NATURAL PHILOSOPHY: THE LOGIC OF PHYSICS

Volume Two:
The Quantum Theory of Everything

Marshall Dixon

Adam Hibshman

James Miller

Marshall Dixon has a Ph.D. in Physics from University of Virginia

Adam Hibshman has a B.S. in Physics from Butler University

James Miller has a B.S. in Physics from Rose-Hulman Institute of Technology

*This book is dedicated to the students
of Cathedral High School who helped us develop
this course. Your drive and enthusiasm made this
venture to be an experience of joy.*

God bless all of you.

Preface:

"I shall instruct thee in the way which thou shall go" Psalm 32

The first volume of NATURAL PHILOSOPHY: THE LOGIC OF PHYSICS is used as the textbook for the first two years of the physics program at Cathedral High School in Indianapolis, Indiana. But the students have requested one or two additional years for a four-year physics majors program at Cathedral High School. This present volume is written to serve as the textbook for the second half of the physics program.

The foundation for physics in the twenty first century is quantum mechanics; hence this second volume is devoted to developing that subject. Within that context, the physics of condensed matter is of great importance in 21st century technology. Thus, this book significantly stresses topics in condensed matter physics.

In studying this book, certain topics can be stressed.

Electromagnetic Theory:
chapters 8 and 12 of volume one and chapter 13 of volume two

Atomic and Nuclear Physics:
chapters 16, 17, and 18 of volume two

Condensed Matter:
chapters 13, 14, 20, and 21 of volume two

Quantum Optics:
chapters 14 and 18 of volume two

Chemical Physics:
chapters 19 and 20 of volume two

The Physical Meaning of Quantum Mechanics:
chapter 22 of volume two

"Back to their source the Holy rivers turn their tide. Order and the universe are being reversed." Euripedes, Medea

All required mathematics is developed when needed in the context of physics, but in doing so, all theorems are rigorously proved. The authors attempt to introduce new research topics at the time of writing this book. Instructors are encouraged to continue to discuss research topics. Physics students should learn what is not understood as well as understood.

In spite of intensive editing of this book, some errors and misprints may still exist. If you find any, let us know at: millerjh10@gmail.com

Table of Contents:

Notice: Each chapter is paged separately. For example, page 4-17, is the 17th page of chapter 4.

Volume One: Describing the World with Mathematics

Chapter 6: Wave motion. From standing waves and interference, it goes into physical optics up to diffraction gratings and spectroscopy.

Chapter 7: Astronomy with black holes. This includes astronomical distance measurements up to parallax methods.

Chapter 8: Electricity. This starts with transport phenomena as electroplating as an introduction to electric fields and Ohm's law. Coulomb's law is derived from this in spherical coordinates. An appendix offers an extensive introduction to electrical measurements.

Chapter 9: Elementary theory of hydrogen atom and special relativity. From Bohr's theory, it moves to de Broglie's rule and a discussion of a quantized particle in a box. Relativity emphasizes proper time, world lines and relativistic energy. Introduction to Feynman graphs and the ideas of quantum field theory.

Chapter 10: Thermodynamics and engine cycles. Starting with experiments on gases, thermodynamics is developed to the derivation of formulas for efficiency of Diesel, Otto, Brayton cycles.

Chapter 11: Angular momentum and torque

Chapter 12: Magnetism up to nuclear magnetic resonance

Constants of Physics:

Velocity of Light in Vacuum	$c = 3 \times 10^8$ meter/second
Electronic Charge	$e = 1.6 \times 10^{-19}$ coulomb
Planck's Constant	$h = 6.625 \times 10^{-34}$ joule-second
Permittivity of Free Space	$\varepsilon_o = 8.85 \times 10^{-12}$ farad/meter
Permittivity of Free Space	$1/(4\pi\varepsilon_o) = 9 \times 10^9$ newton-meter^2/coulomb^2
Permeability of Free Space	$\mu_o = 4\pi \times 10^{-7}$ henry/meter
Gas Constant	$R = 8.3$ joule/mole-kelvin
Avogadro's Number	$N_A = 6 \times 10^{23}$ molecules/gram-mole
Absolute Zero Temperature	$-273°$ Celsius
Electron Mass	9.1×10^{-31} kilogram
Proton Mass	1.67×10^{-27} kilogram
Gravity Field at Earth's Surface	$g = 9.8$ newton/kilogram
Gravitational Universal Constant	$G = 6.67 \times 10^{-11}$ newton-meter^2/coulomb^2
Distance from Earth to Sun	1.5×10^{11} meter
Mass of Sun	2×10^{30} kilogram
Radius of Earth	6.4×10^3 meter
Mass of Earth	6×10^{24} kilogram

Useful Conversions

1 foot = 0.3048 meter	1 mile = 1.6 km
1 meter = 3.28 feet	1 kilometer = 0.6214 mile
1 pound (weight) = 0.4536 kilogram (mass)	
1 kilogram (mass) = 2.2046 pounds (weight)	

CHAPTER THIRTEEN: ELECTROMAGNETISM IN CONDENSED MATTER, INDEX OF REFRACTION,
MAGNETISM, SUPERCONDUCTIVITY, RADIATION

This chapter covers a lot of very important physics. It requires that we
develop a lot of new mathematics which will be developed in the context of
the physics which requires it. This mathematics comes up in many other areas
of physics. Go slowly in reading this chapter, thinking carefully about each
step in the development of ideas. In the long run it will be worth the
effort.

It is assumed that you have studied chapters 8, 11, and 12. Material from
all of these will be required for this chapter. For reference, we repeat the
theory of the hydrogen atom. Particle wave functions and the quantization of
angular momentum will be necessary ideas in this chapter.

The hydrogen atom has an electron with charge, $-e$, in a stable orbit around
a much heavier proton of charge $+e$, where $e = 1.6 \times 10^{-19}$ coulomb. In the
chapter on electricity, an object with charge, q, at a distance, r, from a
charge, Q, has an energy, $E = \frac{1}{2} M v^2 + \left(\frac{q Q}{4 \pi \epsilon_0 r} \right)$.

In that development, we could not avoid using the letter, E, for both
electric force field and for total energy. The context must determine the
sense in which the letter is used. The resulting energy formula is good for
motion with velocity, v, being in any direction. In the following
development we will simplify the formulas by using,

$\frac{1}{4 \pi \epsilon_0} = k = 9 \times 10^9$ Newton-meter^2/coulomb^2 in a vacuum.

We will now consider an atom of the simplest
element, hydrogen, consisting of the electron
going in a circle about a nucleus consisting of
a single nuclear particle, or proton. Each
particle carries one unit of charge being
$e = 1.6 \times 10^{-19}$ coulomb. The electron carries
$q = -e$ and the proton carries $Q = +e$. The force
attracting the electron to the proton,

$F = k \frac{(-e)(+e)}{r^2} = -\frac{k e^2}{r^2}$, is a centripetal force in

Newton's second law, $F = M a$, with centripetal

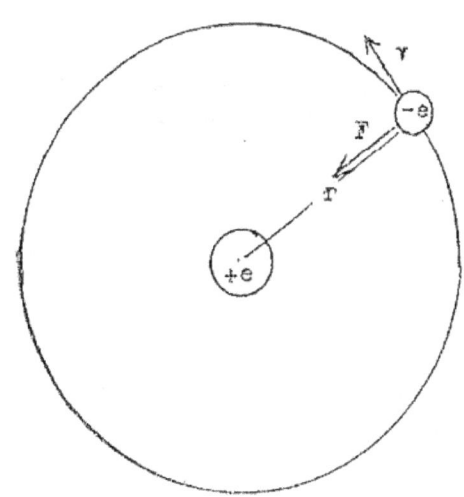

acceleration, $a = -r \omega^2$, first seen in chapter 4, where the velocity was

$v = r \omega$. Hence, $a = -\frac{v^2}{r}$ also.

From $M\,a = F$, we get $-M\dfrac{v^2}{r} = -\dfrac{k\,e^2}{r^2}$, giving $M\,v^2 = \dfrac{k\,e^2}{r}$. If energy is $E = \dfrac{1}{2}M\,v^2 - \dfrac{k\,e^2}{r}$, then $E = -\dfrac{k\,e^2}{2\,r}$.

For reference, we describe some background physics. Although formal quantum mechanics will not be introduced in this chapter, some things like the quantization of the energy of a light wave will be required.

You may ask how is it that we can determine the electron's energy or orbital radius. This is done with light which involves waves of electric and magnetic force fields. On being emitted and absorbed and on passing through matter, the vibrating fields of a light wave interact with the electrical structure of the atoms of matter. Hot solid surfaces, when emitting light give off all colors of the spectrum. Excited gases and vapors give off only certain colors and no others. Each element has a unique set of colors which its excited vapor gives off. If the light from a particular source is passed through a diffraction grating in a spectrometer, the wavelengths of the emitted colors can be determined. Below are given examples of several emission spectra.

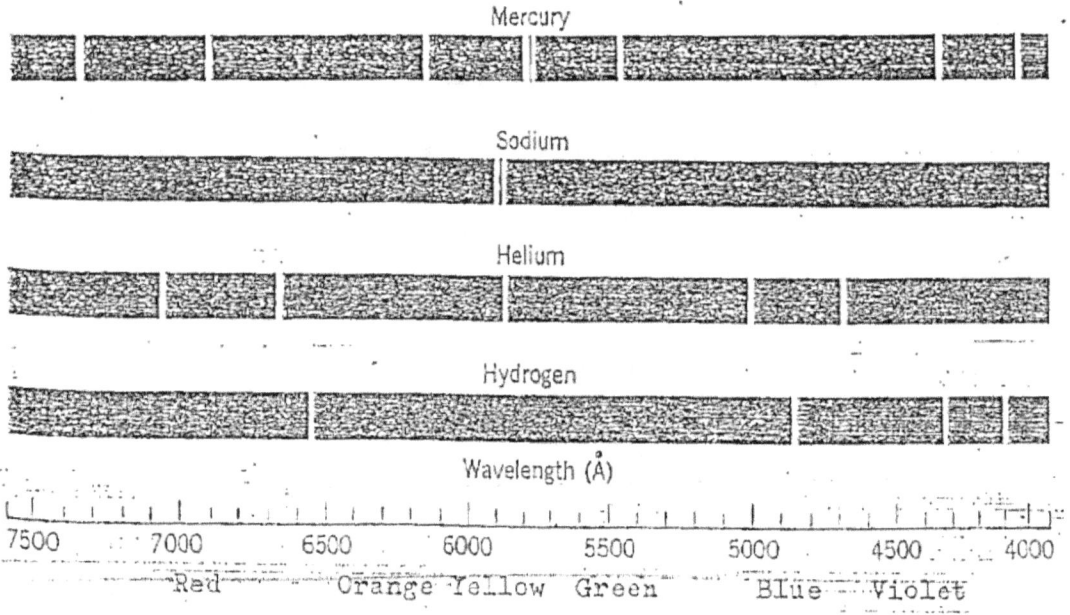

For all of these sources, there are other emissions with wavelengths longer than red, called infrared, and with wavelengths shorter than violet, called ultraviolet. For the visible emissions from exited hydrogen, the wavelengths are: $\lambda_{red} = 6563\,\text{Å}$; $\lambda_{blue-green} = 4861\,\text{Å}$; $\lambda_{violet} = 4340\,\text{Å}$; in which one angstrom unit is $1.0\ \text{Å} = 10^{-10}$ meter.

Expressing the emitted wavelengths from hydrogen in meters, the wavelengths can be described by the formula, $\frac{1}{\lambda} = R\left[\frac{1}{2^2} - \frac{1}{n^2}\right]$, where $n = 3$ for red, $n = 4$ for blue-green, and $n = 5$ for violet. R is a constant called the Rydberg constant. Using the wavelengths given above in meters, we can calculate a number for the Rydberg constant, $R = 1.097 \times 10^7$ meters^{-1}. Including ultraviolet and infrared emissions, all components of the hydrogen spectrum can be described by this same formula if the first term in the bracket is also $1/1^2$ or $1/3^2$ or $1/4^2$ and so forth.

Other spectra do not lend themselves to such a simple description. The spectral data all-in-all suggests that there is something very specific and limited about the structure of various atoms. Only certain states of some properties of atoms are permitted by Nature to exist which, when they are excited, give off only certain colors of light. It was Niels Bohr (1885-1962) who first worked out what particular properties were limited in what particular way.

Before we go further with light from atoms, we must consider light given off by hot surfaces. A hot glowing coal gives off red light. Using a prism or a grating, we know that white light is made up of all colors from red to violet. Why is there a relation between the quality of emitted light and the temperature of the emitting surface?

Notice from the graph on the right that the brightest emitted color associates increasing temperature with increasing frequency. One Hertz is one vibration per second.

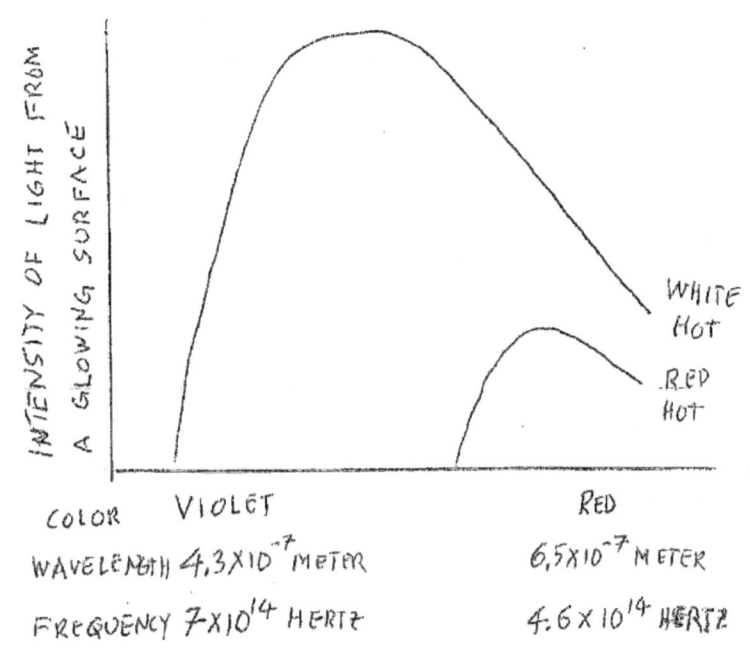

COLOR   VIOLET   RED

WAVELENGTH 4.3×10^{-7} METER 6.5×10^{-7} METER

FREQUENCY 7×10^{14} HERTZ 4.6×10^{14} HERTZ

In the year 1900, Max Planck came up with the right idea. Although light is mostly thought of as involving waves. Its energy is carried in definite sized bundles of energy which we now call photons. Planck reasoned that the higher the frequency of the light wave, the greater the size of the bundle of its energy. Coming off a hot surface, a photon is emitted locally from a small piece of the surface area. Hence, the greater the energy the photon takes, the greater the energy that must be in that piece of area. This is measured by the concentration of energy or energy per unit area on the surface.

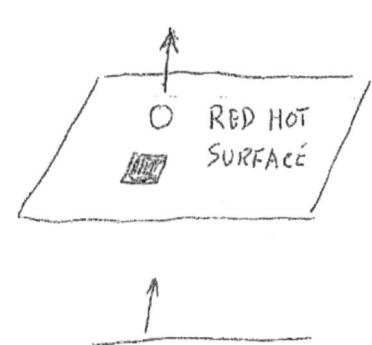

The thermal energy per unit area of a hot surface increases with the temperature. Hence, to get photons of higher energy, and light of increasing frequency, it is necessary to increase the temperature of the surface. As the temperature of the surface increases, the surface can emit light of frequency increasing from red towards the violet, which is what we observe. Obviously, the energy of a photon is proportional to the frequency and we write $dE = h f$, where h is Planck's constant. Measurements give $h = 6.6525 \times 10^{-34}$ joule-second.

For a light wave, $\lambda f = c$, where the velocity of light is $c = 3 \times 10^8$ m/sec. Thus, a photon has energy, $dE = \dfrac{h c}{\lambda}$. Now, recall the Rydberg formula for the hydrogen spectra. A photon of one of these has $dE = h c R \left[\dfrac{1}{2^2} - \dfrac{1}{n^2}\right]$. A hydrogen electron in an orbit of radius, r, had energy, $E = -\dfrac{k e^2}{2 r}$. Maybe the orbit is limited to having only certain radii.

At this point, assume that the energy of the emitted photon is the difference between two orbits in which the electron might exist. Let n be the number of an orbital. Let E_n be the energy of the nth orbital. Recall the electron energy is negative. Take the photon energy above to be the difference between the energies of the nth and 2nd orbits.

$E_n - E_2 = h c R \left[\dfrac{1}{2^2} - \dfrac{1}{n^2}\right]$, then $E_n = -\dfrac{h c R}{n^2}$. $h c R = 2.18 \times 10^{-18}$ joule. Dividing this by $e = 1.6 \times 10^{-19}$, gives $h c R = 13.6$ electron volt.

Equating E_n to $E_n = -\dfrac{k e^2}{2 r_n}$ gives $r_n = \dfrac{k e^2}{2 h c R} n^2$ or $r_n = (5.28 \times 10^{-11}) n^2$. Next look at the picture on page 1. The electron has angular momentum, $L = M r v$. But, $M v^2 = \dfrac{k e^2}{r}$. $L = k e^2 \sqrt{\dfrac{M}{2 h c R}} n$ or $L = (1.06 \times 10^{-34}) n$. Now, $h/2\pi = 1.05 \times 10^{-34}$.

This allows us to write, $L_n = n\dfrac{h}{2\pi}$, $n = 1, 2, 3\ldots$

This result, the quantization of angular momentum, is a general property of quantum mechanics. As magnetic moment rides on angular momentum, this will be important in the theory of magnetism. For simplicity we write haitch-bar, $\hbar = \dfrac{h}{2\pi}$. Then, $L_n = n\hbar$, $n = 1, 2, 3\ldots$

From a standard (before 1900) physics point of view, the limitation of radii to only certain numbers had no reason. This didn't make sense because in classical physics, the only thing limited to certain integers were the frequencies of standing waves. Now, if the light waves had particles, photons, could particles have waves? Let us assume that they do and for an electron its orbit was integer number of wavelengths. $2\pi a n^2 = n\lambda_n$ or $\lambda_n = 2\pi a n$.

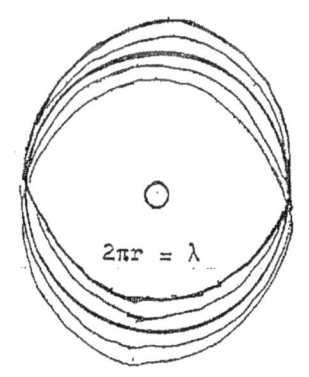

$2\pi r = \lambda$

This then gives $r_n = \dfrac{a^2 n^2}{a} = \dfrac{\lambda_n^{\,2}}{4\pi^2 a}$, $a = 5.28 \times 10^{-11}$.

Now, $M v^2 = \dfrac{k e^2}{r_n}$, where for the electron, $M = 9.1 \times 10^{-31}$ kg.

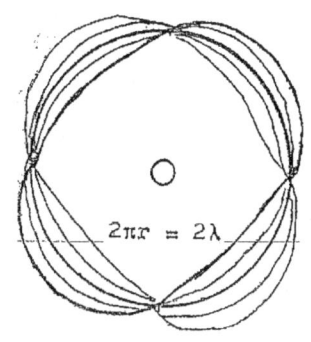

$2\pi r = 2\lambda$

Focusing on momentum, $M^2 v^2 = \dfrac{M k e^2 4\pi^2 a}{\lambda^2}$ or $M^2 v^2 \lambda^2 = M k e^2 4\pi^2 a$, giving $M v \lambda = 6.6 \times 10^{-34}$ looking a lot like $(M v)(\lambda) = h$.

Louis de Broglie came up with the idea in 1924. He suggested that any particle with mass, M, and velocity, v, would have a wave associated with it with $\lambda = \dfrac{h}{M v}$ as it wavelength. These waves for a beam of electrons were observed directly by G. P. Thomson, C. J. Davisson and L. H. Germer in 1927.

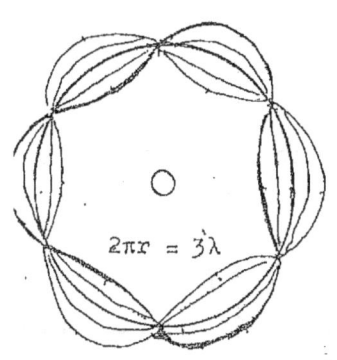

$2\pi r = 3\lambda$

DIGRESSION: MATHEMATICS FOR PHYSICS IN TWO AND THREEE DIMENSIONS

I shot an arrow into the air,
It fell to earth, I know not where; H. W. Longfellow

Although the complex plane, $z = x + i\,y = r\,e^{i\,\theta}$, is perfect for describing
motion in two dimensions, the idea cannot be extended to things in three
dimensions. Another description of space is necessary, but we continue using
complex numbers in mathematics. Physics has quantities having direction as
well as magnitude. Such quantities are called VECTORS. Examples of such
quantities are the position, velocity, acceleration, force, momentum, and
other things to be introduced on. Other quantities in physics are adequately
specified by a single number for their magnitude. These quantities are
called SCALARS with examples being time, mass, temperature, density, etc.

Vector quantities are described by reference to a coordinate system. This
requires an origin, as many axes as there are independent dimensions, and a
unit of magnitude. The coordinate systems that we shall be using will have
no more than three axes and we shall consider only cases with mutually
perpendicular axes. In standard print, vector quantities are denoted by
symbols in bold-faced type. Sometimes we shall speak only of the magnitude
of a vector without reference to the direction. This is, then, merely a
number and will be designated by the symbol without the line over it. For
example, the magnitude of F (bold) is F (not bold), etc.

Once an origin and measure of the magnitude of the vectors in space have
been picked, the directions of the coordinate axes are denoted by a set of
unit length vectors, one of which specifies the direction of each coordinate
axis. These unit vectors are called BASE VECTORS and the set is called a
BASIS for the vector space. Ordinary physical space is generally described
by three mutually perpendicular axes labeled by x, y, and z. The direction
of the x-axis is denoted by a unit vector, i, the y-axis by j, and the z-axis
by k. Any point can be gotten by three mutually perpendicular displacements,
going in the direction of i a distance, x, in the direction of j a distance,
y, and in the direction of k a distance, z. If the vector from the origin to
the point is denoted by r, then the point is specified by $r = i\,x + j\,y + k\,z$.

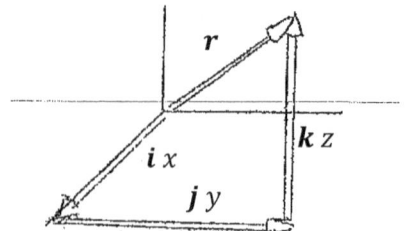

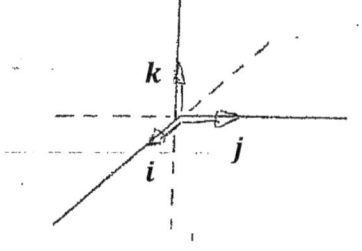

The magnitude of the length of r is easily gotten. The displacements, x and y, form a right triangle with the hypotenuse, $\sqrt{x^2 + y^2}$, and this with the z displacement forms a right triangle with hypotenuse, $\sqrt{x^2 + y^2 + z^2}$, which is a length, r, of the vector, r. Hence, when the axes are mutually perpendicular, the length of a vector is obtained from the three-dimensional Pythagorean rule, $r^2 = x^2 + y^2 + z^2$.

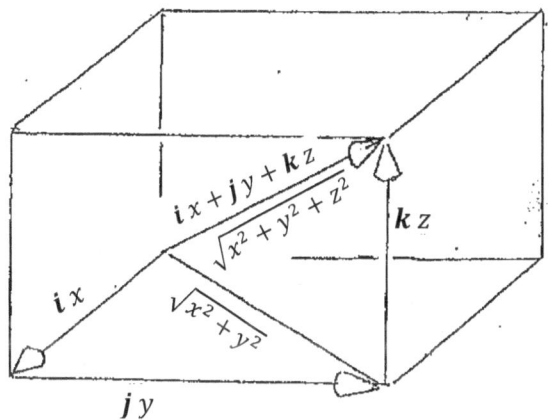

The letters x, y, and z are used as subscripts to distinguish the respective component of a vector that is labeled by some other symbol, for example:
$$A = i\,a_x + j\,a_y + k\,a_z \quad \text{and} \quad B = i\,b_x + j\,b_y + k\,b_z$$
When two vectors are added, their components are added and when two vectors are subtracted their components are subtracted.
$$A + B = i\,(a_x + b_x) + j\,(a_y + b_y) + k\,(a_z + b_z)$$
$$A - B = i\,(a_x - b_x) + j\,(a_y - b_y) + k\,(a_z - b_z)$$

A picture of what this involves can be gotten easily by using two dimensional vectors drawn as arrows on a plane. Adding the two vectors is equivalent to adding the two arrows tail to tip, keeping their lengths and directions constant. The difference between two vectors, for example, $A - B$, is an arrow drawn from the head of one, B, to the head of the other, A.

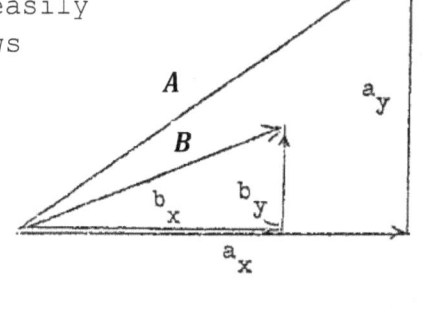

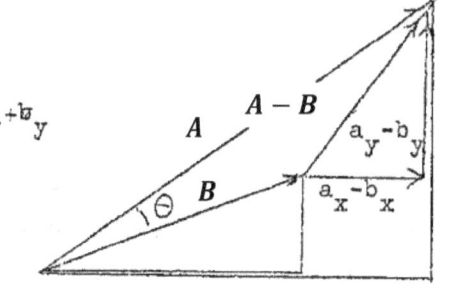

If A is the length of A and B is the length of B and θ is the angle between A and B, then C, the length of $C = A - B$, is given by the law of cosines to be $C^2 = A^2 + B^2 - 2\,A\,B\cos\theta$. To find lengths A and B, use: $A^2 = a_x^2 + a_y^2 + a_z^2$ and $B^2 = b_x^2 + b_y^2 + b_z^2$.

A^2, B^2, and C^2 can be explained in terms of the squares of the components of A, B, and C:

$$(a_x - b_x)^2 + (a_y - b_y)^2 + (a_z - b_z)^2 = a_x^2 + a_y^2 + a_z^2 + b_x^2 + b_y^2 + b_z^2 - 2(a_x b_x + a_y b_y + a_z b_z)$$

which reduces to yield, $a_x b_x + a_y b_y + a_z b_z = A B \cos \theta$.

This result is sufficiently useful and important to be given a name, or rather three names. It is written as the product of vectors A and B with a dot between them to denote the product operation.

$$A \cdot B = a_x b_x + a_y b_y + a_z b_z$$

It is called the INNER PRODUCT of A and B, or the DOT PRODUCT as the product is designated by a dot, or the SCALAR PRODUCT as its result is merely a scalar (number) and not a vector (no direction). Numerically it is equal to the length of A times the length of B times the cosine of the angle between them. If the two vectors are perpendicular to each other, $\theta = \pi/2$ or $90°$ and $\cos \frac{\pi}{2} = 0$. Thus, if A is perpendicular to B, $A \cdot B = 0$. This test for the mutual perpendicularity of two vectors will be used often in this work.

The importance of the dot product lies in its ability to give us the projection of one vector onto another. Let A and B be two vectors with an angle, θ, between them. Using A as its hypotenuse, construct a right triangle with one of its shorter sides lying on B. If A is the length of A and θ is the angle between A and B, the side of the right triangle lying on B is $A \cos \theta$, and the side of the triangle perpendicular to B is $A \sin \theta$. The side of the triangle lying on B is called the PROJECTION OF A ONTO B. Then, if $A \cdot B = A B \cos \theta$.

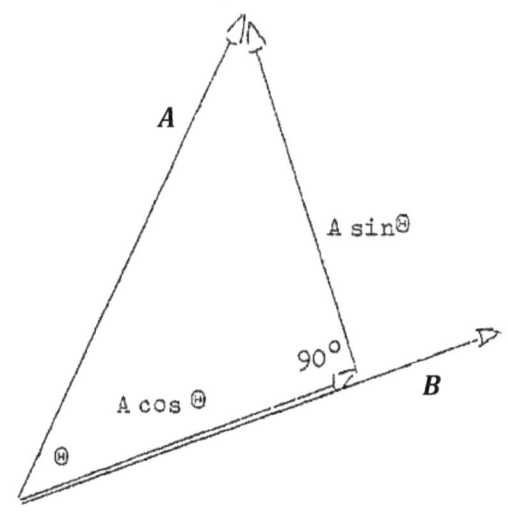

It is evident that the dot product of A and B equals the length of B times the length of the projection of A onto B. A similar construction would show that it is also equal to the length of A times the length of the projection of B onto A.

If i, j, and k are mutually perpendicular unit length vectors, it is evident from the preceding discussion that $i \cdot i = j \cdot j = k \cdot k = 1$ and $i \cdot j = i \cdot k = j \cdot k = 0$.

The following geometry problem serves as an example of the use of the dot product. The problem is: Let $\boldsymbol{A} = \boldsymbol{i}\, a_x + \boldsymbol{j}\, a_y + \boldsymbol{k}\, a_z$ be a vector drawn from the origin to a point in space. Through that point is to pass a plane to which plane $\boldsymbol{A}$ is perpendicular. What is the equation for the plane?

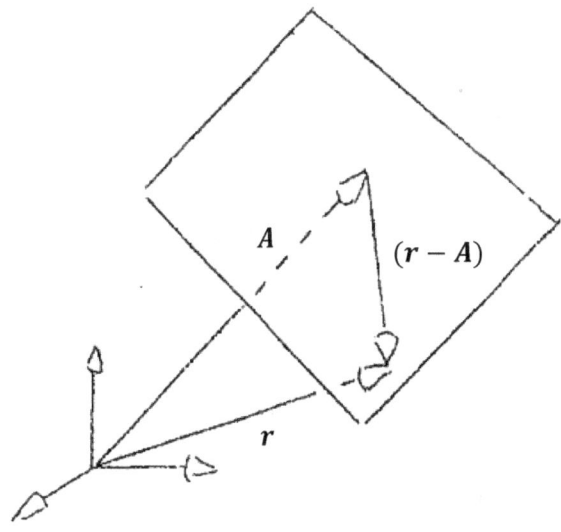

Solution: Let $\boldsymbol{r} = \boldsymbol{i}\, x + \boldsymbol{j}\, y + \boldsymbol{k}\, z$ be any point on the plane. $\boldsymbol{A}$ is a point on the plane by definition. Hence, $(\boldsymbol{r} - \boldsymbol{A})$ is a vector lying on the plane. As $\boldsymbol{A}$ is perpendicular to the plane by defition, $(\boldsymbol{r} - \boldsymbol{A})$ and $\boldsymbol{A}$ are mutually perpendicular vectors and $\boldsymbol{A} \bullet (\boldsymbol{r} - \boldsymbol{A}) = 0$, as the dot product is distributive. Hence, the equation for the plane is,
$$a_x\, x + a_y\, x + a_z\, z = a_x{}^2 + a_y{}^2 + a_z{}^2 .$$

CALCULUS OF FUNCTIONS OF SEVERAL VARIABLES

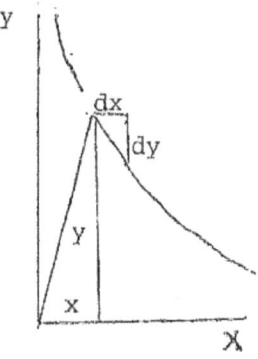

To start, take the curve, $y = \dfrac{C}{x} = C\, x^{-1}$, $C = $ constant.

Now, $\dfrac{dy}{dx} = -C\, x^{-2} = -\dfrac{C}{x^2}$, which is also $\dfrac{dy}{dx} = -\dfrac{y}{x}$.

Now, look at this another way. Suppose $f(x,y) = x\, y$ is a function of both x and y, having a different value at every point on the x-y plane. There will be certain points where $f(x,y)$ is a constant. For those points, $x\, y = c$ gives the curve as shown to the right.

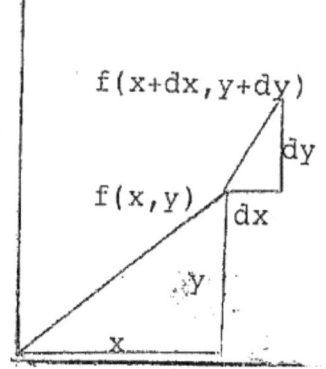

When location changes by dx and dy, the function's value changes. It changes separately by dx and then by dy. A rounded off dee as ∂f means a change of only one variable, the rest staying constant. With a change of dx, $d_x f = \left(\dfrac{\partial f}{\partial x}\right) dx$ and with dy, $d_y f = \left(\dfrac{\partial f}{\partial y}\right) dy$. The function's total change is $df = \left(\dfrac{\partial f}{\partial x}\right) dx + \left(\dfrac{\partial f}{\partial y}\right) dy$. For $f(x,y) = x\, y$; $\dfrac{\partial f}{\partial x} = y$, $\dfrac{\partial f}{\partial y} = x$, and $df = y\, dx + x\, dy$. On the curve where $f(x,y)$ is constant, $df = 0$ and $y\, dx + x\, dy = 0$ gives $\dfrac{dy}{dx} = -\dfrac{y}{x}$, as we got above.

Using i, j and k as unit-length base vectors defining the x-, y- and z-axes, any point in space can be defined by the vector, $r = i x + j y + k z$. Let $f(x, y, z)$ be any smoothly varying function of position in that space. If position changes by, $dr = i\, dx + j\, dy + k\, dz$, the change in the functions will be in three parts,

$d_x f = \left(\frac{\partial f}{\partial x}\right) dx$, $d_y f = \left(\frac{\partial f}{\partial y}\right) dy$, and

$d_z f = \left(\frac{\partial f}{\partial z}\right) dz$. Adding the three changes

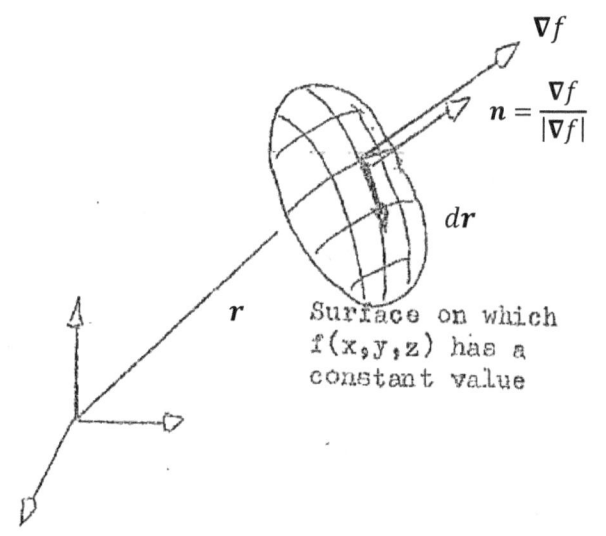

gives $df = \left(\frac{\partial f}{\partial x}\right) dx + \left(\frac{\partial f}{\partial y}\right) dy + \left(\frac{\partial f}{\partial z}\right) dz$ to be the total change of the function, $f(x, y, z)$, when the position changes by $dr = i\, dx + j\, dy + k\, dz$. The change, df, looks like the inner (dot) product of the vector, dr, and another vector, $\nabla f = i \left(\frac{\partial f}{\partial x}\right) + j \left(\frac{\partial f}{\partial y}\right) + k \left(\frac{\partial f}{\partial z}\right)$, which we call the gradient of the function, f. $df = (\nabla f) \bullet (dr)$.

Through the point $r = i x + j y + k z$ there will be a surface of points at which $f(x, y, z)$ has the same value that it did at designated point, r. Thus, f has a constant value on this surface. Let dr be a displacement from r to another point on this surface. If f has a constant value on this surface, $df = (\nabla f) \bullet (dr) = 0$, which means that ∇f is perpendicular to any dr on the surface of constant f and in the direction of increasing f. If $|\nabla f|$ is the size or magnitude of ∇f, then $n = \frac{\nabla f}{|\nabla f|}$ is a unit length (normalized) vector perpendicular to (normal to) the surface of constant f.

With $\nabla f = i \left(\frac{\partial f}{\partial x}\right) + j \left(\frac{\partial f}{\partial y}\right) + k \left(\frac{\partial f}{\partial z}\right)$, taking derivatives of the function are operations done on the function and the symbol for doing the operations is called an operator. Hence, we call $\boldsymbol{\nabla} = \boldsymbol{i}\left(\frac{\partial}{\partial x}\right) + \boldsymbol{j}\left(\frac{\partial}{\partial y}\right) + \boldsymbol{k}\left(\frac{\partial}{\partial z}\right)$ the gradient operator.

Let us see how motion is described in our new coordinates. Position of a moving object is $\boldsymbol{r} = \boldsymbol{i}\,x + \boldsymbol{j}\,y + \boldsymbol{k}\,z$ and velocity is $\boldsymbol{v} = \boldsymbol{i}\,\frac{dx}{dt} + \boldsymbol{j}\,\frac{dy}{dt} + \boldsymbol{k}\,\frac{dz}{dt}$. With force, $\boldsymbol{F} = \boldsymbol{i}\,F_x + \boldsymbol{j}\,F_y + \boldsymbol{k}\,F_z$, $M\frac{d\boldsymbol{v}}{dt} = \boldsymbol{F}$ is the law of motion.

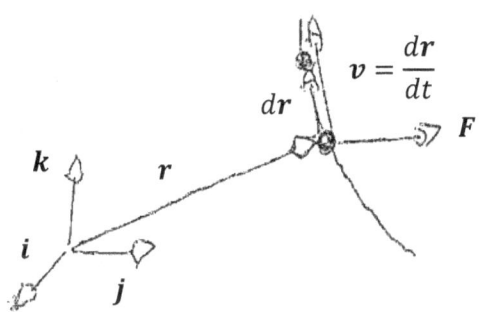

The law of motion can be rewritten, $M\,d\boldsymbol{v} = \boldsymbol{F}\,dt$. Take the dot product with $\boldsymbol{v} = \frac{d\boldsymbol{r}}{dt}$, giving $M\,\boldsymbol{v}\bullet d\boldsymbol{v} = \boldsymbol{F}\,dt\bullet\frac{d\boldsymbol{r}}{dt}$. Let us look at something like $\boldsymbol{v}\bullet d\boldsymbol{v}$. Take $r^2 = x^2 + y^2 + z^2$, from which $d(r^2) = 2\,x\,dx + 2\,y\,dy + 2\,z\,dz = 2\,\boldsymbol{r}\bullet d\boldsymbol{r}$. So, write $M\,\boldsymbol{v}\bullet d\boldsymbol{v} = M\,d\left(\frac{1}{2}v^2\right)$, giving $d\left(\frac{1}{2}M\,v^2\right) = \boldsymbol{F}\bullet d\boldsymbol{r}$.

$\frac{1}{2}M\,v^2$ is kinetic energy and $\boldsymbol{F}\bullet d\boldsymbol{r}$ is work done by the force during the displacement, $d\boldsymbol{r}$. Suppose there is a constant total energy, $E = \frac{1}{2}M\,v^2 + V(x,y,z)$, where the potential energy, $V(x,y,z)$, is a function of position. If E is constant, $dE = 0$. Then, $0 = d\left(\frac{1}{2}M\,v^2\right) + dV$ or $d\left(\frac{1}{2}M\,v^2\right) = -dV$.

This says that potential energy and force are related by $dV = -\boldsymbol{F}\bullet d\boldsymbol{r}$ under the condition of constant total energy. But, $dV = (\boldsymbol{\nabla}V)\bullet(d\boldsymbol{r})$. Hence, $\boldsymbol{F} = -\boldsymbol{\nabla}V$.

Let's try this on an atom with the nucleus at the origin. If $\boldsymbol{r}$ is the position of an electron $\frac{\boldsymbol{r}}{r}$ is a unit vector in the direction of an electron. The force on the electron is $\boldsymbol{F} = k\,\frac{(-e)(+e)}{r^2}\left(\frac{\boldsymbol{r}}{r}\right)$, with $k = \frac{1}{4\,\pi\,\epsilon_0}$. Now, $d(r^2) = 2\,r\,dr = 2\,x\,dx + 2\,y\,dy + 2\,z\,dz$ giving $dr = \frac{x}{r}\,dx + \frac{y}{r}\,dy + \frac{z}{r}\,dz = \left(\frac{\boldsymbol{r}}{r}\right)\bullet d\boldsymbol{r}$.

Then, $\boldsymbol{\nabla}r = \boldsymbol{i}\frac{x}{r} + \boldsymbol{j}\frac{y}{r} + \boldsymbol{k}\frac{z}{r}$ for a function, $f(x,y,z)$, $df = (\boldsymbol{\nabla}f)\bullet(d\boldsymbol{r})$. Now, $d(r^n) = n\,r^{n-1}\,dr = n\,r^{n-1}\left(\frac{\boldsymbol{r}}{r}\right)\bullet d\boldsymbol{r}$, then, $-r^{-2}\left(\frac{\boldsymbol{r}}{r}\right) = \boldsymbol{\nabla}(r^{-1})$. If $\boldsymbol{F} = -\boldsymbol{\nabla}V$, then $\boldsymbol{F} = \boldsymbol{\nabla}\left(\frac{k\,e^2}{r}\right)$. The potential energy of the electron is $V(r) = -\frac{k\,e^2}{r}$.

The gradient operator, and some other tricks involving it, will be found all through physical science, especially in electromagnetic theory, fluid dynamics, heat transport and related subjects.

From time-to-time we will need other tricks with the gradient operator. If electromagnetism and fluid mechanics use the same mathematics, it will be convenient to use fluid mechanics as the context for developing these mathematical tools. This will also enhance our intuition as to the physical interpretation of them. The first context will be a conserved flowing fluid. Since both mass and charge are conserved in normal physics, this will apply to either one.

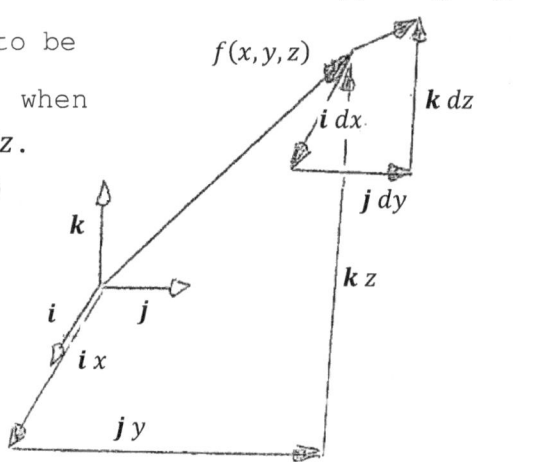

We have found $df = \left(\frac{\partial f}{\partial x}\right) dx + \left(\frac{\partial f}{\partial y}\right) dy + \left(\frac{\partial f}{\partial z}\right) dz$ to be the total change of the function, $f(x, y, z)$, when the position changes by $dr = i\, dx + j\, dy + k\, dz$. The change, df, looks like the inner (dot) product of the vector, dr, and another vector, $\nabla f = i\left(\frac{\partial f}{\partial x}\right) + j\left(\frac{\partial f}{\partial y}\right) + k\left(\frac{\partial f}{\partial z}\right)$, which we call the gradient function.
$df = (\nabla f) \cdot (dr)$.

Let us look at a (non-viscous) fluid flowing in a conduit whose cross-sectional area is A and where the fluid has a velocity, v, which is the same at all points on the area. In a time, t, a volume, $A(vt)$, of fluid will flow across any cross section, giving a volume-per-unit-time,
$\frac{dV}{dt} = A\,v$.

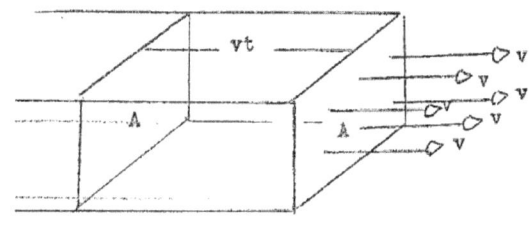

Now, let the fluid flow from the conduit across a beveled surface whose area is $S = dx\, dl$. The cross-section of the flow is $A = dx\, dz$, where $dz = dl\cos\theta$, as shown in the figure below. The volume flow is now,
$$\frac{dV}{dt} = A\,v = (dx\, dz)\,v = (dx\, dl \cos\theta)\,v = S\,v\cos\theta.$$

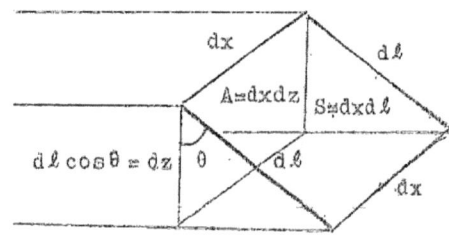

Let v be the velocity vector perpendicular to A and let n be a unit length vector perpendicular to S. If θ is the angle between A and S, then θ is the angle between v and n so $v \cdot n = (v)(1) \cos \theta = v \cos \theta$, giving

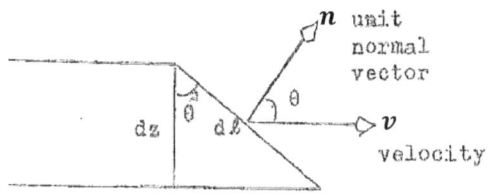

$$\frac{dV}{dt} = S\, v \cdot n.$$

If ρ (rho) is the density or mass-per-unit volume, dV has a mass $dM = \rho\, dV$, and the mass-per-unit time is $\frac{dM}{dt} = \rho\, v \cdot n\, S$. Multiplying the velocity vector by density gives a flow, or flux, vector, $J = \rho\, v$ whose units are: (kilograms/meter^3)(meters/second) = (kilograms/second)/meter^2.

Now construct a wire-edged box whose edges are parallel to the coordinates. The flux vector, $J(x, y, z) = \rho V$, varies with x, y, and z throughout the space. The mass inside the box is $dM = \rho\, dx\, dy\, dz$.

If the flux vector is $J = i J_x + j J_y + k J_z$, only the component of the flux vector perpendicular to a surface will involve flow across that surface. Flow parallel to a surface will slide over it but not across it. The mass per unit time across the surfaces of the cube is as follows:

Top $\left(\dfrac{dM}{dt}\right)_T = J_z(x, y, z + dz)\, dx\, dy$

Bottom $\left(\dfrac{dM}{dt}\right)_{Bo} = -J_z(x, y, z)\, dx\, dy$

Right $\left(\dfrac{dM}{dt}\right)_R = J_y(x, y + dy, z)\, dx\, dz$

Left $\left(\dfrac{dM}{dt}\right)_L = -J_y(x, y, z)\, dx\, dz$

Back $\left(\dfrac{dM}{dt}\right)_{Ba} = -J_x(x, y, z)\, dx\, dy$

Front $\left(\dfrac{dM}{dt}\right)_F = J_x(x + dx, y, z)\, dx\, dy$

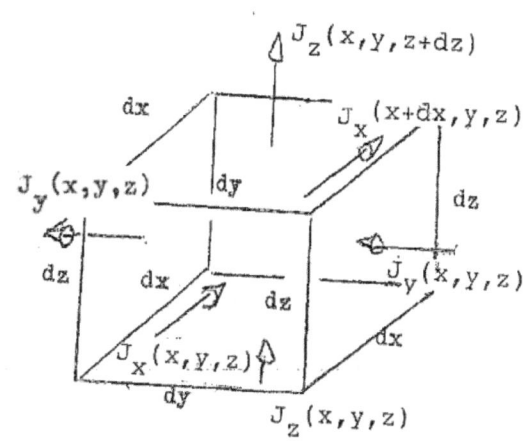

Adding the flows across the separate faces gives the total flow across the boundaries of box.

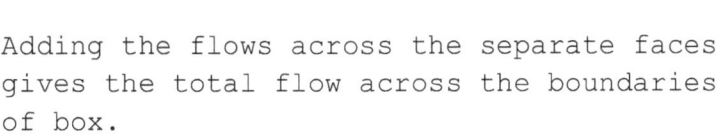

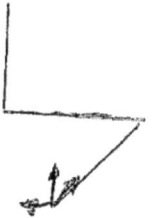

$$\frac{dM}{dt} = \left[J_x(x + dx, y, z) - J_x(x, y, z) \right] dy\, dz +$$
$$\left[J_y(x, y + dy, z) - J_y(x, y, z) \right] dx\, dz +$$
$$\left[J_z(x, y, z + dz) - J_z(x, y, z) \right] dx\, dy$$

This can be written,

$$\frac{dM}{dt} = \left[\frac{\partial J_x}{\partial x}\right] dx\, dy\, dz + \left[\frac{\partial J_y}{\partial y}\right] dy\, dx\, dz + \left[\frac{\partial J_z}{\partial z}\right] dz\, dx\, dy = \left[\frac{\partial J_x}{\partial x} + \frac{\partial J_y}{\partial y} + \frac{\partial J_z}{\partial z}\right] dx\, dy\, dz.$$

With $\dfrac{dM}{dt} = \left[\dfrac{\partial J_x}{\partial x} + \dfrac{\partial J_y}{\partial y} + \dfrac{\partial J_z}{\partial z}\right] dx\, dy\, dz$, we are talking about the rate that mass is crossing the boundaries of the box. If this is positive, the mass of fluid inside the box is decreasing; and if this is negative, the mass inside the box is increasing. The density, ρ, of the fluid can vary with time, t, and with position x, y, and z. The mass inside the box is $\rho\, dx\, dy\, dz$ and as the size of the box is not changing, the mass inside the box changes as $\dfrac{d}{dt}\left[\rho\, dx\, dy\, dz\right] = \left(\dfrac{d\rho}{dt}\right) dx\, dy\, dz$, which is positive if the flow is negative and negative if the flow is positive.

Thus, $\left[\dfrac{\partial J_x}{\partial x} + \dfrac{\partial J_y}{\partial y} + \dfrac{\partial J_z}{\partial z}\right] dx\, dy\, dz = -\dfrac{d\rho}{dt} dx\, dy\, dz$ or $\left[\dfrac{\partial J_x}{\partial x} + \dfrac{\partial J_y}{\partial y} + \dfrac{\partial J_z}{\partial z}\right] = -\dfrac{d\rho}{dt}$, is the local statement for the conservation of a flowing fluid. As the term on the left looks like the inner (dot) product of the gradient operator, $\boldsymbol{\nabla} = \boldsymbol{i}\left(\dfrac{\partial}{\partial x}\right) + \boldsymbol{j}\left(\dfrac{\partial}{\partial y}\right) + \boldsymbol{k}\left(\dfrac{\partial}{\partial z}\right)$, and the vector, $\boldsymbol{J} = \boldsymbol{i}J_x + \boldsymbol{j}J_y + \boldsymbol{k}J_z$, we go ahead and write it that way. $\boldsymbol{\nabla} \boldsymbol{\cdot} \boldsymbol{J} = \dfrac{\partial J_x}{\partial x} + \dfrac{\partial J_y}{\partial y} + \dfrac{\partial J_z}{\partial z}$.

When $\boldsymbol{\nabla} \boldsymbol{\cdot} \boldsymbol{J}$ is positive, fluid is flowing out of the element of volume, or as we might say diverging from it. Thus, we call $\boldsymbol{\nabla} \boldsymbol{\cdot} \boldsymbol{J}$, the divergence of $\boldsymbol{J}$. Thus, $\boldsymbol{\nabla} \boldsymbol{\cdot} \boldsymbol{J} = -\dfrac{d\rho}{dt}$, where $\boldsymbol{J} = \rho\, \boldsymbol{v}$ is the local statement for a conserved fluid.

WITH THE NEW TOOL, RETURN TO ELECTRICITY

Recall the discussion of the motion of charged
particles in a path of cross section area, A,
filled with a medium of conductivity, σ, where
an electric field of strength, E, produced a
charge current, $I = \sigma A E$.

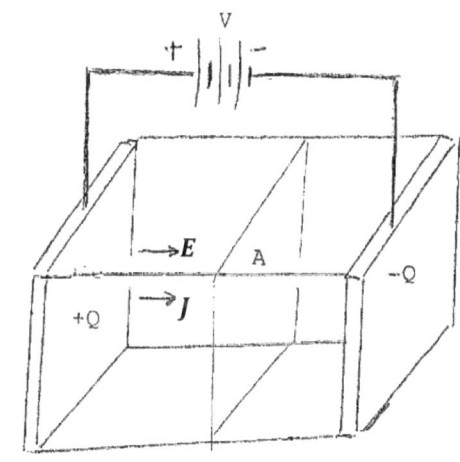

The current density is $J = I/A$ or expressed as
vectors, $J = \sigma E$.

At the left plate, $E = \dfrac{Q}{\epsilon A}$, with ϵ being

permittivity. Putting this into $I = \sigma A E$, gives $I = \dfrac{\sigma Q}{\epsilon}$. But the

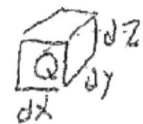

current is the time rate of charge leaving the plate, $I = -\dfrac{dQ}{dt}$.

Hence, $\dfrac{dQ}{dt} = -\dfrac{\sigma Q}{\epsilon}$. If ρ is charge density and $dx\,dy\,dz$ is a volume, the
charge in the volume is $Q = \rho\,dx\,dy\,dz$, and its rate of change is
$\dfrac{dQ}{dt} = \dfrac{d\rho}{dt} dx\,dy\,dz$. But, $\dfrac{dQ}{dt} = -\dfrac{\sigma}{\epsilon}\rho\,dx\,dy\,dz$. Now, $\nabla \cdot J = -\dfrac{d\rho}{dt}$ and $J = \sigma E$.
Therefore, $\sigma \nabla \cdot E = -\dfrac{d\rho}{dt}$. But, $\dfrac{d\rho}{dt} = -\dfrac{\sigma}{\epsilon}\rho$, resulting in $\nabla \cdot E = \dfrac{\rho}{\epsilon}$.

We will discover on the next page that the magnetic field, B, behaves just
like the electric field, E, except that there is no such thing as magnetic
charge. This gives us $\nabla \cdot B = 0$.

These are two of the four equations which James Clerk Maxwell (1831-1879)
derived being the basis of electromagnetic theory. The rest of this theory
is carefully developed at the end of this chapter.

Previously, we showed that if a force has a related potential energy (which
here we write U), then $F = -\nabla U$. In electricity, a charge, q, in a space
where there was an electric field, E, experienced a force, $F = q E$. In the
same space there will be a voltage function, V, which we call electric
potential, from which the charge gets its potential energy as $U = qV$. From
this, the electric field is $E = -\nabla V$.

Then, $\nabla \cdot E = \dfrac{\rho}{\epsilon}$, gives $\nabla \cdot (\nabla V) = -\dfrac{\rho}{\epsilon}$, but what does this look like?
$\nabla V = i\left(\dfrac{\partial V}{\partial x}\right) + j\left(\dfrac{\partial V}{\partial y}\right) + k\left(\dfrac{\partial V}{\partial z}\right)$, then, $\nabla \cdot (\nabla V) = \dfrac{\partial^2 V}{\partial x^2} + \dfrac{\partial^2 V}{\partial y^2} + \dfrac{\partial^2 V}{\partial z^2}$.

The gradient operator is often called DEL and the sum of second derivatives is called DEL SQUARED. It is also called the LAPLACIAN operator. The voltage function, V, then satisfies $\nabla^2 V = -\frac{\rho}{\epsilon}$.

We now have the most general form of electrostatics. There are books and books dealing with solutions of these equations in all sorts of coordinates. The equation that we just derived is called the Poisson equation.

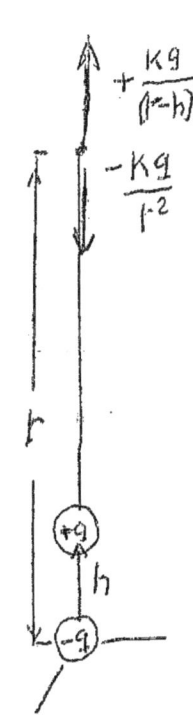

Although a molecule may appear electrically neutral, it does have charges and they may not be distributed in the same manner. One end may be a positive charge (+q) and the other end may have a negative charge (−q) and these, taken as points, may be separated by a distance, h. Calling charges poles, such an object is called a dipole.

$$E = \frac{q\,h}{2\pi\epsilon_0\,r^3}$$

At a distance, r, from the dipole (as shown) each charge produces a component of the electric field (as shown). As they are in line, we add them to get the net electric field. For short, $k = \frac{1}{4\pi\epsilon_0}$.

$$E = \frac{k\,q}{(r-h)^2} - \frac{k\,q}{r^2} \qquad E = k\,q\left[\frac{r^2 - (r-h)^2}{r^2\,(r-h)^2}\right] \qquad E = \frac{k\,q\left(2\,r\,h - h^2\right)}{r^2\,(r-h)^2}$$

If h is of molecular dimensions when compared to r, it can be ignored. Substituting for k we get the net electric field, shown in the above, right-hand picture.

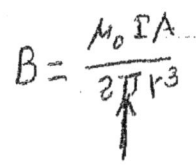

$$B = \frac{\mu_0\,I\,A}{2\pi\,r^3}$$

With our picture of an atom, it was found that charges move in circles giving loops of current. Take such a loop carrying a current, I, in a loop of radius, a, giving an area, $A = \pi\,a^2$.

$$B = \frac{\mu_0\,I\,A}{2\pi\,r^3}$$

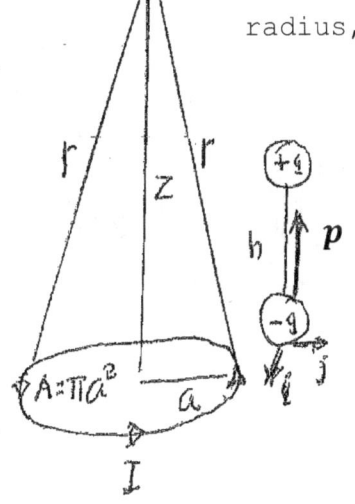

In Volume One's Chapter 12 on magnetism, it was found that such a current carrying loop would produce a magnetic field vector as shown in the left-hand picture. When the loop has atomic dimensions, this gives the right-hand picture. Comparing the electric and magnetic pictures, notice how the formulas are similar.

$$\mu = I\,A\,k$$

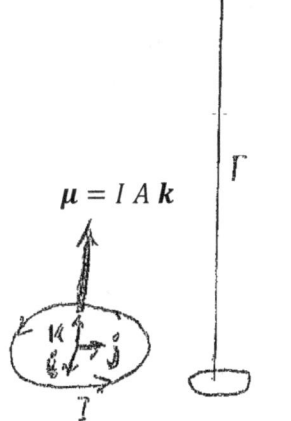

For the electric dipole, we call $p = q\,h$, the electric dipole moment and give it a vector direction as shown. k in the picture on the bottom right is the unit vector for the upward direction. For the current loop, $I\,A$ serves the same purpose. So, we define a magnetic dipole moment, $\mu = IA$, and give it the direction as shown. This suggests that current carrying loops will have the same behavior with magnetic fields that a pair of separated charges have with electric fields. This analogy will simplify many discussions in this chapter.

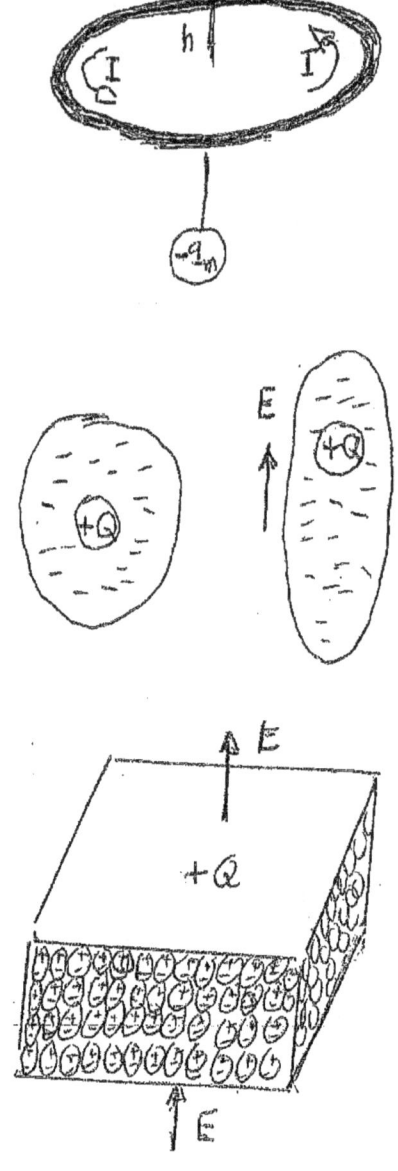

This gives us an idea of a mental artifact or fiction. Suppose there were such a thing as magnetic charge in addition to electric charge and give it the symbol, q_m or Q_m.

Instead of a magnetic moment of IA for a current going around an area, A, suppose that in our thinking we replace it by a magnetic dipole, as shown in the picture above. Let $q_m h = IA$. This might make it easier to carry some of the results from electrostatics over to talk about similar things in magnetostatics.

But remember that this is merely a mental idea. There is no such real thing as magnetic charge although some physicists get carried away talking about magnetic dipoles and start looking for magnetic monopoles. They haven't found any.

We now will talk about electric dipoles. Most atoms and molecules are electrically symmetric showing no charge behavior at all. Take an atom with a positive nucleus with a uniformly distributed electron cloud around it. Now slap on an electric field and the nucleus is pulled in the direction of the electric field and the electrons are distorted in the opposite direction. The resulting dipole moment is proportional to the electric field strength, $qh = \alpha E$, where α is the molecule's polarizability.

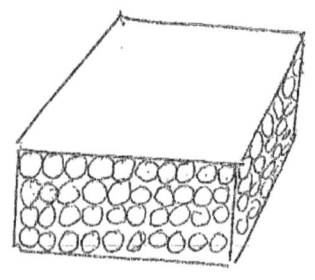

Now look at the molecules in a slab of solid matter (see figure on left). No electric properties are observed. Next slap on an electric field (see figure above). The molecules are distorted. There is now a layer of positive charge on the top face and a layer of negative charge on the bottom. Inside layers of positive and negative charge cancel each other.

13-18

Consider each dipole to have charges, $\pm q$, separated by h. Each molecule has a dipole moment, qh. If A is the area of the top and bottom surfaces, the volume of the molecular layers at these surfaces is Ah. If N is the molecular density, that is MOLECULES PER UNIT VOLUME, then Nah is the number of molecules at each surface. If each molecule contributes a charge, q, to the surface, then the charge (plus or minus) on each surface is $Q = N(qh)A$.

Now we will create the electric field by putting the slab between the plates of a charged capacitor (discussed in chapter 8). Recall that for a charged surface, the electric field at the surface was proportional to the charge per unit area, dQ/dA. For a charge, Q, uniform on a surface, A, $E = \frac{1}{\epsilon_o}\left(\frac{Q}{A}\right)$, where $\epsilon_o = 8.85 \times 10^{-12}$ in empty space.

Between the plates of a charged capacitor whose charge is Q, we call the charges induced on the top and bottom surfaces now Q-prime (Q').

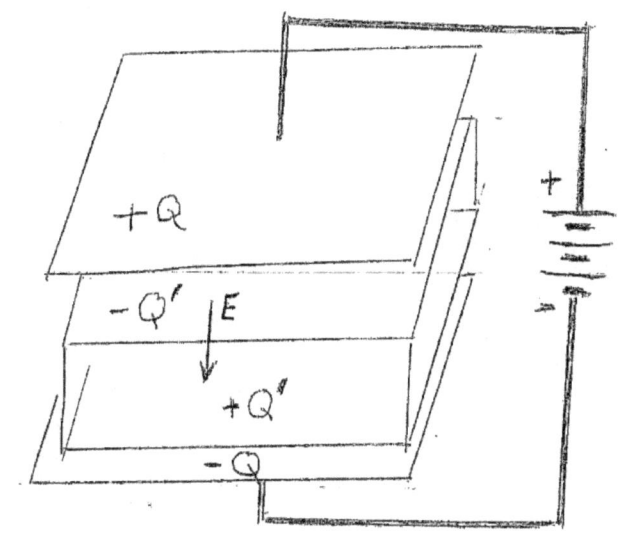

$Q' = N(qh)A$. But, $qh = \alpha E$.
$Q' = N\alpha EA$.
We go to the formula at the bottom of the previous page and for empty space we write ϵ and ϵ_o.

But inside the slab the field is produced by a net charge $Q - Q'$. The formula above is now, $E = \frac{1}{\epsilon_o}\left(\frac{Q-Q'}{A}\right)$.

But, $Q' = N(qh)A$, and thus, $E = \frac{Q}{\epsilon_o A} - \frac{N\alpha}{\epsilon_o}E$. This gives, $\epsilon_o E + N\alpha E = \frac{Q}{A}$.

N is the number of molecules per unit volume in the slab and α is the polarizability of each molecule. As they are properties of the material, we can use another symbol to describe the same properties. We use another Greek letter, χ (chi), called the susceptibility in the following way: $N\alpha = \chi\epsilon_o$.
$[1 + \chi]\epsilon_o E = \frac{Q}{A}$.

As (epsilon) ϵ_o is the permittivity of free space, we now have the permittivity of the material of the slab, $\epsilon = [1 + \chi]\epsilon_o$. $E = \frac{Q}{\epsilon A}$.

At this point, let us make a digression and play with some numbers. $\epsilon_o = 8.85 \times 10^{-12}$. This is a quantity that must be measured. In magnetic formulas there is another quantity, permeability which for free space is defined exactly as $\mu_o = 4\pi \times 10^{-7}$.

The reason that this is defined exactly is that magnetic measurements involving current are easier to do precisely than electrostatic measurements involving charge. The basic quantity in electrical measurements is the ampere and the coulomb is defined in that the ampere is a coulomb per second.

Now, notice the product $\mu_o \epsilon_o = 1.112 \times 10^{-17}$ and $\sqrt{\dfrac{1}{\mu_o \epsilon_o}} = 3 \times 10^8$.

This is a rather suspicious number, especially if you chase down its units and find that they are in meters per second. This raises the question, is light an electromagnetic phenomenon? Later on in this chapter we show that there are waves in electric and magnetic fields and they move with the velocity of light. Let us assume that these can be light waves and see if this helps us understand some things in optics. To start, consider the bending of the direction of a light beam as it crosses the boundary between two media.

In chapter 2, we found that a ray of light broke its direction on going from air into glass and broke it again going from glass into air. This was called refraction (breaking and breaking again). In chapter 6 there were reasons to think that light involved some sort of wave motion and a description of refraction was given involving waves. Let us look at that idea again.

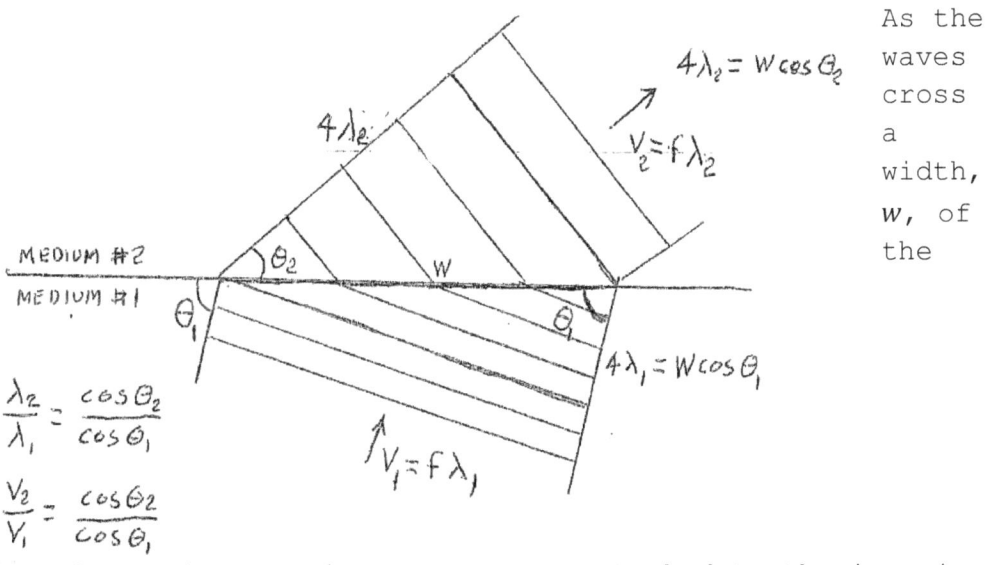

$$\frac{\lambda_2}{\lambda_1} = \frac{\cos\theta_2}{\cos\theta_1}$$

$$\frac{V_2}{V_1} = \frac{\cos\theta_2}{\cos\theta_1}$$

As the waves cross a width, w, of the

boundary, the outgoing wave stays attached to the incoming wave. This requires that the frequency does not change. The wave is interacting with some property of matter to make a change in its velocity. There are only two properties of matter that can be involved, mass and charge. Interactions with mass involve gravity which is ruled out. The wave must be interacting with charge which would involve the wave having electric fields. Later in this chapter we will derive this description of such a wave, shown in the picture on the right. The wave is moving from the bottom to the top of the page. An electric field is vibrating in and out of the page and a magnetic field is vibrating from left to right. All that we are interested in is that as the wave passes it creates a vibrating electric field. Now, look at the picture at the bottom of this page. An electric field distorts the molecule.

Take a hydrogen atom for simplicity. In the absence of fields, we have a spherical nucleus surrounded by spherical electron charge. The atom is electrically neutral. With an electric field present, there are equal and opposite forces on the nucleus and electron. The electron charge, being more flowable, will distort as shown below. Assume that the effective displacement of electron charge is z and that the distortion of the electron orbital gives rise to a

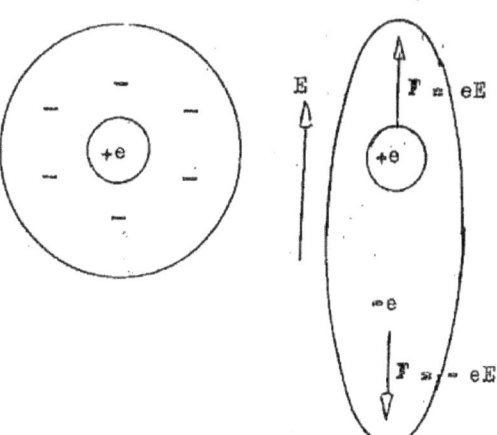

force, $-k\,z$, on the electron attempting to restore symmetry.

In the absence of external effects, a distorted electron would be governed by $m\dfrac{d^2z}{dt^2} = -k\,z$, giving $z = A\,sin\,(\omega_o\,t)$ with $\omega_o{}^2 = \dfrac{k}{m}$.

Now, introduce a vibrating electric field of a passing wave, $\boldsymbol{E} = \boldsymbol{E_o}\,e^{i\,\omega\,t}$, and assume that a vibrating atom would experience energy losses as if it were subject to a drag force, $-b\,v$. Then, $m\dfrac{d^2z}{dt^2} = -k\,z - b\,\dfrac{dz}{dt} + e\,E_o\,e^{i\,\omega\,t}$.

If the atomic charge locks into the driving frequency of the electric field, then $z = A e^{i \omega t}$ and $-m \omega^2 z = -k z - i \omega b z + e E_o e^{i \omega t}$. Rearranging and using $k = m \omega_o^2$, gives $z = \dfrac{e E_o e^{i \omega t}}{m (\omega_o^2 - \omega^2) + i \omega b}$.

The distorted atom becomes an induced charge DIPOLE with equal and opposite charges, $\pm q$, separated by a distance, L. This has the property of a DIPOLE MOMENT, $p = q L$, and if the charges are changing, there will be a current, $I = dq/dt$, between them.

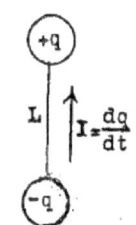

Recall from page 13-20 that in empty space, $c = \sqrt{\dfrac{1}{\mu_o \epsilon_o}}$, was the velocity of electromagnetic waves. Then it is obvious that $v = \dfrac{1}{\sqrt{\epsilon \mu}}$ is the velocity of electromagnetic waves in matter. Then if $\epsilon = \epsilon_o (1 + \chi_e)$, and $\mu = \mu_o$, we get $v = \dfrac{c}{\sqrt{1 + \chi_e}}$.

Now, if the index of refraction of a material is $n = c/v$, we get the index of refraction to be $n = \sqrt{1 + \chi_e}$. Snell's Law is now $\dfrac{\cos \theta_1}{\cos \theta_2} = \dfrac{n_1}{n_2}$. We found that a vibrating electric field $E = E_o e^{i \omega t}$ from an electromagnetic wave caused charges in a molecule with a resonant frequency, ω_o, to be displaced by $z = \dfrac{e E}{m (\omega_o^2 - \omega^2) + i \omega b}$, producing a dipole moment, $p = e z$.

If there are N molecules per unit volume, the volume polarization is $P = N p$. But in terms of the polarizing electric field, $P = \epsilon_o \chi_e E$. Thus, $\chi_e = \dfrac{N e^2}{\epsilon_o [m (\omega_o^2 - \omega^2) + i \omega b]}$ is the electric susceptibility with the index of refraction being $n = \sqrt{1 + \chi_e}$.

The variation of the index of refraction with frequency is called DISPERSION. The figure on the right shows its behavior in the neighborhood of a resonance frequency.

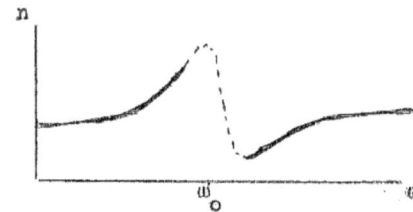

A resonance is where light is absorbed by a quantum transition. In optical materials we are not near such a point. In the graph as frequency increases, wavelength decreases. The horizontal axis is going from long to shorter wavelengths, that is, from red to violet when going from left to right. But as index of refraction increases, velocity decreases.

Let us follow the logic of this result. For light waves, f is frequency, λ is wavelength, and $v = f\lambda$ is velocity. For light waves, when n is the index of refraction, $v = c/n$. In air, $n = 1$ and $v = c$.

In the picture on the right, $\cos\theta_G = \dfrac{\cos\theta_A}{n}$.

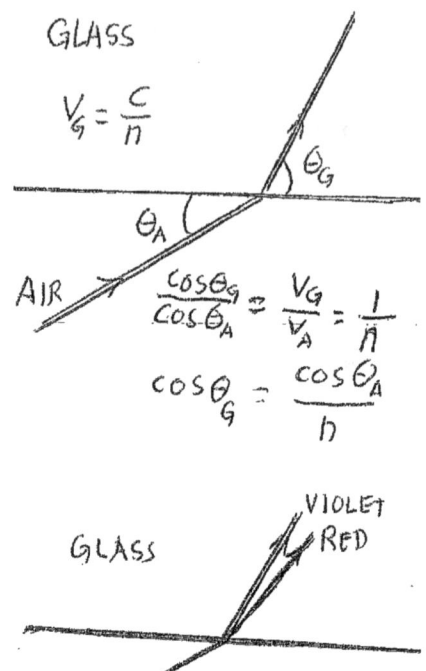

With light, as frequency increases we go from red to violet. In the plot on the previous page, as frequency increased, the index of refraction increased. But in Snell's law, that makes $\cos\theta_G$ decrease. As $\cos\theta_G$ decreases, θ_G increases.

Therefore, as light goes from air to glass its direction bends away from the surface. As we go from red to violet, the direction bends more and more.

Now consider light coming on to a prism. Violet bends away from the surface more than red, giving a spectrum.

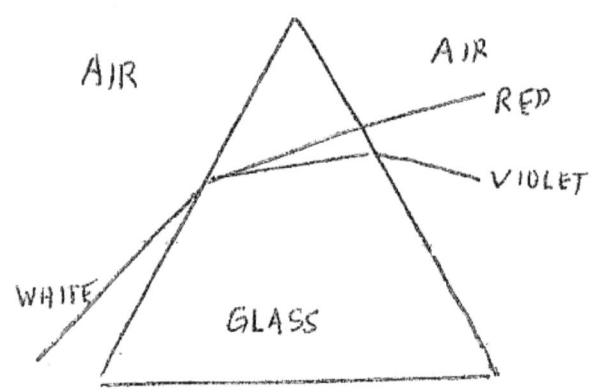

MAGNETISM IN CONDENSED MATTER

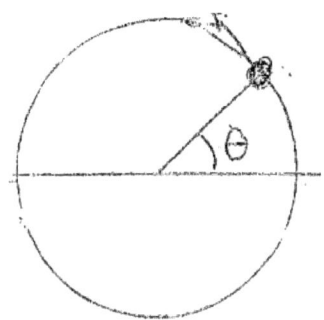

Although magnetism has been known from very ancient times, the theory of it is one of the most difficult topics in physics. From the discussion of the hydrogen atom at the start of this chapter, we found that the electron had angular momentum. We shall soon find that it is spinning, like the earth rotating as it revolves about the sun. But as it has a charge going 'round and 'round, it also has a magnetic moment.

It ends up that all of the basic particles have both angular momentum and magnetic moments and the magnetism comes from the magnetic moments. At the start of chapter 11, an electron going in a circle had angular momentum,

$$L = M r^2 \left(\frac{d\theta}{dt}\right).$$

Quantum physics gives us numbers for the states of angular momentum. Thus, we need a relation between magnetic moment and angular momentum.

Think of a ring with radius, R, having a mass, M, and carrying a charge, Q. It is rotating with $\frac{d\theta}{dt} = \frac{2\pi}{T}$ as its angular velocity. T is the time for one revolution. Its angular momentum is $L = M r^2 \left(\frac{2\pi}{T}\right)$. If the charge, Q, goes around in a time, T, it has a current, $I = \frac{Q}{T}$. The magnetic moment, $m = I A$, becomes $m = \frac{Q}{T}\pi R^2$.

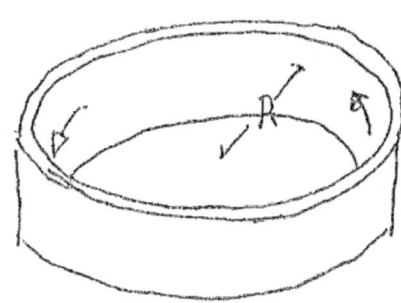

The ratio of magnetic moment to angular momentum is $\frac{m}{L} = \frac{Q}{2M}$.

In the text that follows, we will be describing things that change in three dimensions. All steps in the derivation will be here, but there are a lot of steps. Go through the steps slowly and carefully, thinking about each step as you go along. This derivation is important in physics. The gradient operator, $\boldsymbol{\nabla} = \boldsymbol{i}\frac{\partial}{\partial x} + \boldsymbol{j}\frac{\partial}{\partial y} + \boldsymbol{k}\frac{\partial}{\partial z}$, has been discussed.

It is often called by its short name DEL and is treated as vector. We will reproduce some of the identities which were developed. If $\boldsymbol{r} = \boldsymbol{i}x + \boldsymbol{j}y + \boldsymbol{k}z$ is the radius vector to any point in our region of interest, let $\boldsymbol{F}(\boldsymbol{r}) = \boldsymbol{F}(x,y,z)$ and $\psi(\boldsymbol{r}) = \psi(x,y,z)$ be a vector and a scalar function of position.

In addition to the gradient of ψ, $\boldsymbol{\nabla}\psi = \boldsymbol{i}\frac{\partial\psi}{\partial x} + \boldsymbol{j}\frac{\partial\psi}{\partial y} + \boldsymbol{k}\frac{\partial\psi}{\partial z}$, which is often called DEL PSI, we also have the divergence, $div\,\boldsymbol{F} = \boldsymbol{\nabla}\boldsymbol{\cdot}\boldsymbol{F} = \frac{\partial F_x}{\partial x} + \frac{\partial F_y}{\partial y} + \frac{\partial F_z}{\partial z}$, which gives a number.

We have discussed the gradient of a number (scalar) function which gives a vector, and the divergence of a vector function which gives a number function. But what about the cross product of the gradient with a vector function? We can do it but what is its use in physics and why do we call it a curl?

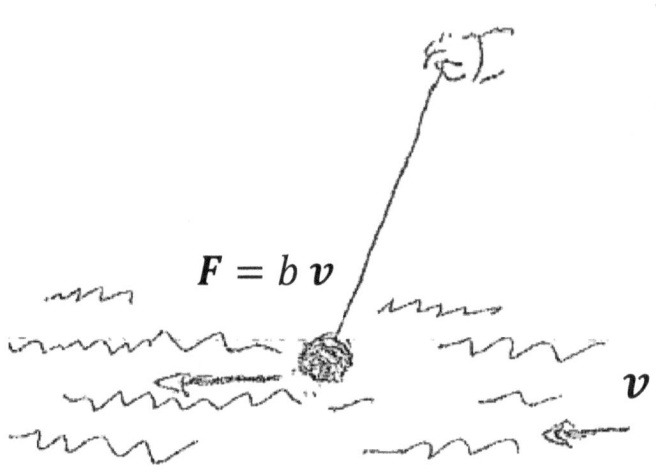

$$\boldsymbol{F} = b\,\boldsymbol{v}$$

$$\boldsymbol{v}$$

To answer that question, we again go to the example of flowing water. You are holding on a string tied to a ball that floats in a stream of water flowing with a velocity, $\boldsymbol{v}$. There is a drag force, $\boldsymbol{F} = b\,\boldsymbol{v}$, on the ball which has a drag coefficient, b.

If the ball moves a short distance, $d\boldsymbol{r}$, work, $dW = b\,\boldsymbol{v}\boldsymbol{\cdot}d\boldsymbol{r}$ is done. If the ball is carried around a closed path, the total work done is:

$$W = \oint b\,\boldsymbol{v}\boldsymbol{\cdot}d\boldsymbol{r}$$

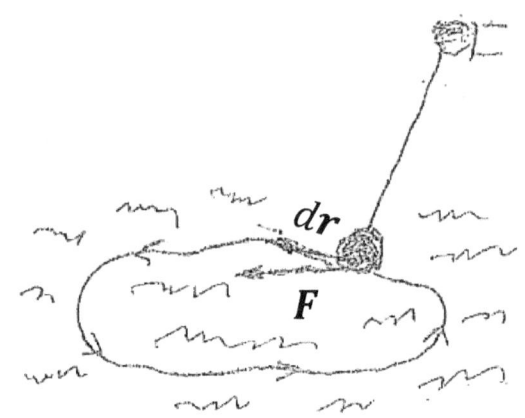

$$d\boldsymbol{r}$$

$$\boldsymbol{F}$$

Going around a closed path is indicated by a circle on the summation sign.

Let us move the ball very slowly along some contour. With any displacement, $dr = i\,dx + j\,dy$, the applied force will do work, $F \cdot dr$.

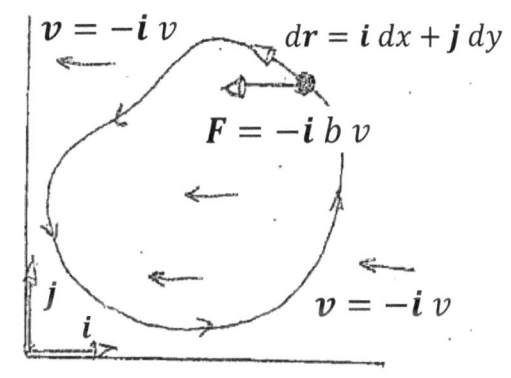

We consider the work on the ball carrying it around a closed contour, ending up wherever we began. Convention takes the positive direction for going around a contour to be counter-clockwise, as seen from above the xy-plane. Such a contour sum around a closed contour is represented by:

$$\oint_C F \cdot dr$$

In a region where the velocity is constant in direction and magnitude, say with v being constant, the applied force is $F = -i\,b\,v$ and $F \cdot dr = -b\,v\,dx$. Going around a closed contour, the work is:

$$\oint F \cdot dr = \oint -b\,v\,dx = -b\,v \oint dx = 0$$

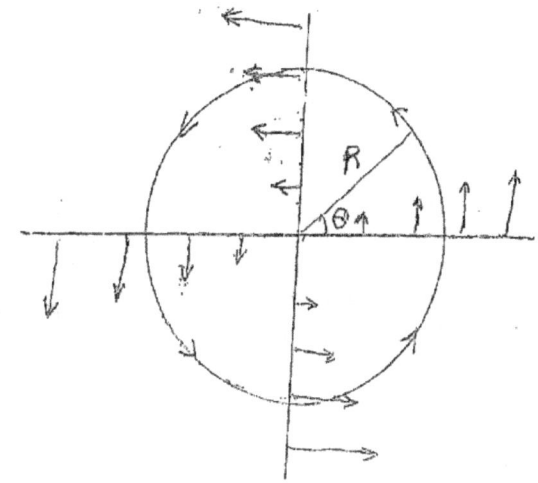

As in going around a closed contour, the stopping value of x is the starting value of x giving a net change of zero. In a constant velocity of flow, zero work around a closed contour is no surprise.

Next take a whirlpool giving a force, $F = -i\,c\,y + j\,c\,x$. In displacement, $dr = i\,dx + j\,dy$, work is $dW = F \cdot dr = -c\,y\,dx + c\,x\,dy$.

In summing the work done around a closed path there is no loss of generality if the path is a circle of radius, R. Then, in a direction, θ, $x = R\cos\theta$ and $y = R\sin\theta$, with $dx = R\sin\theta\,d\theta$ and $dy = R\cos\theta\,d\theta$. This gives $dW = c\,R^2 \sin^2\theta\,d\theta + c\,R^2 \cos^2\theta\,d\theta$ or $dW = c\,R^2\,d\theta$. Going around a closed path, $W = \oint c\,R^2\,d\theta$ or $W = c\,R^2 \oint_{\theta=0}^{2\pi} d\theta = 2\pi\,c\,R^2$, which is no surprise.

From all this we conclude that $\oint F \cdot dr = 0$ is a condition for what we might call streamline flow. When the sum is not zero, the flow involves some sort of eddying or rotational flow. To complete the picture, we will next prove the most general theorem for such a summation around a closed path. The result of that is called Green's theorem.

This suggests that when a fluid is rotating in an eddy or vortex, the integral of the velocity around a closed contour is not-zero. When the integral of the velocity around a closed contour is zero, we call the flow laminar or irrotational.

For any fluid flowing across a plane, we will now develop a general test as to whether the flow is rotational or irrotational. Let the velocity vector be a function of position, $v = i\,v_x(x,y) + j\,v_y(x,y)$ where the components: $v_x(x,y)$ and $v_y(x,y)$ are differentiable functions of position.

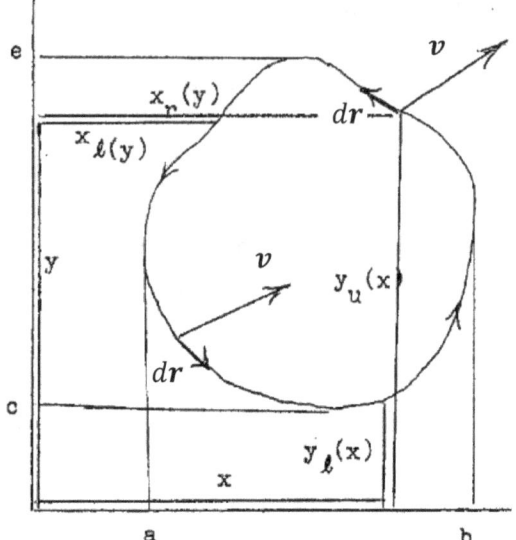

The integral, $\oint_C v \cdot dr = \oint_C v_x(x,y)\,dx + \oint_C v_y(x,y)\,dy$, done counterclockwise around a closed contour gives us two separate integrals. For the integral over the variable, x, we again divide the contour into an upper half and a lower half, as we did before.

$\oint_C v_x(x,y)\,dx = \int_b^a v_x\big(x,y_u(x)\big)\,dx + \int_a^b v_x\big(x,y_l(x)\big)\,dx$ gives us,

$\oint_C v_x(x,y)\,dx = -\int_a^b \big[v_x\big(x,y_u(x)\big) - v_x\big(x,y_l(x)\big)\big]\,dx$.

But, $v_x\big(x,y_u(x)\big) - v_x\big(x,y_l(x)\big) = \int_{y_l}^{y_u}\left(\dfrac{\partial v_x}{\partial y}\right)dy$ and

$\oint_C v_x(x,y)\,dx = -\int_a^b\int_{y_l}^{y_u}\left(\dfrac{\partial v_x}{\partial y}\right)dy\,dx = -\iint_{Area\ inside\ C}\left(\dfrac{\partial v_x}{\partial y}\right)dy\,dx$.

The integral over y is then, $\oint_C v_y(x,y)\,dy = \int_c^e v_y(x_r(y),y)\,dy + \int_e^c v_y(x_l(y),y)\,dy$, dividing the contour into right and left halves. This leads to:

$$\oint_C v_y(x,y)\,dy = \int_c^e \big[v_y(x_r(y),y) - v_y(x_l(y),y)\big]\,dy = \int_e^c\int_{x_l}^{x_r}\left(\dfrac{\partial v_y}{\partial x}\right)dx\,dy = \iint_{Area\ inside\ C}\left(\dfrac{\partial v_x}{\partial y}\right)dy\,dx$$

Putting the two integrals together:

$$\oint_C v \cdot dr = \iint_{Area\ inside\ C}\left[\dfrac{\partial v_y}{\partial x} - \dfrac{\partial v_x}{\partial y}\right]dx\,dy$$

This identity is known as Green's Theorem on a plane.

Green's theorem goes as follows: The condition for $\oint_C \boldsymbol{v} \cdot d\boldsymbol{r} = 0$ around any closed contour is that $\dfrac{\partial v_y}{\partial x} - \dfrac{\partial v_x}{\partial y} = 0$ everywhere in the region of flowing fluid.

It has been shown that for a particle with momentum, $m\,v$, and wavelength, λ, they were related by $m\,v\,\lambda = h$. If the orbit of the hydrogen electron contained n (integer) wavelengths, then $\lambda = \dfrac{2\,\pi\,r}{n}$ and $m\,v\,\dfrac{2\,\pi\,r}{n} = h$.

Therefore, $m\,v\,r = n\,\dfrac{h}{2\,\pi}$.

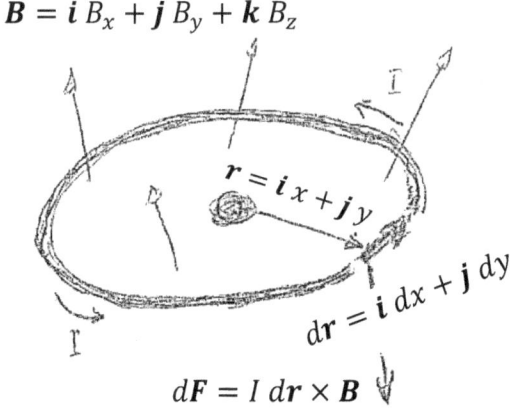

$$\boldsymbol{B} = \boldsymbol{i}\,B_x + \boldsymbol{j}\,B_y + \boldsymbol{k}\,B_z$$

$$\boldsymbol{r} = \boldsymbol{i}\,x + \boldsymbol{j}\,y$$

$$d\boldsymbol{r} = \boldsymbol{i}\,dx + \boldsymbol{j}\,dy$$

$$d\boldsymbol{F} = I\,d\boldsymbol{r} \times \boldsymbol{B}$$

But, $m\,r\,v$ is angular momentum. Giving angular momentum the letter, L, we see that the electron's orbit has a quantized angular momentum, $L_n = n\,\dfrac{h}{2\,\pi}$.

If the electron with a charge, $-e$, goes around its orbit in a time, T, it constitutes a current of $I = -\dfrac{e}{T}$ amperes. The following discussion will involve the energy of such a current loop in a magnetic field.

Look at a current carrying loop in a region with a magnetic field. The field may have differing values at differing locations. The force on a current, I, (taken in the direction positive charge would move) in an arc, $d\boldsymbol{r}$, of the loop is $d\boldsymbol{F} = I\,d\boldsymbol{r} \times \boldsymbol{B}$ and the sum, $\boldsymbol{F} = \oint I\,d\boldsymbol{r} \times \boldsymbol{B}$, is the total magnetic force.
$$d\boldsymbol{r} \times \boldsymbol{B} = \boldsymbol{i}\left(dy\,B_z - dz\,B_y\right) + \boldsymbol{j}\left(dz\,B_x - dx\,B_z\right) + \boldsymbol{k}\left(dx\,B_y - dy\,B_x\right).$$

In the case we have taken, $dz = 0$. The summation of things around closed loops is done on page 13-27, except this time we have components of magnetic field rather than of velocity.

$$\oint B_z\,dy = \iint \left(\frac{\partial B_z}{\partial x}\right) dx\,dy \qquad \oint B_z\,dx = -\iint \left(\frac{\partial B_z}{\partial y}\right) dx\,dy$$

$$\oint B_y\,dx = -\iint \left(\frac{\partial B_x}{\partial y}\right) dx\,dy \qquad \oint B_x\,dy = \iint \left(\frac{\partial B_x}{\partial x}\right) dx\,dy$$

The double summations are over the pieces $dx\,dy$ of the area bounded by the loop. If the rates of change (derivatives) of the magnetic field are constant over the area, then in each double summation they can be factored out and $\iint dx\,dy = A$, the area of the loop.

The total force on the current loop is:

$$F = I\,A\left[i\left(\frac{\partial B_z}{\partial x}\right) + j\left(\frac{\partial B_z}{\partial y}\right) - k\left(\frac{\partial B_x}{\partial x} - \frac{\partial B_y}{\partial y}\right)\right]$$

Recall from page 13-15, $\nabla \cdot B = 0$ giving $\dfrac{\partial B_x}{\partial x} + \dfrac{\partial B_y}{\partial y} = -\dfrac{\partial B_z}{\partial z}$.

The force is then $F = I\,A\left[i\left(\frac{\partial B_z}{\partial x}\right) + j\left(\frac{\partial B_z}{\partial y}\right) + k\left(\frac{\partial B_z}{\partial z}\right)\right]$. But that is the gradient operating on B_z. Therefore, we can rewrite F, $F = I\,A\,\nabla B_z$.

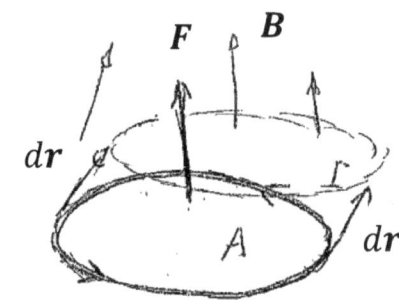

With a force of $F = I\,A\,\nabla B_z$ acting, if the ring is displaced, dr, work, $dW = F \cdot dr$, is done on page 13-11, which is $dW = I\,A\,\nabla B_z \cdot dr$. But, $\nabla B_z \cdot dr = dB_z$, the change is of B_z with the change of location, dr. This gives $dW = d(I\,A\,B_z)$.

If the force has an associated potential energy, $V(x,y,z)$, the work done by the force equals a loss of potential energy, $dW = -dV$. Potential energy changes by $dV = -d(I\,A\,B_z)$, suggesting that $V = -I\,A\,B_z$. Now, $B = i\,B_x + j\,B_y + k\,B_z$; and $k \cdot i = 0$, $k \cdot j = 0$, and $k \cdot k = 1$.

We can write $V = -I\,A\,k \cdot B$. Hanging $I\,A$ onto k gives a vector, $\mu = I\,A\,k$, which we call the magnetic moment of the current loop. Yes, the Greek letter mu is used for too many things in the physics of magnetism, but I am not making this up as I go along, this is standard physics language.

Let the ring be made of something with mass, M, and charge, q, and let it all go around in a time, T. As all the charge passes any point in a time, T, the current is $I = \frac{q}{T}$. If the ring is a circle with radius, R, the area is $A = \pi R^2$. This gives $\mu = \frac{q}{T}\,\pi R^2$ as the size of the magnetic moment. The rotating mass has an angular momentum, $L = M\,R\,v$, but the velocity is $v = \frac{2\pi R}{T}$, giving $L = \frac{M\,2\pi R^2}{T}$, from which $\frac{\pi R^2}{T} = \frac{L}{2M}$.

As result, we hang the magnetic moment onto the angular momentum. $\mu = \frac{q}{2M}\,L$. And for the hydrogen electron, $L_n = n\frac{h}{2\pi}$.

The energy of a magnetic dipole (call energy E here) in a magnetic field is $E = -\boldsymbol{\mu} \cdot \boldsymbol{B}$. Then, $E = -\frac{q}{2M} \boldsymbol{L} \cdot \boldsymbol{B}$.

When a particle moves, it moves in the direction of decreasing energy so if the particle is in a region where the strength of the magnetic field varies with location, and if angular momentum is in the same direction as the field, then the particle will move in the direction of stronger field. If angular momentum is opposite to the field direction, a particle will move in the direction of a weaker field. This brings us to an important experiment done by Stern and Gerlach in 1922.

When a beam of silver atoms passed between magnet poles shaped to give a field of varying strength (as shown in the right-hand picture), the beam split into two distinct beams; one pulled in the direction of stronger field and one pulled in the direction of weaker field. Some property of the atoms had two states of angular momentum.

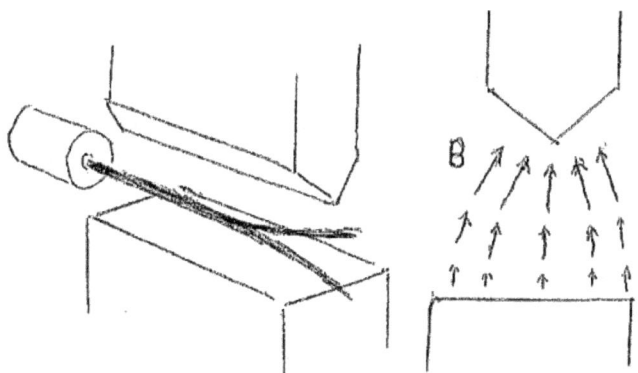

We are now ready to study MAGNETISM IN CONDENSED MATTER. Using pictures, we will try to give you physical intuition about the subject. Because the interactions between particles are so much stronger than in the gas phase, the mathematics involved is much more difficult and on top of that, we are not sure that we understand all the physics of what we are about to discuss. On the other hand, the potential applications are so important that industry from computers to medical equipment to electric power distribution is following every bit of research in these areas.

Magnetic moment, $\boldsymbol{\mu}$, was related to angular momentum, $\boldsymbol{L}$, by $\boldsymbol{\mu} = \frac{q}{2M} \boldsymbol{L}$ and $L = n\frac{h}{2\pi}$ with h being Plank's constant and the quantum number changed by integer jumps. In a magnetic field the particle had energy, $E = \boldsymbol{\mu} \cdot \boldsymbol{B}$.

For the basic particles, the charge is always plus or minus $e = 1.6 \times 10^{-19}$ coulomb. The charge and mass of the particles do not have the same distribution giving rise to a g-factor. $\boldsymbol{\mu} = g\left(\frac{e}{2M}\right)\boldsymbol{L}$. For all particles, the angular momentum has the same quantization and as the proton has about **2000** time the mass of the electron, magnetism is a function of electron behavior.

The result of the Stern-Gerlach experiment on the previous page holds for
all elementary particles. With only two states of what we call spin and the
quantum numbers differing by unity, the spin quantum numbers are plus or
minus 1/2.

The electron has two states of angular momentum and in a magnetic field it
is either in the direction or in the opposite direction to the field. In the
same direction, the energy is negative and in the opposite direction, the
energy is positive. Given a choice, the negative energy is the stable state.

Now, take a material (maybe iron) where each atom has an electron whose
magnetic moment is free to do anything. Normally the moments will happen to
be pointing in random directions and add up to give no magnetic properties
seen from the piece of material. But slap a magnetic field onto the
material, and the electron moments line up with the field giving the whole
thing a big magnetic moment, behaving as what we call a magnet. This is
called PARAMAGNETISM.

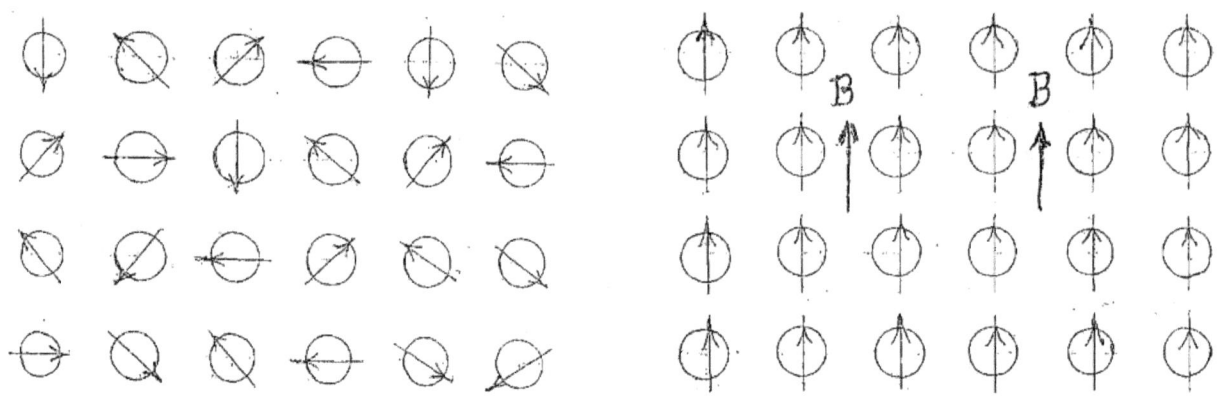

But with iron, the moments may still be lined up, even after taking away the
magnetic field. When we have a permanent magnet, this is called
FERROMAGNETISM and is due to interactions between nearby magnetic moments.

To get some idea as to what is going on here, we now look at interactions
between nearby magnetic moments. Doing so we use the identical behavior of
electric and magnetic moments. It is more simple to talk about electrical
things than magnetic things.

Use Φ (capital phi) for the electric potential (voltage). What is its formula due to a charge dipole at the coordinate origin with a dipole moment of $p = Q h k$?

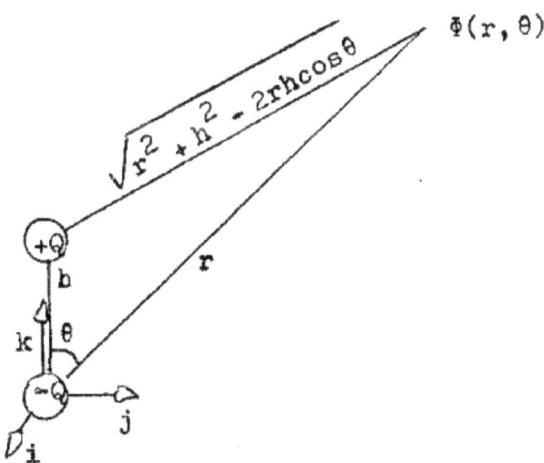

At a direction θ from the z-axis, go a distance, r (spherical coordinates), and the electric potential due to the two charges is:

$$\Phi(r,\theta) = \frac{+Q}{4\pi\epsilon_o \sqrt{r^2 + h^2 - 2rh\cos\theta}} + \frac{-Q}{4\pi\epsilon_o r}$$

$$\Phi(r,\theta) = \frac{Q}{4\pi\epsilon_o r}\left[1 + \left(\frac{h}{r}\right)^2 - 2\left(\frac{h}{r}\right)\cos\theta\right]^{-1/2} - \frac{Q}{4\pi\epsilon_o r}$$

Ignoring $\left(\frac{h}{r}\right)^2$ and using the tricks from the box:

$$\left[1 + \left(\frac{h}{r}\right)^2 - 2\left(\frac{h}{r}\right)\cos\theta\right]^{-1/2} = 1 + \frac{h}{r}\cos\theta$$

Giving the electric potential, $\Phi(r,\theta) = \frac{Q h \cos\theta}{4\pi\epsilon_o r^2}$.

$$\Phi(r,\theta) = \frac{Q h}{4\pi\epsilon_o r^2}\left(\frac{z}{r^3}\right) \text{ and } E = -\nabla\Phi.$$

If E is electric field, call energy, U. Let the second dipole be parallel to the first.

$$U = -(q h k)\bullet E \text{ or } U = q h \frac{\partial\Phi}{\partial z}.$$

$$U = \frac{(q h)^2}{4\pi\epsilon_o}\frac{\partial}{\partial z}\left(\frac{z}{r^3}\right) \text{ and }$$

$$\frac{\partial}{\partial z}\left(\frac{z}{r^3}\right) = \frac{1}{r^3} + z\frac{\partial}{\partial z}(r)^{-3}$$

$$\frac{\partial}{\partial z}(r)^{-3} = \frac{\partial}{\partial r}(r)^{-3}\frac{\partial r}{\partial z} = -3r^{-4}\left(\frac{z}{r}\right) = \frac{-3z}{r^5}$$

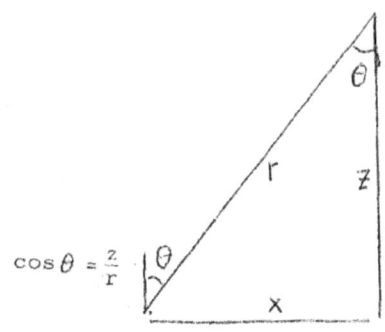

$\cos\theta = \frac{z}{r}$

Copy everything on the previous page over on a piece of paper and think about each step. It will take a while but if you do it with care, you will understand the logic.

Then, putting it all together, the energy is $U = \dfrac{(q\,h)^2}{4\,\pi\,\epsilon_o}\left[\dfrac{1}{r^3} - \dfrac{3\,z^2}{r^5}\right]$.

But, $z = r\cos\theta$. The mutual energy of a pair of dipoles is $U = \dfrac{(q\,h)^2}{4\,\pi\,\epsilon_o}\left[\dfrac{3\cos^2\theta - 1}{r^3}\right]$.

For the magnetic case, replace $\dfrac{(q\,h)^2}{4\,\pi\,\epsilon_o}$ by $\dfrac{\mu_o\,(I\,A)^2}{4\,\pi}$.

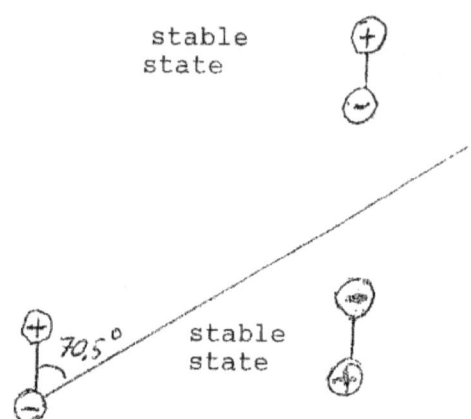

This assumes that the dipoles are in the same direction and parallel. This is stable (or negative) when $\cos\theta$ is greater than $\sqrt{1/3}$ or θ is less than 70.5 degrees. When θ is greater than 70.5 degrees this gives a positive number (unstable). For these directions the stable state is when the dipoles are in opposite directions.

With magnetism in matter, the interacting dipoles are electron spins and the interacting electrons are valence electrons of atoms. Hence, atomic orientations determine magnetic properties of materials. We now look at the way atoms are arranged in solid matter.

THE STRUCTURE OF SOLID MATTER

When there are very many atoms in the solid phase, they most often arrange themselves in an orderly structure called a crystal lattice. Some basic pattern of atoms, sometimes simple but often quite complicated, is repeated over and over again in an orderly manner. An explanation of this involves the shapes of atomic electron orbitals developed in chapter 17.

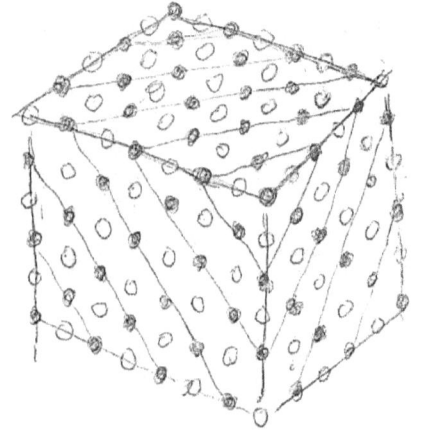

An example of a crystal is MnO. The black atoms are manganese and the white atoms are oxygen in the picture.

A structure that is repeated over and over again is called a unit cell. The following pictures on the next page are two variations on a cubic cell.

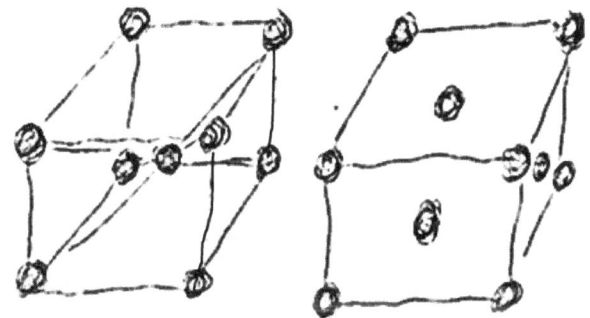

On the left is an example of a body centered cubic lattice, such as iron.

On the right is a face centered cubic lattice (shown as a solid).

A couple two dimensional structures can be stacked to form a three dimensional solid.

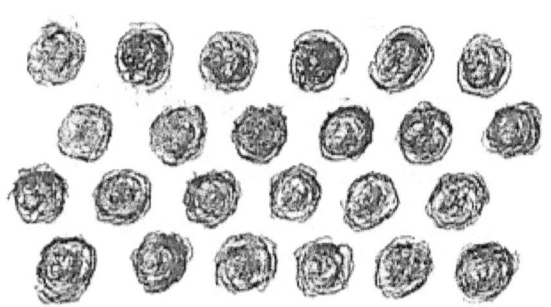

Hexagonal close packed

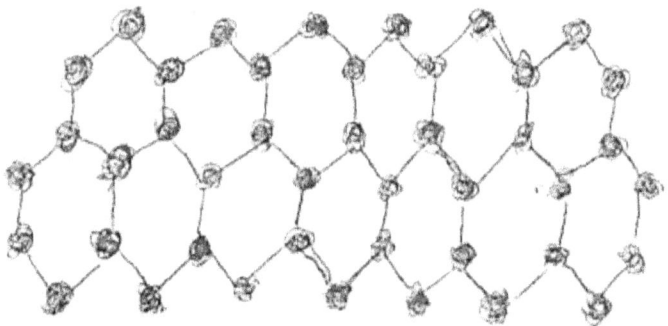

Graphene is a vast extension of benzene rings. When rolled into a cylinder it makes carbon nanotubes. When stacked in layers it makes graphite.

It is evident that through any atom in a crystal there pass several planes of atoms and that, depending on the unit cell, these planes of atoms define the type of crystal. Most commonly, the set of planes is observed with x-rays.

Electrons hit a hard metal surface. The energy lost creates photons.

Electrons are accelerated through voltage, V, and those passing through a hole in a baffle come out in a narrow beam.

Electrons are emitted from a hot metal surface.

The accelerating voltage, V, gives each electron an energy, $E = hf$, where $f\lambda = c$. The resulting wavelength is $\lambda = \frac{hc}{eV}$. An accelerating voltage of 10^4 volts produces waves with wavelength, $\lambda = 1.24 \times 10^{-10}$ meter, which is on the order of the size of atoms.

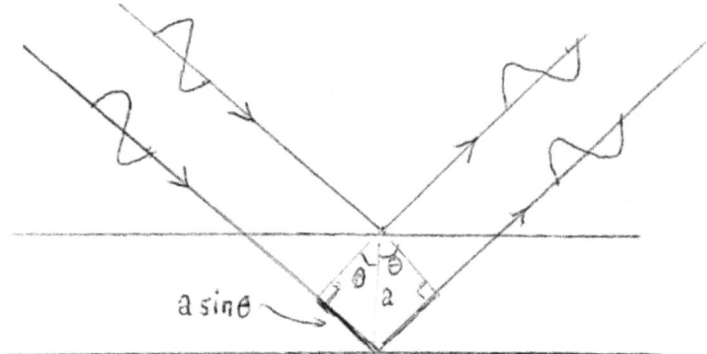

The two horizontal lines in the figure on the left represent two successive planes of atoms.

Waves reflected from the lower plane go this distance before aligning with waves reflected from the upper plane.

The distance between the atomic planes is a. Then incoming x-ray waves making an angle, θ, with the atomic planes will be reflected off adjacent atomic planes. The wave reflected from the lower plane goes a distance, $2\,a\sin\theta$, before aligning with the waves reflected from the upper plane. If that distance is an integer plus one-half wavelengths, the reflected waves will cancel each other. Hence, the directions of x-ray beams coming off a crystal will give us the orientation of atomic planes in the crystal.

In 1927, C.J. Davisson and L. Germer got the same effect with a beam of electrons, experimentally confirming that electrons had waves.

We are now able to think about two types of permanent magnetic alignment, FERROMAGNETIC and ANTIFERROMAGNETIC. Ferromagnetism gives us magnets and always has been of interest.

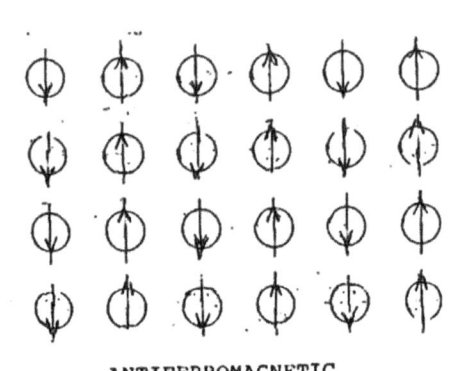

FERROMAGNETIC   ANTIFERROMAGNETIC

Antiferromagnetic materials have no overall magnetic behavior and up to recently have not been of much interest. But the discovery of high-temperature superconductivity, as we shall see later in this chapter, has made antiferromagnetism all the rage in 21st century physics. The dipole-dipole energy formula may help us understand the differences between these two forms of magnetism.

In actual atoms, there are numerous electrons in each Bohr orbit (now called a shell) and as their spins are paired (up and down), electrons in filled shells make no magnetic contribution. It is the valence electrons in unfilled shells that are of interest. It is the type of crystal lattice that is important. If each atom has more nearby atoms above and below than on the sides, the result may be ferromagnetism. In antiferromagnetic materials atoms above and below each other may have spins in the same direction. Atoms beside each other may have spins in opposite directions, being overall antiferromagnetic.

There is now the consideration of temperature and thermal energy. As temperature rises, atoms in the lattice vibrate more and more. In a later chapter, we will find that the energy of lattice vibrations is carried by particles called phonons (similar to photons in light energy). Phonons bumping into electrons keep them from lining up.

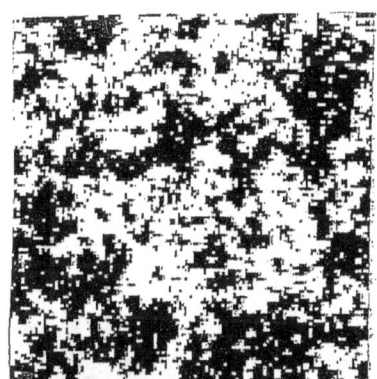

As the material is cooled, there are less and less phonons and the collisions are reduced to the point where electron spins can line up. The theory of abrupt phase transitions from disorder to order is a serious topic in physics. With magnetism the temperature of disorder to order is called the Curie temperature, T_c, after Pierre Curie (1859-1906), who started the serious study of magnetism. With antiferromagnetism, it is the Néel temperature, T_c, after Louis E. F. Néel (1904-2000), who made major contributions to the study of antiferromagnetism. Some Curie temperatures are 1043 K for iron, 1394 K for cobalt, and 613 K for nickel. Some Néel temperatures are 300 K for CuO 185 K for FeO, 122 K for MnO.

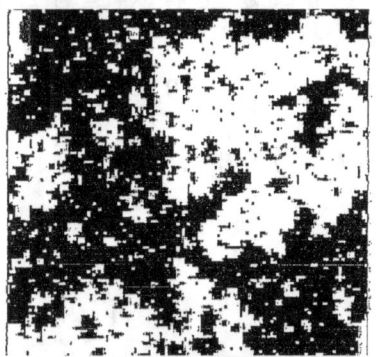

The pictures on the right show a computer simulation of a disorder to order phase transition. In black areas, the spin quantum number is $+1/2$ and in the white areas, the spin quantum number is $-1/2$. If T_c is the Curie temperature, in the top picture, $T = 1.2\,T_e$. Here, above the Curie temperature, we see small domains of aligned spins which are randomly distributed. Overall, there is no effective long-range magnetization.

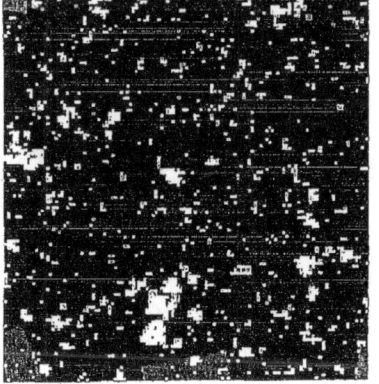

The middle picture shows the system at its Curie temperature. Now large domains with magnetization spin up and spin down are forming.

In the bottom picture, $T = 0.95\,T_e$. Now, an over-all magnetization with spin, $S = +1/2$, dominates the system.

What we have here is spontaneous magnetization. From the outside we apply a magnetic field, B. The domains will be stimulated to produce over-all magnetization which will then persist after the external field is removed. This produces what is called hysteresis.

In a region where there is a cloud of charge, not dominated by spin, another type of magnetism occurs. This is called DIAMAGNETISM and it is the opposite of paramagnetism. Before we think about diamagnetism, let us return to paramagnetism.

With permanent dipole moments, IA, let the average amount that they are in line with the field, B, be the magnetic polarizability, α, letting $IA = \alpha B$. If N dipoles PER UNIT VOLUME is the dipole density of particles, we define the volume magnetization vector as $M = N(IA)$ or $M = N\alpha B$. If B_0 is the applied field from the outside, the aligned dipole moments create their own field and the total field is $B = B_0 + M$. If $M = N\alpha B_0$, then the magnetic field in the material is:

$$B = B_0 + N\alpha B_0 = (1 + N\alpha)B_0$$

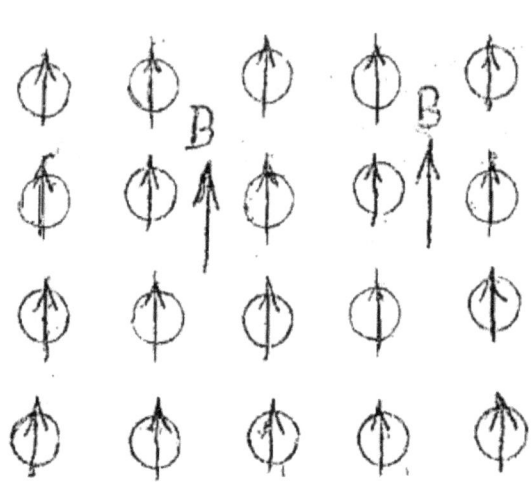

$N\alpha = \chi_m$ is called the magnetic susceptibility.

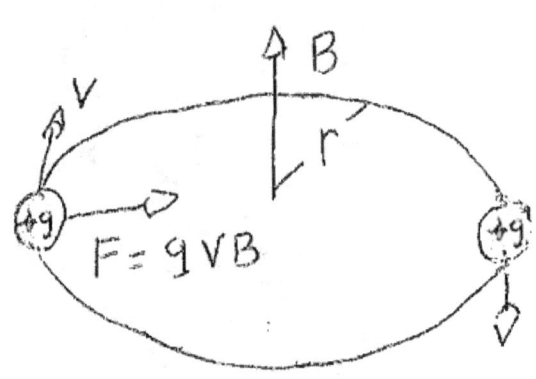

Now, consider a gas of particles each with a charge, $+q$, and mass, m. Because of their thermal energy a particle will have energy, $E = \frac{1}{2}mv^2$. Introduce a magnetic field, B, and a force, $F = qvB$, which is perpendicular to both velocity and field due to the definition of the cross product. $F = q\,v \times B$ leads to a magnitude of qvB. A force perpendicular to velocity makes the particle go in a circle. Let m be a single particle's mass.

Mass times centripetal acceleration gives $E = m\dfrac{v^2}{2} = qvB$ or $mv = qBr$. Now, $v = \dfrac{2\pi r}{T}$, where T is time for one revolution. $\dfrac{2\pi m r}{T} = qBr$.

Cancelling r gives $\dfrac{2\pi m}{T} = qB$, from which $\dfrac{1}{T} = \dfrac{qB}{2\pi m}$. If charge q goes around once in time, T, the current is $I = \dfrac{q}{T}$.

The induced magnetic moment is $(I\,A) = \left(\dfrac{q}{T}\right)\pi\,r^2$ giving $(I\,A) = \left(\dfrac{q^2\,B\,r^2}{2\,m}\right)$.

But let's look at this result. If the current is going counterclockwise and the dipole moments are directed down, the volume magnetization is directed down, opposite to the applied field, B_o. If the resulting field due to induced dipoles is $B' = M$. The total field in the material is $B = B_o - M$ and the susceptibility in this case is negative.

The primary place where diamagnetism is found is in the cloud of charge that constitutes a molecular electron orbital. Here it causes the chemical shift in nuclear magnetic resonance, a major spectroscopy spectrum for the chemists.

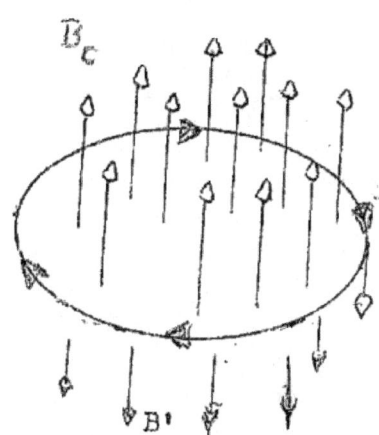

At very low temperatures, near absolute zero, materials behave in ways that are strange (and even goofy). There seems to be a whole new physics acting at low temperatures, unlike what we are accustomed to. We will start with the fact that the ability of certain materials to conduct electricity involves their resistivity going to zero. For that, we go back now and think about normal electrical resistance.

We shall now develop a simple-minded theory of electrical conduction: Consider a gas of free electrons whose overall charge is neutralized by a background lattice of positive ions, with which they do not interact. Let them be subject to collisions.

An electric field, E, will accelerate electrons in a direction opposite to itself, creating an electric current. But the electrons will be bombarded by phonons knocking electrons out of the current. Assume that the average effect of a phonon collision is to reduce the electron's velocity in the current to zero and so the field has to start all over again to bring the electron into the current. If t is the average time between collisions, the electron will move a distance, $x = \frac{1}{2}\left(\frac{eE}{m}\right)t^2$, with an average velocity,

$$\bar{v} = \frac{x}{t} = \frac{eE}{2m}t.$$

By the law of Dulong and Petit, the heat capacity of a metal is $C = 3R$, R being the gas constant. Hence, the number of phonons should be proportional to the temperature, T, giving collisions per unit time, $v = cT$, where c is some constant. The time between collisions is $t = \frac{1}{v}$ and the average velocity,

$$\bar{v} = \frac{eE}{2mv} = \frac{eE}{2mcT}.$$

Let the current be flowing in a wire of cross-section area, A. Let there be N electrons per unit volume. An average electron goes a distance, $\bar{v}t$, in a time, t,

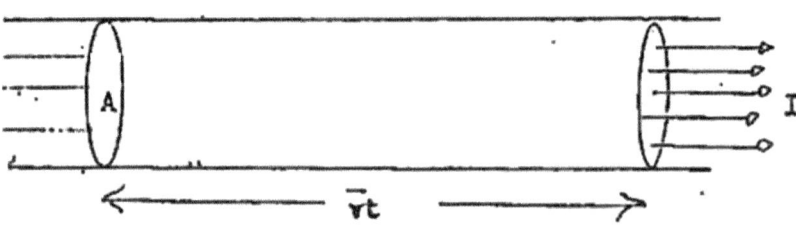

and all electrons before it will cross the far cross-section in that time. All electrons in a volume, $A\bar{v}$, will pass that point in a unit time. This gives an electric current, $I = NeA\bar{v} = \left(\frac{Ne^2}{2mcT}\right)AE$.

$\sigma = \dfrac{N\,e^2}{2\,m\,c\,T}$ is the conductivity and the

electrical resistance is $R = \dfrac{L}{\sigma A}$ for a path

of length, L. $R = \left(\dfrac{2\,m\,c\,L}{e^2\,N\,A}\right) T$ is proportional
to temperature, as measurements of tungsten
wire will confirm.

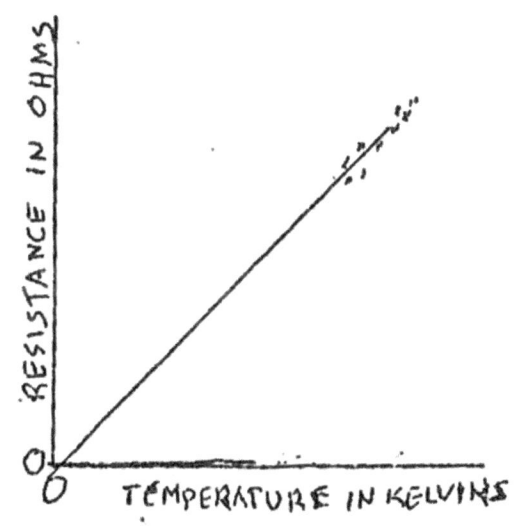

Although talking about collisions of
particles gives some answers, we found at
the start of this chapter that particles
had waves. We next look at the wave of a
particle in a box.

The wave will be a standing wave with an
integer number of wavelengths in the length
of the box. We found that a particle with momentum, $m\,v$,
had a wave of wavelength, λ, satisfying $(m\,v)\,(\lambda) = h$,
h being Planck's constant. For integer number of
wavelengths in the length, L, of the box, $L = n\,\lambda$ for

$n = 1,\ 2,\ 3,\ 4$ etc. The energy, $E = \dfrac{m^2\,v^2}{2\,m}$, gives

$E_n = \dfrac{n^2\,h^2}{2\,m\,L^2}$.

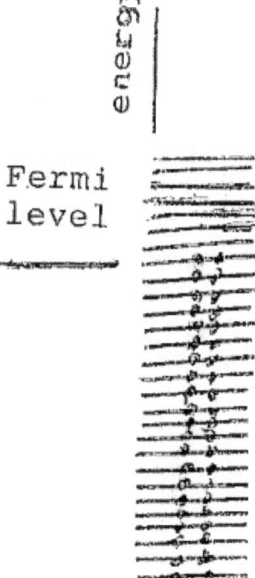

Fermi
level

Electrons have two spin states for each energy.
If there is only one electron in each quantum state, at
low temperature, all states are full to a maximum called
the Fermi level.

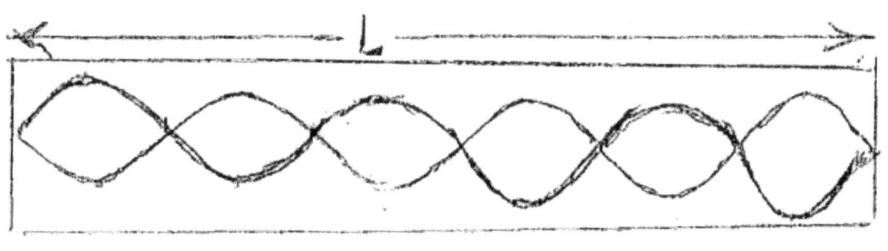

Go back to page 13-30 and review the Stern-Gerlach
experiment. We now know more about atoms than was
known in 1922. A silver atom is made up of electrons,
protons and neutrons. Electrons have a charge, $-e$, and protons have a
charge, $+e$, where $e = 1.6 \times 10^{-19}$ coulomb, while neutrons have no charge. The
electron mass is 9.1×10^{-31} kilogram, the mass of the proton and neutron
being 1.67×10^{-27} kilogram.

The silver atom has **47** protons and either **60** or **62** neutrons in its nucleus and **47** electrons in the region around its nucleus. Although all of the particles may have the same angular momentum values, if the mass of the nuclear particles is around a thousand times the electron, the electron's magnetic moment will govern the atom's magnetic behavior. Having an odd number of electrons, we might assume that the odd (**47**[th]) electron will be responsible for the Stern-Gerlach results. If the magnetic moment has two directions, then the angular momentum is quantized into two directions in the magnetic field. If angular momentum has values that are multiples of $\frac{h}{2\pi}$, we give this angular momentum property a quantum number, s. Now s has two values, plus and minus. If angular momentum quantum numbers go from one to the next by unit steps, then $(+s) - (-s) = 1$, requiring $s = +1/2$ or $s = -1/2$. Orbital angular momentum had integer quantum numbers, so this must be an additional rotation type of motion which we call spin. Like the earth having a rotation about its axis in addition to orbiting around the sun, the electron has what we call intrinsic spin with two quantum values for s. Then we find that the proton and neutron also have the same quantized spin.

Putting all the particles into an atom, the total spin is merely the sum of the spins of the particles. If there is an even number of particles, the total spin will be an integer; if there are an odd number of particles, the total spin will be half of an odd integer.

For example, ^{4}He has **2** electrons, **2** protons and **2** neutrons. The spins plus and minus add up to zero. ^{3}He has **2** electrons and **2** protons but only **1** neutron. Its spins add up to be $s = 1/2$.

SPIN AND STATICS

Kamerlingh Onnes (1853-1926) at the University of Leiden managed to liquify helium at 4.2 degrees Kelvin in 1908.

Then in 1938, Pyotr L. Kapitza cooled liquid helium to below 2.2 K (degrees above absolute zero are in Kelvin, K) and the stuff began behaving really goofy. For example, when a beaker was partly immersed in it, it would climb up over the walls into the beaker; but when the beaker was removed, it would climb back over the wall into the pool. Other behavior was just as wacky. Later, when enough ^{3}He was obtained, a liquid of that isotope remained a normal liquid at those temperatures. Why was it that only ^{4}He and not ^{3}He behaves this way?

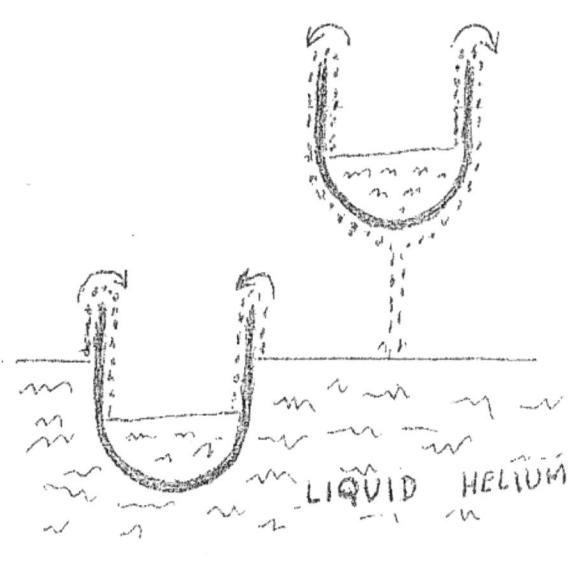

LIQUID HELIUM

One clue lies in the heat capacity, which you will recall from chapter 10 is the heat energy necessary to change temperature by one degree. Near the point where ^{4}He goes from being a normal liquid to being a SUPERFLUID, the heat capacity has a very big increase. With normal adding or subtracting of heat, molecules are jumping between nearby energy quantum states with small changes in quantum

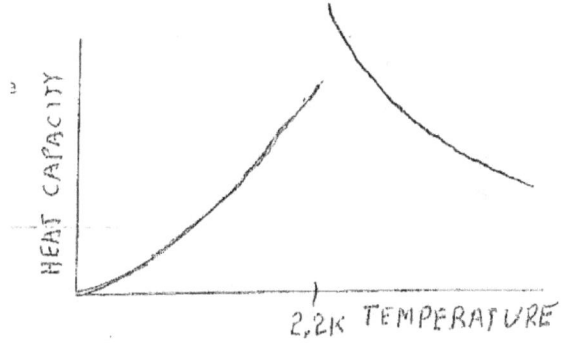

2.2K TEMPERATURE

numbers. The interpretation of this point (called the lambda point as it is shaped like the Greek letter, λ) is that at this point the atoms cease being distributed among a number of quantum states and lose a lot of energy dropping into the lowest energy quantum state. You will recall that with electrons, their behavior required that any quantum state be either empty or occupied by a single electron. But in the case of ^{4}He, it appears that the atoms can all crowd into the single lowest quantum state. Electrons obeyed the Pauli exclusion principle (after Wolfgang Pauli 1900-1958) of no more than one particle in any quantum state. Furthermore, if the Stern-Gerlach experiment is done with a beam of electrons, they would split into two beams giving them spin 1/2 like ^{3}He atoms which don't have a superfluid state, where ^{4}He does. Now ^{4}He has spin 0 so it appears that particles with spin 0, 1, 2, etc. can crowd into single quantum states, where particles with spin

1/2, 3/2, 5/2, etc. are limited to one particle per state. The theory of distributing particles among quantum states is called statistical thermodynamics (see chapter 19) and the types of statistics are named after the physicists who developed the theories. Particles that obey the Pauli exclusion principle are described by Fermi-Dirac statistics and are called fermions. Particles which do not obey the Pauli exclusion principle obey Bose-Einstein statistics and are called bosons. Just why the difference in spin should make such a difference in the behavior of matter is not easily explained IF IT CAN BE EXPLAINED AT ALL.

NOW, SUPERCONDUCTIVITY

For the moment, look at the atomic model at the start of this chapter. With an atom with numerous electrons, the orbits (now called orbitals) have several, but only a fixed number of quantum states. On filling up orbitals with electrons, when the fixed number of states in any orbital are occupied, the next electrons, must go into the next orbital. Only one electron seems to be allowed in any quantum state. Wolfgang Pauli (1900-1958) stated this as general principle for particles with two spin states (spin $\pm 1/2$). This includes electrons, protons, neutrons and other particles mentioned in chapter 17. They are called fermions.

Chapter 14 has details of electrons in crystal lattices of atoms. Here we give a simplified picture of electron-atom collision processes.

Let the electrons be in a lattice of atoms (reality is three dimensional, but we will think in one dimension). At low temperatures and low phonon density, we will consider that most collisions will be electrons bouncing off of atoms. Taking the average of the collisions shown in the picture, assume that an electron stops its motion in the current and has to start in the current all over again from rest.

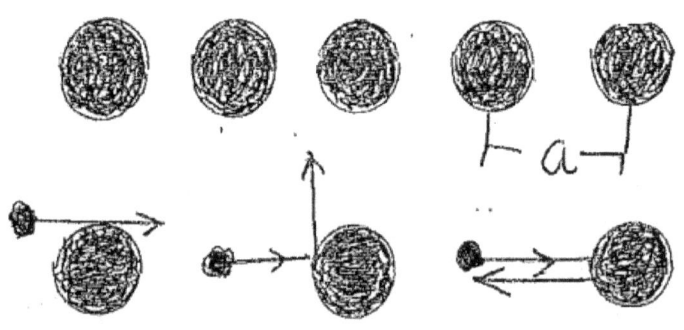

In Chapter 8, we showed that an electric field accelerates a free electron by $m\left(\frac{dv}{dt}\right) = eE$. If t is the average time between collisions, the electron will move a distance $x = \frac{1}{2}\left(\frac{eE}{m}\right)t^2$, with an average velocity, $\bar{v} = \frac{x}{t} = \frac{eE}{2m}t$.

Let the current be flowing in a wire of cross-section area, A. Let there be N electrons per unit volume. An average electron goes a distance, $\bar{v}t$, in a time,

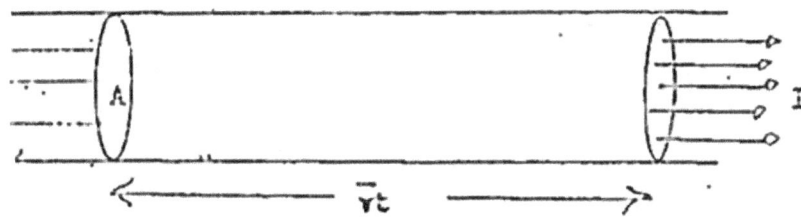

t, and all electrons before it will cross the far cross-section in that time. All electrons in a volume, $A\bar{v}$, will pass that point in a unit time. This gives an electric current, $I = NeA\bar{v}$. Recall $I = \sigma AE$. Here,

$$I = \left(\frac{Ne^2 t}{2m}\right)AE.$$

In this model, assume that an electron has some sort of collision with each atom. If an electron is moving with a velocity, v, and the atom spacing is a then the time between collitions is $t = \frac{a}{v}$. The conductivity is $\sigma = \frac{Ne^2 t}{2m}$ giving $\sigma = \frac{Ne^2 a}{2mv}$. But a particle with momentum, mv, has a wavelength, λ. $(mv)(\lambda) = h$. This gives conductivity, $\sigma = \frac{Ne^2 a\lambda}{h}$.

In the next chapter, it is shown that the electrons involved in conduction are near the Fermi level, and these electrons have wavelengths, $\lambda = a$, for the material we are discussing. "a" is the space between atoms. This gives a normal conductivity, $\sigma_n = \frac{Ne^2 a^2}{h}$.

Now suppose that it were possible to get all electrons into their lowest quantum energy state. In that case their wavelength would be the dimension of the box, $\lambda = L$. As it will turn out to be, this will be a superconducting state. Its conductivity is $\sigma_{sc} = \frac{Ne^2 aL}{h}$.

The resistance varies inversely with the conductivity so the ratio of the superconducting resistance to the normal resistance will be $\frac{R_{sc}}{R_n} = \frac{\sigma_n}{\sigma_{sc}} = \frac{a}{L}$. If the lattice spacing is $a = 1\,\text{Å} = 10^{-10}$ meter and the dimension of the box is $L = 1\,\text{cm} = 10^{-12}$ meter, we get $\frac{R_{sc}}{R_n} = 10^{-8}$, which is rather close to zero.

But if electrons are going to be responsible for all the things that depend on the band theory of semiconductors, they must satisfy the condition of only one electron per quantum state. So, how can they do all that and still be able to be superconducting at low temperatures?

Electrons by themselves are Fermions obeying the Pauli exclusion principle. If there can only be one electron in a quantum state, they cannot all pile up in the lowest quantum state. If it were possible for electrons with opposite spins to combine, they would have zero spin and would become bosons, not subject to the Pauli exclusion principle. Now, opposite spins with opposite magnetic moments attract if they are parallel, next to each other. But electrons have charge and like charges repel each other more strongly than opposite spins attract. If we are to have superconductivity, how do we get electrons to pair with each other. IT DOES HAPPEN, but what is going on?

Up to 1986, the highest temperature at which superconductivity had been observed was $23\,$K. Then J. G. Bedborz and K. A. Muller at the IBM research laboratory in Zurich, Switzerland achieved an important breakthrough. Before discussing their result, it is necessary that we say something about the material they were studying. They made a non-conducting MOTT INSULATOR to be superconducting. (Who ever said that the real world made any sense?)

In an insulator, electrons remain stuck on their atoms, but valence electrons on nearby atoms can interact with their magnetic properties. In certain materials such as transition metals like iron, cobalt and nickel, when the jiggling motion is reduced below a certain temperature called the Curie temperature the electron magnetics line up in the same direction adding up to give the effect of one big magnet. Because this is mainly noticed with iron in nature, it is called ferromagnetism. In other crystals, with different relative positions of the atoms, the same forces cause the electrons to line up with spins in opposite directions. These materials are called antiferromagnetic and Mott insulators are of this type.

These lattices are alternate layers of oxygen and first row metal atoms, such as Fe, Co, Ni, Cu, etc. Much work has been done with copper although iron is now being studied. Valence electrons in metal ions line up in antiferromagnetic order. Normally, these materials are insulators and their electrons cannot move around. But adjacent spins are opposite, and if brought together, they cancel, forming bosons with no spin. Can we get the electrons to move around and cancel their electrostatic repulsion?

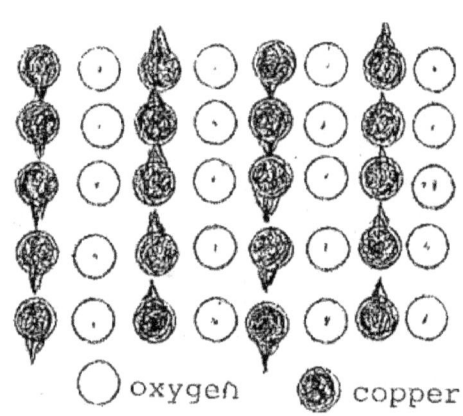

◯ oxygen   ⊛ copper

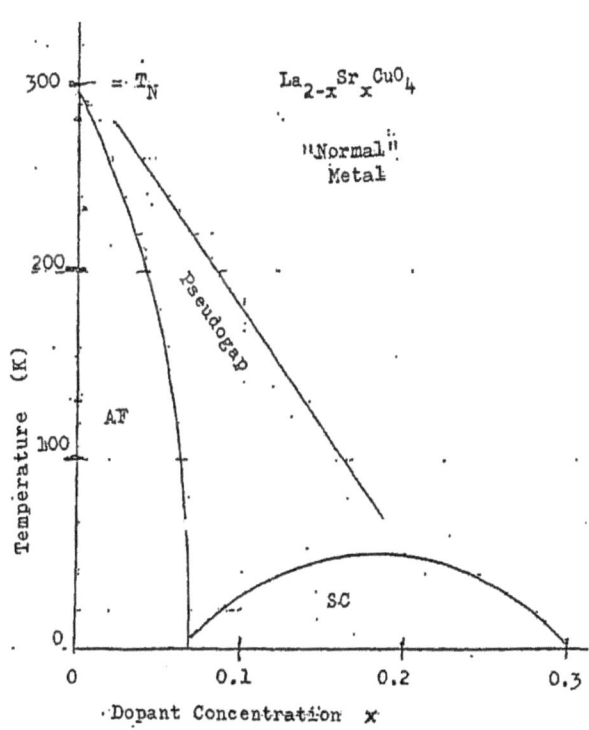

Now, we return to the Mott insulator where we have seen the possibility of pairing spins as with the Copper pair. Now that we have bosons, how can they move around in an insulator? You are referred to the discussion of holes in semiconductors in chapter 14. Take CuO_4 and introduce trivalent lanthanum. Then replace a fraction, x, of La with divalent strontium to form $La_{2-x}Sr_xCuO_4$. This is designed to provide holes so that the bosons can move around.

The results are shown in the figure to the left. The AF region is antiferromagnetic. The SC region is superconducting. If these are taken to be phases, there is not a normal phase transition between them, but rather a strange pseudo gap.

This does achieve superconductivity at temperatures higher than attained with metals giving rise to high-temperature superconductivity.

An idea, although not accepted, might give some feeling as to what is going on. If we start off with a lattice of electrically neutral atoms and each gives an electron to a sea of conducting electrons, the atoms having lost electrons, are now positively charged. If this positive charge acts as if it is smeared out uniformly over the entire crystal, it would produce an atmosphere of positive charge throughout the space occupied by the electrons, thereby shielding them from their electrical repulsion. When the motion of the electrons was reduced by lowering the temperature, the weak attraction of their magnetic moments would pull them together to form pairs with spin 0. (Remember that each little spinning electron acts like a little magnet with north and south poles)

Physicists in the Soviet Union (now Russia) considered superfluidity and superconductivity as part of the same phenomenon. V. L. GINZBURG AND L. D. Landau (1950) Zh. Eksp. Teor. Fiz. U.S.S.R. 20, 1064 presented a theory for both superphenomena. Leon Cooper (1956) Phys. Rev. 104, 1189 by theory allowed two spin $1/2$ electrons to combine to form a Cooper pair which was a boson, as the Ginzburg-Landau theory required. This allowed Bardeen, Cooper and Schrieffer (1957) Phys. Rev. 108, 1175 to come up with their famous theory of superconductivity.

We next compare the phases of two doped cuprates, the one on the right introducing holes into the lattice, the one on the left introducing electrons.

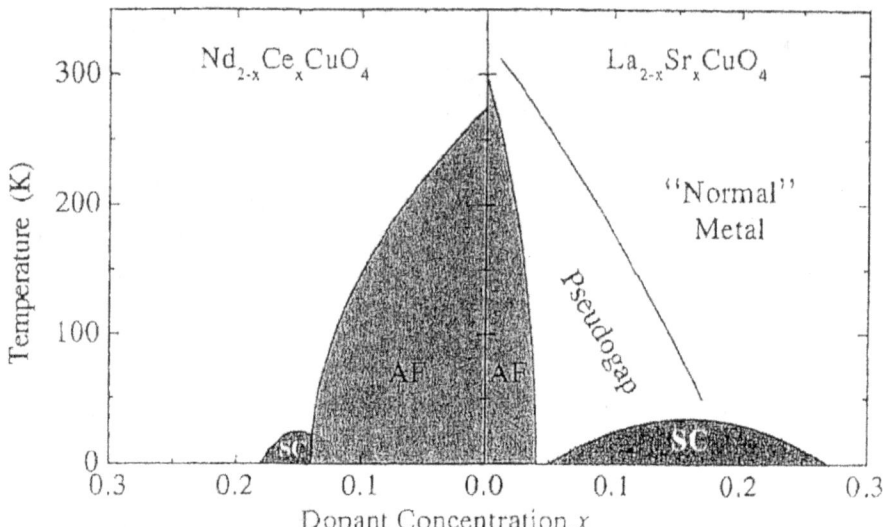

These data tell us several things. Somehow or other there is electron pairing to produce bosons. The CuO_4 is a nonconducting Mott insulator, with conduction similar to pure semiconductors. The introduction of impurities with holes allows the bosons to move about, just like inserting donor atoms to make a p-type semiconductor. Impurities introducing electrons do not lead to conduction as the carriers are electron pairs, not single electrons as with n-type semiconductors, and the impurity atoms are too far from each other for their electrons to pair up.

The $La_{2-x}Sr_xCuO_4$ of Bednorz and Müller had a critical temperature of $T_C = 38$ K (the maximum at which it was superconducting). Soon after, $YBa_2Cu_3O_7$ was found to have $T_C = 92$ K and $HgBa_2Ca_2Cu_3O_{8+x}$ was found to have $T_C = 135$ K. As liquid nitrogen boils at $T_C = 77$ K, this was significant because liquid helium as a coolant is expensive while liquid nitrogen as a coolant is almost free. The trouble is that these materials are exotic and rather brittle materials for being drawn out into wires needed for most applications of superconducting materials.

This is an active realm of research. As of this writing (2015) what is going on in the pseudo gap is a bit of a mystery. Unfortunately, textbook writers stick to things we understand, and the fun things in physics are those things we do not understand. For reference to areas of current research you will have to go to the research journals. The summary articles in the Reviews of Modern Physics are the best places to start. The Ginzburg-Landau Theory of Superconductivity is discussed in detail in chapter 20.

The next important effect to be considered is the MEISSNER EFFECT. Here magnetic flux will not penetrate a superconducting region and flux, which existed in a conducting region, will be ejected when that region becomes superconducting. The effect was reported by W. Meissner and R. Ochsenfeld in Naturwiss. 21,787 (1933) and analyzed by Fritz and Heinz London Proc. Roy. Soc. A149.71 (1935).

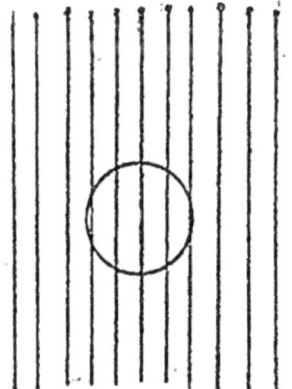

Magnetic flux with a
conducting cylinder
to the left.

Magnetic flux around
a superconducting
cylinder to the right.

MEISSNER EFFECT

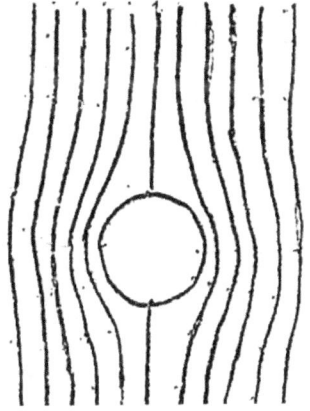

The simplest explanation of this uses diamagnetism, discussed on page 13-38. An external magnetic field, increasing with time, causes a current to flow in a conducting ring, which flows in such a direction as to produce a reactive magnetic field that opposes the increasing applied field.

Look at this more closely. On page 13-39 we saw that an electric field is what causes current to flow. If the conducting ring has a cross section area, A, and its material has a conductivity, σ, an electric field, E, will cause a current, I, where $I = \sigma A E$.

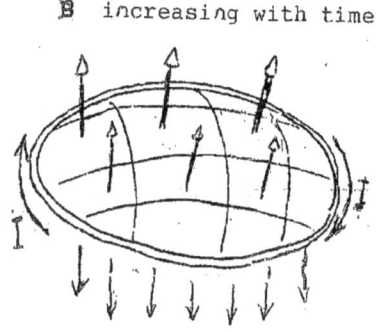

B increasing with time

Reactive field
produced by the
induced current

From this we conclude that the increasing magnetic field produces electric fields that go 'round and 'round as shown in the picture. If the entire space is filled with a conducting material, currents with current densities, $\frac{I}{A} = \sigma E$, flow that produce reactive magnetic fields that oppose the increasing applied field.

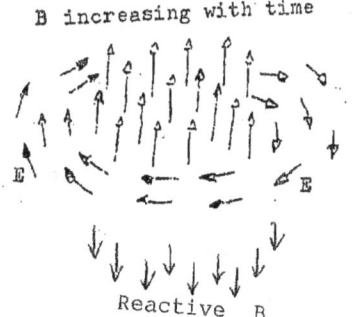

B increasing with time

Reactive B

But in a superconducting space the conductivity is infinite, producing induced currents that are also infinite, as are the reactive magnetic fields. Things don't get to the infinite. Things increase until the reaction cancels the applied magnetic field, not allowing it to exist in the superconductor.

In chapter 10, it was seen that the energy at temperature, T, in one gram molecular weight of a substance is approximately RT, where $R = 8.3$ joules per mole per kelvin. Then if there are 6.02×10^{23} molecules in a mole, one molecule would have about $k_B T$ joule of energy where the Boltzmann constant is $k_B = 1.38 \times 10^{-23}$ joule per molecule per kelvin. This leads to an interesting calculation for atoms in a gas. If we set $\frac{1}{2} M v^2 = k_B T$, then $(M v)^2 = 2 M k_B T$, giving a temperature of a single atom.

But with the momentum, $M v$, there is a wavelength, λ, where $M v = \frac{h}{\lambda}$. Then, $\left(\frac{h}{\lambda}\right)^2 = 2 M k_B T$ gives a temperature, $T = \frac{h^2}{2 M k_B \lambda^2}$. Now, $h = 6.625 \times 10^{-34}$ joule-sec and if the atomic mass number, A, equals the sum of the protons plus the neutrons, the mass of one atom is $M = A \times 1.67 \times 10^{-27}$ kilogram. Then, $T = 9.52 \times 10^{-18} / A \lambda^2$.

For sodium, ^{23}Na, $T = 4.14 \times 10^{-19} / \lambda^2$. A sodium atom confined to a space of 10^{-6} meter has a temperature of 4.14×10^{-7} K!

HOW CAN WE SO CONFINE A GASEOUS ATOM?

On page 12-11 a magnetic field was derived at the center of a pair of current-carrying coils called Helmholtz coils. If you use the same formulas to derive the magnetic field at other points on the axis, you will find that the field strength is a maximum at the center of the pair of coils.

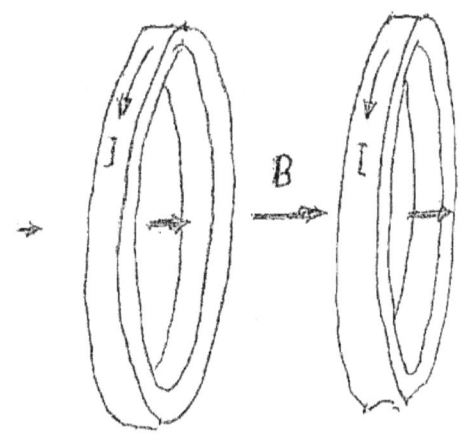

Then, on page 13-30 a particle with a magnetic moment, μ, in a magnetic field, B, had an energy, $E = -\mu \cdot B$.

If μ is in the direction of B the particle's energy is a minimum at the center of the coils. If μ is in the direction opposite to B the energy is a maximum at the center of the coils.

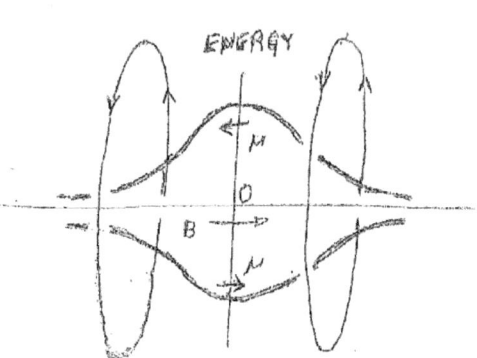

The experiment to be described is done in an evacuated (all air pumped out) tube into which a vapor of the atoms to be studied are inserted. Atoms with moments opposite to *B* will lose energy by moving away from the center and will be ejected from the region. Atoms with moments in the same direction as *B* will lose energy by going toward the center and will be trapped there if they have a low kinetic energy.

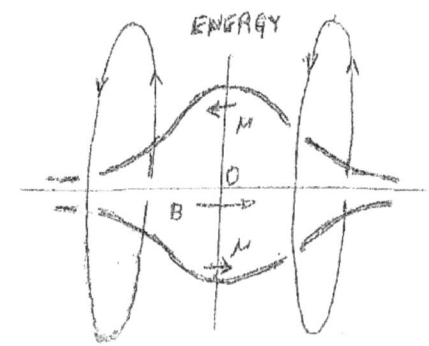

Let us insert ^{23}Na into the region. Sodium is the next element above neon. ^{20}Ne has ten of each: protons, neutrons and electrons, giving it zero spin. ^{23}Na has 11 protons and electrons and 12 neutrons giving it integer spin but remember that the magnetic moment of an electron is a thousand times that of a proton.

^{23}Na is thus a boson but with a magnetic moment. Because of its moment, it experiences a magnetic energy well. Those atoms with kinetic energy higher than the limits of the well will escape from the region.

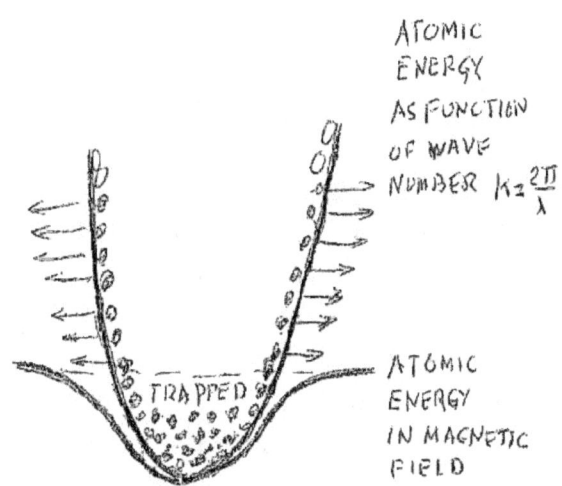

Those with kinetic energy below the limits are trapped in the magnetic field, like being in a bottle without walls, this is called magnetic trapping. A similar trapping can be achieved using the light pressure of laser beams.

Because ^{23}Na is a boson at low temperature, there is a Bose-Einstein condensation resulting in many atoms piling in the magnetic trap.

With clever probes, it is now possible to study cold atoms floating in the middle of an evacuated space.

SCANNING TUNNELLING MICROSCOPE

The materials we have discussed have a complicated mixture of atoms. We can get some idea of their crystal structure by studying the order of atoms on their surface with a scanning tunneling microscope, which is an important instrument in condensed matter research. In quantum mechanics, tunneling is a phenomenon where particles can exist in a region where the total energy is less than the potential energy. Being in such a region is forbidden in classical physics. We have seen that electrons have waves where momentum, $p = m v$, and wavelength, λ, are related by $(m v)(\lambda) = h$. If particles and waves obey the same physics, electron energy and frequency are related by $E = h f$.

Consider a wave, $\psi(x, t) = A \cos(k x - \omega t)$. A point of constant value must move so that $(k x - \omega t)$ keeps a constant value. Let this be zero. Then $(k x - \omega t) = 0$, giving a velocity, $v = \frac{\omega}{k}$. But, $v = \lambda f$. We have seen that $\omega = 2 \pi f$. Hence, $\frac{2 \pi f}{k} = \lambda f$, from which $k = \frac{2 \pi}{\lambda}$. But $\lambda = \frac{h}{p}$ and $f = \frac{E}{h}$. The wave is $\psi(x, t) = A \cos 2 \pi (p x - E t)/h$. We can write a generalized wave function as an exponential function.

First, let us review the phenomenon of tunneling. $\psi(x, t) = A e^{i(p x - E t)/h}$ is the wavefunction for a particle with a constant potential energy, V, having energy, E, and momentum, $p = \sqrt{2 m (E - V)}$.

To start, let a particle with energy, E, be moving toward the right in a region where its potential energy is zero. From $x = 0$ to $x = L$, let there be a region where its potential energy is a constant, V, which is less than E after which it emerges into a region where $V = 0$ again.

Let $k_o = \sqrt{2 m E}/h$ be its wavenumber where, $V = 0$, and $k = \sqrt{2 m (E - V)}/h$ be the wavenumber between $x = 0$ and $x = L$. Take time to be at $t = 0$. At each discontinuity there will be a reflected wave. For reference, let the incoming wave have unit amplitude.

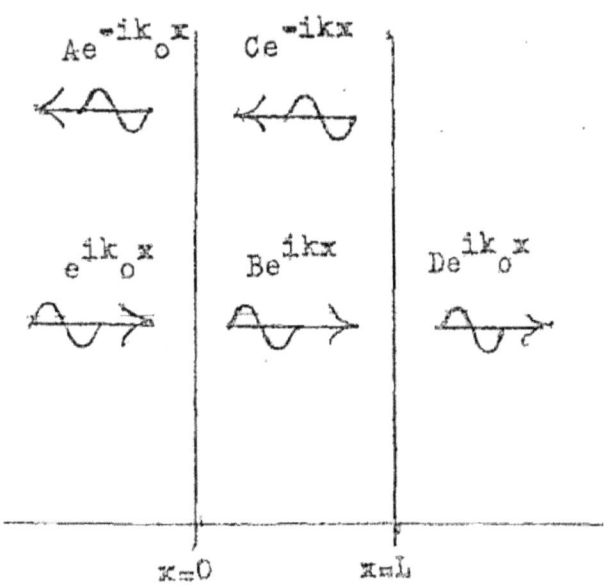

When $x < 0$, $\psi = e^{ik_o x} + A e^{-ik_o x}$. When $0 < x < L$, $\psi = B e^{ikx} + C e^{-ikx}$.
When $x > L$, $\psi = D e^{ik_o x}$.

That the Schrodinger equation involves a second derivate of ψ requires that ψ and its first derivative be continuous at boundaries. We shall relate the transmitted wave, $\psi = D e^{ik_o x}$, to the incoming wave, $\psi = e^{ik_o x}$, for $V < E$ and then modify the result for $V > E$ to give a tunneling current.

If ψ is continuous, $1 + A = B + C$ at $x = 0$, $B e^{ikx} + C e^{-ikx} = D e^{ik_o x}$ at $x = L$.
If $\dfrac{d\psi}{dx}$ is continuous, $k_o (1 - A) = k (B - C)$, and $k \left(B e^{ikx} - C e^{-ikx} \right) = k_o \left(D e^{ik_o x} \right)$.
Eliminating B and C gives $1 + \dfrac{k_o}{k} + \left(1 - \dfrac{k_o}{k} \right) A = \left(1 + \dfrac{k_o}{k} \right) D e^{i(k_o - k) L}$ and

$1 - \dfrac{k_o}{k} + \left(1 + \dfrac{k_o}{k} \right) A = \left(1 - \dfrac{k_o}{k} \right) D e^{i(k_o + k) L}$. Then, $D = \dfrac{4 \dfrac{k_o}{k} e^{-ik_o L}}{\left(1 + \dfrac{k_o}{k} \right)^2 e^{-ikL} - \left(1 - \dfrac{k_o}{k} \right)^2 e^{ikL}}$.

This is the transmitted waves amplitude when the potential energy of the barrier is less than the electron's total energy. But what about a case where the potential energy of the barrier is greater than the electron's total energy? An example would be two metal surfaces with a thin vacuum gap between them. Take an electron at the Fermi level, $E = E_F$. It needs an extra energy called the work function to be emitted from the metal surface by thermionic-emission, or photo-emission, call this w (or often Φ).

In the metal, momentum is $h k_o = \sqrt{2 m E}$. In the vacuum, $V = E_F + w$ and $h k = \sqrt{2 m (E - V)}$ or $h k = \sqrt{-2 m w} = h i g$ with $g = \sqrt{-2 m w}/h$. This gives,

$D = \dfrac{4 \dfrac{k_o}{i g} e^{-ik_o L}}{\left(1 + \dfrac{k_o}{i g} \right)^2 e^{g L} - \left(1 - \dfrac{k_o}{i g} \right)^2 e^{-g L}}$ in which we will ignore the term with the $e^{-g L}$ in

the denominator. In that case, $D = \dfrac{4 \dfrac{k_o}{i g} e^{-ik_o L}}{\left(1 + \dfrac{k_o}{i g} \right)^2 e^{g L}}$.

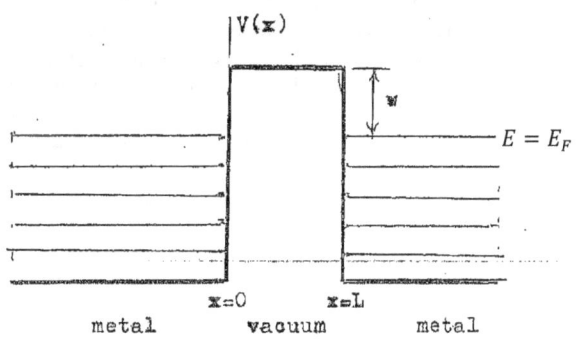

The product of wave function with its complex conjugate gives a real positive number. We use this as the probability density of the particle being at a point.

Current is density times velocity. The rightward moving transmitted current

is $J = D^* D$ with $D^* D = \dfrac{16 \left(\frac{k_0}{g}\right)^2 e^{-2gL}}{\left(1 + \frac{k_0}{ig}\right)^2 e^{gL}}$. Notice $e^{-2gL} \cdot \left(\dfrac{k_0}{g}\right)^2 = E/w$.

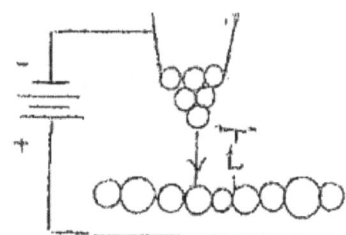

In the case shown on the previous page there will also be a leftward moving tunneling current. To get a net tunneling current toward the right, we tilt the potential curve with an applied voltage.

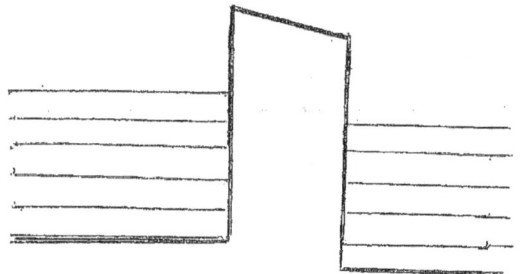

With the scanning tunneling microscope the vacuum gap is between the sharp tip of a probe and the surface of a conducting material whose atoms may be of different elements with differing sizes.

At this point we shall go to the account by Gerd Binning and Heinrich Rohrer in SCANNING TUNNELING MICROSCOPY – "From Birth to Adolescence" RMP 59, 615 (1987) and let Heinrich Rohrer describe how their gadget works.

Note that the STM can only scan surfaces of conducting materials which connect a closed circuit. Scanning a nonconducting surface can be done with an atomic force microscope, for which we refer you to the literature.

For further study, we refer you to:
F.J GIESSIBI ADVANCES IN ATOMIC
    FORCE MICROSCOPY RMP 75,949,
    (2003)
W.A. HOFER, A.S. Foster, A.L.
    Shluger THEORIES OF SCANNING
    PROBE MICROSCOPIES AT ATOMIC
    SCALE RMP 75, 1287 (2003)

Constant Current Mode

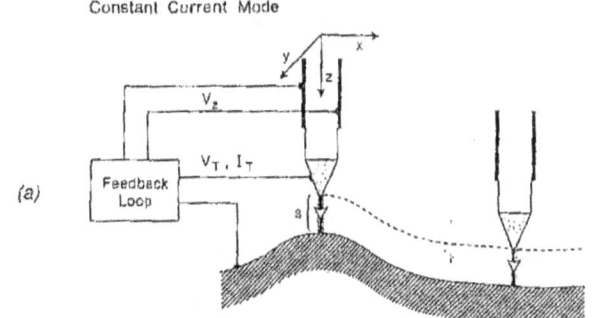

(a)

$V_z (V_x, V_y) \dashrightarrow z(x,y)$

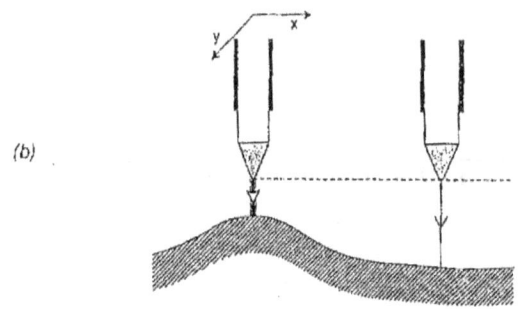

(b)

$\ln I (V_x, V_y) \dashrightarrow \sqrt{\overline{\Phi}} \cdot z(x,y)$

. Imaging. (a) In the constant current mode, the tip is scanned across the surface at constant tunnel current, maintained at a preset value by continuously adjusting the vertical tip position with the feedback voltage V_z. In the case of an electronically homogeneous surface, constant current essentially means constant s. (b) On surface portions with denivellations less than a few Å—corresponding to the dynamic range of the current measurement—the tip can be rapidly scanned at constant average z position. Such "current images" allow much faster scanning than in (a) but require a separate determination of $\sqrt{\phi}$ to calibrate z. In both cases, the tunnel voltage and/or the z position can be modulated to obtain, in addition, $d \ln I /dV$ and/or $d \ln I /ds$, respectively.

Fischer, M. Kugler, I. Maggio-Aprile
    SCANNING TUNNELING SPECTROSCOPY
    OF HIGH TEMPERATURE
    SUPERCONDUCTORS RMP 79, 353
    (2007)

Hofer, Foster and Shluger cite Wouda,
et al Surf. Sci 959. 17 (1996) to show
a very nice STM scan of PtRh (100)
surface. Rh atoms possess a different
electronic structure due to alloying
with adjacent Pt atoms, they are
therefore clearly discriminated by
their apparent height (see detail on
right). Measurements of this type are
the first real-space visualizations of
chemical atoms.

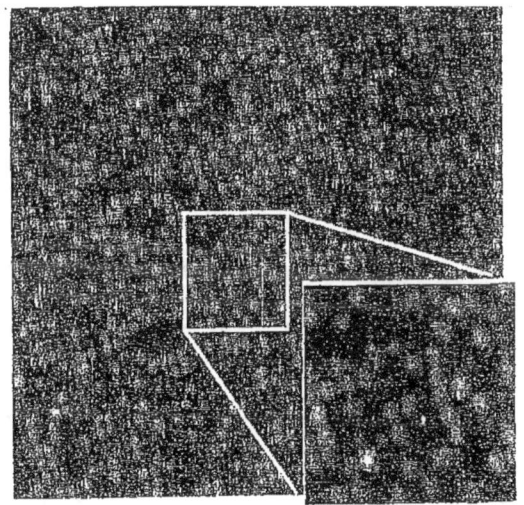

There are other phenomena in magnetism in condensed matter that have not
been covered and some of the topics here could have been covered in more
detail. We shall conclude our discussion at this point. By now you have
learned that the ultimate references are not textbooks but rather to reports
in the research journals. By referring to the Reviews of Modern Physics we
suggest that that is the best place to start a study of a topic in physics.
From there you will be guided to articles in other journals covering
specifics more deeply. Have fun!

APPENDIX: MAXWELL'S EQUATIONS AND RADIATION

The interaction of matter with electromagnetic radiation (light) was discussed earlier in this chapter. Having come this far with electromagnetic theory, a few more pages will allow us to look at the nature of electromagnetic radiation itself.

On page 13-15 we found that Coulomb's law could be restated as the divergence of the electric field from the density (coulombs per cubic meter) of charge. This is the first Maxwell equation that we will derive.

Distributions of charge produce an electric field, E, in the region around them. The direction of the electric field is more or less, away from the positive source charge, and toward the negative source charge.

$$\nabla \cdot E = \frac{\rho}{\epsilon_0}$$

On page 13-17 the similarity between electric and magnetic fields was shown suggesting that equations involving either one could be used to describe the other. The main difference was in their source. Electric fields were produced by charge and magnetic fields came from loops of current. When we introduced the idea of magnetic charge, it was a fiction which allowed us to get some equations easily. Actually, there is no such thing as magnetic charge. Bring the relation we just now discussed over into magnetism, ρ is zero. This gives us the second of Maxwell's equations, $\nabla \cdot B = 0$.

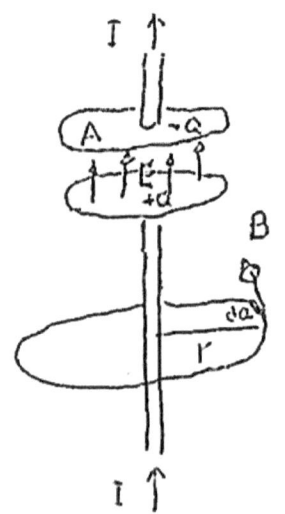

Again, let a current, I, flow in a long wire, being interrupted at one point by a capacitor. At the bottom of page 8-14, It was shown that the time rate of change of the electric field acted as a continuation of the current. $I = \epsilon A \frac{dE}{dt}$.

Then at the bottom of page 12-8 it was shown that a magnetic field of strength, $B = \frac{\mu I}{2 \pi r}$, exists in a direction perpendicular to a radius out from the wire or tangent to a circle about the wire.

This gives a result that looks as if it might be interesting. B has the same value at all points on a contour of radius, r, and is tangent to the contour. Multiply B by the circumference of the contour. Now assume that either a current or a changing E could exist. We get

$$B \, 2 \, \pi \, r = \mu \left[I + \epsilon \, A \frac{\partial E}{\partial t} \right].$$

Carry this a bit farther. Let the wire have a cross section area, A, and let J be the current per unit area in the wire. $I = J \, A$. Then let da be an element of arc length of the contour.

This leads to another idea which is actually correct. In a region of space let there be a current density, J, and a time-varying E. For the moment, let them be perpendicular to an area. Summing around the closed contour (C),

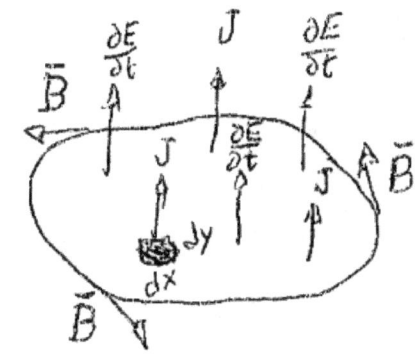

$$\oint_C B \, da = \iint_{Area\ inside\ C} \left[J + \epsilon \frac{\partial E}{\partial t} \right] dx \, dy$$

At this point we are tempted to review some old mathematics and develop some new mathematics. Go back to page 13-27 and replace velocity with magnetic field. We proved:

$$\oint_C \boldsymbol{B} \cdot d\boldsymbol{r} = \iint_{Area\ inside\ C} \left[\frac{\partial B_y}{\partial x} - \frac{\partial B_x}{\partial y} \right] dx \, dy$$

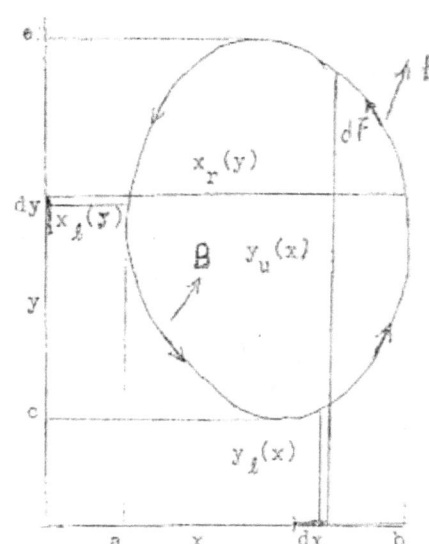

For a function, $f(x,y,z)$, with only numbers, we defined the gradient of f as,
$\boldsymbol{\nabla} f = \boldsymbol{i} \left(\frac{\partial f}{\partial x} \right) + \boldsymbol{j} \left(\frac{\partial f}{\partial y} \right) + \boldsymbol{k} \left(\frac{\partial f}{\partial z} \right)$. Then $df = \boldsymbol{\nabla} f \cdot d\boldsymbol{r}$, with the gradient operator, $\boldsymbol{\nabla} = \boldsymbol{i} \left(\frac{\partial}{\partial x} \right) + \boldsymbol{j} \left(\frac{\partial}{\partial y} \right) + \boldsymbol{k} \left(\frac{\partial}{\partial z} \right)$.
$\boldsymbol{\nabla} \cdot \boldsymbol{F} = \frac{\partial F_x}{\partial x} + \frac{\partial F_y}{\partial y} + \frac{\partial F_z}{\partial z}$ was the divergence of a vector function.

For a reference, we will review the vector mathematics and then add a few new things.

First, let $F = i F_x + j F_y + k F_z$ have length F where $F^2 = F_x{}^2 + F_y{}^2 + F_z{}^2$ and $G = i G_x + j G_y + k G_z$ have length, G, given by $G^2 = G_x{}^2 + G_y{}^2 + G_z{}^2$. Recall the dot product, $F \bullet G = F_x G_x + F_y G_y + F_z G_z = F G \cos \theta$, and the cross product, $F \times G = i [F_y G_z - F_z G_y] + j [F_z G_x - F_x G_z] + k [F_x G_y - F_y G_x]$.

Now, substitute B for G and ∇ for F:

$$\nabla \times B = i \left[\frac{\partial B_z}{\partial y} - \frac{\partial B_y}{\partial z}\right] + j \left[\frac{\partial B_x}{\partial z} - \frac{\partial B_z}{\partial x}\right] + k \left[\frac{\partial B_y}{\partial x} - \frac{\partial B_x}{\partial y}\right].$$

Now, $i \bullet k = 0$ and $j \bullet k = 0$, and $k \bullet k = 1$.

Then, $(\nabla \times B) \bullet k = \frac{\partial B_y}{\partial x} - \frac{\partial B_x}{\partial y}$.

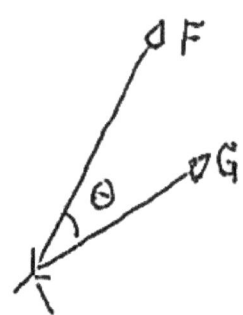

$$\oint_C [B_x \, dx + B_y \, dy] = \iint_{xy \text{ inside } C} [(\nabla \times B) \bullet k] \, dx \, dy$$

$$\oint_C [B_x \, dx + B_y \, dy] = \mu_o \iint_{xy \text{ inside } C} \left[J_z + \epsilon_o \left(\frac{\partial E_z}{\partial t}\right)\right] dx \, dy$$

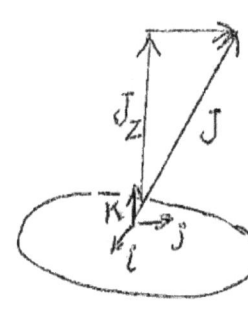

Now k is a unit vector perpendicular to the x-y plane and J is current density in any direction. $J_z = J \bullet k$ gives the component of J perpendicular to the x-y plane. Likewise, $E_z = E \bullet k$. Putting things together:

$$\iint [(\nabla \times B) \bullet k] \, dx \, dy = \mu_o \iint \left[J + \epsilon_o \left(\frac{\partial E}{\partial t}\right)\right] \bullet k \, dx \, dy$$

This gives $\nabla \times B = \mu_o J + \mu_o \epsilon_o \left(\frac{\partial E}{\partial t}\right)$.

What do the new Maxwell equations mean? $\nabla \bullet B = 0$ means that the traces of magnetic field vectors don't start and stop anywhere but go around current vectors. A picture of all this is as follows.

A current of charge designated by the vector symbol, J, produces a magnetic field, or flux density, in the region about it. The field arrows go 'round and 'round the current.

$\nabla \bullet B = 0$ $\qquad$ $\nabla \times B = \mu J$

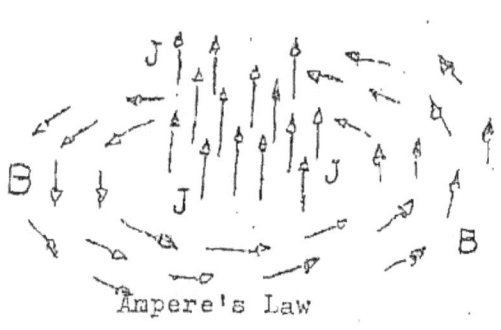

Ampere's Law

The new equation was $\nabla \times B = \mu J + \mu \epsilon \frac{\partial E}{\partial t}$.

What did the new term mean? It is evident that in the absence of currents, i.e. $J = 0$, the new equation looks like the old one, except for a constant coefficient, $\mu\epsilon$, and the lack of a minus sign. Hence, a time varying electric field could induce a magnetic field in the same region of space.

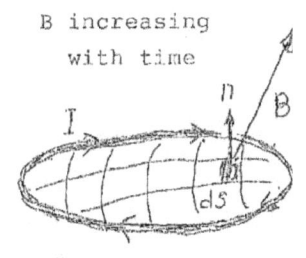

E, increasing with time

$$\nabla \times B = \mu\,\epsilon\,\frac{\partial E}{\partial t}$$

$\mu_o = 4\pi \times 10^{-17}$ is the permeability of empty space or matter without dipoles. It is defined that way because magnetic measurements involving current are easier than electric measurements with charge. The permittivity of empty space must be measured. $\epsilon_o = 8.85 \times 10^{-12}$. Now, $\mu_o\,\epsilon_o = 1.11 \times 10^{-17}$.

$$\frac{1}{\mu_o\epsilon_o} = 9 \times 10^{16} = c^2.$$

For the fourth Maxwell equation, we go to page 12-28. There we found that if the magnetic field on a surface (as shown) is increasing with time, a voltage is induced in a conducting wire surrounding the surface, making a current flow, as shown.

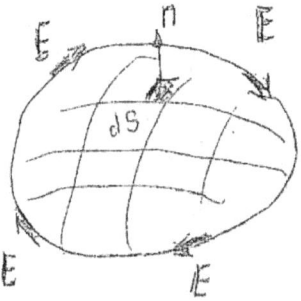

B increasing with time

$$V = -\frac{d}{dt} \iint\limits_{S\,in\,C} B \cdot n\,dS$$

In chapter 8 it was shown that if there is a voltage difference between two points in space, there is an electric field in that space. If the space contains a conducting medium with movable charges, there will be a current flowing. On page 12-30 we pointed out that the current would flow in such a direction as to produce an effect that opposed whatever the changing magnetic field was doing. A change of voltage, dV, in a gap, dr, produces an electric field, E, with $dV = E\,dr$.

The statement above becomes:

$$\oint\limits_{C} E \cdot dr = -\frac{d}{dt} \iint\limits_{S\,in\,C} B \cdot n\,dS$$

$$\oint_C \mathbf{E} \cdot d\mathbf{r} = \iint_{S\,in\,C} (\nabla \times \mathbf{E}) \cdot \mathbf{n}\, dS$$

If $\iint (\nabla \times \mathbf{E}) \cdot \mathbf{n}\, dS = -\dfrac{d}{dt} \iint \mathbf{B} \cdot \mathbf{n}\, dS$, then

$\nabla \times \mathbf{E} = -\dfrac{\partial \mathbf{B}}{\partial t}$.

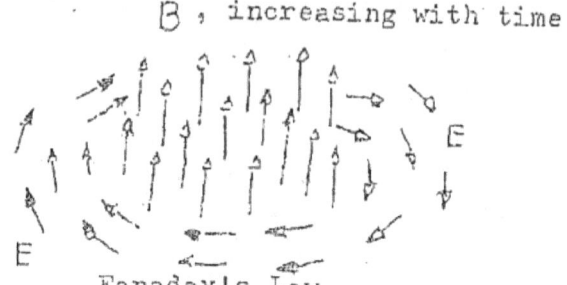

This is the fourth Maxwell equation and its effect in space is shown in the picture on the right.

If the magnetic flux density, **B**, is changing with time, an electric field, **E**, is produced in the same region of space. The electric field arrows go 'round and 'round the changing magnetic field arrows.

We now summarize the relations between **E** and **B**.

$$\nabla \cdot \mathbf{E} = \frac{1}{\epsilon_o} \rho \qquad\qquad \nabla \cdot \mathbf{B} = 0$$

$$\nabla \times \mathbf{E} = -\frac{\partial \mathbf{B}}{\partial t} \qquad\qquad \nabla \times \mathbf{B} = \mu_o \mathbf{J} + \mu_o \epsilon_o \frac{\partial \mathbf{E}}{\partial t}$$

These are called Maxwell's equations after James Clerk Maxwell (1831-1879) after studying the experimental work of Michael Faraday (1791-1867) and others.

The gradient operator, $\nabla = \mathbf{i}\left(\dfrac{\partial}{\partial x}\right) + \mathbf{j}\left(\dfrac{\partial}{\partial y}\right) + \mathbf{k}\left(\dfrac{\partial}{\partial z}\right)$, is often called by its short name, DEL.

Let $\mathbf{F}(\mathbf{r}) = \mathbf{F}(x,y,z)$ and $\psi(\mathbf{r}) = \psi(x,y,z)$ be a vector and a scalar function of position. In addition to the gradient of ψ, $\nabla\psi = \mathbf{i}\left(\dfrac{\partial\psi}{\partial x}\right) + \mathbf{j}\left(\dfrac{\partial\psi}{\partial y}\right) + \mathbf{k}\left(\dfrac{\partial\psi}{\partial z}\right)$, which is often called DEL PSI, we also have the divergence and curl of $\mathbf{F}$.

$div\, \mathbf{F} = \nabla \cdot \mathbf{F} = \dfrac{\partial F_x}{\partial x} + \dfrac{\partial F_y}{\partial y} + \dfrac{\partial F_z}{\partial z}$ gives a number.

$curl\, \mathbf{F} = \nabla \times \mathbf{F} = \mathbf{i}\left[\dfrac{\partial F_z}{\partial y} - \dfrac{\partial F_y}{\partial z}\right] + \mathbf{j}\left[\dfrac{\partial F_x}{\partial z} - \dfrac{\partial F_z}{\partial x}\right] + \mathbf{k}\left[\dfrac{\partial F_y}{\partial x} - \dfrac{\partial F_x}{\partial y}\right]$ gives a vector.

Often, we must take the change of a function with one variable and then change it with another variable. Notice that in changing location on a plane, we can go by dx and then dy or we can go by dy and then dx. Thus, for a function of x and y, $f(x,y)$, we have $\frac{\partial}{\partial y}\left(\frac{\partial F}{\partial x}\right) = \frac{\partial}{\partial x}\left(\frac{\partial F}{\partial y}\right)$.

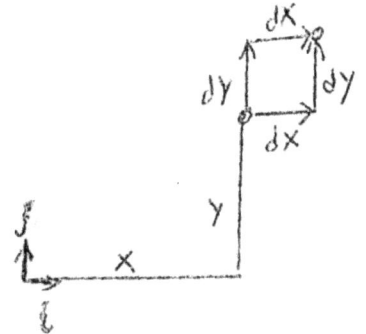

The development of electromagnetic theory is much less complicated than what Maxwell did if we use several vector identities. These make more sense if they have been carefully proved. We will now prove several very useful identities which involve using the gradient twice. The curl of the gradient of a scalar function is zero and the divergence of the curl of a vector function is zero. These proofs are:

$$\nabla \times (\nabla \psi) = i\left[\frac{\partial}{\partial y}\left(\frac{\partial \psi}{\partial z}\right) - \frac{\partial}{\partial z}\left(\frac{\partial \psi}{\partial y}\right)\right] + etc. = 0 \quad \text{or} \quad curl(grad\ \psi) = 0.$$

$$\nabla \cdot (\nabla \times F) = \frac{\partial}{\partial x}\left(\frac{\partial F_z}{\partial y} - \frac{\partial F_y}{\partial z}\right) + \frac{\partial}{\partial y}\left(\frac{\partial F_x}{\partial z} - \frac{\partial F_z}{\partial x}\right) + \frac{\partial}{\partial z}\left(\frac{\partial F_y}{\partial x} - \frac{\partial F_x}{\partial y}\right) = 0 \quad \text{or} \quad div(curl\ F) = 0.$$

cancel

cancel

cancel

This immediately gives us a bright idea. Recall $\nabla \cdot B = 0$.

Maybe there can be another vector function, $A\,(x,y,z)$, such that $B = \nabla \times A$. This would be a vector potential. What can we do with it?

Put together, $\nabla \times E = -\frac{\partial B}{\partial t}$ and $B = \nabla \times A$.

This suggests a magnetically related electric field, E_m. Two rates of change (derivatives) can be taken in either order.

If $\nabla \times E_m = -\frac{\partial B}{\partial t}$ and $B = \nabla \times A$, then $\nabla \times E_m = -\nabla \times \frac{\partial A}{\partial t}$, suggesting $E_m = -\frac{\partial A}{\partial t}$.

The vector potential, A, gives electric and magnetic fields, E and B, which exert on a particle carrying a charge, q, and moving with a velocity, v, a force, $F = qE + qv \times B$.

ELECTROMAGNETIC RADIATION

We now investigate the relation between fields in empty space. There is zero charge density, $\rho = 0$, and zero currents, $J = 0$.

We will need one more identity on the vector potential. $\boldsymbol{E} = -\frac{\partial \boldsymbol{A}}{\partial t}$ and in empty space, $\boldsymbol{\nabla} \bullet \boldsymbol{E} = 0$. Then, $\boldsymbol{\nabla} \bullet \left(-\frac{\partial \boldsymbol{A}}{\partial t} \right) = -\frac{\partial}{\partial t} (\boldsymbol{\nabla} \bullet \boldsymbol{A}) = 0$ giving $\boldsymbol{\nabla} \bullet \boldsymbol{A} = 0$.

Now, let $\boldsymbol{A}$ be a function of z and t. $\boldsymbol{A} = \boldsymbol{i} A_x(z,t) + \boldsymbol{j} A_y(z,t) + \boldsymbol{k} A_z(z,t)$. Then, $\boldsymbol{\nabla} \bullet \boldsymbol{A} = \left(\frac{\partial A_x}{\partial x} \right) + \left(\frac{\partial A_y}{\partial y} \right) + \left(\frac{\partial A_z}{\partial z} \right)$ or $\boldsymbol{\nabla} \bullet \boldsymbol{A} = 0 + 0 + \frac{\partial A_z}{\partial z} = 0$. In this case, A_z is a constant, let it be zero. If $\boldsymbol{A}$ has only x- and y- components, let $\boldsymbol{A} = \boldsymbol{i} A_x(z,t)$ or just $\boldsymbol{A} = \boldsymbol{i} A(z,t)$. This gives $\boldsymbol{E} = -\boldsymbol{i} \left(\frac{\partial A}{\partial t} \right)$.

If $\boldsymbol{B} = \boldsymbol{\nabla} \times \boldsymbol{A}$, then $\boldsymbol{B} = \boldsymbol{i} \left[\frac{\partial A_z}{\partial y} - \frac{\partial A_y}{\partial z} \right] + \boldsymbol{j} \left[\frac{\partial A_x}{\partial z} - \frac{\partial A_z}{\partial x} \right] + \boldsymbol{k} \left[\frac{\partial A_y}{\partial x} - \frac{\partial A_x}{\partial y} \right]$ or $\boldsymbol{B} = \boldsymbol{j} \frac{\partial A}{\partial z}$. $B_y = \frac{\partial A}{\partial z}$.

$$\boldsymbol{\nabla} \times \boldsymbol{B} = \boldsymbol{i} \left[\frac{\partial B_z}{\partial y} - \frac{\partial B_y}{\partial z} \right] + \boldsymbol{j} \left[\frac{\partial B_x}{\partial z} - \frac{\partial B_z}{\partial x} \right] + \boldsymbol{k} \left[\frac{\partial B_y}{\partial x} - \frac{\partial B_x}{\partial y} \right] \qquad \boldsymbol{\nabla} \times \boldsymbol{B} = -\boldsymbol{i} \frac{\partial^2 A}{\partial z^2}$$

Next, if $\boldsymbol{\nabla} \times \boldsymbol{B} = \mu_o \epsilon_o \frac{\partial \boldsymbol{E}}{\partial t}$ and $\boldsymbol{E} = -\frac{\partial \boldsymbol{A}}{\partial t}$, $\frac{\partial^2 A}{\partial z^2} = \mu_o \epsilon_o \frac{\partial^2 A}{\partial t^2}$ or $\frac{\partial^2 A}{\partial z^2} = \frac{1}{c^2} \frac{\partial^2 A}{\partial t^2}$.

Let us guess an exponential solution: $A = A_o e^{i \omega t} e^{-i k z}$

$\frac{\partial^2 A}{\partial z^2} = -k^2 A$ and $\frac{\partial^2 A}{\partial t^2} = -\omega^2 A$ which works if $k^2 = \frac{\omega^2}{c^2}$. Now, $A = A_o e^{i(\omega t - k z)} = A_o \cos(\omega t - k z) + i A_o \sin(\omega t - k z)$. Now, $(\omega t - k z)$ is constant, say zero, if $\omega t = k z$ or $\frac{z}{t} = \frac{\omega}{k}$. But, if $k = \frac{\omega}{c}$, the wave velocity is $v = \frac{z}{t} = c$.

Let $\theta = (\omega t - k z)$. If $\boldsymbol{A} = \boldsymbol{i} A_o \cos \theta$, then $\boldsymbol{E} = -\frac{\partial \boldsymbol{A}}{\partial t}$ or $\boldsymbol{E} = -\frac{\partial \boldsymbol{A}}{\partial \theta} \frac{\partial \theta}{\partial t}$.

$\boldsymbol{E} = \boldsymbol{i} \omega A_o \sin(\omega t - k z)$. If $\boldsymbol{B} = \boldsymbol{\nabla} \times \boldsymbol{A}$, then $\boldsymbol{B} = \boldsymbol{i} \left[\frac{\partial A_z}{\partial y} - \frac{\partial A_y}{\partial z} \right] + \boldsymbol{j} \left[\frac{\partial A_x}{\partial z} - \frac{\partial A_z}{\partial x} \right] + \boldsymbol{k} \left[\frac{\partial A_y}{\partial x} - \frac{\partial A_x}{\partial y} \right]$.

$\boldsymbol{B} = \boldsymbol{j} \frac{\partial A_x}{\partial \theta} \frac{\partial \theta}{\partial z}$ or $\boldsymbol{B} = \boldsymbol{j} k A_o \sin(\omega t - k z)$.

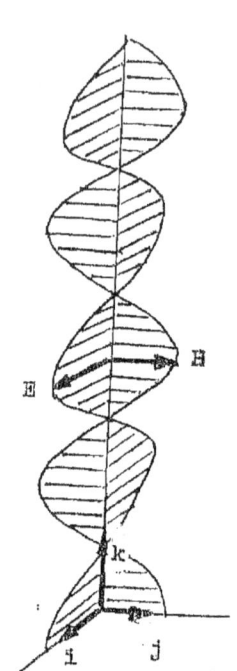

The wave is moving along the z-axis and the fields are perpendicular to the direction of motion. This is a called transverse wave.

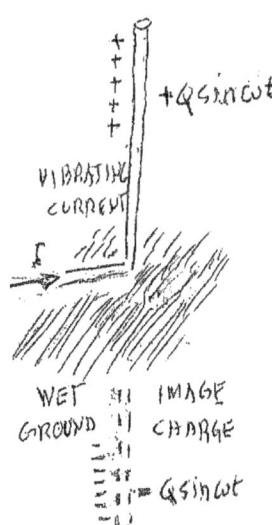

If a wire carrying a vibrating current ends in a vertical wire, the current will end there with a vibrating charge. If the vertical wire is above a damp ground, there will appear to be an image charge in the ground.

If a voltage wave comes down a co-axial cable (cylinder with a wire down its center) there will be an oscillating charge in the stub of wire at the end.

In both cases, there will be an oscillating charge at the end of a wire carrying an oscillating current.

Oscillating changes and currents produce oscillating electric and magnetic fields which move away from their source with the velocity of light.

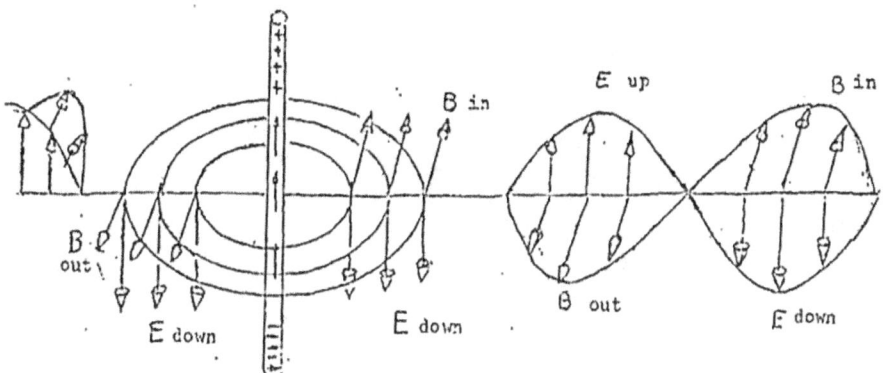

ENERGY IN ELECTROMAGNETIC FIELDS AND RADIATION

If electromagnetic fields can exert forces and move things, they can do work; and that work must come from energy somewhere. We shall now show that the energy is stored in the fields themselves. Special physical cases will be used to develop equations for electromagnetic theory. For energy in an electric field, we will discuss a capacitor being charged. Let the plates have area, A, and be separated by a distance, L.

Take a circuit with no electrical activity. Close the switch and a current, $I = dQ/dt$, will flow. If $E = \frac{Q}{\epsilon A}$, work, $dW = \frac{L \, Q \, dQ}{\epsilon A}$, must be done moving a charge, dQ, from one plate is $W = \frac{L}{\epsilon A} \int Q \, dQ$ or $W = \frac{L}{2 \epsilon A} Q^2$. But $Q = \epsilon A E$, so $W = \frac{1}{2} L A \epsilon E^2$. Dividing energy (work stored) by the volume between plates, we get an energy density, $\frac{W}{L A} = \frac{1}{2} \epsilon E^2$, in joule per meter^3.

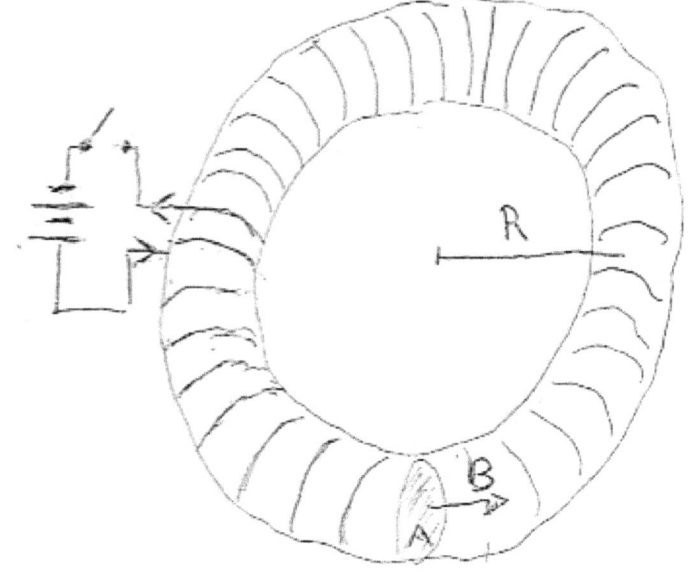

The energy of an isolated charge is stored in its electric field.

To discuss the energy stored in a magnetic field, take a coil of wire, formed as a torus to give a uniform magnetic field. Let each turn have a radius, A, and let there be N turns over the length, $L = 2 \pi R$. Chapter 12 gave the magnetic field to be $B = \frac{\mu_o N I}{L}$.

The magnetic flux through each turn is $B A$, and if that is changing, there is a voltage, $V = A \frac{dB}{dt}$, in each turn opposing the increase of flux.

This must be overcome by an externally applied voltage. Thus, $N A \frac{dB}{dt}$ must be the applied voltage to establish a current in a coil of N turns. This is $\frac{\mu_o N^2 A}{L} \left(\frac{dI}{dt} \right)$. Multiply this by the current to get the power, or rate of doing work, used to put a current into the coil. Then, $\frac{dW}{dt} = \frac{\mu_o N^2 A}{L} I \left(\frac{dI}{dt} \right)$.

Recall, $\int_0^I I\, dI = \frac{1}{2} I^2$. Then, $W = \frac{\mu_o N^2 A}{2 L} I^2$ is the work expended to establish a current in the coil. This work is stored in the magnetic field in the coil. From the previous page, $N I = B L / \mu_o$. From this, $W = \frac{B^2 A L}{2 \mu_o}$. Then, $\frac{W}{A L} = \frac{B^2}{2 \mu_o}$ is the work per unit volume (density) stored in the magnetic field.

$\rho_{EM} = \frac{1}{2} \epsilon_o E^2 + \frac{1}{2 \mu_o} B^2$ is the density of electromagnetic energy. This result is better (more general) than the special cases used to derive it.

ENERGY OF ELECTROMAGNETIC RADIATION

Look at the EM wave on page 13-61. The electric field wave has a magnitude, $E = \omega A_o \sin(\omega t - k z)$, and the magnetic field wave has a magnitude, $B = k A_o \sin(\omega t - k z)$. Notice, $B = \frac{k}{\omega} E$.

Now, $\omega = 2 \pi f$ and $k = \frac{2 \pi}{\lambda}$ giving $\frac{k}{\omega} = \frac{1}{c}$. $c = f \lambda$ being wave velocity. Recall now that $\mu_o \epsilon_o = \frac{1}{c^2}$. Hence, $\left(\frac{k}{\omega}\right)^2 = \mu_o \epsilon_o$.

The magnetic wave has an amplitude, $B = \frac{k}{\omega} E$. The electromagnetic wave has an energy density, $\rho_{EM} = \frac{1}{2} E^2 \left(\epsilon_o + \frac{1}{\mu_o}\left(\frac{k}{\omega}\right)^2\right)$. But, $\frac{1}{\mu_o}\left(\frac{k}{\omega}\right)^2 = \epsilon_o$ giving $\rho_{EM} = \epsilon_o E^2$.

Recall that fluid density times velocity gives fluid per unit time crossing a unit area. $\boldsymbol{J} = \rho \boldsymbol{v}$. Hence, for an EM wave, energy per unit time crossing a unit area is $\boldsymbol{J}_{EM} = \epsilon_o E^2 c \boldsymbol{k}$, $\boldsymbol{k}$ giving the direction of energy flow.

Electromagnetic theory has another way of writing this. Recall that $\boldsymbol{k} = \boldsymbol{i} \times \boldsymbol{j}$ and $\epsilon_o c = \frac{1}{\mu_o}\left(\frac{k}{\omega}\right)$ and $\frac{k}{\omega} E = B$. Then, $\boldsymbol{J}_{EM} = \frac{1}{\mu_o}[(E\,\boldsymbol{i}) \times (B\,\boldsymbol{j})]$ or simply,

$\boldsymbol{J}_{EM} = \frac{1}{\mu_o} \boldsymbol{E} \times \boldsymbol{B}$. In EM theory, $\boldsymbol{J}_{EM}$ is given the symbol, $\boldsymbol{S}$, and is called the Poynting vector.

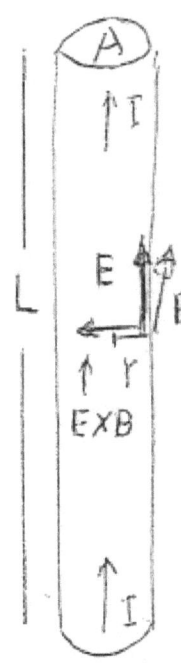

In chapter 8, we found that an electric field, E, in a wire of cross section area, A, and of a material with conductivity, σ, made a current to flow where $I = \sigma A E$. In chapter 12 we found that if a current, I, flowed in a wire, a magnetic field, $B = \frac{\mu_0 I}{2\pi r}$, existed at a radius, r, from the wire.

The electric and magnetic fields meet at the surface of the wire. In the picture, the electric field is up, in the direction of the current. The magnetic field is into the picture. The combination gives a power flow vector, $\frac{1}{\mu_0} E \times B$, directed into the wire. This is $\frac{1}{\mu_0}\left(\frac{I}{\sigma A}\right)\left(\frac{\mu_0 I}{2\pi r}\right) = \frac{I^2}{2\pi \sigma r A}$.

This is in joules per second (watts) per square meter. When this is multiplied by the surface area, $2\pi r L$, of the wire, we get the total power flow into the wire from the surrounding space, $P = \left(\frac{I^2}{2\pi \sigma r A}\right)(2\pi r L)$ or $P = \left(\frac{L}{\sigma A}\right)I^2$.

In chapter 8, we found that $\frac{L}{\sigma A} = R$ is the resistance of the wire and that $R I = V$ is the voltage from one end to the other of the wire. $P = I V$ is exactly the power loss to the resistance of the wire.

This tells us that the energy lost in the resistance is flowing in from the space outside the wire. Who said reality made any sense?

THE QUANTIZATION OF ELECTROMAGNETIC RADIATION

Next, we shall show that the energy of an EM wave is carried in particles called photons. In doing this, we refer to the quantum mechanics of a harmonic oscillator developed in chapter 16.

If A is the magnitude of a wave in the vector potential, $E = \frac{dA}{dt}$ and $B = k A$.

From this, $\rho_{EM} = \frac{1}{2}\epsilon_o \left(\frac{dA}{dt}\right)^2 + \frac{1}{2\mu_o}k^2 A^2$, but $\frac{k^2}{\mu_o} = \epsilon_o \omega^2$.

The energy density is now $\rho_{EM} = \frac{1}{2}\epsilon_o \left(\frac{dA}{dt}\right)^2 + \frac{1}{2}\epsilon_o \omega^2 A^2$.

A harmonic oscillator (object of mass, M, on a spring of stiffness, k, vibrating with a displacement, x, and angular frequency, ω, where $\omega^2 = \frac{k}{M}$) has an energy, $E = \frac{1}{2} M \left(\frac{dx}{dt}\right)^2 + \frac{1}{2} M \omega^2 x^2$.

Replace M by ϵ_o and x by A and the electromagnetic energy has the same form. In chapter 16 it is shown by quantum mechanics that the energy of an oscillator is given by $E = \left(n + \frac{1}{2}\right) \hbar \omega$; $n = 1, 2, 3, 4 \ldots$ n is called a quantum number. As $dn = 1$, $dE = \hbar \omega$. As $\omega = 2\pi f$ and $\hbar = \frac{h}{2\pi}$, $\hbar \omega = hf$.

It is evident that quantum mechanics quantizes electromagnetic energy of a wave into definite amounts, $dE = \hbar \omega = hf$.

These pieces of EM energy act as particles and are called PHOTONS.

CHAPTER FOURTEEN: BAND THEORY OF SEMICONDUCTORS: LIGHT EMITTING DIODES, LASERS, AND FIELD EFFECT TRANSISTORS

It is the goal of theoretical physics that we may be able to construct a constant mathematical scheme which provides a unified description of the measurable appearance of the universe. To accomplish this aim, we attempt to formulate axioms from which we may derive mathematically as conclusions both the results of past measurements and predictions of future measurements to be done in the laboratory. The development of a theory begins with a body of experimental evidence – in this case the observation of the nature of light emitted from hot solids and excited gases. After much enlightened guesswork, some bright people stumble onto a set of statements from which a description of the experiments may be derived. As a bonus for good mathematics, we may expect predictions of phenomena which have not yet been observed - e.g. antimatter, mesons, semiconductors, lasers, etc. Unfortunately, in going from known physics to unknown physics, our mathematics may carry no physical intuition along with it. Quantum mechanics is an extreme case of this; it tells you nothing physical about what it is doing until the point when it says that if you go into the laboratory and do such-and-such then this and that will be the results. As a result, we are totally dependent of the rigor of our mathematics to hold everything together until that point where it suddenly predicts the result of an experiment.

At this point you might review what was said about atoms, particles and waves in chapter 13. Max Plank (1901) postulated that a light wave with a frequency, f, had its energy carried in packets (quanta) or particles (photons) of energy, $E = hf$. Robert A. Millikan (1916) measured Planck's constant to be 6.624×10^{-34} Joule second. Planck's postulate was directly observed by Arthur H. Compton (1923) when he scattered x-rays off of electrons. Louis de Broglie (1923) postulated that particles with momentum, $p = mv$, would have associated waves whose wavelength was $\lambda = h/mv$. These waves for electrons were directly observed by C. J. Davisson, L. H. Germer and G. P. Thompson in 1927.

Assume now, that a free particle with no interactions experiences the Planck and de Broglie wave-particle relationships. We use the Greek letter, ψ, for the wave that is associated with a particle with momentum, $p = mv$, and energy, E. You will recall that either $\psi = A \sin(kx - \omega t)$ or $\psi = A \sin(kx - \omega t)$ will describe a wave. ψ will have a constant value if $(kx - \omega t)$ has a constant value. Let $(kx - \omega t) = C$, a constant. Take the time derivative, $k\frac{dx}{dt} - \omega = 0$, or $\frac{dx}{dt} = \frac{\omega}{k}$ is the velocity of the moving wave. Putting the two waves together gives the generalized wave, $\psi = e^{i(kx - \omega t)}$.

If λ is the wavelength, $k(x+\lambda) = kx$, requiring $k\lambda = 2\pi$. If the wave has a time, T, for one vibration, $\omega(t+T) = \omega t$, requiring that $\omega T = 2\pi$. Now, $\frac{1}{T} = f$, the frequency of vibration. This gives $\psi = A e^{i2\pi\left(\frac{x}{\lambda} - ft\right)}$. But suppose that $f = \frac{E}{h}$ and $\frac{1}{\lambda} = \frac{p}{n}$ with $p = mv$. Then $\psi = A e^{i\frac{2\pi}{h}(px - Et)}$. In all the discussions that follow, Planck's constant will always be divided by 2π, so we will follow a suggestion by P. A. M. Dirac and define an haitch-bar as $\hbar = \frac{h}{2\pi}$. Then our wavefunction is $\psi = A e^{i\left(\frac{px - Et}{\hbar}\right)}$.

If $\psi = A e^{i\left(\frac{px}{\hbar}\right)} e^{-i\left(\frac{Et}{\hbar}\right)}$, then $\frac{\partial \psi}{\partial x} = i\frac{p}{\hbar}\psi$ and $\frac{\partial^2 \psi}{\partial x^2} = -\frac{p^2}{\hbar^2}\psi$.
Similarly, $\frac{\partial \psi}{\partial t} = -i\frac{E}{\hbar}\psi$ and $\frac{\partial^2 \psi}{\partial t^2} = -\frac{E^2}{\hbar^2}\psi$.

But $E = \frac{p^2}{2M}$ for a particle with no interactions, where $p = Mv$. Multiplying this by ψ gives $E\psi = \frac{p^2}{2M}\psi$, and substitution gives $i\hbar\frac{\partial \psi}{\partial t} = \frac{\hbar^2}{2M}\frac{\partial^2 \psi}{\partial x^2}$. For a particle whose interaction has a potential energy, $V(x)$, $E = \frac{p^2}{2M} + V(x)$. If we multiply this by ψ and substitute, we get $i\hbar\frac{\partial \psi}{\partial t} = \frac{\hbar^2}{2M}\frac{\partial^2 \psi}{\partial x^2} + V(x)\psi$. This result is the time-dependent Schrödinger equation.

Returning to $\psi = A e^{i\left(\frac{px - Et}{\hbar}\right)}$, $\frac{\hbar}{i}\frac{\partial \psi}{\partial x} = p\psi$ and $i\hbar\frac{\partial \psi}{\partial t} = E\psi$. Substituting $i\hbar\frac{\partial \psi}{\partial t} = E\psi$ into the time-dependent Schrödinger equation gives the time-independent Schrödinger equation, $-\frac{\hbar^2}{2M}\frac{\partial^2 \psi}{\partial x^2} + V(x)\psi = E\psi$.

Yes, this is more of an excuse for the Schrödinger equation than a derivation, but a rigorous derivation might confuse things rather than clarify them. Given the Schrödinger equation, we will justify it by the answers we get in using it.

In the following discussion we will consider how electrons behave in a lattice of atoms. Although real materials such as metals, semi-conductors and insulators exist in three dimensions, we will simplify the picture by considering one dimensional lattices. These actually exist in such exotic structures like quantum wires and Luttinger liquids.

To start, consider an electron wave in a confined one-dimensional box. Such a confined particle exists in a standing wave with nodes (zero points) at the end walls. The length of the box will be taken to be an integer number of wave lengths.

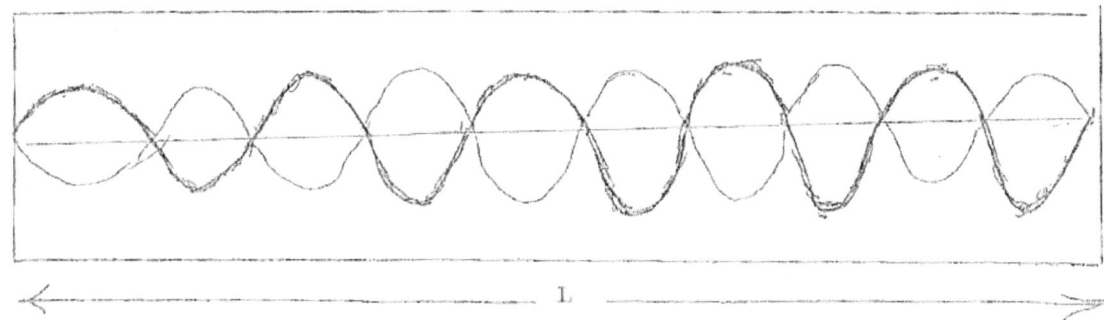

In the picture, the length, L, of the box contains five wavelengths of a standing wave. For the "particle" related to the wave, its momentum, $p = m v$, is $p = \frac{h}{\lambda}$ and its energy is $E = \frac{1}{2} m v^2 = \frac{p^2}{2m}$. If there are n wavelengths in the length, L, then $\lambda = \frac{L}{n}$ and $p = \frac{nh}{L}$. The particle's energy is $E = \frac{n^2 h^2}{2 m L^2}$.

Before going on, let us look at $p = \frac{nh}{L}$. If n is an integer, we only know the momentum in jumps of $\Delta p = \frac{h}{L}$. But with quantum mechanics, we only knew that the particle's location is somewhere in a length, L. The particle is somewhere in a span of position, $\Delta x = L$.

Taking the product of the uncertainties, $\Delta p \, \Delta x = h$.

This is a general fact of quantum mechanics called the HEISENBERG UNCERTAINLY PRINCIPLE. The closer that we know the position of a particle, the less we know how fast it is going and the less known of its location, the better we can specify its velocity.

Return now, to $p = \frac{h}{\lambda}$ and multiply this and divide it by 2π. $p = \frac{2\pi}{\lambda} \frac{h}{2\pi}$ and recall $\hbar = \frac{h}{2\pi}$. Define wave number, $k = \frac{2\pi}{\lambda}$. Then $p = k \hbar$.

With quantized systems, the wave number is often used instead of p. The electron's energy was $E = \frac{p^2}{2m} = \frac{k^2 \hbar^2}{2m}$. If $\lambda = \frac{L}{n}$, then $k = \frac{2\pi n}{L}$.

Just as we found with Niels Bohr's theory of atomic electrons, the electron's energy can have only certain values. Just as water flows downhill rather than uphill, so does a system of particles seek their lowest energy. One mystery of multielectron atoms is that all of the electrons don't pile up in the lowest energy state. In developing a theory of atomic electron behavior, it was evident that for particles such as electrons, only one particle out of many could exist in any quantum state. As Wolfgang Pauli first thought of this, we call it the PAULI EXCLUISION PRINCIPLE.

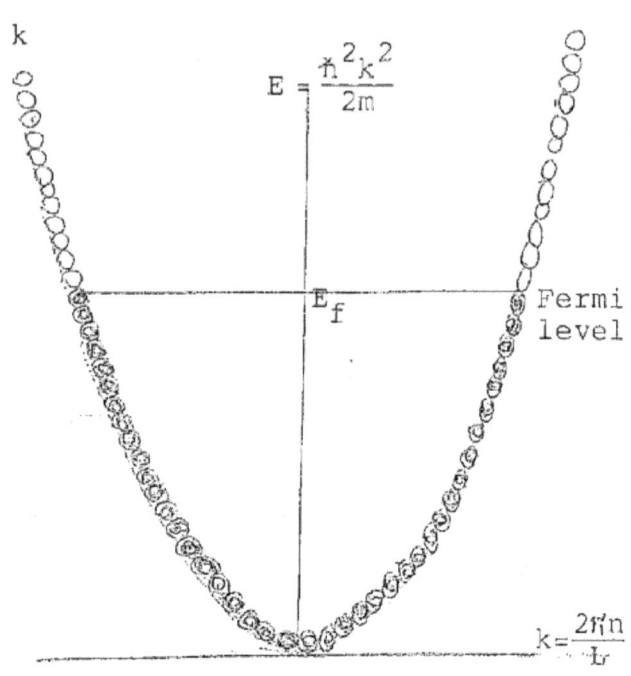

If the electron energy is plotted against momentum as wave number, we get a parabola. As only certain values of this parabola represent electron states, we will represent these by little circles. The Pauli exclusion principle says that at the lowest energy, all quantum states starting at the lowest are filled up until we have accounted for all of the electrons. We represent filled states by black circles.

When all of the quantum states have been filled, we come to the maximum filled energy level. After Enrico Fermi, this is called the Fermi level and its energy is called the Fermi energy, E_f.

If an electric field is created in the region, electrons will be pulled into momentum states moving in a direction opposite to the electric field with more electrons moving in one direction than the other, we now have an electric current. But there is a limit to how many electrons will be pulled into the current, and at this point we must be aware of the atoms in our box. For each electron that goes into the system, an ionized atom is left in the lattice. At first, we will consider only the atoms. Then we will return and consider them as ions.

In this picture, an electric field toward the right makes a shift of negative electrons into left moving momentum states, causing a current of negative charge toward the left.

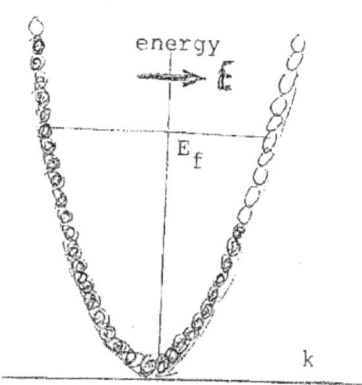

Now, put the atoms back in a lattice. Let a be the distance between adjacent atoms. Here, consider atoms to be hard balls. A wave hitting a hard surface will be partially reflected, and when the surface is hard, the reflected wave will have the opposite phase from the ongoing wave, that is, if the ongoing wave is zigging the reflected wave will be zagging.

In most cases of reflection, the wavelength is irrationally related to atomic spacing, and the reflected waves will interfere with each other in random or destructive interference and in the total cancel themselves. But, in the next picture, when the wavelength equals twice the atomic spacing, the reflected waves are all doing the same thing and add up to give a net left moving wave, just as if they had been reflected off a mirror.

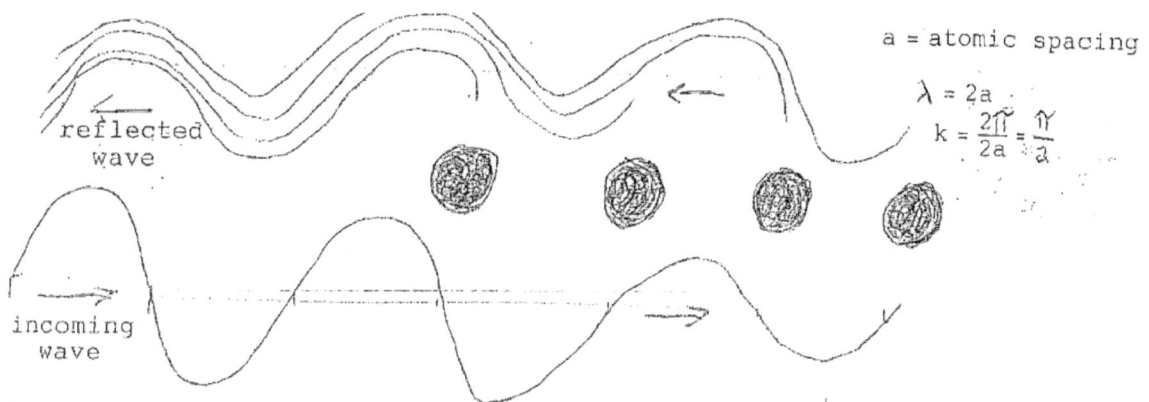

Because of this, the electrons are limited to momentum states between $k = -\frac{\pi}{a}$ and $k = +\frac{\pi}{a}$. These are called Brillouin zone boundaries, and the range of momentum states between them is called the first Brillouin zone after Leon Brillouin.

The next boundaries are at $k = \pm\frac{2\pi}{a}$ giving the second zone.

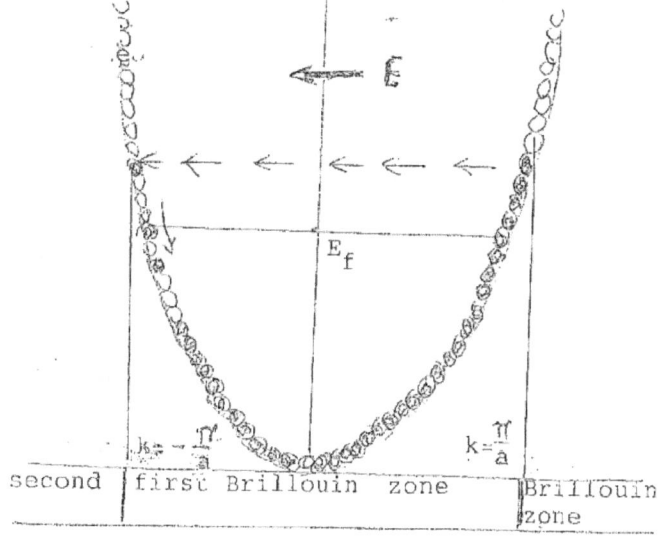

Now, an atom having given up an electron to wander in the box is then left with a positive charge becoming an ion. A negative charge, $-q$, a distance, r, from a positive charge, $+Q$, has a potential energy, $V(r) = \frac{(-q)(+Q)}{4\pi\epsilon_0 r}$.

With the potential energy of an electron interacting with an ion, we are ready to apply the Schrödinger equation to electrons in periodic lattices. The time-independent Schrödinger equation was

$$-\frac{\hbar^2}{2M}\frac{\partial^2\psi}{\partial x^2} + V(x)\,\psi = E\,\psi.$$

An electron sees a Coulomb potential energy in the neighborhood of each, which is repeated over and over as shown.

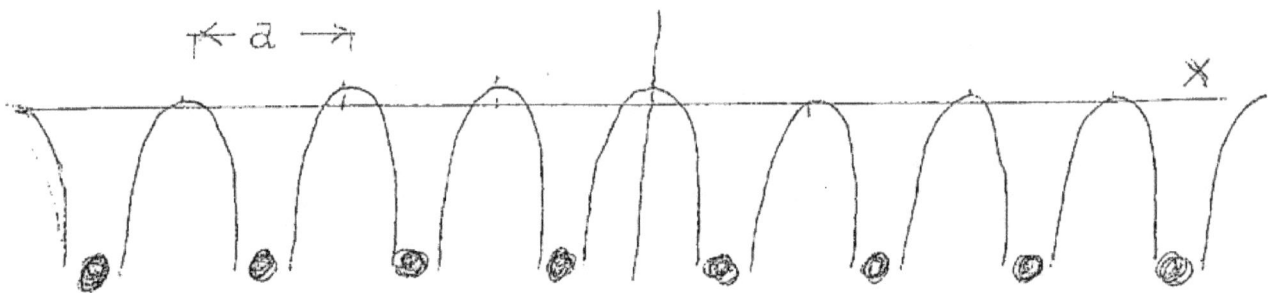

A simpler potential function of the same shape would be a sine or cosine potential with the atomic spacing as wavelength. $V(x) = V\frac{sin}{cos}\left(\frac{2\pi x}{a}\right)$.

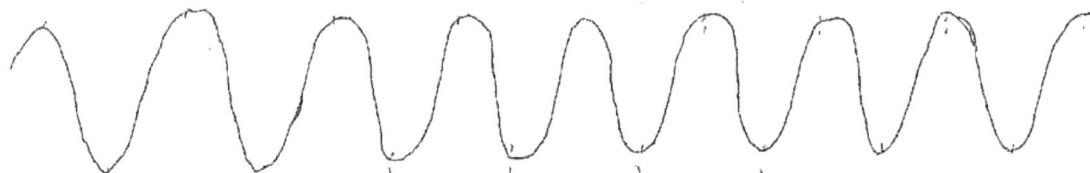

As all we want is the shape, it doesn't matter if it is sine or cosine. Therefore, we can use a generalized potential, $V(x) = V\,e^{i\frac{2\pi x}{a}}$.

Now strongly interacting with lattice ions, the electron is no longer in standing waves. Some new state function, $\psi(x)$, must result from this potential energy. To construct it, quantum mechanics proceeds as follows.

The standing waves in the length, L, had wavelengths, $\lambda_n = \frac{L}{n}$, for n wavelengths. Then, $e^{i\frac{2\pi x}{L}}$ would be a generalized wave of that sort.

When a simple quantum mechanical problem becomes complicated so that an exact solution for the state function, $\psi(x)$, can't be derived, we assume that whatever it is, it can be described in terms of a sum of functions holding in the related simple case. We assume the countable summation:

$$\psi(x) = \sum_{n=1}^{\infty} A_n \frac{sin}{cos} \left(\frac{2\pi n x}{L} \right)$$

The amplitudes are interpreted to mean that an electron in the new state has a probability of $A_n{}^2$ of having $E_n = \frac{n^2 h^2}{2 M L^2}$ as its energy.

To simplify things, we will let $g = \frac{2\pi}{a}$ and $k_n = \frac{2\pi n}{L}$.

Putting all this into the Schrödinger equation we will primarily be interested in the effect on the energy states.

Let us, somewhat arbitrarily, write $\psi = 2B\cos(kx) - 2C\sin(kx)$. This can be written, $\psi = B[e^{ikx} + e^{-ikx}] + iC[e^{ikx} - e^{-ikx}]$ which is $\psi = (B + iC)e^{ikx} + (B - iC)e^{-ikx} = A e^{ikx} + A^* e^{-ikx}$, where $A = (B + iC)$ and $A^* = (B - iC)$, its complex conjugate. This means that $A_{-n} = A_n^*$

Then we write,

$$\psi = \sum_{n=-\infty}^{+\infty} A_n e^{i k_n x}$$

Put this with the potential energy, $V e^{i\frac{2\pi x}{a}} = V e^{igx}$, into the Schrödinger equation, $-\frac{\hbar^2}{2M}\frac{\partial^2 \psi}{\partial x^2} + V(x)\psi = E\psi$.

$$\sum_{n=-\infty}^{+\infty} \left(\frac{\hbar^2 k^2}{2M} - E \right) A_n e^{i k_n x} + \sum_{n=-\infty}^{+\infty} V A_n e^{i(k+g)x} = 0$$

You can sum over n's like so, or you can just sum from $k = -\infty$ to $k = +\infty$, the integer quantum number, n, being understood.

Look at $\sum_{k=-\infty}^{+\infty} V A_k e^{i(k+g)x}$. If k goes from $-\infty$ to $+\infty$, then $k + g$ also goes from $-\infty$ to $+\infty$ also. Replace $k + g$ by just k and k by $k - g$.

With the new k we get $\sum_{k=-\infty}^{+\infty} V A_{k-g} e^{ikx}$. This allows us to write the Schrödinger equation:

$$\sum_{k=-\infty}^{+\infty} \left[(\frac{\hbar^2 k^2}{2M} - E) A_k + V A_{k-g} \right] e^{ikx} = 0$$

This requires $(\frac{\hbar^2 k^2}{2M} - E) A_k = -V A_{k-g}$.

Consider this at the first Brillouin zone boundary. $\lambda = 2a$. $k = \frac{2\pi}{\lambda} = \frac{\pi}{a}$ and $g = \frac{2\pi}{a}$ or $k = \frac{g}{2}$. Let $E_0 = \frac{\hbar^2 k^2}{2M}$. At the first Brillouin zone boundary, $(E_0 - E) A_{g/2} = V A_{-g/2}$.

But, $A_{-g/2} = A_{g/2}^*$. $(E_0 - E) A_{g/2} = V A_{g/2}^*$. Take the complex conjugate of this, $(E_0 - E) A_{g/2}^* = V A_{g/2}^{**}$. But, $A_{g/2}^{**} = A_{g/2}$, so $(E_0 - E) A_{g/2}^* = V A_{g/2}$.

Multiply $(E_0 - E) A_{g/2} = V A_{g/2}^*$ by $(E_0 - E)$ giving $(E_0 - E)^2 A_{g/2} = V (E_0 - E) A_{g/2}^*$.

Then $(E_0 - E)^2 A_{g/2} = V^2 A_{g/2}$, if $(E_0 - E)^2 = V^2$ or $(E_0 - E) = \pm V$. This is at the first Brillouin boundary.

Without an interaction with atoms, the electron quantum states lie on a parabola. But after they interact with ionized atoms, the parabola has an energy gap at the first Brillouin boundary.

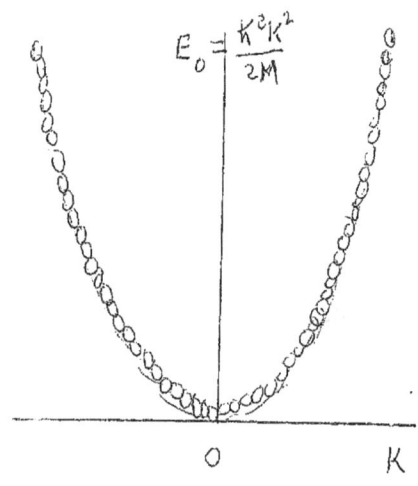
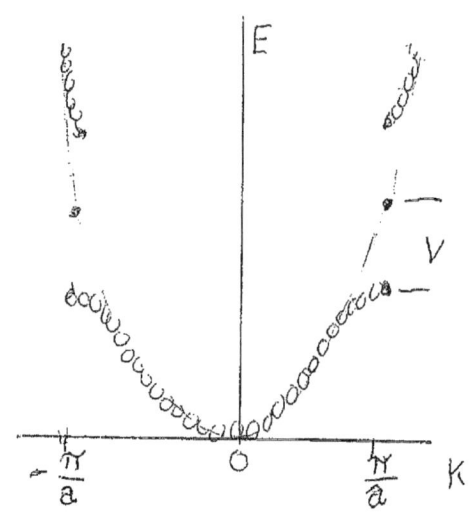

If the free electron energy is $E_o = \dfrac{\hbar^2 k^2}{2M}$, when $k = \dfrac{\pi}{a}$ the energy is $E = E_o - V$ for states below E_o. The parabola is distorted as shown at the first Brillouin zone boundary. A similar energy gap occurs at all other Brillouin boundaries. Thus, we have the BAND THEORY of the electrons in solid lattices.

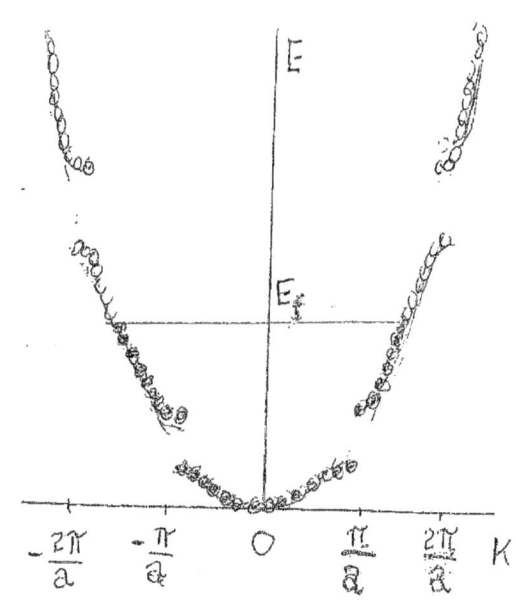

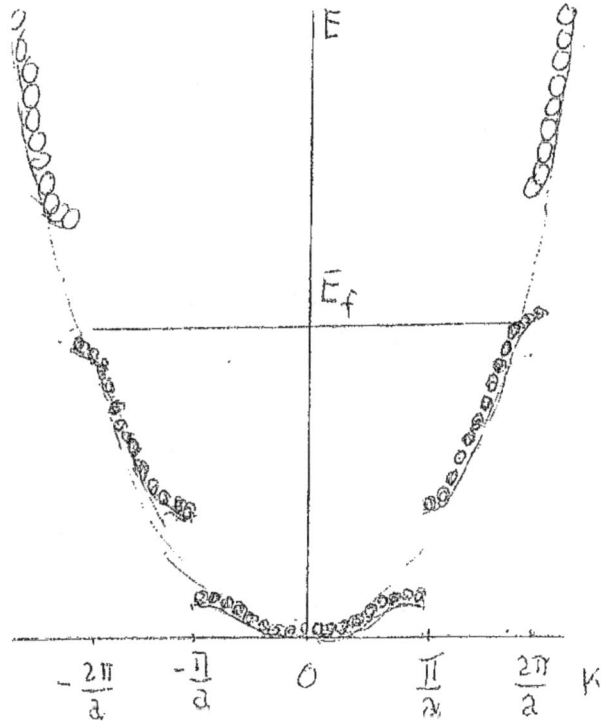

Now, start filling quantum states with electrons. With the Pauli exclusion principle allowing only one electron per quantum state, at the lowest system energy all states are filled up to the Fermi energy. If the Fermi level only comes up halfway in the second band (see above), there are empty states into which electrons can shift with an electric field, allowing more electrons moving in one direction than the other allowing a current to flow. Such a material is a CONDUCTOR, usually a metal. If the Fermi level is at a Brillouin zone boundary, the electrons cannot shift into empty quantum states without jumping the energy gap. Such a material is either an insulator or SEMICONTUCTOR.

To get the physics of semiconductors explained as simply as possible, we are using a one-dimensional crystal. Let us stop a moment but real crystals are three-dimensional. Brillouin boundaries were at momentum states where electron waves were reflected by constructive interference. In three-dimensions, a Brillouin zone boundary is a plane surface off a polyhedron determined by the structure of the crystal lattice of atoms.

In our discussion, we have plotted electron energy against wave number, k, rather than momentum, $p = \hbar k$, getting the parabola as shown. When the Fermi level was not near a Brillouin zone boundary, the energy states lay on the original parabola. But near Brillouin zone boundaries the parabola was distorted by energy gaps. When the Fermi level was at a Brillouin zone boundary, the energies were not on the parabola.

In a three-dimensional box, the electron is a three-dimensional standing wave, each dimension of the box being an integer number of wavelengths in that dimension. Momentum is now a three-dimensional vector, each point in a three-dimensional vector space having energy value, $E = \frac{1}{2M}\left[p_x^2 + p_y^2 + p_z^2\right]$ or $E = \frac{\hbar^2}{2M}\left[k_x^2 + k_{,y}^2 + k_z^2\right]$. The Fermi level becomes a Fermi surface, all points having the same energy of maximum occupied quantum states.

When the Fermi level is not near a Brillouin zone boundary, these points from a sphere in momentum space. When the Fermi level is near a Brillouin zone boundary, that sphere is distorted, as shown in the picture below.

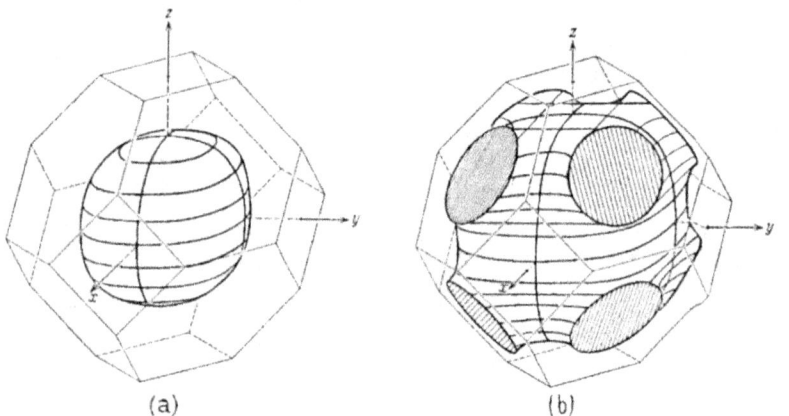

(a) (b)

Fig. 26. Surfaces of constant energy in k-space, face-centred cubic lattice.†

(a) Zone nearly empty. (b) Zone nearly full.

Many electron properties of solids are explained by the topology of Fermi surfaces in the neighborhood of Brillouin zone boundaries. The mathematics needed for further discussion of the topic is more than can be used in this discussion and will be found in advanced books on solid state theory. One such reference is The Theory of the Properties of Metals and Alloys by Mott and Jones.

Which elements of the periodic
table have electrons in which
band? A reference to the Bohr
model will help. Although Bohr
considered only hydrogen, and
can't describe any other atom's
structure, it does form a basis
for discussion. Considering atoms
with more electrons they can be
thought to go into quantum states
in the Bohr orbits. The formal
quantum mechanics developed in
the later chapters will tell us
how many quantum states are in
any Bohr orbit.

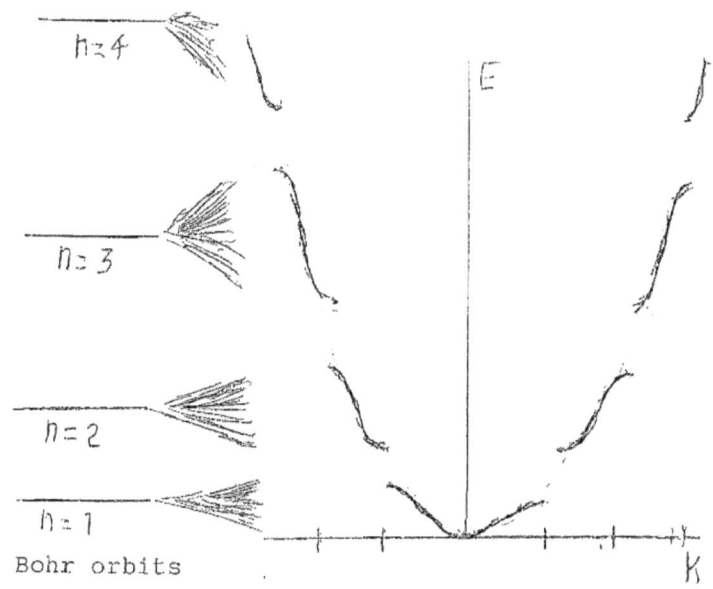

Bohr orbits

Two states are in the first orbit, eight in the second, and so on. Each Bohr
quantum state is split up into the multiple states on combining atoms into
molecules and when the atoms are in crystalline matter, the multiple states
fuse into the energy bands. Thus, the highest energy band with any electrons
in it involves valence electrons and is called the VALENCE BAND. Any
electrons that can get into the next band can move around easily. The next
band is called the CONDUCTION BAND.

It may be evident by now that the electrical nature of any material will be
determined by the relation of the Fermi level to some Brillouin zone
boundary. For descriptive purposes, we will stick to a one-dimensional
model. Actual matter is in three dimensions and then the discussion involves
the topology of Fermi surfaces in relation to Brillouin zone boundaries.

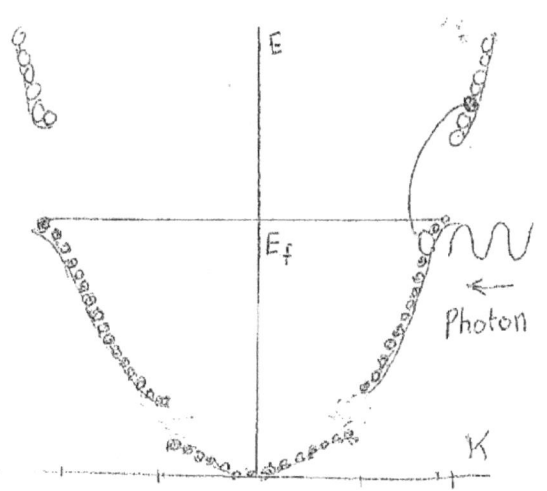

Consider now a case where the Fermi level
is at a Brillouin zone boundary and a
normal electric field will not cause
conduction. Let an electron absorb a photon
with enough energy to put the electron into
the conduction band with empty quantum
states.

The electron in the conduction band is free
to respond to an electric field and there
is a VACANCY in the valence band allowing
some electron to respond to an electric
field. For a light wave of frequency of the
photon energy, $E = hf$, must be greater than
the band gap energy, E_g.

The difference between an insulator, where an electric field strong enough to make it conduct would cause the material to break down, and a semiconductor, which can be made to conduct with only a bit of difficulty, is somewhat arbitrary. For photons, or light waves, we might take it to be in the near ultraviolet wavelengths around $\lambda = 3000\,\text{Å}$, which is about 4 volts. We might take semiconductors to have band gap energies less than that.

Presumably to save space, it is standard practice to fold the bands of energy so that they all appear over the first Brillouin zone.

Take a case where the Fermi level is at the top of the second energy band. A photon whose energy is greater than the band gap energy may be absorbed by an electron in the valence band, knocking it up into the conduction band. The electron can then bounce down to the lowest energy in the conduction band and fall back to the valence band emitting a photon of definite energy and frequency, $h f = E_g$.

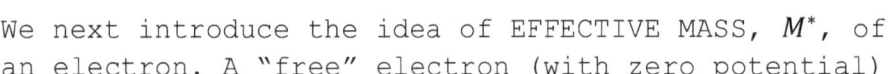

Photons of varying frequency can be absorbed, but emitted photons have only a specific frequency.

We next introduce the idea of EFFECTIVE MASS, M^*, of an electron. A "free" electron (with zero potential) has energy, $E_o = \dfrac{\hbar^2 k^2}{2M}$, where $k = \dfrac{2\pi}{\lambda}$. Notice that

$$\frac{d^2 E}{dk^2} = \frac{\hbar^2}{M}.$$ Remember, $h k = p = m v$.

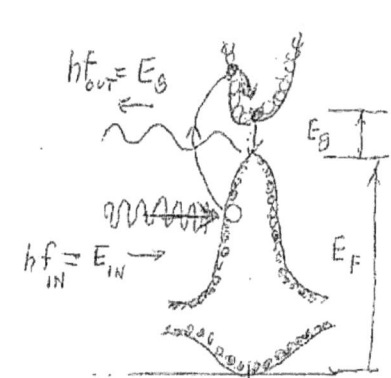

Hence, $+k$ is a motion toward the right, and $-k$ is a motion toward the left.

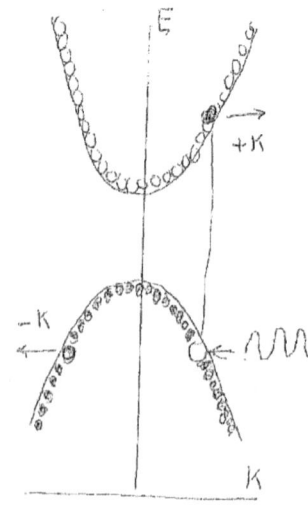

With the band energies we are plotting E against k. Carried over into this situation, the effective mass is defined as $\dfrac{d^2 E}{dk^2} = \dfrac{\hbar^2}{M^*}$, M^* now does some strange things.

Electron states in the conduction band have energies, $E = E_f + E_g + b k^2$, where b is some constant, as the conduction band is an upright parabola.

Then, $\frac{d^2E}{dk^2} = 2b$ or $M^* = \frac{\hbar^2}{2b}$ is positive as expected. In the valence band,

$E = E_f - ck^2$, $\frac{d^2E}{dk^2} = -2c$ and $M^* = -\frac{\hbar^2}{2c}$. What does this mean?

The electron in the conduction band has velocity, $+v$, mass, $+M^*$, and charge, $-e$. Its momentum is $(+M^*)(+v) = M^* v$ and current, $(-e)(+v) = -ev$. It is a normal electron.

In the valence band, things are a little more complicated. An electron in a rightward moving state is missing, leaving an un-balanced electron in a leftward moving state with velocity, $-v$. This electron has charge, $-e$, and a negative effective mass which can be written, $-M^*$. It has a momentum, $(-M^*)(-v) = M^* v$, as if it had positive mass moving rightward. Its electric current is $(-e)(-v) = +ev$. This is the current of a positive charge moving toward the right!

Thus, when there is a missing electron (called a HOLE) in the valence band, the remaining electron at that energy acts as an electron with positive mass and positive charge moving in the opposite direction.

When a photon knocks an electron from a filled valence band into the conduction band, the electron acts now as a negative electron, but the hole left behind acts as a positive electron.

In series with a zig-zag shaped strip of cadmium-sulfide, CdS, or cadmium-selenide, CdSe, place an ammeter for measuring electric current and a battery source of voltage, or electrical energy, V. Limited by the resistance, R, of the strip an electric current, I, in ampere = coulomb/sec, will flow, where

$I = \frac{V}{R}$.

Now, place a light source above the strip and move your hand back and forth between the light source and the strip. When the light on the strip is bright, the current goes up and when the light on the strip becomes dim, the current goes down. When the light is cut off all together, the current goes to zero.

The resistance of the strip is dependent on the intensity of light falling on it, decreasing as the brightness increases.

Wires in light bulbs are usually tungsten, but bulbs with carbon wires exist. Using a variable voltage source, measure the current through a bulb of each type. Plot the resistance, $R = \frac{V}{I}$, against temperature. There is no relation between the two curves.

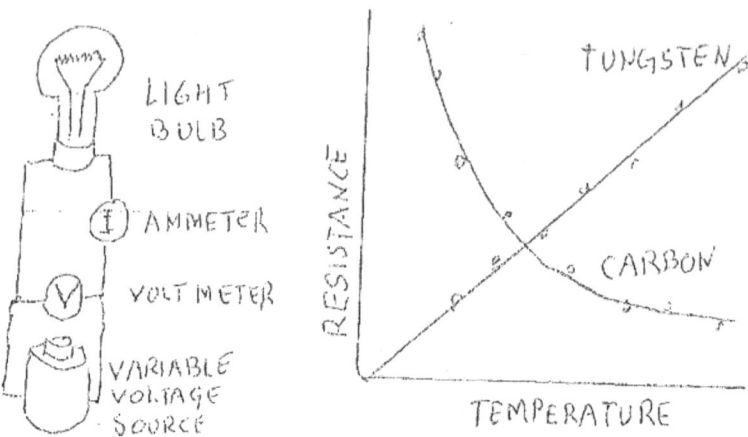

Thermal (heat) energy is distributed among both the electrons and the lattice atoms. The lattice atoms collectively vibrate in standing waves. In a later chapter, we show that the energy of mechanical waves is carried by particles of energy called phonons, just as the energy of light waves is carried by photons. As with photons, the phonons of a wave of frequency, f, is $E = hf$. It is the interaction between a phonon liquid and electron liquid that accounts for many properties. (Note: we used to call them gasses, but the word liquid is used today.)

We shall now develop a simple-minded theory of electrical conduction. Consider a gas of free electrons whose overall charge is neutralized by a background lattice of positive ions, with which they do not interact. Let them be subject to collisions with phonons or lattice vibration energy.

An electric field, E, will accelerate electrons in a direction opposite to itself, creating an electric current. But the electrons will be bombarded by phonons knocking electrons out of the current. Assume that the average effect of a phonon collision is to reduce the electron's velocity in the current to zero and so the field has to start all over again to bring the electron into the current. If t is the average time between collisions, the electron will move a distance, $x = \frac{1}{2}\left(\frac{eE}{M}\right)t^2$, with an average velocity, $\bar{v} = \frac{x}{t} = \left(\frac{eE}{2M}\right)t$.

By the law of Dulong and Petit, the heat capacity of metal is $C = 3R$, R being the gas constant. Hence, the number of phonons should be proportional to the temperature, T, and the rate of phonon-electron collisions should be proportional to T, giving collisions per unit time, $v = cT$, where c is some

constant. The time between collisions is $t = 1/v$ and the average velocity is
$$\overline{v} = \frac{eE}{2Mv} = \frac{eE}{2McT}.$$

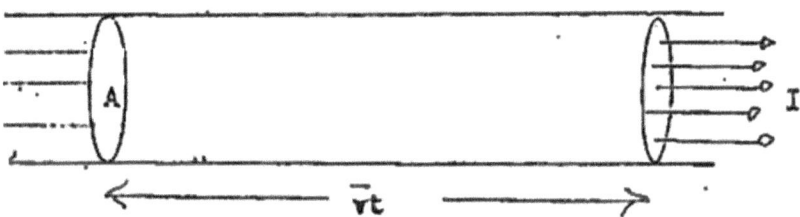

Let the current be flowing in a wire of cross-section area, A. Let there be N electrons per unit volume. An average electron goes a distance, $\overline{v}\,t$, in a time, t, and all electrons before it will cross the far cross-section in that time. All electrons in a volume, $A\overline{v}$, will pass that point in a unit time. This gives an electric current, $I = NeA\overline{v} = \left(\frac{Ne^2t}{2m}\right)AE.$

$\sigma = \left(\frac{Ne^2t}{2m}\right)$ is the conductivity and the electrical resistance is $R = \frac{L}{\sigma A}$, for a path of length, L. $R = \left(\frac{2McL}{e^2NA}\right)T$, proportional to temperature as measurements with a tungsten wire will confirm. $\rho = \frac{1}{\sigma}$ is called resistivity.

With tungsten, the Fermi level is in the middle of a conduction band and can behave in the manner just now described. With carbon, the Fermi level is at the top of the valence band. There are no empty quantum states for them to shift into and at normal temperatures carbon is a non-conductor. Rising temperature produces more energetic phonons which can kick carbon electrons into the conduction band, allowing carbon to conduct. The higher the temperature, the more phonons and the more conduction electrons and holes.

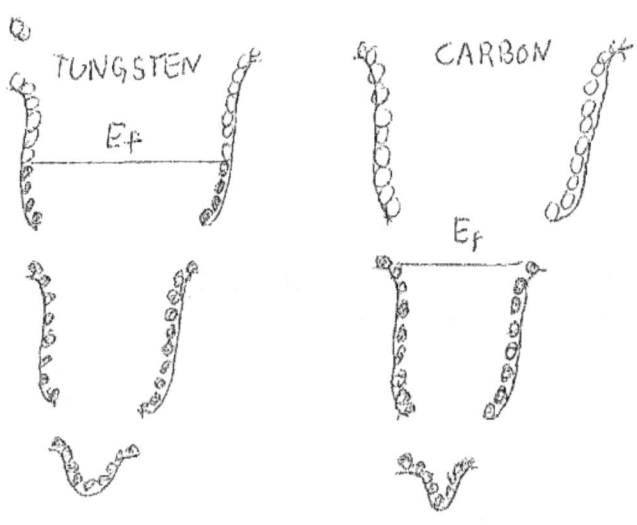

	Rb	Sr	Y	Zr	Nb	Mo	Te	Ru	Rh	Pd	Ag	Cd	In	Sn	Sb	Te	I	Xe
	K	Ca	Sc	Ti	V	Cr	Mn	Fe	Co	Ni	Cu	Zn	Ga	Ge	As	Se	Br	Kr
Na	Mg												Al	Si	P	S	Cl	Ar
Li	Be												B	C	N	O	F	Ne
H																		He

In the following discussion, we will ignore the first Bohr orbit electrons, the lowest band being second Bohr orbit electrons, and so on. We will focus on silicon (Si) whose Fermi level comes at a Brillouin zone boundary. With pure silicon at normal temperature, the electrons have no freedom to move around and it is a non-conductor, which is an insulator.

By adding "impurity" atoms it can be made to conduct in a limited or controlled manner. If we add phosphorus atoms, each phosphorus has one more electron than silicon. The extra electron is forced up into the conduction band where it can move around freely. Being able to conduct with negative carriers, silicon "doped" with phosphorus is called an n-type semiconductor.

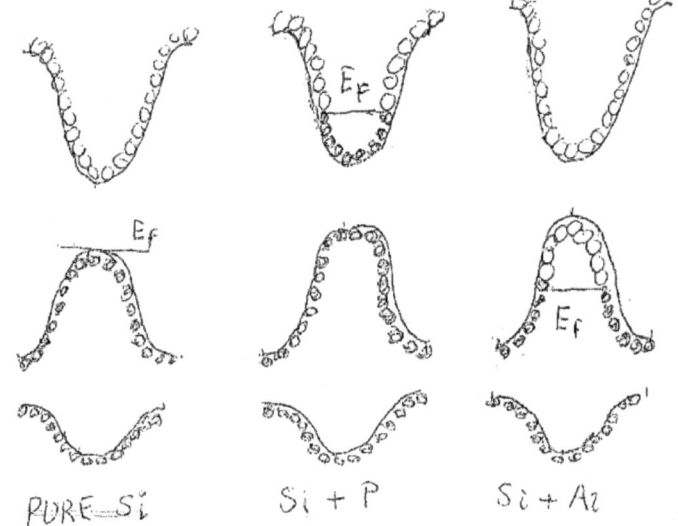

PURE Si   Si + P   Si + Al

Next, let silicon be "doped" with aluminum "impurities" where aluminum has one less electron than silicon. Now there are too few electrons and the Fermi level is lowered creating vacancies or holes. Because holes act as if they have positive charge, this material is called a p-type semiconductor. Recent work is concentrating on column three-five compounds as Gallium-Arsenide or Gallium-Nitride. We discuss more about exotic materials later in this chapter.

The game is played with pieces of p-type abutting pieces of n-type semiconductors. Let us begin with a simple sandwich of p- and n- types. With an applied voltage as shown on the right, positive carriers go from plus to minus voltage and negative carriers go from minus to plus. When carriers reach the boundary, electrons fall into holes. The battery pumps electrons from the p-type up into the n-type, and current flows in the circuit.

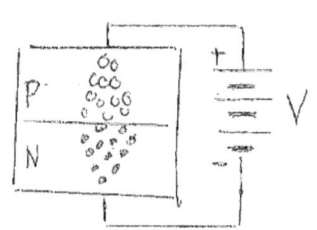

When the voltage is reversed, carriers are pulled
away from the boundary. New carriers can be produced
now only if electrons can be lifted across the gap
between the valence band and the conduction band.
This can be done only if enough energy can be
concentrated at the P-N boundary to pull electrons
across the band gap energy, E_g.

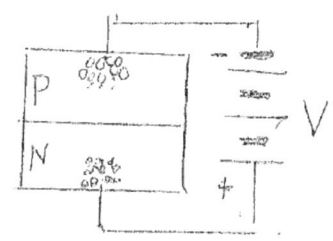

We have been discussing the semiconducting
properties of fourth column elements, such as
carbon and silicon. But if you combine third
and fifth column elements as gallium and
arsenic or second with sixth column elements
as zinc and selenium, you get fourth column
behavior. Now take ZnSe strip and connect it
in a circuit with a voltage source and an
ammeter. If there is no light falling on the
strip, there is no current flowing. But when
light falls on the strip, a current flows in
the circuit. With no light, the Fermi level is
at the top of the valence band and the
electrons cannot shift into states with a net
momentum in any direction. Light falling on
the strip excites electrons into the
conduction band, leaving holes in the valence
band allowing current to flow.

Next, put a piece of p-type material abutting
a piece of n-type material, with the voltage
polarity in the direction making current to
flow. As electrons and holes approach the
boundary, the electrons fall down into the
holes, yielding band gap energy, which if
the material is transparent goes out as
light. This gives a light-emitting-diode,
LED. The color (wavelength) of the light
is determined by the band gap energy.

This gives rise to an idea. With an LED,
all of the photons leave as soon as they
are created, but suppose they were
somehow trapped in the region of electron
recombination? Could the trapped photons
make the electrons jump back into the
conduction band?

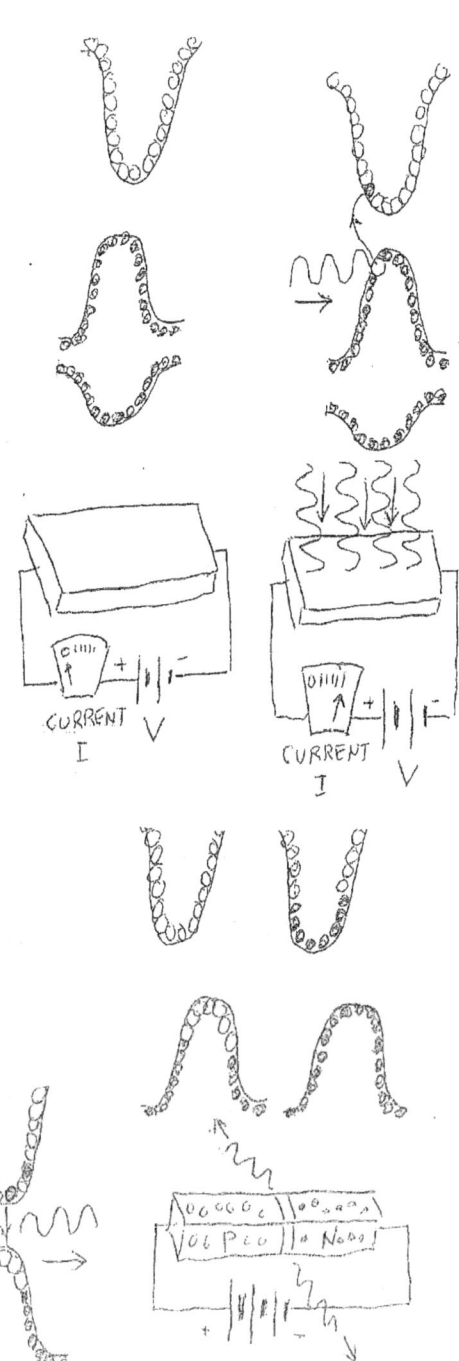

Take now, a block of n-type and a block of p-type materials separated by a non-conducting layer which limits current and electron-hole recombination to a channel, as shown. Now, put a mirror surface at each end of the channel (one of the mirrors can be partially coated to let some light out). Standing waves of light will build up in the channel, giving a high photon density.

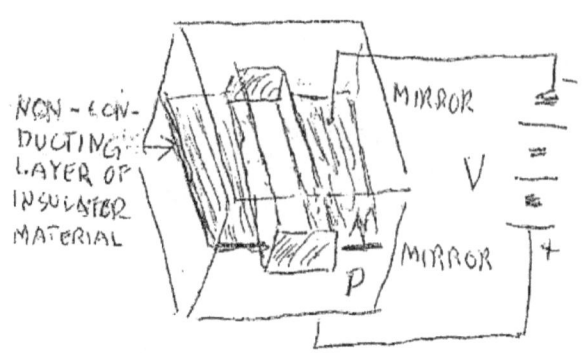

In the external circuit with the voltage source, current will flow replacing electrons in the conduction band as they fall down into holes in the valence band, and then pulling them out of the valence band to keep that full of holes. Electron-hole recombination produces photons which are trapped in the channel with increasing concentration. If a photon from the outside can cause electrons to jump from the valence band to the conduction band, a photon already in the channel can do the same thing. This would provide more electrons to then create more photons when they fall back to the valence band. Enhancing the normal LED performance, this is called STIMULATED EMISSION OF RADIATION and as it involves light, we call the device a LASER.

Electrons bouncing between the valence and conduction band are more or less oscillating, and this makes us think of a mechanical analogue. Take a mass bouncing on a spring. M is the mass and k is the stiffness of the spring. A negative force proportional to velocity represented drag and reduced vibrations. Now, introduce a positive force proportional to velocity to represent a positive feedback.

$M \dfrac{dv}{dt} = -k\,x + b\,v$. Multiply both sides by $v = \dfrac{dx}{dt}$.

$M\,v\,\dfrac{dv}{dt} = -k\,x\,\dfrac{dx}{dt} + b\,v^2$ giving $\dfrac{d}{dt}\left[\dfrac{1}{2}M\,v^2 + \dfrac{1}{2}k\,x^2\right] = b\,v^2$.

The energy (kinetic plus potential) of the oscillator is increasing with time at a time rate that increases quadratically with the velocity of oscillation. In the case of the LASER the extra energy must be demanded from the voltage source. But there is a limit to the rate that the voltage source can deliver energy, and that puts a clamp on the increasing amplitude.

In an actual Light Amplification by Stimulated Emission of Radiation (LASER) device, the amplification (gain) is many times that shown in the following figure.

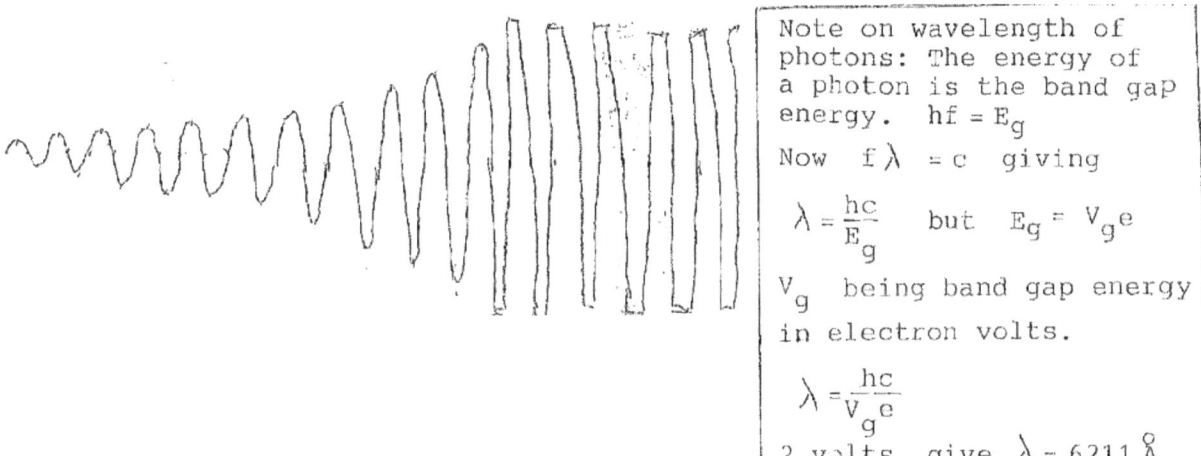

LEDs, LASERs, QUANTUM WELLS, QUANTUM DOTS

If the wavelength of emitted light depends on the gap energy, V_g, what determines the gap energy? We have shown that $\lambda = \frac{hc}{V_g e}$ which is $\lambda = 1.24 \times 10^{-6}/V_g$ meters. Where $V_g = 2$ volts gave $6211\,Å$ which is in the yellow range. Band gap energies and lattice spacings for two types of semiconductors, nitrides and arsenides, are shown in the display below.

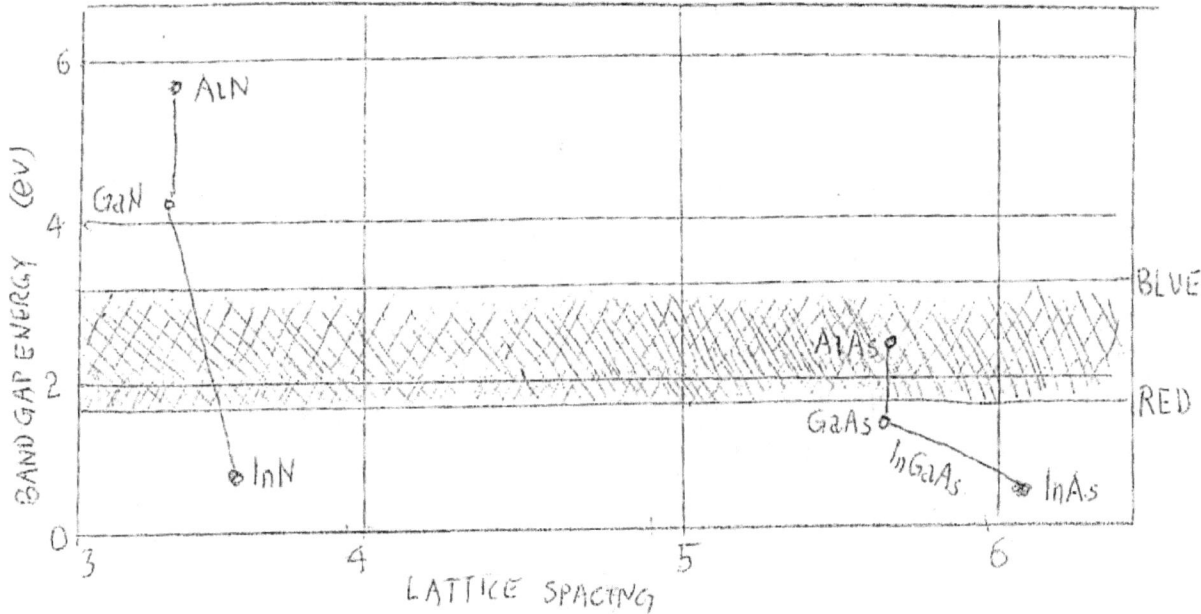

The shaded band shows those gap energies that give wavelengths in the visible range of light. We see that crystals of Al$_x$Ga$_{1-x}$As will emit in the red. This is why most seven segment digital readout devices glow in the red. On the other hand, In$_x$Ga$_{1-x}$N can be made to emit across the visible spectrum.

Although this property of the gallium nitrides was predicted as early as 1960, it wasn't until the late 1990's that these rather difficult materials could be handled industrially.

We will attempt to give some intuitive reasons for the behavior of these materials. We will use the Bohr atom as a background for reference, knowing that quantum mechanics will give quite different numbers.

Let us put the electrons in Bohr orbits. Label each orbit by the element whose valence electrons would be in that orbit.

B, C, N   Al, Si, P   Ga, Ge, As

Obviously, the atomic spacing in lattices of gallium and arsenic will be much bigger than the atomic spacing in lattices of aluminum and nitrogen.

Giving some intuitive excuse for the band gap energies is a bit more tricky. Refer to the Bohr model. Assume that the energy gap between the valence band and the conduction band is the difference between the atom's valence band and its ionization level. (Obviously way over simplified)

Ionization level

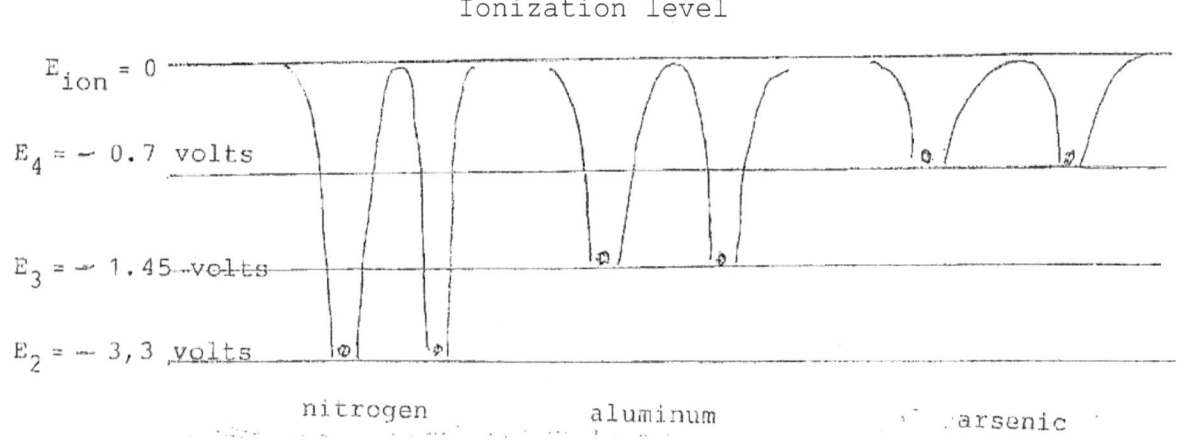

$E_{ion} = 0$

$E_4 = - 0.7$ volts

$E_3 = - 1.45$ volts

$E_2 = - 3,3$ volts

nitrogen   aluminum   arsenic

Although the quantum mechanics of many-electron atoms will give a far more complicated picture, this simple Bohr model will give some intuitive understanding of the data at the bottom of the previous page.

Given the basic materials of $Al_xGa_{1-x}As$ or $In_xGa_{1-x}N$, we must "dope" them with impurities to produce p-type and n-type semiconductors. For p-type we need a deficiency of electrons. This could be done with beryllium or zinc. For n-type we need a surplus of electrons. This might be done with selenium or tellurium.

The ultrapure form of a semiconductor is referred to as an intrinsic semiconductor. Because the valence band is full, and the conduction band is empty, this material is a nonconductor with very high resistivity. Any conductivity depends on thermal excitation of electrons from the valence to the conduction band.

This bring us to the PIN led or laser, where a thin layer of intrinsic material is put between the p-type and n-type layers.

For gallium-arsenic structures, the intrinsic layer is GaAs whose $V_g = 1.43$ volts with $\lambda = 870$ nm.

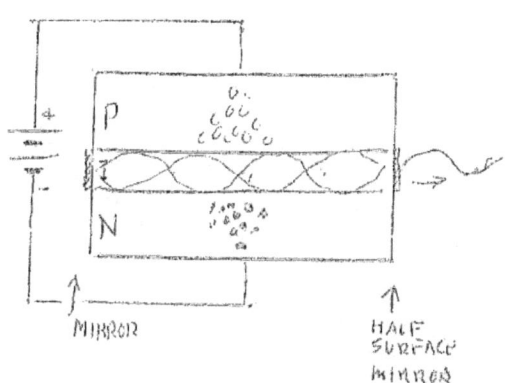

The p- and n-type layers will be doped $Al_xGa_{1-x}As$ with V_g between 1.43 and 1.97 volts, giving wavelengths between 870 nm and 630 nm. Visible light is between 660 nm (red) and 430 nm (violet).

GaN at 300° K has $V_g = 3.4$ volts giving $\lambda = 3.65$ nm. $In_xGa_{1-x}N$ gives wavelengths all the way from red to ultraviolet.

With the PIN laser shown above, the V_g of the intrinsic layer, here taken to be GaAs, is in the non-visible infrared. Between the p- and the n-type layers, V_g for $Al_xGa_{1-x}As$ can be in the red. But the energy jump creating light is in the GaAs layer. Because of the much higher resistance of the intrinsic layer, most of the voltage drop occurs in this intrinsic layer concentrating energy here and raising the V_g to that between the p- and n-type layers.

The wavelength of emitted light is governed by the gap energy between the p- and n-type materials. Thus, a GaAs PIN laser can be made to emit red light which you see in seven segment digital readouts.

Such a thin sheet sandwiched between layers of higher
band gap energy is called a quantum well. What does
that mean? Let us look at a sheet with length and width
being $L_x = L_y = 1.0$ mm and thickness, $D = 10$ nm.

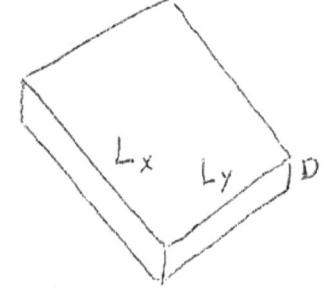

If each dimension is an integer number of standing
wavelengths, then the quantized kinetic energy is

$$E = \frac{h^2}{2M}\left[n_x^2/L_x^2 + n_y^2/L_y^2 + n_z^2/D\right].$$ When $n_x = 1$, $E_{x1} = \frac{h^2}{2ML_x^2}$

is the energy of motion in the x-direction. When $n_z = 1$, $E_{z1} = \frac{h^2}{2MD^2}$ is the

energy of motion in the z-direction. Take the ratio, $\frac{E_{z1}}{E_{x1}} = \frac{L_x^2}{D^2}$. If $L_x = 10^{-3}$ m

and $D = 10^{-8}$ m, $\frac{E_{z1}}{E_{x1}} = 10^{+10}$.

At $300°$K the x- and y- energies are closely spaced that the electrons are
spread over a wide range of energy states as if they constituted an energy
continuum. But in the z-direction, the jump from the first to the second
quantum state is so big that, in that dimension, the electrons stay in their
ground state. This gives what is called a two-dimensional electron LIQUID.
Graphene is already such a liquid.

Let us now construct a PIN laser with an intrinsic layer of GaAs, $V_g = 1.43$
volts, between p- and n- type regions of AlAs with $V_g = 2.15$ volts.

A plot of gap volts across the transistor is shown. The dip across the
intrinsic layer is known as a QUANTUM WELL.

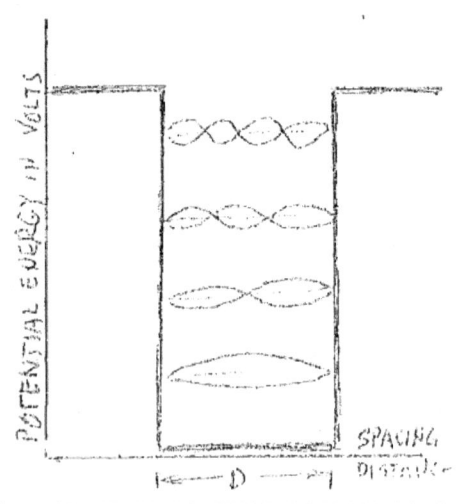

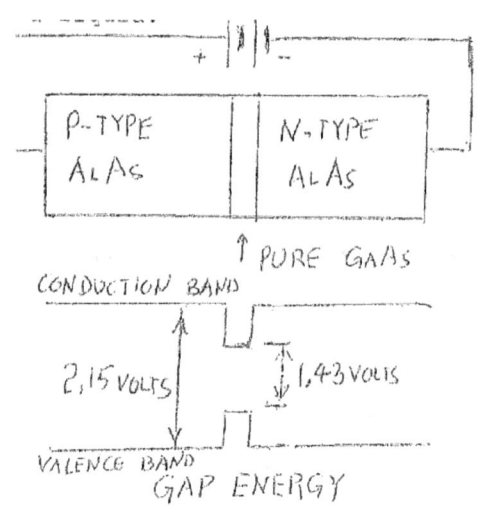

If D is the thickness of the intrinsic layer, then the energy of standing waves in the potential well is given by the above analysis. If the well is of finite potential depth, the wave functions will tunnel into the walls of the well and the broadenings of them will reduce the energy levels.

The quantized energy levels in the potential wells are in our favor. In addition to the normal gap energy, V_g, of the intrinsic layer we have jumps V_g' between quantized energy levels. Remember that the greater the gap energy, the shorter the resulting wavelength of emitted light. For the GaAs PIN laser this helps pull the emitted light from the infra-red into the visible red.

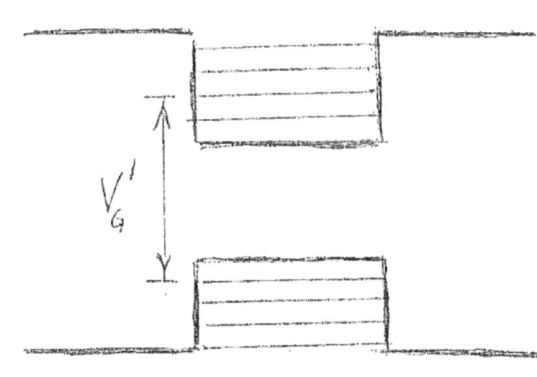

The next picture shows a PIN laser geometry.

For further reading, the primary reference is LASER PHYSICS by Simon Hooker and Colin Webb, Oxford University Press.

Although this may be the book on lasers, it is written for university graduate students as preparation for doing research in the field.

When we go from a two-dimensional quantum well to a one-dimensional electron liquid, this is called a quantum wire, also known as a Luttinger Liquid.

Restricting one more dimension gives a quantum dot. With large jumps between all of its energy levels, this begins to behave as an artificial atom. Quantum dots are being used more often in engineering. Refer to "quantum dot" on Wikipedia for more.

P-TYPE

N-TYPE

NON-CON
DUCTING
LAYER

INTRINSIC
LAYER
WITH MIRRORS
AT BOTH ENDS

As soon as we start to restrict the dimensions of electron motion, electron behavior becomes stranger and stranger, especially when in a magnetic field. None of the rules for electron behavior in three dimensions hold, and in some cases electrons disappear to be replaced by exotic quasiparticles. Unfortunately, these exiting areas of research are more than we can cover here.

MEASUREMENT OF BAND GAP ENERGIES

The picture shows the experimental setup for measuring band gap energy for various semiconductors. The importance of such measurements is evident in the previous discussion.

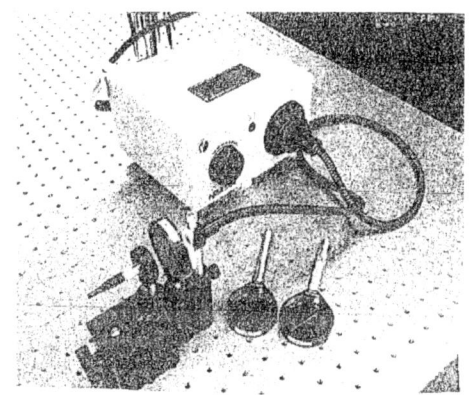

The figure below shows details of what is in the box.

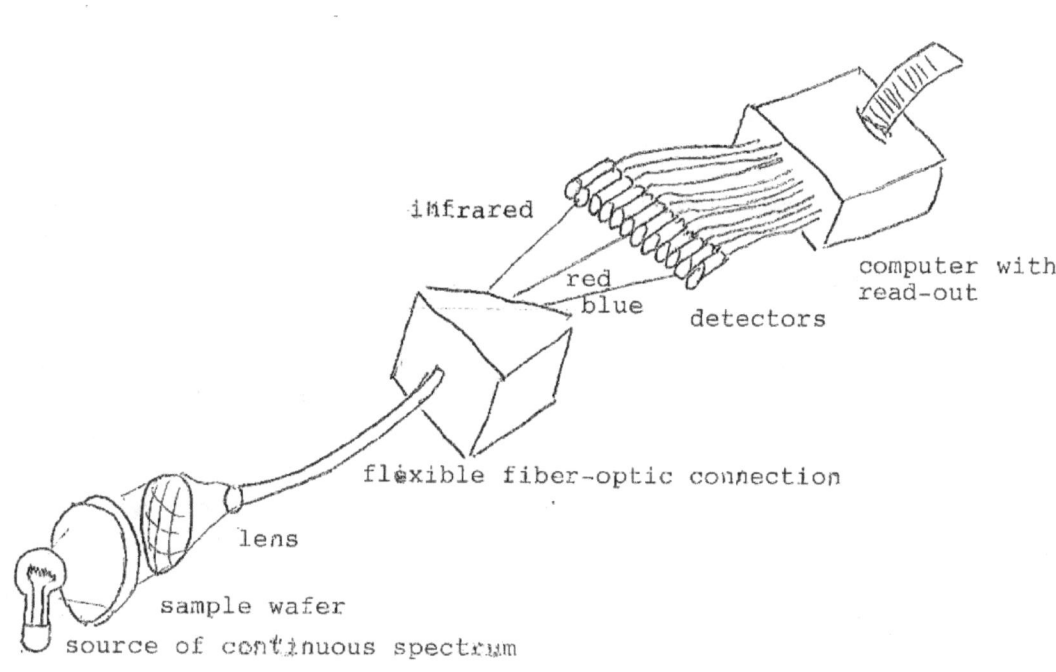

A light source gives off all colors in the visible and infrared spectra. The semiconductor material in the wafer will pass all photons having energies less than the band gap energy. These photons cannot excite electrons to jump from the valence to the conduction band. All photons having energy greater than the band gap energy will be absorbed by electrons jumping from the valence to the conduction band and such wavelengths will not pass through the sample. The prism breaking up light into different colors must have previously been calibrated with known spectral wavelengths. The computer reads out a spectrum of which wavelengths were passed, and which were absorbed by the semiconductor sample.

J. H. Miller was kind to provide the following spectrum. The high points are
wavelengths passed by the wafer. The low points are wavelengths absorbed.

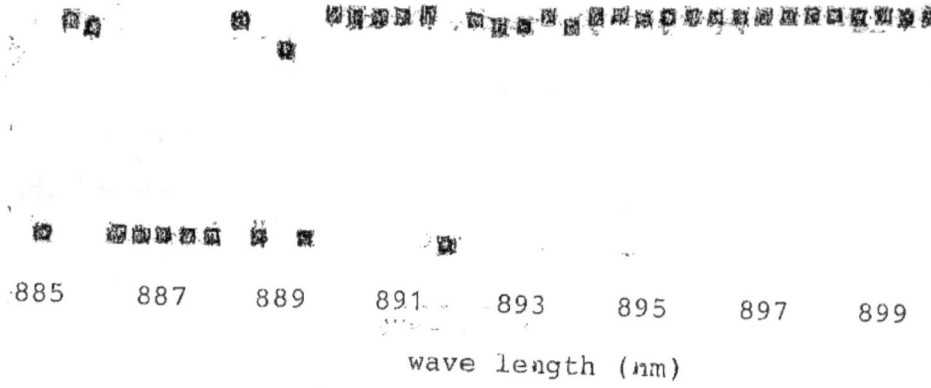

| 885 | 887 | 889 | 891 | 893 | 895 | 897 | 899 |

wave length (nm)

The wavelength, $\lambda = 889$ nanometer gives the band energy. The formula,
$V_g = 1.24 \times 10^{-6}/\lambda$ with λ in meters gives V_g in electron volts. $\lambda = 889$ nm gives
$V_g = 1.39$ eV.

The erratic data points near 886 nm and 892 nm represent noise. But the jumpy
points near 889 nm may represent some interesting physics. The electron
excitation from the valence band to the conduction band by just simply
absorbing a photon is called direct absorption. But it has been found that
the process may be more complicated than that. If the photon doesn't have
quite enough energy, the electron may grab a photon or lattice vibration
energy to complete the jump. Or if the photon has a bit too much energy, the
extra may be given off to create a phonon of lattice vibration energy. This
is called indirect absorption. Theory of electron-phonon interactions are a
bit beyond the level of this discussion but are important in condensed
matter physics, especially in Leon Cooper's theory of electron pairing in
superconductivity.

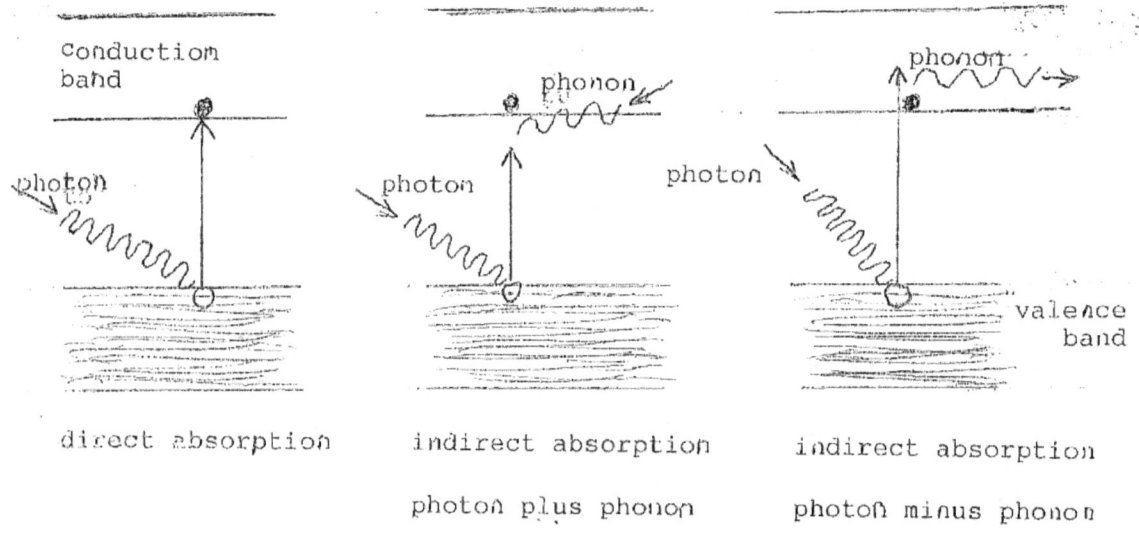

direct absorption   indirect absorption   indirect absorption

photon plus phonon   photon minus phonon

14-26

THE HALL EFFECT was discovered by Edwin Hall in 1878.

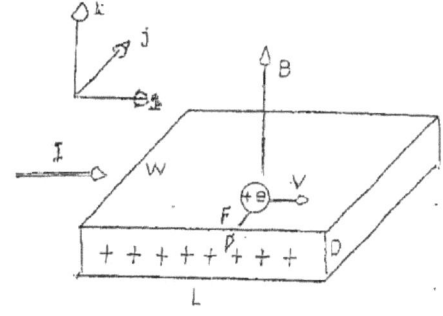

Take a piece of p-type semiconductor as shown, having a magnetic field perpendicular to it in the z-direction and having a current in the x-direction. The positive carriers will experience a force in the negative y-direction making the near face positive and the far face negative. The separated charges will produce an electric field in the positive y-direction. A steady state configuration will occur when the electric force balances the magnetic force.

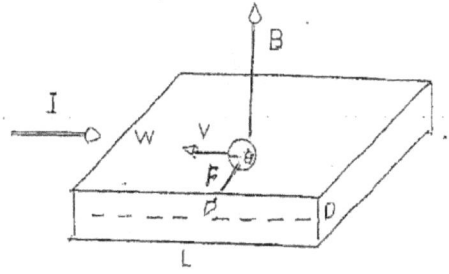

$e E_y = e v_x B_z$ or $E_y = v_x B_z$. The Hall voltage is $V_H = E_y W$ across the width, W.

If there are N carriers per unit volume and each moving with a velocity, v, we showed that the current across the area, $W D$, is $I_x = e N W D v_x$ giving

$$v_x = \frac{I_x}{e N D W} \quad \text{and} \quad V_H = \frac{I_x B_z}{N e D}.$$

For negative carriers the opposite polarization of charge occurs.

Using a strip of gold foil. Hall placed a galvanometer across the width, W, and measured the current due to the voltage, V_y. Taking the ratio of that current, call it I_y, to the original current, he plotted the ratio, $\frac{I_y}{I_x}$, against the strength of B_z and got the result shown in the figure.

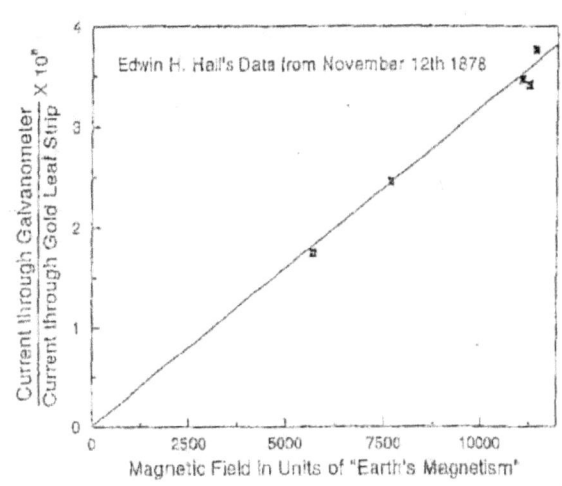

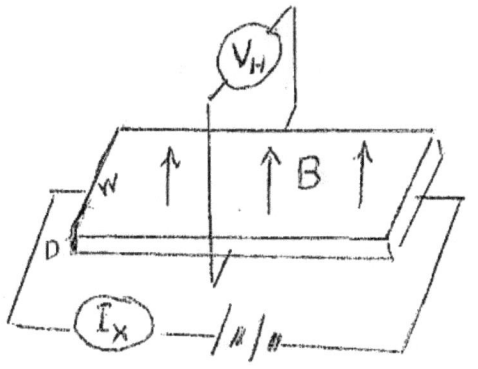

For the picture on the left, refer to the formula for V_H above, if this and I_x are measured and the thickness, D, of the strip has been measured, then either the magnetic field, B, or the charge density, N_e are left to be measured.

For semiconductor research, the magnetic field will have been measured. The polarization of the strip and the direction of the Hall voltage will tell us if we have electrons or holes as carriers. If B is known, the carrier density Ne can be determined.

Otherwise, if NeD has been determined, the hall effect gives us a probe for measuring magnetic field strength. This is of great importance for engineering applications.

THE FIELD-EFFECT TRANSISTOR (FET)

In our discussion we will build up this device by pieces. Start off with a channel of n-type (silicon with phosphorus impurities) where the extra phosphorus electrons must go in the conduction band. A voltage is put between contacts called the source and the drain. Electrons are emitted from the source and move up the channel to the drain. We will measure the drain current in the direction that positive charge would flow.

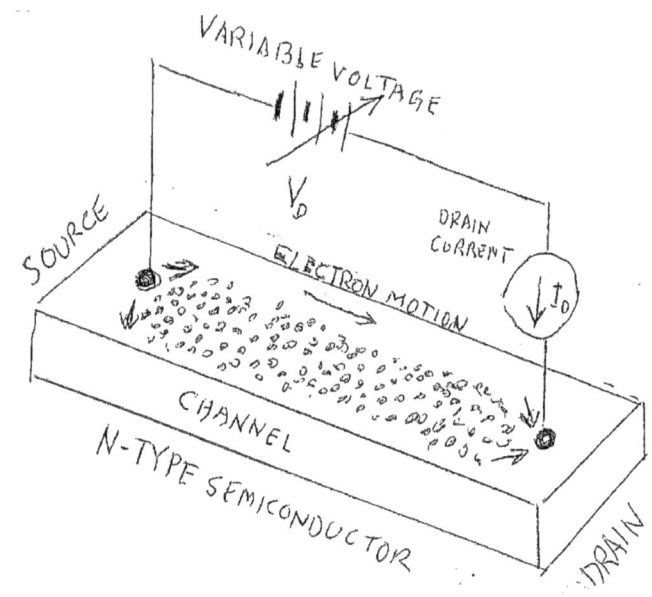

The important thing is that because the channel is a semiconductor it does not obey Ohm's law. When drain current is measured against drain volts, it saturates as shown in the picture below. This should be no surprise. With a limited number of electrons in the conduction band when they are all moving in the same direction, that is all the current you are going to get.

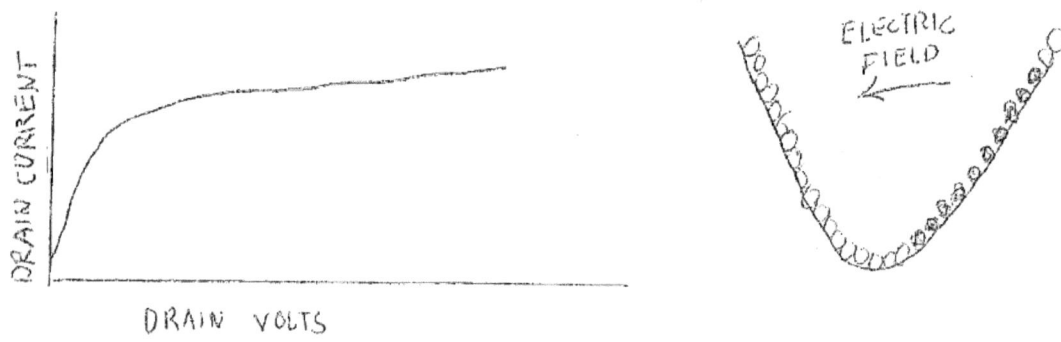

In the picture on the following page, the n-type channel is embedded in the non-conducting silicon of a wafer which will contain a lot of transistors. A layer of a non-conducting silicon of a wafer which will contain a lot of transistors. A layer of a non-conduction oxide is placed

above the channel to electrically insulate the channel from a conducting
aluminum layer called the base. Plugs of aluminum through the oxide layer
allow electrical contact at the source and the drain. Because of the layers
of metal, oxide and semiconductor making a field-effect transistor, this is
called a MOSFET.

The field effect comes in when the gate is given a positive or negative
voltage in reference to the source. This puts a positive or negative charge
on the gate and it is the effect of the electric field due to this charge
that controls the current in the channel. A negative charge on the gate
repels the negative electrons in the channel. A positive charge on the gate
attracts them.

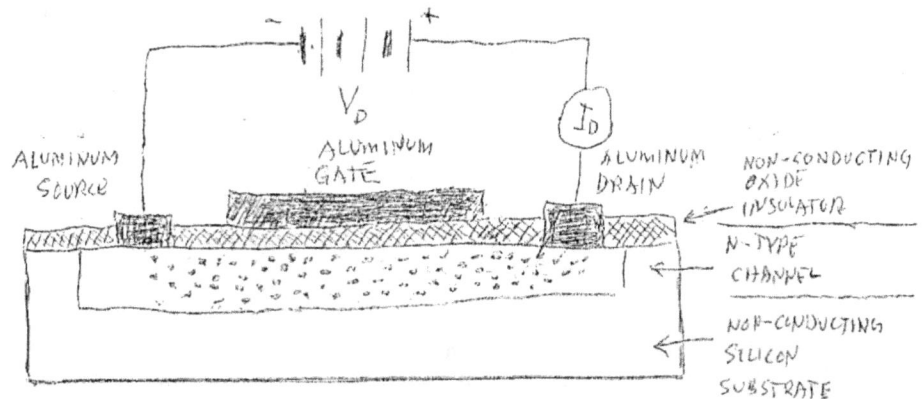

A negative voltage at the gate puts
negative charge on the gate which
repels the electrons in the channel
reducing the current. This is called
a depletion mode MOSFET. The net
effect is that a varying negative
voltage at the gate makes a variation
of the current at the drain.

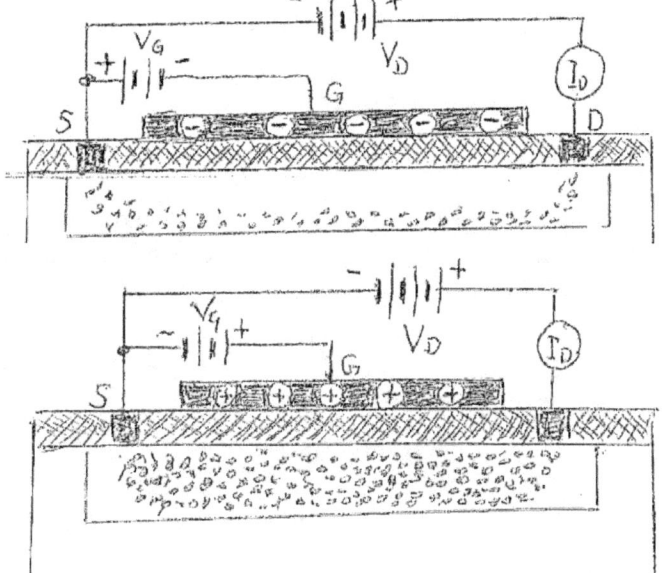

A positive voltage at the gate puts a
positive charge on the gate, drawing
more electrons into the channel. As
this increases the current, this is
called an enhancement mode MOSFET.

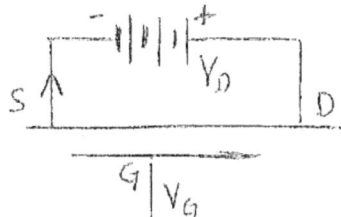

The figure on the left gives the symbol for an n-
channel MOSFET. An arrow at the source indicates the
direction of the positive equivalent current at the
source.

APPENDIX TO CHAPTER 14: INTRODUCTION TO SOLID STATE ELECTRONICS

Most of this textbook deals with the theory of this and the theory of that with references to experiments. At this point we do something different. Having developed an understanding of a transistor, it is necessary that you have some idea of what to do with it. In almost any area of science or technology, if you must design something, it will involve electronics and you may not be able to find what you want available at the store. You may have to design it yourself.

Information is encoded in voltage. If the voltages are continuously changing with time, we have ANALOG ELECTRONICS. If the voltages are encoded in constant values, a lot of which code things, we have DIGITAL ELECTRONICS. This appendix is not a course in electronics. It is intended to make you comfortable with the subject so that you are ready for a course in electronic design. Even so, it would be nice if your instructor could develop a lab so that you could get your hands dirty. Commercially, there are training kits available to teach it to people who need it for their work.

We will start off with analog electronics. You are already familiar with amplifiers. They take weak voltage wiggles and make big voltage wiggles. We will take you very slowly through the design of one stage of a signal amplifier. It would be nice if you had a lab bench to play with this.

Next, we take you into digital design. This, of course is the electronics of a computer, but you may have to design with it in your scientific work. Many physical quantities we have studied come in continuous functions in time. When we detect them, we get an analog signal. But you may have to convert this into a digital form to get it into a computer. This involves an analog to digital converter.

We start off with a very important component of electronic systems. It is a POWER SUPPLY. Electrical energy comes with sine wave voltages which allow us to use transformers to step voltage up or down with little energy loss, but electronic circuits require constant voltage. We don't want to do this with batteries. A power supply is involved in AC-to-DC conversion.

When you get around to designing electronic equipment, an important consideration is the effect of temperature on semiconductors. On page 14-13 the creation of holes and electrons in semiconductors was described. Devices need materials to have either positive or negative carriers, but not both.

Trouble arises with temperature discussed on page 14-14. As temperature rises, lattice vibrations become more vigorous; creating more energetic phonons. These bump into electrons in the valence band kicking them into the conduction band creating both holes and electrons. The distinction between p- and n-type materials is wiped out. Things no longer behave as designed. With decreasing resistance components may even burn up. Pay attention to the temperature on semiconductor electronics.

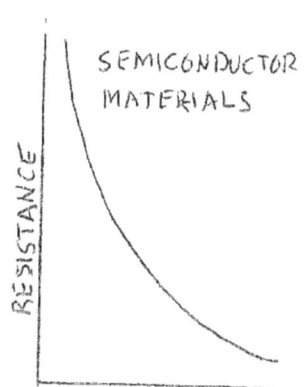

On page 14-16 we discussed how an n-type and p-type current is only one direction.

Such a device is called a DIODE and is depicted by two lines with an arrow in the direction of positive charge flow when the circuit conducts.

A diode is used in the following manner. When a sine-wave, or alternating voltage, is placed across the diode in series with a resistor, current flows in the resistor only, in the direction in which the diode conducts. Only when the current is flowing will there be a voltage, RI, across the resistor.

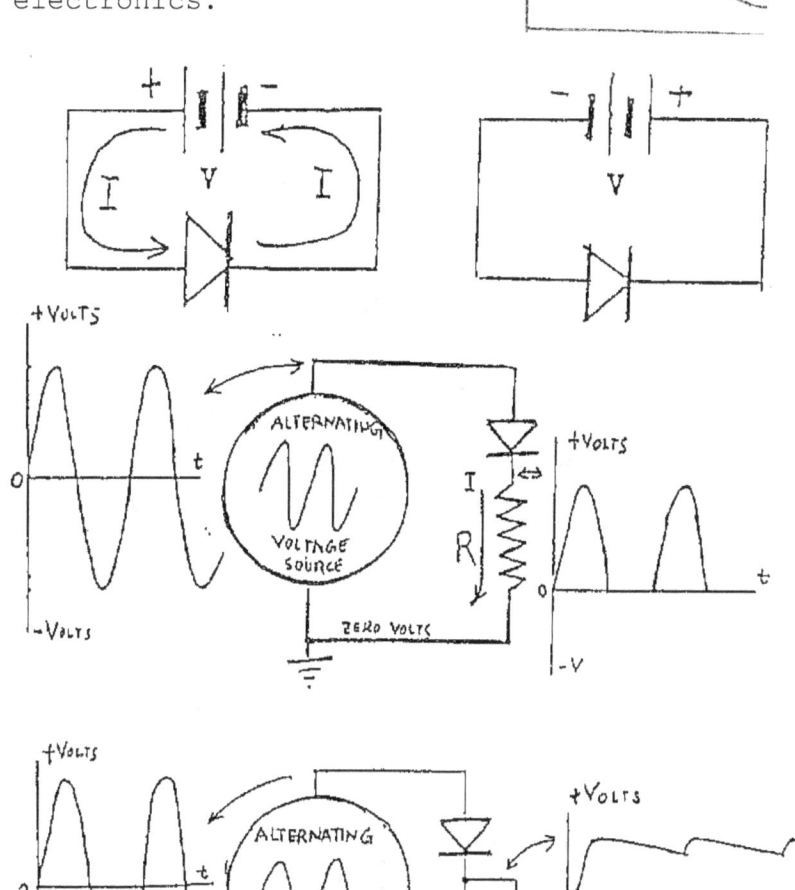

Recall that a capacitor is a device with plates that store charge with an applied voltage but are separated by a non-conducting material allowing no current to flow.

The field-effect-transistor is a negative voltage (at the gate) controlled resistor. Its circuit symbol is a straight line for the channel with contacts at the source and the drain and a diode contact at the gate.

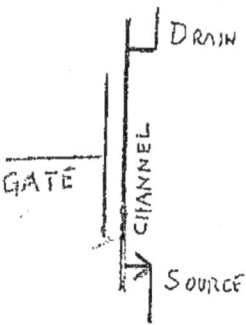

The behavior of the field effect transistor is described by the curves shown below. The gate voltage is kept constant at zero or some negative value. The drain current is measured as the drain voltage is varied. (An arrow through a circuit symbol says that that quantity is a variable.) The source is taken as reference voltage zero indicated by the ground symbol or upside-down Christmas tree.

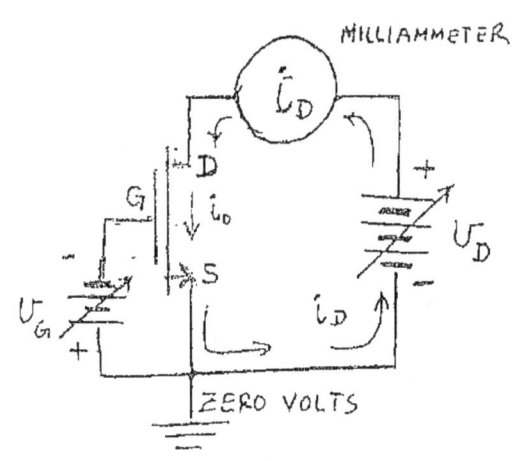

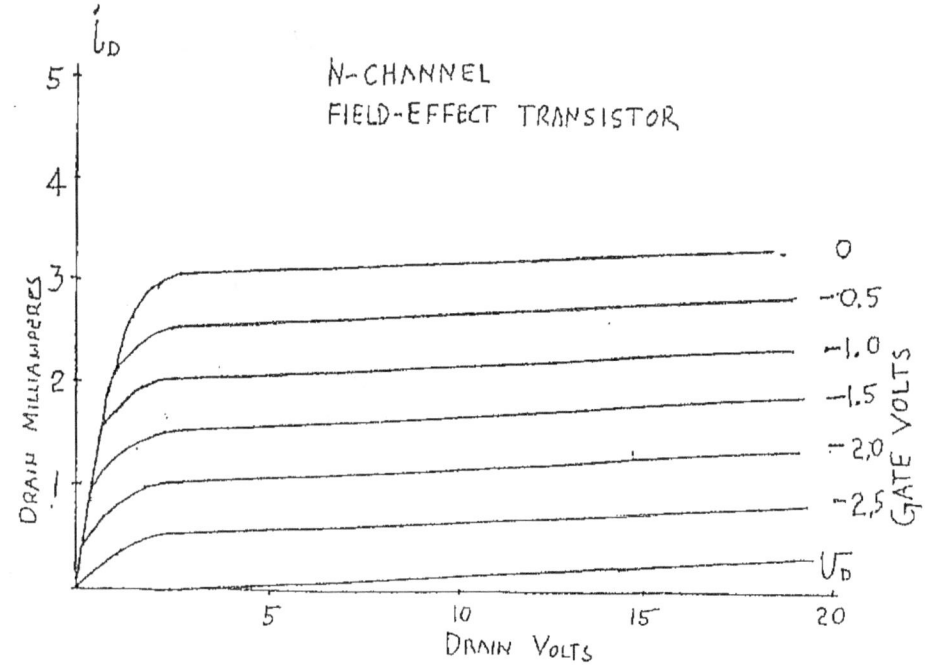

VOLTAGE SIGNAL AMPLIFIER

A signal is a time-dependent voltage which encodes information as varying voltage levels. Sources would include microphones, magnetic tape pick-ups, and keyboards on computer terminals. We shall next see how transistors are used to take weak voltage wiggles and make stronger voltage wiggles.

To do this, the transistor is placed in series with a resistor called the load resistor, R_L. A constant voltage, V_{DD}, is put across this combination producing a current, i_D, passing through R_L and from the drain to the source through the transistor. The weak signal voltage is added to a negative constant voltage and applied at the gate. We take the source to be at reference voltage zero. In the drain circuit the drain will be at a voltage, v_D, above the source and the current, i_D, produces a voltage drop, $R_L i_D$, across the load resistor.

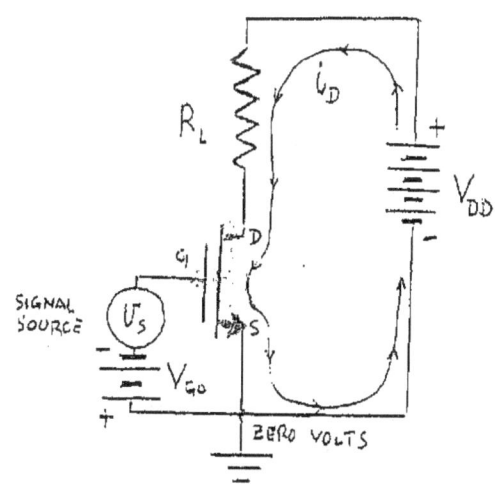

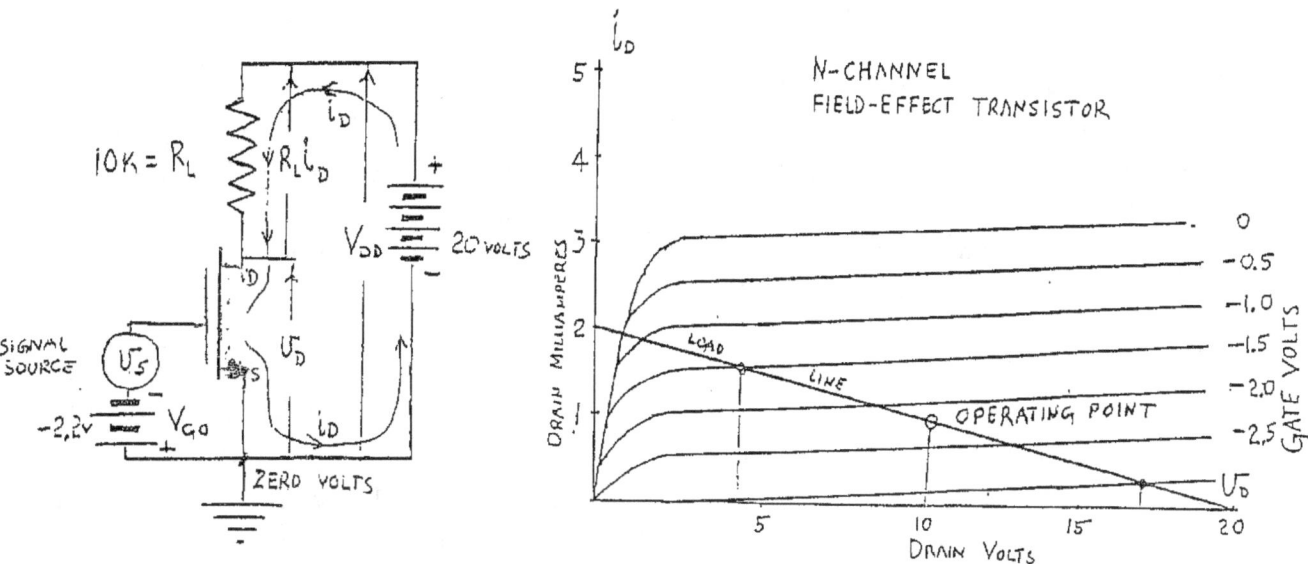

In the drain circuit, the voltage rise, V_{DD}, equals the sum of the voltage drops. Thus, $V_{DD} = v_D + R_L i_D$, which gives us, $i_D = \frac{1}{R_L} V_{DD} - \frac{1}{R_L} v_D$. These are the same drain currents, i_D, and drain volts, v_D, that were measured for the transistor curves on the previous page. Now the current and voltage values are limited to only those values allowed by the external drain circuit. Let us choose $V_{DD} = 20$ volts and $R_L = 10$ k ohms (10 kilohms = 10,000 ohms), then $i_D = \frac{1}{10} 20 - \frac{1}{10} v_D$ in milliamps.

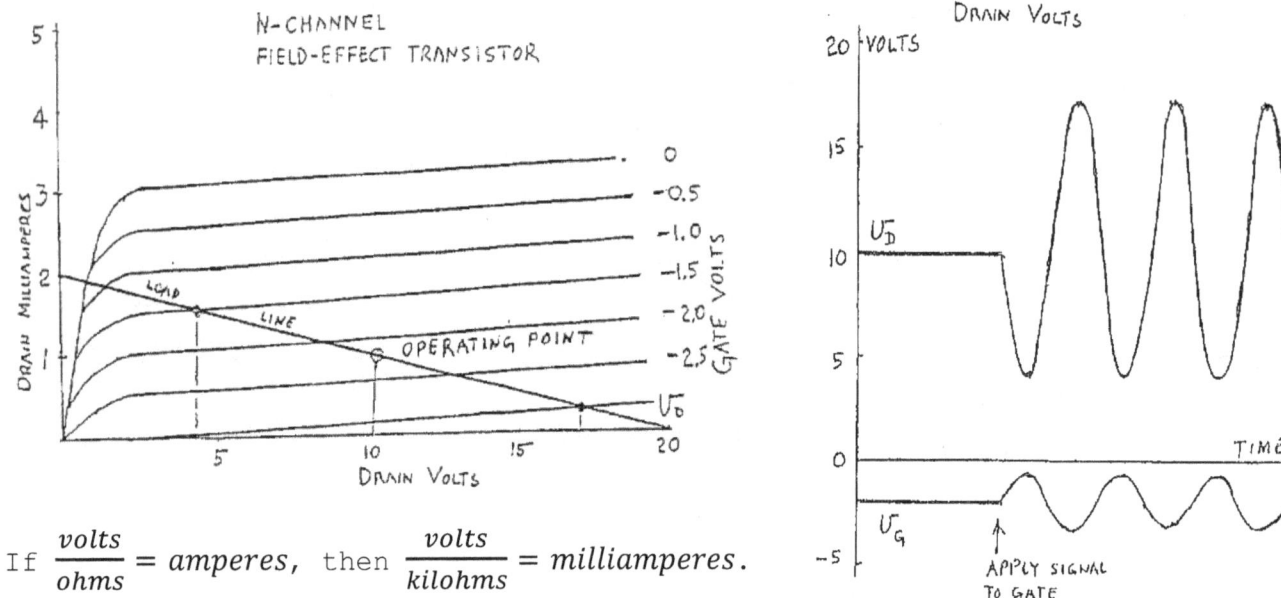

If $\dfrac{volts}{ohms} = amperes$, then $\dfrac{volts}{kilohms} = milliamperes$.

With these components in the external drain circuit, the equation limiting the drain voltage and current values is $i_D = 2 - 0.1\,v_D$. Placed on top of the transistor curves, this gives a straight line going from $i_D = 2$ ma when $v_D = 0$, to $i_D = 0$ ma when $v_D = 20$ volts.

Only these drain voltage and current values will happen when the transistor is in this circuit. Which drain volts and current combination will happen depends now on the negative gate volts. If $v_G = -2.2$ volts, $v_D = 10$ volts, and $i_D = 1.0$ milliampere. If the gate becomes less negative, the drain current rises and the drain voltage drops. If the gate becomes more negative, the drain current drops and drain volts rise.

A constant voltage, -2.2 volts, at the gate thus puts us in the middle of this line which is called the load line. If a time-varying signal voltage with 1.5 volts from bottom-to-top (peak-to-peak) is added to the constant voltage you will notice that when the gate rises to -1.5 volts, the drain falls to 4 volts; and when the gate falls to -3.0 volts, the drain rises to 17 volts.

The resulting amplification of the signal strength is evident when both signals are plotted against time (see top right graph). The ratio of the peak-to-peak voltages is called the voltage gain. Here, a 1.5 volt signal is boosted to a 13 volt signal. The resulting voltage gain is $G = \dfrac{13}{1.5} = 8.6$.

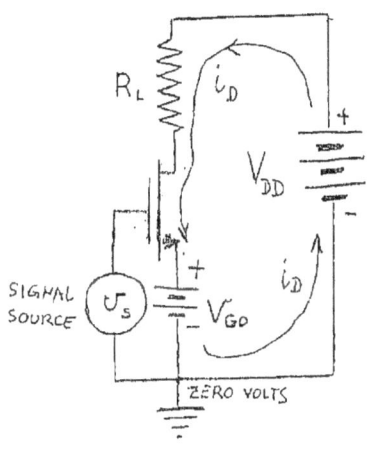

We now consider two modifications in this circuit. The output voltage now vibrates above and below **10** volts and we would like for it to vibrate above and below zero volts. Secondly, we wish to get rid of the need for a second battery for the constant part of the gate volts. We solve the second problem first. So long as the gate operates about $V_{Go} = -2.2$ volts below the source, it makes no difference what point in the system is taken to be at reference voltage zero. Hence, we put zero volts at the connection between $V_{Go} = -2.2$ volts and the signal voltage.

Notice, now, that the drain current is pushed through the battery $V_{Go} = -2.2$ volts. Notice also that at the operating point, when the signal voltage is zero ($v_s = 0$) and the gate is **2.2** volts below the source, the drain current equals **1.0** milliampere. If **1.0** milliampere flows through a **2.2** k = **2200** ohm resistor, the voltage drop will be **2.2** volts, the same as the V_{Go} battery. Thus, if the battery is replaced by a **2.2** k resistor, with the operating point current of **1.0** milliampere, we still get a **2.2** volt drop from the source to the gate. But you will say that this voltage is to be kept constant.

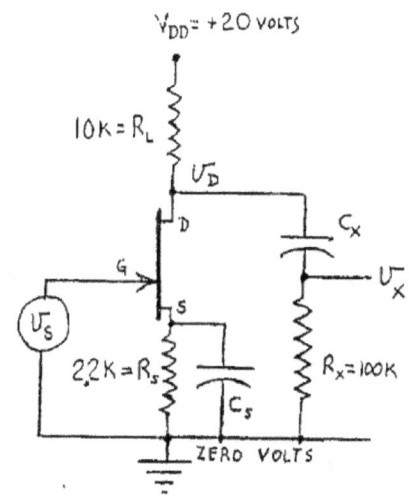

If the signal voltage varies the drain current will vary and the voltage across the resistor, $R_s = 2.2$ k, will wander from being **2.2** volts. Right you are! That is why we put a capacitor acros the R_s resistor with possibly $C_s = 50$ microfarads (for capacitors relatively large).

The variations of the drain current, now coming out of the source, will make small wiggles in the charge on the capacitor but not change the voltage across the resistor which responds now only to the constant or average value of the current.

Now the signal voltage into the gate is vibrating above and below the reference of zero volts; but the output voltage is vibrating above and below the ten volts at the drain.

To get the output signal to vibrate above and below zero volts, we use the following trick. We connect the drain to the zero volt level (ground) through a resistor and capacitor in series, with the resistor connected at its one end to ground. The resistor should be about ten times the load, R_L. $R_x = 100$ k will do. For C_x, 0.1 microfarad might do. When there is no signal variation, the capacitor charges up to 10 volts.

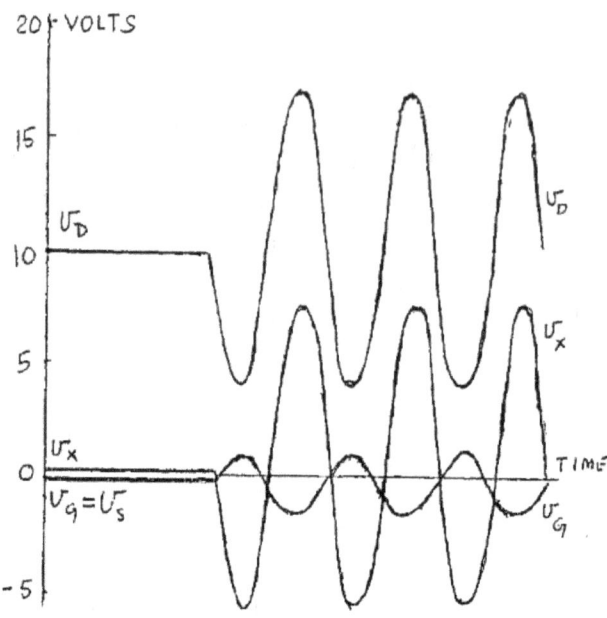

When the drain voltage is varying, the charge on the top plate of the capacitor follows the varying drain voltage. The opposite charge on the bottom plate follows the charge on the top plate. This charge must come from the ground (zero volt) level through the resistor, R_x. The current of charge through R_x produces a voltage drop, $v_x = R_x i_x$, which is identical to v_D varying at the drain, except that v_x vibrates above and below zero volts, as is desired.

The transistor which we have discussed was an n-channel FET which used an n-type channel with a p-type gate. Just as well, we can use a p-type channel FET using a p-type channel and an n-type gate. The channel-to-gate diode boundary would conduct positive current from the channel to the gate and would be non-conducting when the gate is positive with respect to the channel. Thus, a positive charge on the gate will control a current of positive holes moving from the source to the drain in the channel.

P-CHANNEL MOSFET

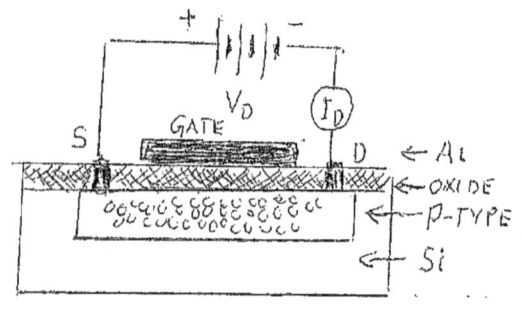

With a p-channel FET, the drain will be at a negative voltage with reference to the source and the gate will be positive. Otherwise it behaves exactly the same as the n-channel FET which we have discussed.

The transistor, described at the bottom of the previous page, is called a depletion-mode FET, because the effect of the gate is to deplete the carriers carrying current in the channel. Next, we discuss an enhancement-mode FET, where the gate pulls carriers into the channel.

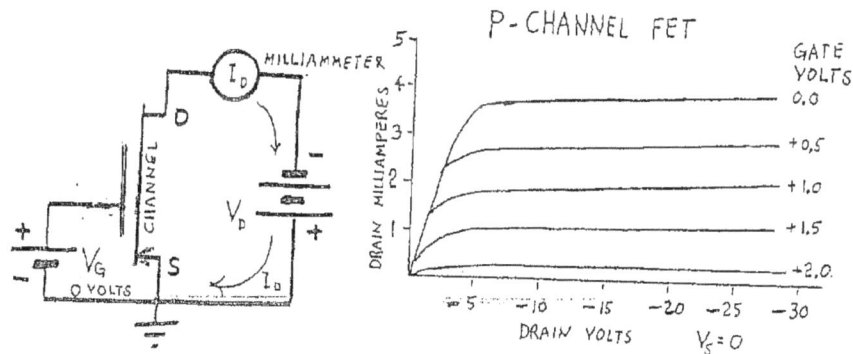

The Enhancement-Mode transistor is called a Metal-Oxide-Semiconductor Field-Effect-Transistor or MOSFET as it involves a metal gate, an insulating oxide, and a semiconductor between source and drain. The device shown here is an n-channel MOSFET as the source and drain are n-type semiconductors. But note that the channel is not connected continuously from the source to the drain. The source and the drain are separated by a gap of p-type semiconducting material which extends up from a p-type region called the substrate.

Because of the non-conducting oxide insulator, no current can flow between the metal gate and the semi-conductor material.

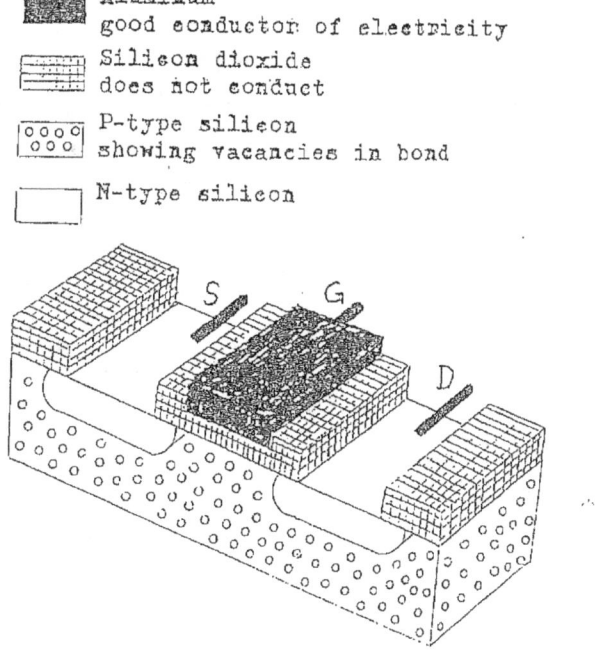

BI-POLAR TRANSISTOR ELECTRONICS

The n-channel MOSFET has an n-type
source and an n-type drain
separated by a p-type substrate. A
metal gate is insulated from the
semi-conductors by an oxide layer.

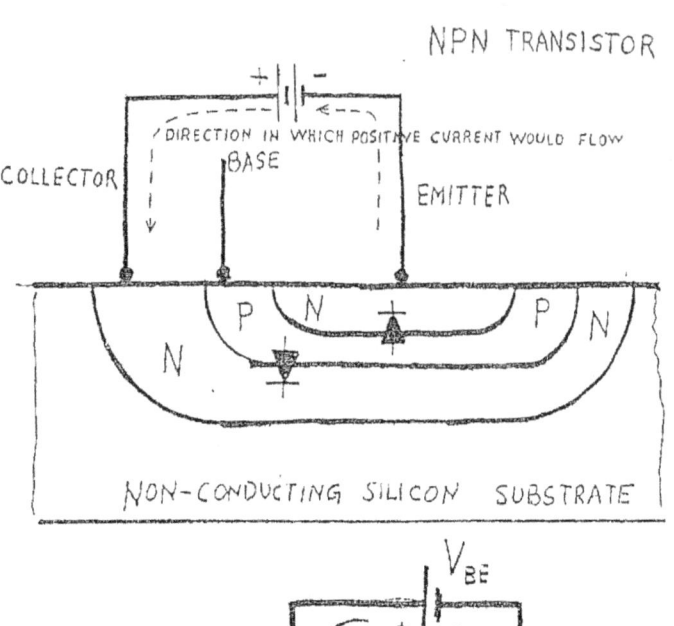

We get another type of transistor
by eliminating the gate and oxide
and rearranging the n- and p-
materials on the silicon chip.
Embedded in the silicon, let there
be a n-type layer above a p-type
layer, as shown in the picture. The
p- layer is now called the base.
The n- layer below it is now called
the collector. This is called an NPN
bipolar transistor and has a circuit
symbol shown here.

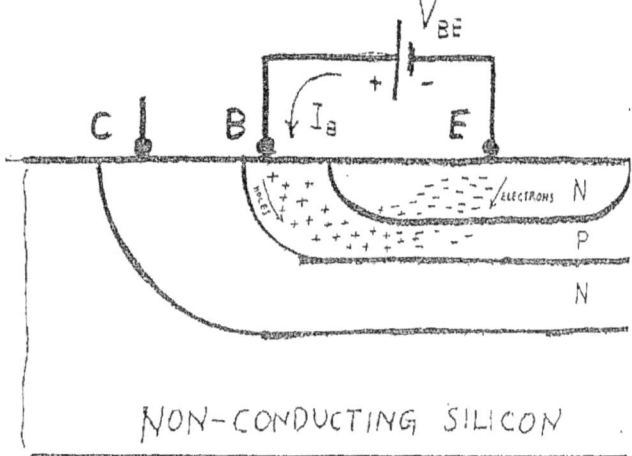

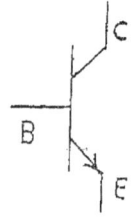

The emitter behaves as did the source and the collector behaves as did the
drain. The base, however, differs from the gate. The P-N boundaries would
conduct positive current from the base into the collector and from the base
into the emitter. A positive voltage at the collector above the emitter
would not conduct across its collector-base boundary. So far, just like an
n-channel MOSFET.

A positive voltage at the base above the emitter would conduct across the
base-emitter boundary, with holes flowing in the base and electrons flowing
in the emitter. Some electrons would pass into the base region.

Now, if we put a positive voltage at the collector above the emitter,
electrons flowing into the base will be drawn into the collector and
continue to flow as normal electrons in the n-type material.

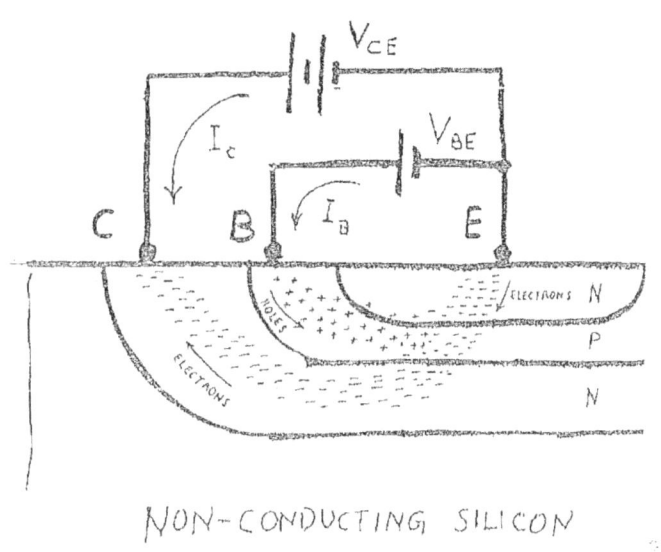

VNON-CONDUCTING SILICON

In fact, the geometry of the base is so designed that most of the electrons flowing toward the base will cross it and continue to flow in the collector. The collector current may be approximately ten times the base current. It is obvious that with this device, a current flowing in the base stimulates a current into the collector. With the FET, a voltage at the gate controlled a current at the drain. With an NPN transistor, a current at the base controls a current at the collector.

In the picture below, we show a signal voltage amplifier using an NPN transistor. Except for the base, it looks just like an amplifier using a FET. The **1.0** megaohm resistor divide the **+20** volts in such a manner that the base will be about **0.5** volt above the emitter. Thus, there will always be currents flowing at the base and collector.

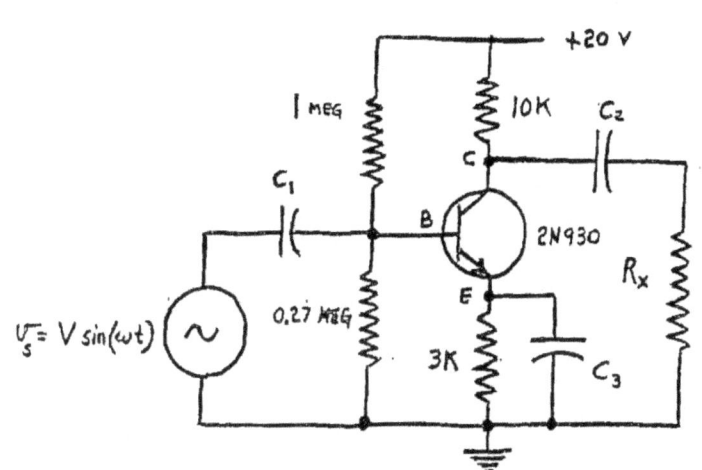

COMMON EMITTER NPN TRANSISTOR AMPLIFIER CIRCUIT

The signal voltage wiggle, v_s, which we wish to amplify will be stuck on top of the voltage at the base using the capacitor, C_1. By making the voltage at the base wiggle, it will make the current into the base wiggle. The wiggling base current will make the current into the collector to wiggle. The **10** kilohm load resistor converts the wiggling colector current into a wiggling collector voltage.

This is exactly like how the FET amplifier operated using C_x and R_x to make the resulting output signal voltage wiggle above and below zero volts at the ground level.

The FET amplifier which we designed earlier in this chapter got a voltage gain of about **20** at best. The amplifier circuit which we have just now described can get a voltage gain to maybe over **200**.

The bipolar transistor can also come as a PNP transistor, having an n-base between p-type emitter and collector. Reversing the materials, it reverses the directions in which the P-N boundaries conduct. The symbol for an NPN transistor showed a diode conducting from the base into the emitter. The symbol for a PNP transistor shows the diode conducting from the emitter into the base.

The amplifier circuit shown above would work for a PNP transistor, except for a negative supply voltage, **−20** volts for the base and collector to operate with voltages below the emitter.

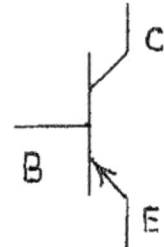

PNP TRANSISTOR

> (1) Every one who is sane can do Logic;
> (2) No lunatics are fit to serve on a jury;
> (3) None of *your* sons can do Logic.
> (4) Hence. . . .

Imagine a system with two switches and two lights. You may be able to imagine even more switches and more lights. When a switch is up, the wire from it is at five volts ground. When a switch is down, the wire from it is at zero volts, which is called ground. When the wire onto a light is at five volts it glows. When the wire into a light is at zero volts, the light is off. Call zero volts LOW or ZERO. The lights are connected to the switches through a black box in which anything involving a mixture of highs and lows can happen. We will call the switches INPUTS to the black box and we will call the light OUTPUTS from the black box.

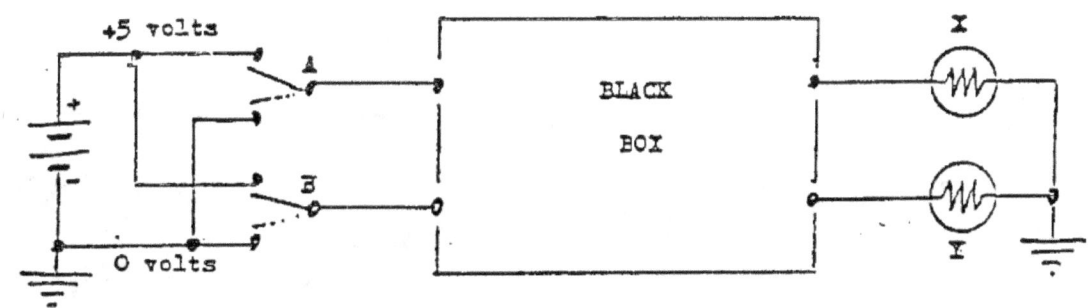

Call the switches A and B. Call the lights X and Y.

There are four states in which two switches can exist:

$$A = 0 \quad A = 1 \quad A = 0 \quad A = 1$$

$$B = 0 \quad B = 0 \quad B = 1 \quad B = 1$$

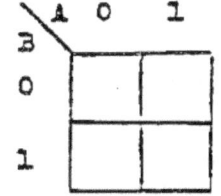

We assign each of these states to one of four boxes in an array of boxes called a KARNAUGH MAP. The columns are labeled by states of A and the rows labeled by states of B. The box in the upper right corner represents the state, $A = 1 ; B = 0$.

Anything that the black box can do to the outputs of switches A and B so that a light, X or Y, is turned on or off can be represented by a Karnaugh map.

Let us use only light X to start with. For whatever state of A and B, X is lighted, the box for that state will be shaded. For whatever state of A and B, X is not lit, the box for that state is unshaded.

If X is connected only to A, then X is lit when A is **1**, and X is unlit when A is **0**, B having no effect. This gives the set of shaded boxes for $X = A$:

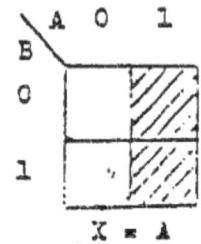

If light X is connected straight to switch B with no effect from switch A, then for $X = B$ we get:

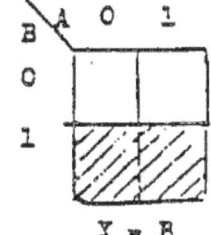

If the black box does something with the input from A so that X is on when $A = 0$, and X is off when $A = 1$, we say that X is NOT A, which is written as $X = \overline{A}$. This gives the set of boxes:

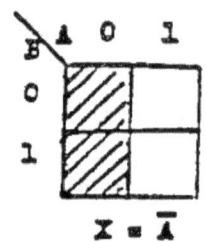 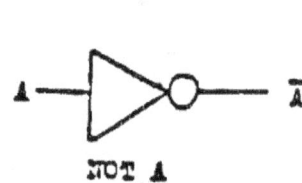

Anything which produces NOT A which is also called the COMPLEMENT OF A is depicted by a triangle with a little circle at the tip of its output.

In the rest of our Karnaugh maps we shall omit the labels of the columns and rows, the labels being understood to be as shown above.

The function NOT B has the Karnaugh map and picture representation as shown below.

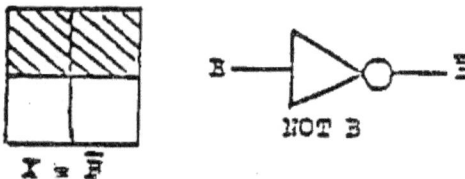

Using orderly bunches of **1**'s and **0**'s, generated by bunches of switches and displayed by bunches of lights we can represent numbers in an encoding called BINARY. Two switches together can generate a binary encoding of the decimal numbers from **0** to **3** as $0 = 00$, $1 = 01$, $2 = 10$ and $3 = 11$. If you want to count past three, add more switches. Each place in a binary number is called a Binary Digit (BIT).

Suppose we give our black box a job to do. Switch A encodes a one bit number and switch B encodes another one bit number. The black box is to add A to B to produce a two bit sum, displayed in lights X and Y. Following are the things which can happen:

```
A=        0       1       0       1
B=        0       0       1       1
Sum=     00      01      01      10
                                 └┴── sum bit   =X
                                  └── carry bit =Y
```

The sum of two one bit numbers gives a two bit number. The bit under the bits added we call the sum bit. The other resulting bit we call the carry or overflow bit. Each bit in the sum is represented by its own Karnaugh map.

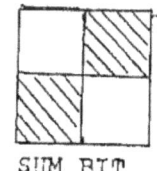

SUM BIT

Our problem now becomes that of finding something to put into the black box to do the job of adding. This will be done with things called integrated circuits. Here we shall be interested only in what they do, not how they do it. We shall perform our logic with a basic electronic unit called a NAND gate. Its output will light a light if either one or both of its inputs is zero.

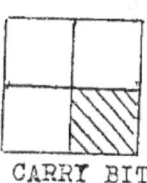

CARRY BIT

It will not light a light if both inputs are one. Its Karnaugh map and symbol are shown here. Starting with switches A and B and using nothing but NAND gates, how shall we produce the sum and carry bits? This requires a form of thinking called BOOLEAN ALGEBRA.

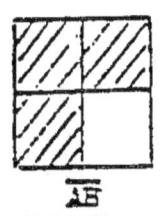

$\overline{AB}$
A NAND B

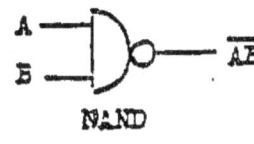

NAND

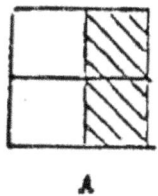

A

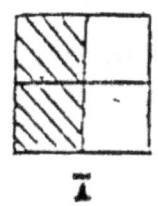

$\overline{A}$

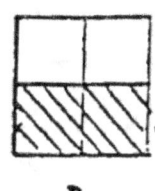

B

$\overline{B}$

Let us start with the maps for A, NOT A, B and NOT B. Consider each map to be a set of shaded boxes. Consider set A and set B. The shaded state-box that is common to both sets is in the lower right corner. The box common to both of two sets is called the INTERSECTION of those sets. Mathematicians write this as $A \cap B$. We shall use an engineering notation and write this as AB. The intersection function is called AND and an electronic component doing this would be called an AND GATE. The function is represented as a half circle, inputs to the diameter. AB is shown on the next page.

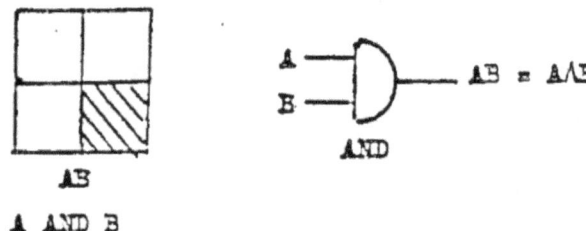

AB

A AND B

We are no longer in arithmetic algebra, we are in Boolean Algebra which is a branch of set theory. Here AB means A and B, not A TIMES B.

The set of all boxes shaded in both sets A and B is called the UNION of the two sets. The union function is called OR and an electronic component doing this is called an OR GATE. Mathematicians write the OR function as $A \cup B$. We shall use the engineering notation of $A + B$. Here this means A OR B, not A PLUS B. The Karnaugh map and the symbol for the OR GATE are as shown.

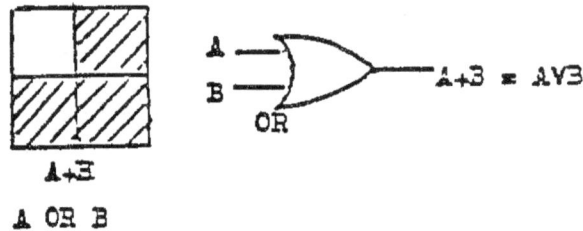

A+B

A OR B

A little circle on the output of an AND gate denotes the opposite or complement of AND. This we call NOT AND, which is shortened to NAND. The Karnaugh map and picture are:

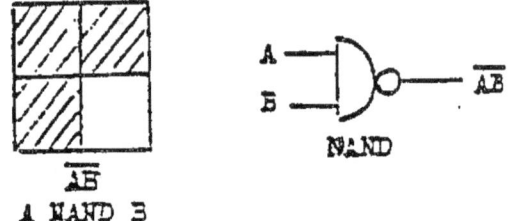

$\overline{AB}$

A NAND B

A little circle on the output of an OR gate denotes the opposite or complement of OR. This is called NOT OR, shortened to NOR. Its Karnaugh map and picture are:

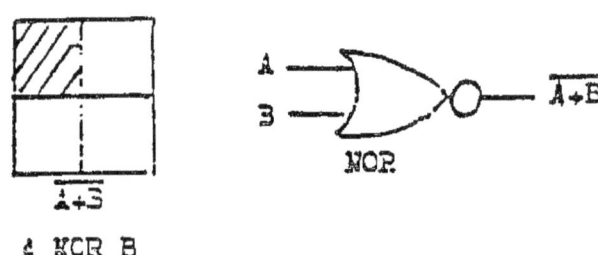

$\overline{A+B}$

A NOR B

You have probably figured out that a line over anything denotes the opposite or complement of that thing. This has nothing to do with the idea of negatives.

On page 14-43, it is obvious that the set of shaded boxes for NOT A AND B (NAND) has a pattern similar to that for A OR B. Looking at the Karnaugh maps at the top of 14-43. It is evident that the NAND function is the UNION of the set NOT A ($\overline{A}$) with the set NOT B ($\overline{B}$). Thus, we have the equivalence, $\overline{AB} = \overline{A} + \overline{B}$.

Again, it evident that NOR has a pattern similar to AND. Looking at the sets, it is evident that NOR is the INTERSECTION of the sets NOT A and NOT B. Thus, $\overline{A + B} = (\overline{A})(\overline{B})$.

These two statements, taken together, are known as DeMORGAN'S THEOREM. They are essential for the logic design of a digital computer.

Before we design an adder, let us look at the Karnaugh map for our one electronic component, the TTL gate. This obeys the function NAND. For the function NOT A AND NOT B, it is obvious that if B=1, the output is shaded (=1) when A=0, and the output is not shaded (=0) when A=1. Thus, if B=1, the NAND gate produces the function NOT A. The state of B=1 can be produced by pulling the input B up to +5 volt. On TTL NAND gates it can be produced by just not connecting B to anything.

Returning to our addition problem, it is evident that the CARRY function is just A AND B. But we have only NAND gates to play with. Well, if you complement something twice, you are right back where you started.

We have,

A AND B = NOT (NOT A AND B)

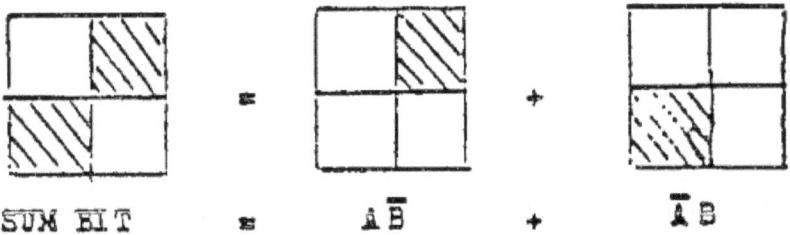

Next, it is evident that the set of shaded boxes describing the SUM BIT is the UNION of the two INTERSECTIONS. We draw the following decomposition.

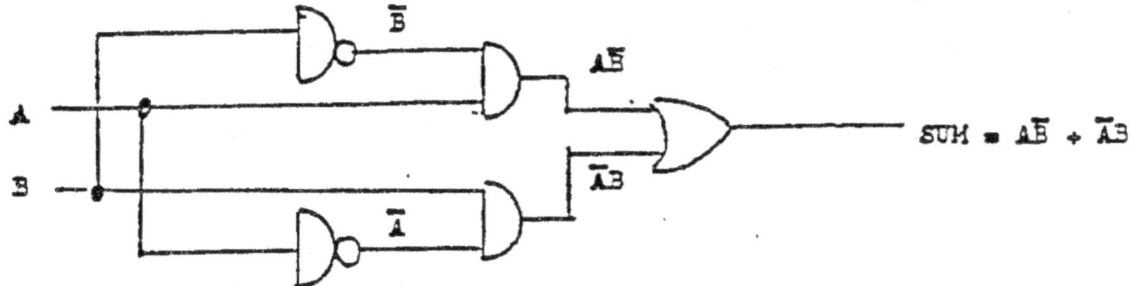

This leads to the set of gates illustrated below.

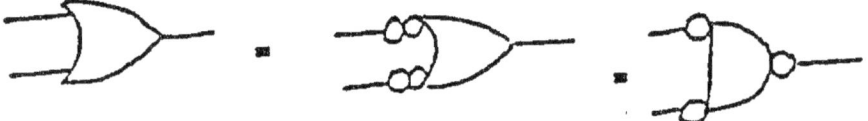

This requires a mixture of types of gates. We have only NAND gates. The problem is solved by using DeMorgan's Theorem along with the fact that a double NOT on a logic line does not change that logic. Thus:

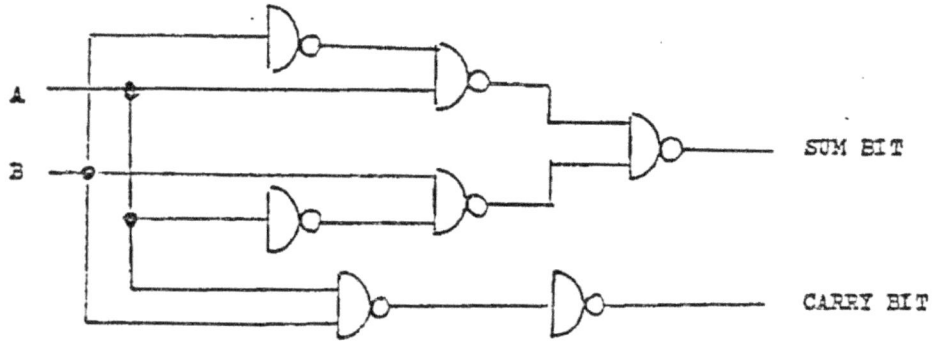

This gives the following logic for adding two one bit numbers.

PROBLEM FOR THE STUDENT: Refer to the Karnaugh map for A NAND B and show
that the SUM function can be described as the UNION of two INTERSECTIONS as:

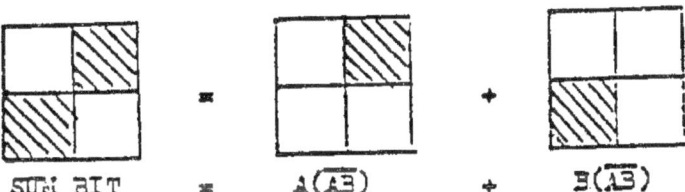

SUM BIT   =   A(AB)   +   B(AB)

Use this result - we must produce a NAND for the carry anyway - and also
DeMorgan's Theorem to show that the sum of two one bit numbers can be done
with the following logic. Note that it uses fewer NAND GATES.

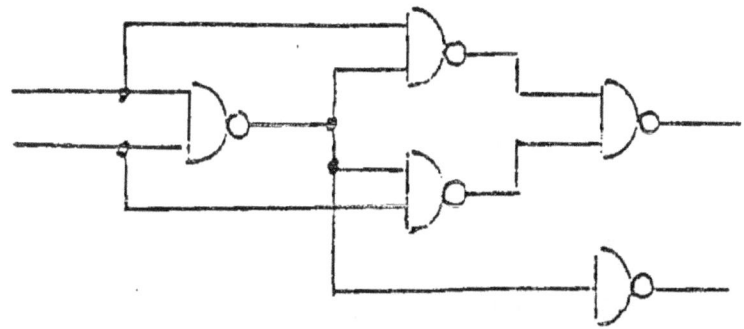

The concept of number encodes the results of counting things. The first
counting was done on fingers (and sometimes toes) giving rise to the term
digit for a number. You are familiar with Roman and Standard (not Arabic as
you will discover in Egypt or Saudi Arabia). In binary encoding, we have
only two symbols to use in each bit - It is either 0 or it is 1. Just as we
must use the standard symbols over and over again as we count past ten and
then past a hundred, we must use binary bits over and over again as we count
past one and then past three, etc. A comparison of these standard, Roman and
binary systems is given below.

Standard	Roman	Binary
0		0000
1	I	0001
2	II	0010
3	III	0011
4	IIII	0100
5	V	0101
6	VI	0110
7	VII	0111
8	VIII	1000
9	VIIII	1001
10	X	1010
11	XI	1011
12	XII	1100
13	XIII	1101
14	XIIII	1110
15	XV	1111

You notice that the Romans did not use a symbol
to encode nothing - zero. You notice that binary
code puts a bit in every place which might
require a one bit in the conning scheme.

Before we talk about adding binary numbers in a
computer or calculator, let us recall how we add
standard numerals - say to add 853 to 548.
Recall that we add each pair of corresponding
digits and then add a carry in front of the pair
to the right and produce a carry out to be added
to the sum of digits on the left.

For example:

```
      5  4  8
      8  5  3
    ───────────
      3  9  1  First Sum
   1  1  1      Carry
   ────────────
   1  4  0  1  Final Sum
```

Now let us add what would be standard or decimal numerals, $A = 13_D$ and $B = 7_D$ but instead of doing it in a decimal, we do it in binary.

In both cases, notice that a carry-in is added to the first sum to produce a final sum; and that the carry-out may come from either the first or final sums.

```
A = 13_D =   1  1  0  1
B =  7_D =   0  1  1  1
           ──────────────
           1  0  1  0    X = First Sum
        1  1  1  1        Carry (in and out)
        ──────────────
S = 20_D = 1  0  1  0  0  S = Final Sum
```

To accomplish the binary addition, we can use the logic from page 14-45, sum-bit, momentarily ignoring the carry-bit logic. First, we get a first-sum-bit, X, for corresponding bits of numbers, A and B. Then we produce, in the same way, a final-sum-bit for X and carry in $C = C_{in}$. That was easy. But carry out can now come from either X or from S, and this requires some logic design.

There are now three input-bits, A, B, and C. There are eight different combinations for three bits – requiring an eight-box Karnaugh map.

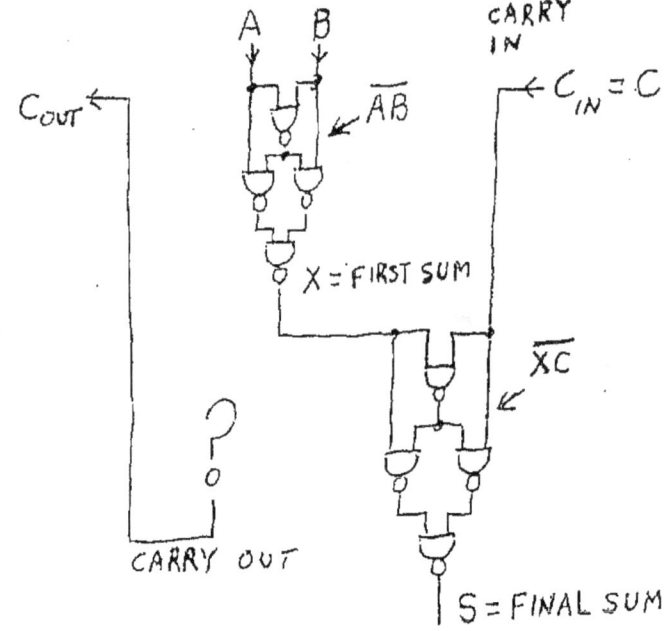

For a Karnaugh map with eight boxes for eight combinations of A, B, and C, we will take four columns labelled – note carefully how – by combinations of A with B. The rows will be labelled by values of carry-in C, each input separately being one will give one of the following sets of the shaded boxes.

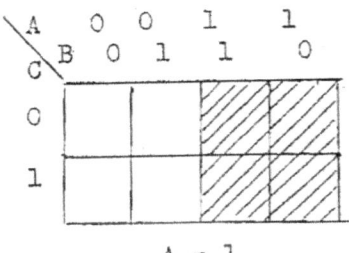

A = 1

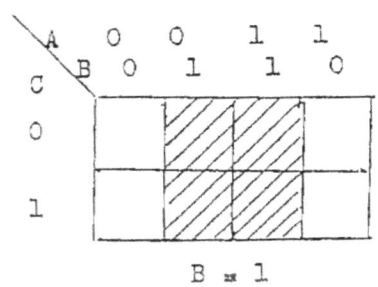

B = 1

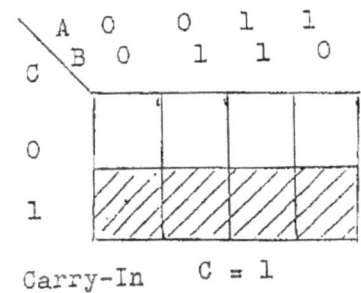

Carry-In   C = 1

You can draw your own Karnaugh maps for NOT-A = $\overline{A}$, NOT-B = $\overline{B}$, and NOT-C = $\overline{C}$. From this you get Karnaugh maps for first-sum, $X = A\overline{B} + \overline{B}A$, and then for final-sum, $S = \overline{C}X + X\overline{C}$.

There is a contribution to carry-out from first-sum when A and B are both one. As at the top of page 2-3, carry-out from first-sum is A AND B = AB.

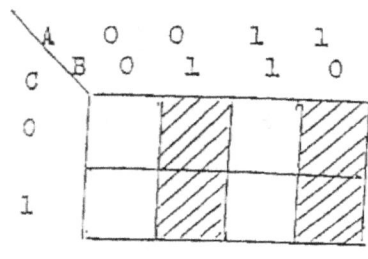

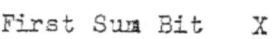

First Sum Bit   X

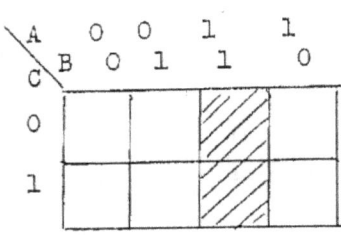

Carry-Out from A and B

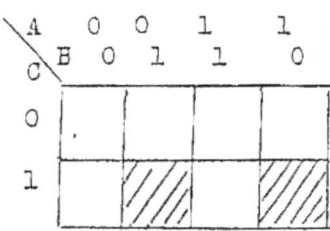

Carry-Out from X and C

Carry-out from the second-sum or final-sum is then, X AND C = XC. The total carry-out is the UNION of these two carry-out bits.

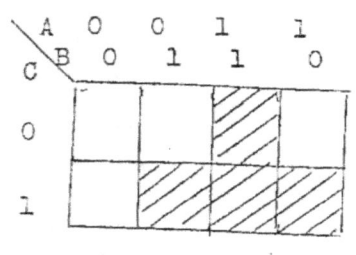

Total Carry-Out = C_{out}

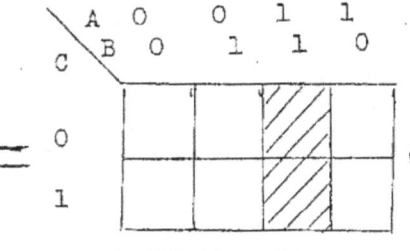

A AND B = AB

OR

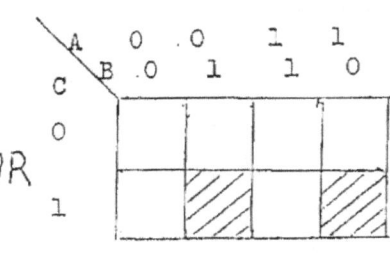

X AND C = XC

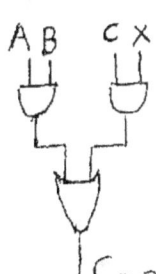

$C_{out} = AB + CX$ would produce gates as shown. But these are not NAND gates. Double-NOTS on AB and CX between the first and second gates change nothing. But, recall from page 14-44…

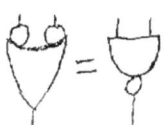

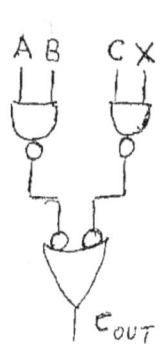

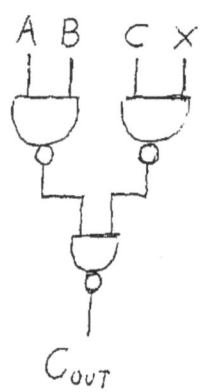

Thus, from A, B, C, and X, we produce a total carry-out using only NAND gates. $C_{out} = AB + CX = \overline{(\overline{AB})(\overline{CX})}$. You will notice that in producing first-sum X, and final-sum S, we have already produced $\overline{AB}$ and $\overline{CX}$ so one more NAND gate gives our carry-out bit.

The preceding logic has developed a unit for adding two corressponding bits of muli-bit numbers - with carry-in for the right and a carry-out to the left. A linked row of four of these will add two four bit numbers. A row of **64** adding units will add two **64**-bit numbers.

This is all the arithmetic logic design needed for a digital computer, as, using some extra logic units help, we subtract by adding, we multiply by adding, and with some cleverness, we divide by adding.

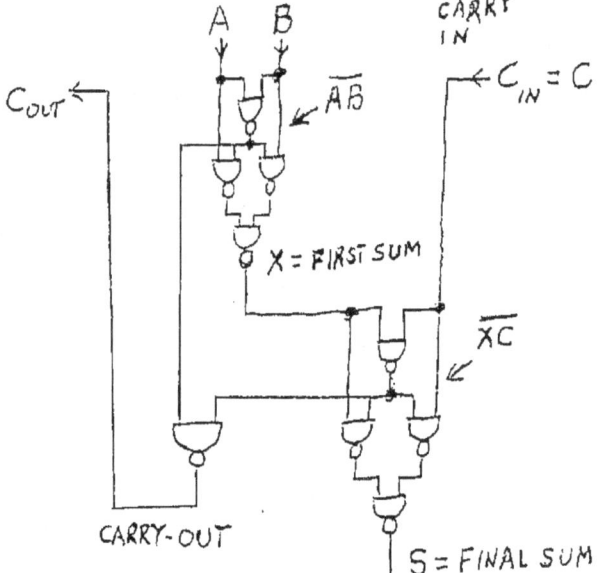

Next, we turn our attention to a DIGITAL CLOCK. The time is given by arrays of Liquid Crystal Displays (LCD in gray) or Light Emitting Diodes (LED in red). The LED was explained earlier in chapter 14. Each numeral is displayed from an array of seven segments which people can read easily. Each numeral is produced by a binary counter which gives a four bit encoding of that numeral. The binary encoding is not easy to read by humans accustomed to standard decimal numerals. Thus, each four bit binary code must be decoded to activate the required segments in the seven-segment display assigned to it. We now turn our attention to this decoding logic.

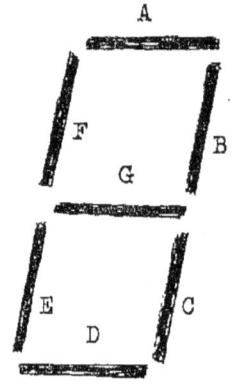

It is important to note that only binary codes from 0000 to 1001 will be produced by the counters, or some sub-set of them. The codes 1010 to 1111 will never happen here. This set of four bit codes is called Binary-Coded-Decimal (BCD) and occurs in all binary treatments of standard decimal numbers. For each BCD code, only the appropriate segments of the related 7-segment display will be activated. Our BCD-to-7-segment display logic unit will have four input lines for BCD codes and seven output lines, each activating a specific segment. A 1 on an output line turns its segment on; a 0 turns it off.

As shown on the previous page, we label the segments - starting at the top and going clockwise - as A, B, C, D, E, F, and G. The BCD bits will be labelled - from left to right - as W, X, Y, and Z. There are 16 possible combinations for W, X, Y, and Z. Although only ten of them are to be used, we contruct a Karnaugh map with 16 boxes for 16 states of W, X, Y, and Z. The rows of boxes will be labeled by combinations of W and X; the columns by Y and Z. Note carefully the order of 0's and 1's for X and Z.

In each box we show the segments that will display that combination of W, X, Y, and Z. Note that one (1) uses the two right-hand segments.

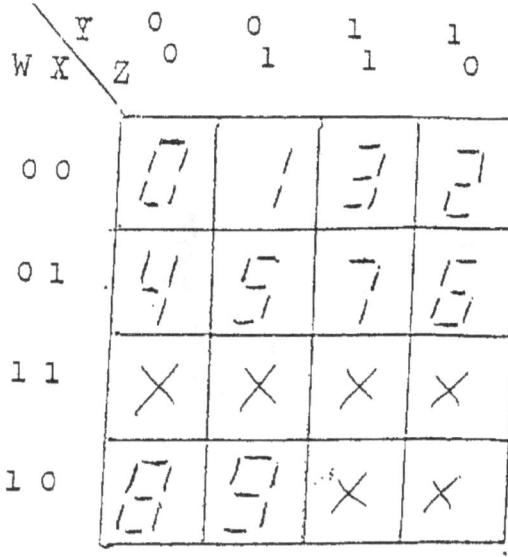

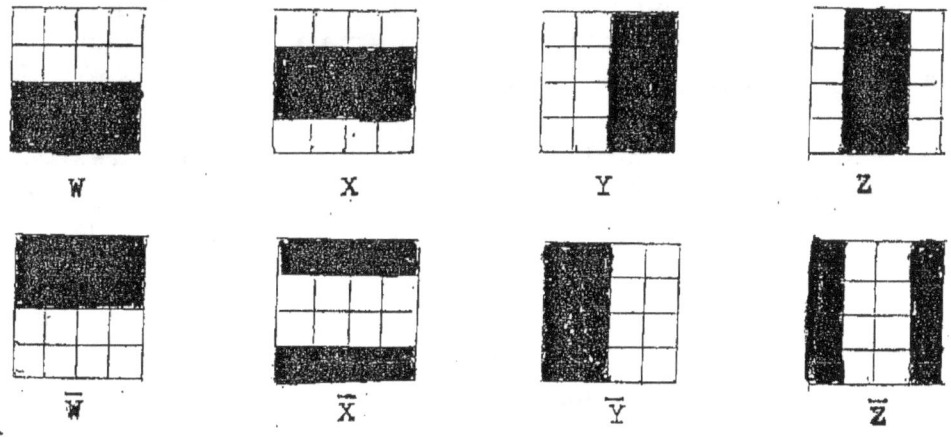

0 0 0 0 = 0
0 0 0 1 = 1
0 0 1 0 = 2
0 0 1 1 = 3
0 1 0 0 = 4
0 1 0 1 = 5
0 1 1 0 = 6
0 1 1 1 = 7
1 0 0 0 = 8
1 0 0 1 = 9

For each BCD, we must produce 0's and 1's on each segment line telling that segment to be off or on. Hence, we must design seven decoding logic gate arrays each driving its particular segment.

Let us look how that is accomplished. The six boxes or states labelled X can be included in any decoding scheme as they never occur to mess things up.

We next display Karnaugh maps, shaded for those states for which each of the BCD bits are one. Then we display maps for not, or the opposites, of the BCD input bits being one.

W   X   Y   Z

$\overline{W}$ $\overline{X}$ $\overline{Y}$ $\overline{Z}$

Now we shade those boxes or states of W, X, Y, and Z for which A is on.

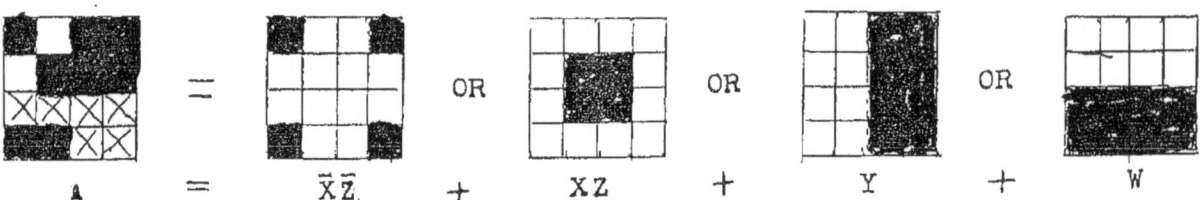

$$A = \bar{X}\bar{Z} + XZ + Y + W$$

All BCD states except $0001 = 1_D$ and $0100 = 4_D$ will require the top segment, A, to be on. Six non-BCD states of W, X, Y and Z will never happen. We can use all states of W, X, Y, and Z in the logic to decode the BCD bits to produce the driver bits for the seven-segment display as only the BCD states will occur and the never-happen states will never happen to mess up the desired result. The Karnaugh map for segment, A, can be represented by the union (OR) of the intersection (AND) of BCD bits and non-BCD bits. The logic is not unique as other logic arrays would do just as well.

In the same manner, we can encode the bit turning on segment 3.

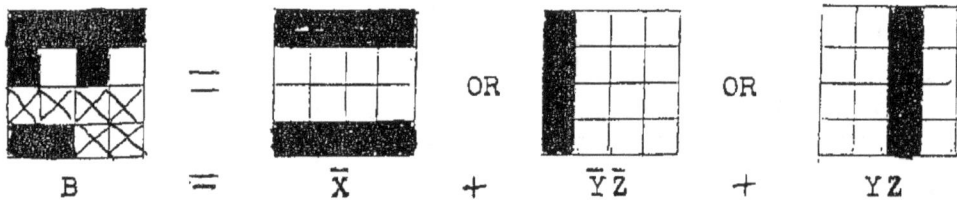

$$B = \bar{X} + \bar{Y}\bar{Z} + YZ$$

We can now start to think about the design of a silicon chip that would accomplish this decoding-encoding logic. With parallel conducting lines we take the bits W, X, Y and Z across one side of the chip. Next, we decode these, using AND and OR gates to implement the logic producing segments A and B. Similar decoding-encoding would produce the five other segments.

But we do not have AND and OR gates on our silicon chip, we only can have NAND and NOR gates, as described earlier in chapter 14. We solve this problem in the same manner as we have solved it before. We put two NOTs on each line to the OR gate. The opposite of an opposite is exactly what you started with – not changing the logic.

The AND gates are converted to NAND gates (see the two figures on following page) and by the DeMorgan's Theorem – page 14-44, and holding for any number of inputs – the OR of NOTs is the NOT of an AND, which is a NAND. Thus, we come to the second implementation of the decode-encode logic, using all NAND gates. The Y, Z, and $\overline{X}$ in the first diagram which did not go through AND gates now want the opposite of what they were at first and in the second diagram are connected to $\overline{Y}$, $\overline{Z}$, and X.

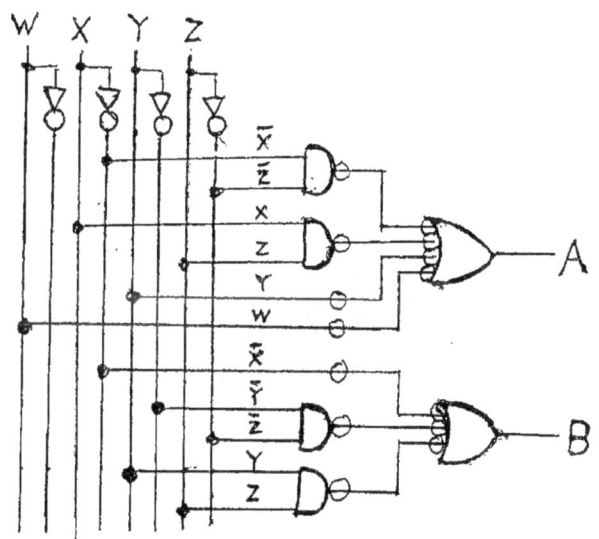
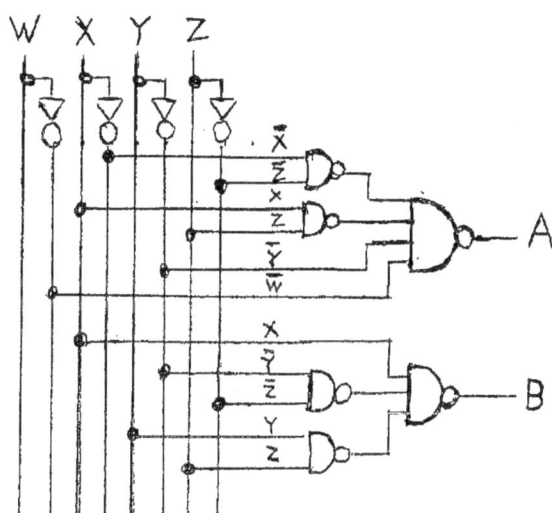

We have taken you through a design problem that is not trivial. You have seen all of the basics principles of the logic design. Go back, now, and study the sequence of logic steps very carefully. In the further design problems there will be new tricks resulting from cleverness, but you now have the basic principles.

The following problems involve encoding the remaining segments. No particular encoding logic is unique. Even where a logic scheme is suggested, you may derive another logic that works just as well.

In each case, draw a Karnaugh map for that segment in states of W, X, Y, and Z.

14-1: Derive logic that would produce segment C.

14-2: Derive logic that would produce segment E.

14-3: Show that the segment D could be produced by the logic,
$$D = W + \overline{X}\,\overline{Z} + \overline{Y}\,\overline{Z} + X\,\overline{Y}\,Z.$$

14-4: Show that the segment F could be produced by the logic,
$$F = W + X\,\overline{Y} + X\,\overline{Z} + \overline{Y}\,\overline{Z}.$$

14-5: Show that the segment G could be produced by the logic,
$$G = W + X\,\overline{Y} + Y\,\overline{Z} + \overline{X}\,Y.$$

Please notice that in this chapter we have been using symbols that looks the same as normal algebra, but where the symbols stood for OR = the union of sets, AND = the intersection of sets, and NOT = the opposite of… In the rest of the book, Algebra will have its normal reference to arithmetic operations on numbers.

The channel does run between the source and the drain, but the n-type source is separated from the n-type drain by a region of p-type substrate. This produces two diode boundaries. If the drain is at a positive voltage above the source, then the positive current will try to flow into the drain and back to the source through the semi-conductors. But the right-hand diode will not allow conduction in that direction. In the top picture, there is no current through the device.

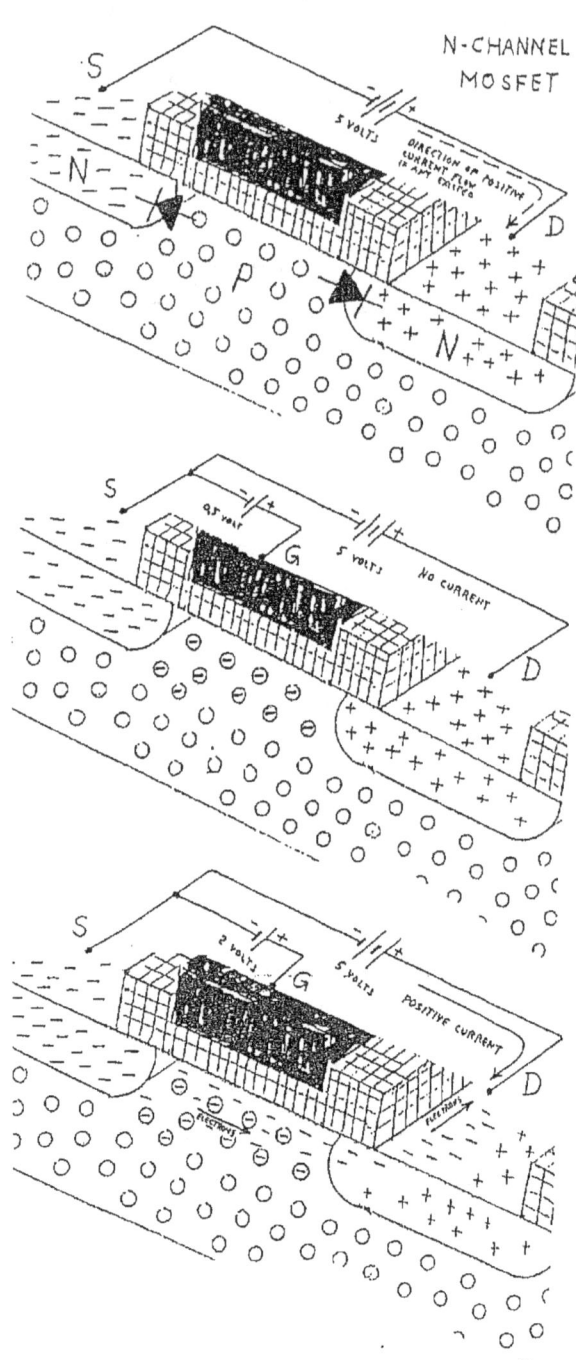

Let us put a small **0.5** volts from the source up to the gate. The aluminum gate becomes slightly positively charged, but charge can-not flow from the gate to the semi-conductor through the insulating silicon dioxide. The positive gate pulls electrons into the unfilled bonds at the impurity sites in the p-type semi-conductor. The region under the gate becomes negatively charged, but as the electrons are tied down in the bond structure, they can-not move. Thus, no current flows.

From the source, let us increase the positive voltage at the gate to **2.0** volts. The more positive gate pulls more electrons from the source into the channel. As the bonds in the channel are filled, these electrons will be free to move in the conduction band. As the **5.0** volts at the drain above the source has pulled the surplus electrons off the impurity sites in the n-type drain, these will now be replaced by the movable electrons in the channel. A current now flows in the channel.

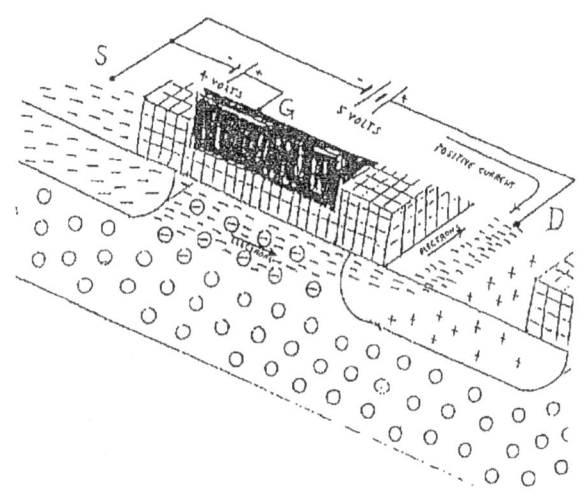

A positive **4.0** volts at the gate pulls more electrons into the channel, producing more current at the drain.

This is now called an enhancement mode FET as the gate now increases the number of carriers in the channel rather than decreasing them as with the type of FET previously discussed. The configuration of materials shown is as they would be built onto the surface of a computer chip, or integrated circuit.

In circuit diagrams, the n-channel MOSFET is given the symbol shown below.

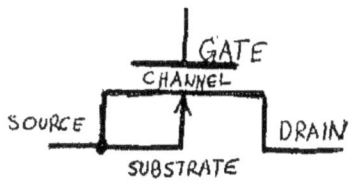

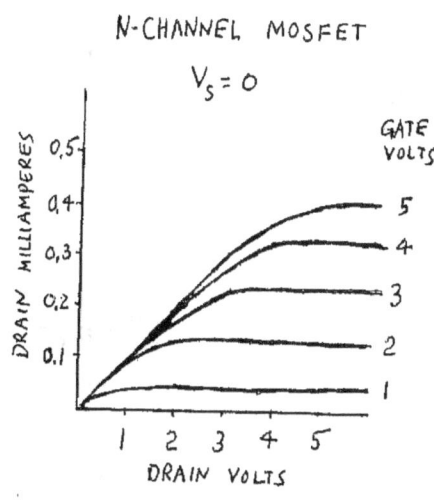

Although the channel is interrupted by the substrate, in drawing circuit diagrams it is common practice to show the channel as if it were continuous from source to drain and to show a diode arrow in the direction in which the channel-base boundary would conduct positive current.

For a typical n-channel MOSFET, drain current may vary with drain voltage for various gate volts as shown in the graph on the right. In the following graph, the drain is kept at **5** volts and the drain current is plotted against gate volts. Notice that the gate volts must get above a certain threshold value before the channel will conduct current.

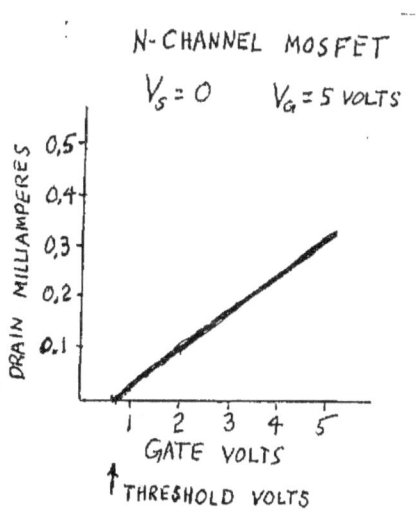

In the figure below, we show a cross-section of a p-channel MOSFET. This device works in exactly the same manner with the only exception that the gate and drain voltages are negative with reference to the source and that positive current now flows from the source to the gate.

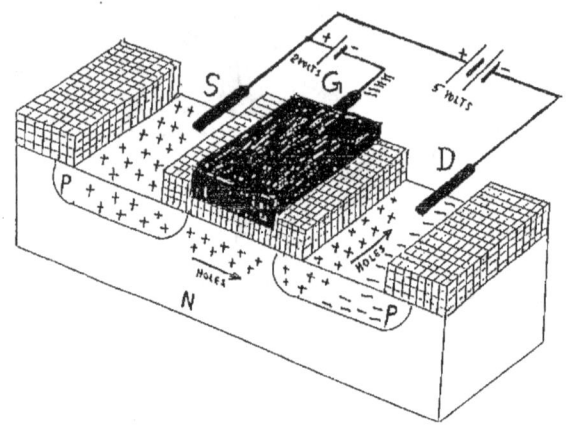

For further reading The Proceedings of the IEEE (Institute of Electrical and Electronics Engineers) has published several issues on the design of VLSI (very-large-scale integrated circuits).

January 1983: VLSI: Problems and Tools, May 1983: Micron and Sub-Micron Circuit Engineering, May 1993: VLSI Reliability.

Now that we have described the components, we can discuss how they are put together on a silicon chip to implement the gate logic with which we design digital computers. Please return to chapter 2 to review your understandings of the principles of gate logic.

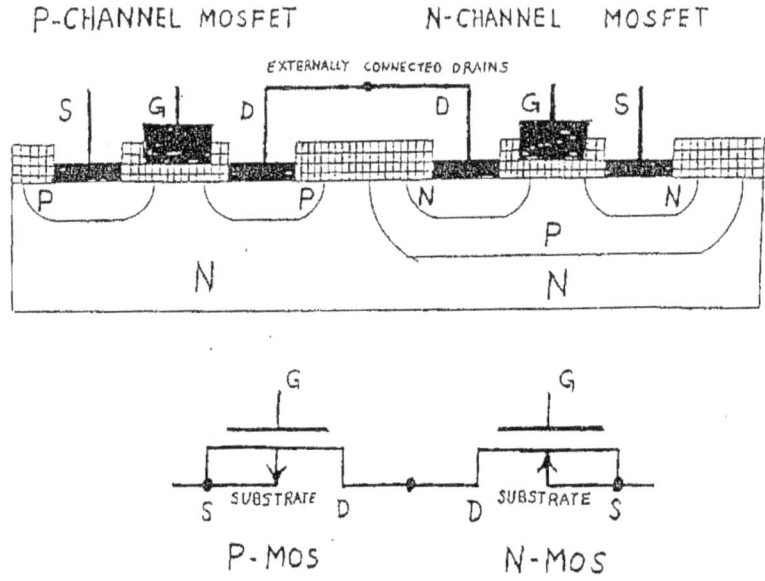

In designing integrated circuits, N-MOS denotes designs using only n-channel MOSFETS, and P-MOS denotes designs using p-channel MOSFETS only. If we place P-MOS and N-MOS transistors back-to-back on an integrated circuit chip and connect their drains, we say that each complements the behavior of the other and we denote by C-MOS, the design with complementary MOSFETs.

We will now implement the logic of this chapter, using C-MOS design. We take the pair of complementing transistors and put the N-MOS source at 0 volts and put the P-MOS source at 5 volts. If the N-MOS gate is at zero volts, its channel will not conduct. The N-MOS channel will conduct only if the gate is higher than about +0.5 volts. If the P-MOS gate is at 5 volts (the same as its source), the P-MOS channel will not conduct. When the P-MOS gate is less than 4.5 volts (negative with reference to its source at 5 volts), its channel will conduct. We now connect the two gates and apply an input voltage (labeled A) to them. The output voltage from the connected drains we call F.

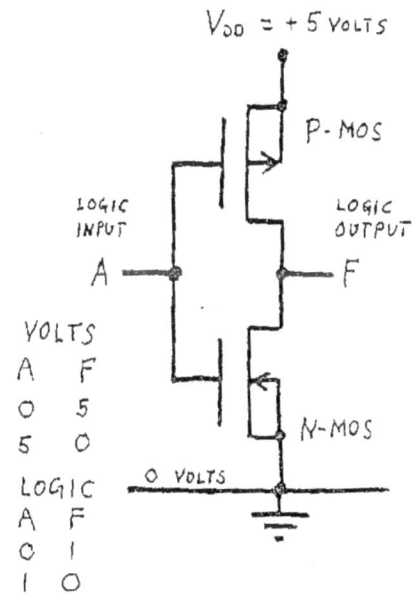

When A is at 5 volts, the N-MOS will conduct but the P-MOS will not conduct. The output, F, is connected to 0 volts. When A is at 0 volts, the bottom N-MOS will not conduct but the top P-MOS will conduct and the output, F, is connected to 5 volts. The voltage at F is the opposite of the voltage at A. This is shown in the table of voltage values.

Now, suppose we are accomplishing the logic of this chapter, using C-MOS electronics. Let 0 volts encode a logic zero (0) and let 5 volts encode a logic one (1). Now, when $A = 0$, $F = 1$, and when $A = 1$, $F = 0$. For an input A, the circuit produces the opposite of A which we shall call not-A ($\overline{A}$) and has the logic symbol shown here and on page 14-41.

$$A \quad \longrightarrow \quad F = \overline{A}$$
$$\text{NOT-A}$$

On page 14-41, we stated that the logic design would be done electronically using a component which performed the logic function NAND or Not-And. We are now able to see how this can be done using C-MOS design. The component will have two logic inputs, A and B, and a logic output being a function, F, of the inputs, A and B.

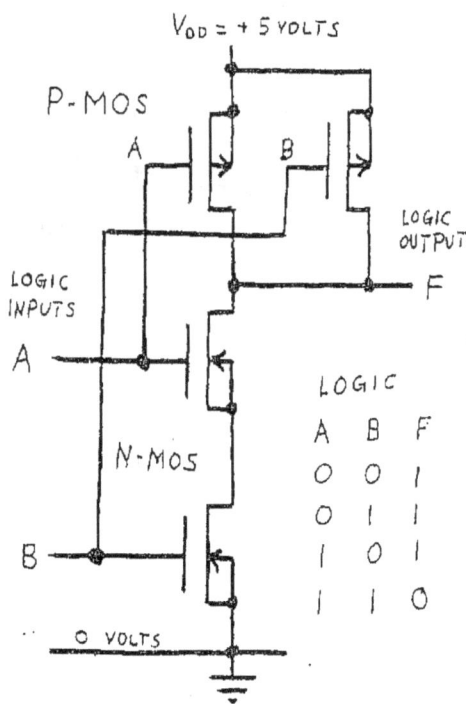

$V_{DD} = +5$ VOLTS

P-MOS

A   B

LOGIC
OUTPUT

LOGIC
INPUTS

A

F

N-MOS

B

0 VOLTS

LOGIC

A	B	F
0	0	1
0	1	1
1	0	1
1	1	0

To accomplish the logic function NAND = Not-And, the output, F, will be connected to zero volts through two N-MOS transistors in series and to 3 volts through two P-MOS transistors in parallel. At any moment, either P-MOS or N-MOS transistors will conduct, but not both types at once. Both N-MOS transistors must be conducting for F to be connected to zero volts. This happens when both A and B are at 5 volts; and when this happens neither of the P-MOS transistors is conducting.

If either $A = 0$ volts or $B = 0$ volts, one or the other of the N-MOS transistors can not conduct and F is dis-connected from zero volts. But, if $A = 0$ volts or $B = 0$ volts, one or the other of the P-MOS transistors will be conducting and F is connected to 5 volts.

If 5 volts encodes logic 1 and 0 volts encodes logic 0, then if A and B are both 1, F will be zero. If either A or B or both of them is zero, the output F will be 1. This result is shown in the logic table, and in the Karnaugh map. The boxes are shaded if the inputs A and B make the output F to be 1. This accomplishes the function NAND, Not A and B.

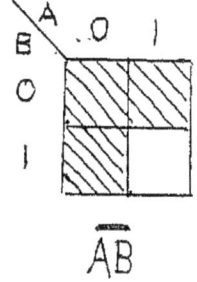

$\overline{AB}$

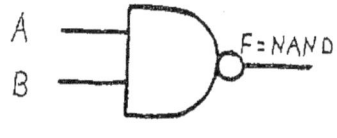

A
B

F = NAND

The function NOR is done by two N-MOS FETs in parallel connecting F to 0 volts and two P-MOS FETs in series connecting F to 5 volts. A and B must be both 0 volts for F to be connected to 5 volts. If either A or B is 5 volts, F will be 0 volts.

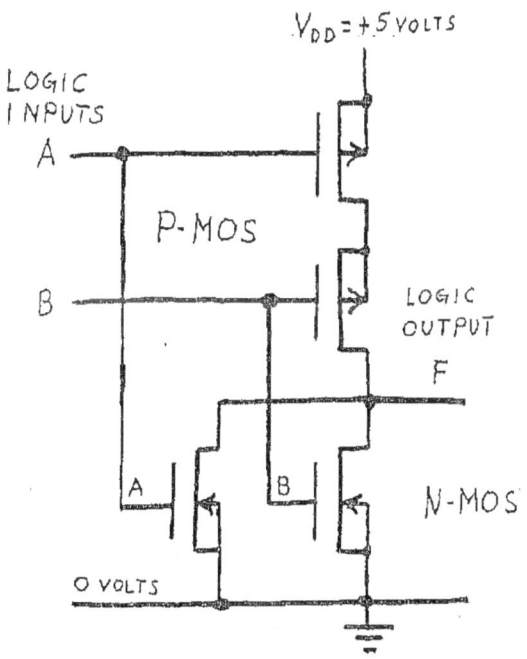

LOGIC

A	B	F
0	0	1
0	1	0
1	0	0
1	1	0

In terms of logic, if A and B are both 0, then $F = 1$. If $A = 1$ or $B = 1$ or both are 1, then $F = 0$. This result is shown in the logic table and gives the function, NOR = Not A OR B, as shown in the Karnaugh map when a box is shaded if $F = 1$.

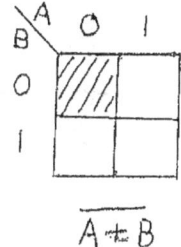

$\overline{A + B}$

An old and much used implementation of the logic operation NAND = NOT AND uses three NPN bi-polar transistors and three resistors. It is called Transistor-transistor-Logic (TTL). The secret of its operation is that the input transistor serves as a sort of two-way switch.

The input transistor has a p-type base, an n-type collector, and two more emitters. Recall that the base-emitter and base-collector boundaries are one-way diodes where positive charge current will flow from the base to the emitter or to the collector if the base is at a higher voltage than either of those. If an emitter is at a higher voltage than the base, current will not flow across that boundary.

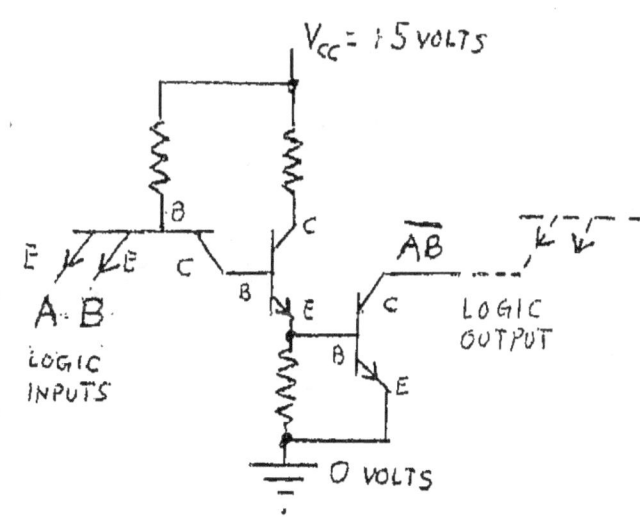

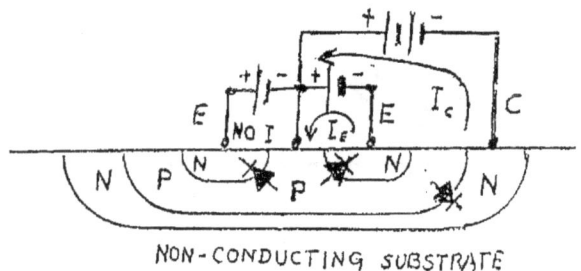

NON-CONDUCTING SUBSTRATE

A low level - logic-zero - at an input emitter will exist when the emitter is connected to zero volts. A high level - logic-one - will exist when the emitter is not connected to anything or is connected to 5 volts, the power supply to the electronic units.

If an emitter is connected to zero volts current will flow from the base through that emitter and take an easy route to zero volts (called ground). Then, current will not flow out through the collector as that is now a not-easy route to zero volts. The next transistors act as normal NPN transistors. If there is no current from the base to the emitter there is no current from the collector to the emitter.

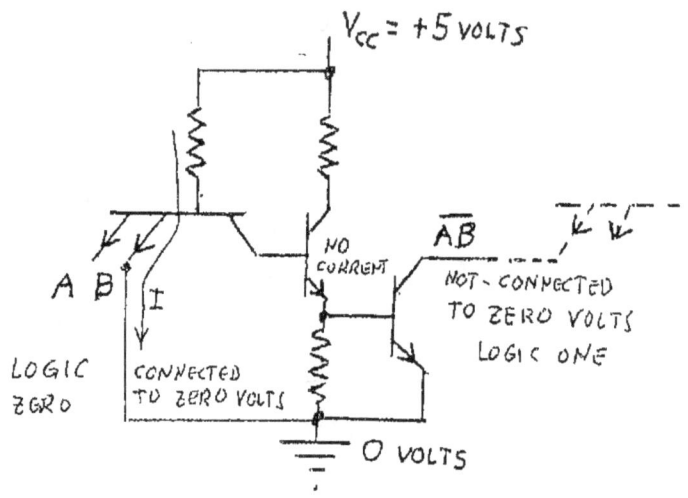

If there is no current into the base of the second transistor then there can be no current into the base of the third transistor. The input emitter to the next NAND gate sees this collector as disconnected from anything.

So, if either or both input emitters are connected to zero volts, the output collector is disconnected from anything. If the input emitters are both either disconnected or pulled up to 5 volts, current can now flow only from the base of the first transistor out of its collector, and then into the second base. The second transistor now conducts and current out of its emitter flows into the third base, and the third transistor conducts. An input emitter connected to it will be at logic zero.

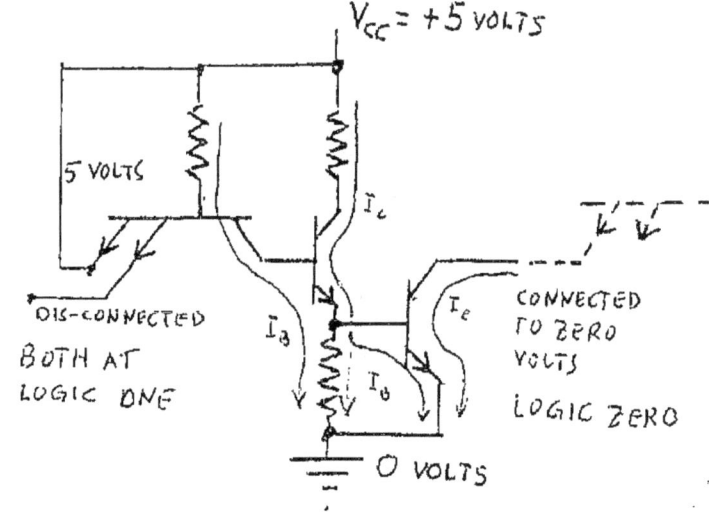

14-60

Chapters 13 and 14 introduced the quantum mechanics of condensed matter. This is a vast and very important area of current research. As this is only an introduction, you may want to continue your study. The Oxford University Press Master Series in Physics are probably the most definitive textbooks available. They are reasonably priced in paperback editions. However, they are written at the university fourth year undergraduate or first year graduate level and are intended to prepare the student for starting research. For a high school student, any one of these books will be the start of a long-term study and growth in the particular branch of physics or engineering.

THE FOLLOWING TITLES ARE AVAILABLE FROM OXFORD UNIVERSITY PRESS:

J. C. Foot: ATOMIC PHYSICS
Martin T. Dove: STRUCTURE AND DYNAMICS
John Singleton: BAND THEORY AND ELECTRONIC PROPERTIES OF SOLIDS
Mark Fox: OPTICAL PROPERTIES OF SOLIDS
Mark Fox: QUANTUM OPTICS, AN INTRODUCTION
Simon Hooker and Colin Webb: LASER PHYSICS
Stephen Blundell: MAGNETISM IN CONDENSED MATTER
James F. Annet: SUPERCONDUCTIVITY, SUPERFLUIDS AND CONDENSATES
Richard A. L. Jones: SOFT CONDENSED MATTER
S. M. Barnett: QUANTUM INFORMATION

THE FOLLOWING TITLES MAY ALSO BE OF INTEREST:

R. P. Huebener: CONDUCTORS, SEMI-CONDUCTORS, SUPERCONDUCTORS: AN
        INTRODUCTION TO SOLID STATE PHYSICS, Springer International (2015)

D. I. Khomskii: BASIC ASPECTS OF THE QUANTUM THEORY OF SOLIDS
        Cambridge University Press (2010)

R. Skomski: SIMPLE MODELS OF MAGNETISM, Oxford University Press (2008)

M. Sharon & M. Sharon (editors) CARBON NANOFORMS AND APPLICATIONS
        McGraw-Hill (2010)

K. Iniewski (editor) NANOELECTRONICS: NANOWIRES, MOLECULAR ELECTRONICS, AND
        NANODEVICES, McGraw-Hill (2011)

Also try searching the internet for "condensed matter physics" and "solid state physics" (Google it).

Quantum mechanics uses a notation called Dirac notation. In this chapter, we will present two mathematical interpretations of this notation. First will be a non-standard form, related to the calculus of the Schrödinger equation. Then we will develop the customary form used in physics. This will involve things called linear operators in abstract many dimensioned spaces.

Let us go back to the start. Most problems involve relations between two sets of numbers. For every number (call it x) there will be another number (call it y) so that for every x there will be a y. For example, take $y = x^2$ for which a graph of their relation gives what is called a parabola.

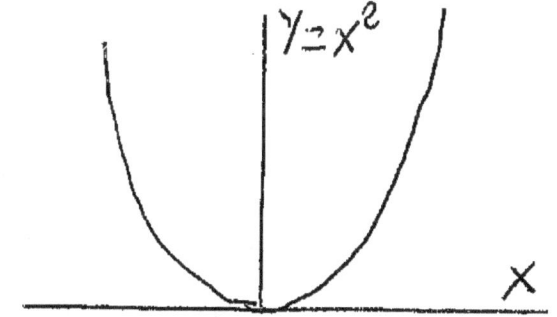

In such a relation, y is called a function of x. We write y is a function of x as $y = f(x)$. For the parabola, $f(x) = x^2$.

If we multiply $f(x)$ by x we get $x f(x) = x^3$, which is another function of x, call it $g(x)$. Then $x f(x) = g(x)$. Anything that does something to give another function is called an OPERATOR. Even in this simple case, x is said to operate on $f(x)$ to give $g(x)$.

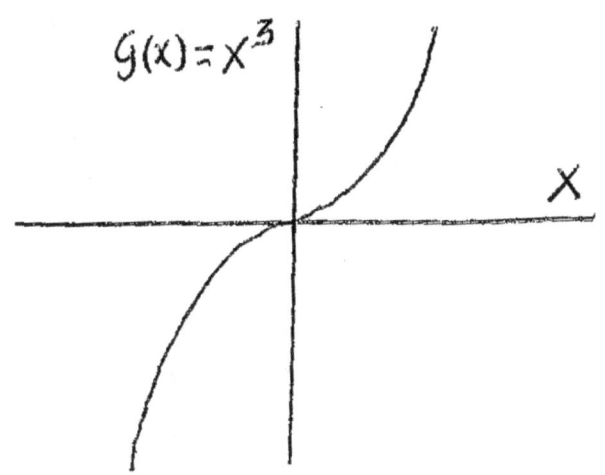

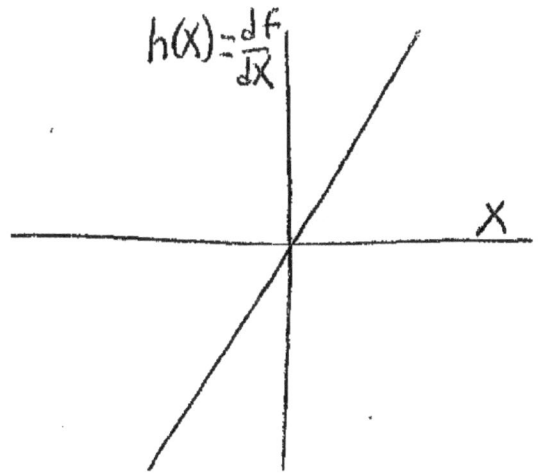

Now, for any function, $f(x)$, we can take its derivate, $\frac{df}{dx} = h(x)$, producing a new function, giving the slope or steepness of $f(x)$'s curve. As the derivative gives a new function, it is represented by an operator symbol, D, which represents whatever we do to $f(x)$ to get $h(x)$. We write $h(x) = D f(x)$.

We next introduce what we shall call a FUNCTION SYMBOL which merely stands for a function and its curve. We write $|f\rangle$, $|g\rangle$, and $|h\rangle$ to identify the curves for $f(x)$, $g(x)$, and $h(x)$. Now the value or number of the function given by a particular value or number of x is written $f(x) = \langle x|f\rangle$ and likewise, $g(x) = \langle x|g\rangle$ and $h(x) = \langle x|h\rangle$.

We introduce a modified idea for an operator which will represent whatever we do to one function symbol to get a related function symbol. $\hat{X}|f\rangle = |g\rangle$ and $\hat{D}|f\rangle = |h\rangle$.

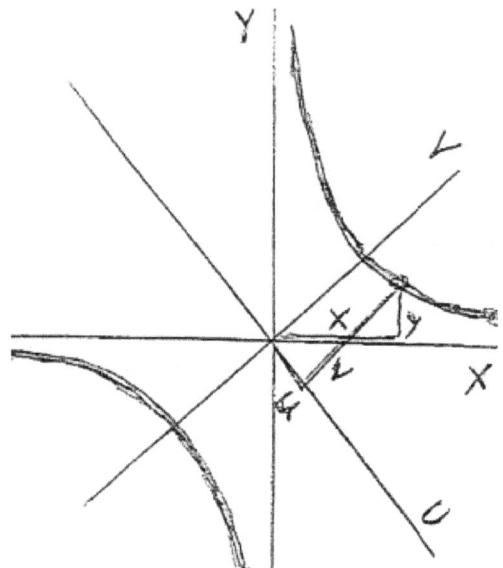

Although we have been talking about normal calculus, we have arrived at the notation for doing formal quantum mechanics.

There is a reason for this notation in quantum mechanics. The condition of a particle is described by its STATE. Its state is described by a function, call it $f(x)$ here. This will describe the value of the particle's state at some position, x. But, in quantum mechanics the same state might be a function of the particle's velocity. This would give another set of numbers for the same state. We would get $f(v)$.

In the picture there is a curve for $x\,y = c^2$ or $y = c^2/x$. But we have another pair of coordinates at $45°$ to the coordinates x and y. In these coordinates, the same curve is given by $v^2 = 2\,c^2 + u^2$. The same curve has two formulas. Now, let $|f\rangle$ stand for whatever formula that describes the curve. Then, $\langle x|f\rangle = c^2/x$ and $\langle u|f\rangle = \sqrt{2\,c^2 + u^2}$.

In quantum mechanics (and elsewhere in applied mathematics) if $f(x)$ and $g(x)$ are two functions (not the ones we have been thinking about) and they have values for x when it has numbers from some a to some number, b, it is necessary that we do the integral of their product on that domain of x. We want:

$$\int_{x=a}^{b} f(x)\,g(x)\,dx$$

15-2

Let $g(x) = \langle x|g \rangle$ and in the new notation $\langle x|f \rangle$ can also be $\langle f|x \rangle$. The integral is now:

$$\int_{x=a}^{b} \langle f|x \rangle \langle x|g \rangle \, dx$$

Now if that isn't strange enough, in the new notation it is then rewritten as shown below.

$$\langle f|g \rangle = \langle f| \left[\int_{x=a}^{b} |x \rangle \langle x| \, dx \right] |g \rangle$$

This emphasizes that the integral is something done with the states $|f \rangle$ and $|g \rangle$. This notation was invented by the outstanding physicist, Paul Dirac. He called $\langle f|$ a BRA symbol and $|g \rangle$ a KET symbol and then $\langle f|g \rangle$ was a BRA-KET. In most books these are called vectors in a Hilbert space, but rather than getting bogged down in the mathematics of abstract vector spaces, we have decided to get the same notation with a simpler algebra.

Integrals of products of functions are used in Chapter 18 and in the first part of Chapter 21.

It often happens that the terms in $\langle x|f \rangle$ are complex such as $\langle x|k_\alpha \rangle = \sqrt{\frac{1}{2L}}\, e^{i\,\alpha\,\pi\, x/L}$, but that we require $\langle f|x \rangle \langle x|f \rangle$ to be a real number. In this case, recall that if $z = x + i\,y$, $z^* = x - i\,y$, and $z\,z^* = x^2 + y^2$, a real number. Then for $\langle x|k_\alpha \rangle$, $\langle k_\alpha|x \rangle = \sqrt{\frac{1}{2L}}\, e^{-i\,\alpha\,\pi\, x/L}$.

This gives the integral:

$$\langle k_\alpha|k_\alpha \rangle = \int_{x=-L}^{+L} \frac{1}{2L} e^{-i\,\alpha\,\pi\, x/L}\, e^{i\,\alpha\,\pi\, x/L}\, dx = 1$$

The explanation of Dirac notation may be enough for you to read chapters 16, 17, and 18. In the most quantum mechanics books, Dirac notation refers to linear operators in abstract spaces, usually called Hilbert Spaces. If you desire to study the mathematics of abstract spaces, continue to study this chapter. For quantum mechanics with Dirac notation, read Paul A. M. Dirac's own The Principles of Quantum Mechanics.

THE FORMAL THEORY OF QUANTUM MECHANICS

"All things transitory but as symbols are sent" J. W. Goethe <u>Faust</u>

As far its ability to describe laboratory data and predict new phenomena, quantum mechanics is the most successful physical theory of all time. There is an old saying that a camel is a horse put together by a committee. Quantum mechanics may indeed appear to be an intellectual camel, held together with scotch tape and paper clips. In fact, it was created by a committee of some of the most brilliant physicists of all time; Schrödinger, Heisenberg, Jordan, Born, Pauli, Wigner and others. It was Paul Dirac who brought all the loose ends together, and it is his version that we shall follow.

Assume now that a free particle, with no interactions, experiences the Planck and de Broglie wave-particle relationships. In chapter 9, we discussed waves of frequency, f, wavelength, λ, and wave velocity, $u = f\lambda$. We can write for them a wave function, $\Psi = A \, \begin{smallmatrix} sin \\ cos \end{smallmatrix} \, 2\pi f\left(\frac{x}{u} - t\right)$.

If either a sine function or a cosine can describe our wave, we can use $e^{i\theta} = \cos\theta + i\sin\theta$ to write a generalized wave, $\Psi = A \, e^{i\,2\pi f\left(\frac{x}{u} - t\right)}$, but $u = f\lambda$ giving $\Psi = A \, e^{i\,2\pi\left(\frac{x}{\lambda} - f t\right)}$. But suppose that $f = \frac{E}{h}$ and $\frac{1}{\lambda} = \frac{p}{h}$ with $p = mv$. This gives us $\Psi = A \, e^{i\frac{2\pi}{h}(px - Et)}$. In all the discussions that follow, Planck's constant will always be divided by 2π, so we will follow a suggestion by P. A. M. Dirac and define an haitch-bar as $\hbar = \frac{h}{2\pi}$.

Then our wavefunction is $\Psi = A \, e^{i\left(\frac{px - Et}{\hbar}\right)}$.

If $\Psi = A \, e^{i\frac{px}{\hbar}} e^{-i\frac{Et}{\hbar}}$, then $\frac{\partial \Psi}{\partial x} = i\frac{p}{\hbar} \Psi$ and $\frac{\partial^2 \Psi}{\partial x^2} = -\frac{p^2}{\hbar^2} \Psi$.

Similarly, $\frac{\partial \Psi}{\partial t} = -i\frac{E}{\hbar} \Psi$ and $\frac{\partial^2 \Psi}{\partial t^2} = -\frac{E^2}{\hbar^2} \Psi$.

But, $E = \frac{p^2}{2M}$ for a particle with no interactions where $p = Mv$.

Multiplying this by Ψ gives $E\Psi = \frac{p^2}{2M}\Psi$, and substitution gives

$$i\hbar \frac{\partial \Psi}{\partial t} = -\frac{\hbar^2}{2M} \frac{\partial^2 \Psi}{\partial x^2}.$$

For a particle whose interaction has a potential energy, $V(x)$, $E = \frac{p^2}{2M} + V(x)$.

If we multiply this by Ψ and substitute we get $i\hbar \frac{\partial \Psi}{\partial t} = -\frac{\hbar^2}{2M} \frac{\partial^2 \Psi}{\partial x^2} + V(x)\,\Psi$,

which is the time dependent Schrödinger equation.

But, $i\hbar \frac{\partial \Psi}{\partial t} = E\,\Psi$ allowing us to write $\left[-\frac{\hbar^2}{2M} \frac{\partial^2}{\partial x^2} + V(x) \right] \Psi = E\,\Psi$.

The quantity $\hat{H} = \left[-\frac{\hbar^2}{2M} \frac{\partial^2}{\partial x^2} + V(x) \right]$ works on a function, taking rates of change and multiplying it by something. Such a quantity is called an OPERATOR and we put a little hat over it to remind us that it is not just another arithmetic multiplier. So, we write $\hat{H} \Psi = E\,\Psi$.

When an operator times a function merely multiplies that function by a number, this is a unique case. The unique function is called an eigenfunction and the unique number is an eigenvalue. Obviously, some German mathematicians had something to do with developing this.

A particle with momentum, p, and energy, E, has an associated wave, $\Psi(x,t) = e^{i(px - Et)/\hbar}$, where $\hbar = \frac{h}{2\pi}$. But let us look at the particle right at the present with $t = 0$. Then, $\Psi(x) = e^{ipx/\hbar}$, giving $\frac{\partial \Psi}{\partial x} = i\frac{p}{\hbar}\Psi$ or $\left[\frac{\hbar}{i} \frac{d}{dx} \right] \Psi = p\,\Psi$.

In this case, $\frac{\hbar}{i}\frac{d}{dx}$ is the momentum operator and the numbers, p, are the momentum eigenvalues. Here the functions, $\Psi(x)$, will be the momentum eigenfunctions.

Finally, quantum mechanics refers to classical (before 1900) physics in this way. Every classical quantity is replaced in quantum mechanics by an operator. If energy is $E = \frac{p^2}{2M} + V(x)$, its operator, called the Hamiltonian is $\hat{H} = \frac{\hat{p}^2}{2M} + V(\hat{x})$, replacing momentum and position by their operators.

The Dirac notation is the standard language of quantum mechanics. The presentation in this discussion is not the standard one. We have modified the usual interpretation to conform better to the Schrödinger picture of quantum mechanics. The normal mathematical context involves linear operators in an abstract space called a Hilbert space. A ket symbol stands for a column vector whose components are written in a column. A bra symbol stands for a row vector, whose components are written in a row. A bra times a ket is an inner product.

As an excuse for developing this form of mathematics and, then, as a means for demonstrating what the results look like, we will consider a light weight string, loaded at uniform intervals with balls all having the same mass, M. The dimension (number of coordinate axes) will be determined by the physical problem. Multiplying a vector (column of numbers) by an array of numbers called a matrix, will give another vector, mapping any point in the vector space onto another point.

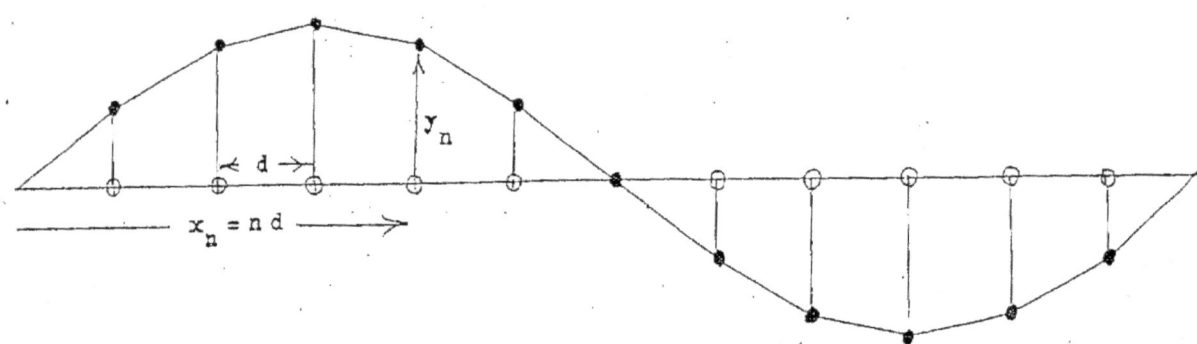

You may wish to review the discussion of moving and standing waves at the start of chapter 6. We now consider a two-loop standing wave in a light-weight string, loaded with equally spaced balls, each of which has a mass, M. Let the integers, n, denote the balls, where in this case n goes from 1 to 11. In general, let N be the total number of balls on the string. Let d be the distance between each pair of balls, and $x_n = n\,d$ be the distance from the left end of the string to the nth ball. When the string is vibrating, let y_n be the displacement of the nth ball. Referring to the description of a vibrating standing wave in chapter 6, $y_n = A \sin\left(\frac{2\pi x_n}{L}\right) \cos(2\pi f_2 t)$, where L is the length of the string and f_2 is the frequency when vibrating with two loops.

Obviously, each ball being in vibration motion has a force acting on it. Let F stand for the tension force in the string. When the string is horizontal, the string on either side of any ball exerts a tension force, and when the string is horizontal and straight, the tension forces cancel each other.

When the objects are displaced, the strings on either side of any ball are not pulling in opposite directions, and the tension forces cannot cancel each other. Consider the effective force on the nth ball. But first we must describe some geometry.

The nth object has a displacement, y_n, while the objects on either side have displacements, y_{n-1} and y_{n+1}. Let θ_n be the angle between the string and the horizontal to the right of the nth ball and θ_{n-1} be the angle to the right of the $(n-1)^{th}$ ball and to the left of the nth ball. If d is the horizontal distance between balls, then $tan\,\theta_n = \frac{y_{n+1} - y_n}{d}$ and $tan\,\theta_{n-1} = \frac{y_n - y_{n-1}}{d}$.

The tension force to the right of the nth ball has horizontal and vertical components, $F_h = F\cos\theta_n$ and $F_v = F\sin\theta_n$. The string to the left has horizontal and vertical components, $F_h = F\cos\theta_{n-1}$ and $F_v = F\sin\theta_{n-1}$.

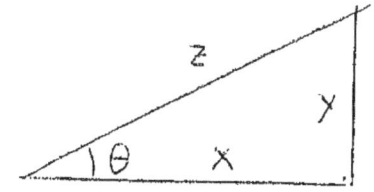

For the triangle shown, $sin\,\theta = \frac{y}{z}$ and $tan\,\theta = \frac{y}{x}$. But when the angle is very small, $x = z$ approximately.

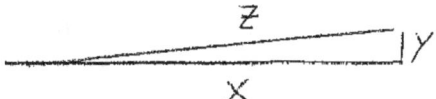

For very small angles $sin\,\theta = tan\,\theta$

Now, for the triangle, $cos\,\theta = \frac{x}{z}$, and for small θ, $cos\,\theta = 1$. Then, for small vibrations, in the standing wave, the horizontal forces on each ball cancel themselves, and the net vertical force (up is positive and down is negative in the picture) will be $F\sin\theta_n - F\sin\theta_{n-1} = F\,tan\,\theta_n - F\,tan\,\theta_{n-1}$. If the net force equals mass time acceleration, we put everything together to get

$$M\frac{d^2y_n}{dt^2} = F\left[\frac{y_{n+1} - y_n}{d} - \frac{y_n - y_{n-1}}{d}\right].$$

We now introduce a new notation. More about it in an appendix to this chapter on matrix algebra. If a point in three dimensions has a_x, a_y, a_z for its location, the numbers can be written in a column, $|A\rangle = \begin{pmatrix} a_x \\ a_y \\ a_z \end{pmatrix}$ or a row,

$\langle A| = \begin{pmatrix} a_x & a_y & a_z \end{pmatrix}$.

We can multiply a column by a row as

$$\langle A|A \rangle = \begin{pmatrix} a_x & a_y & a_z \end{pmatrix} \begin{pmatrix} a_x \\ a_y \\ a_z \end{pmatrix} = a_x{}^2 + a_y{}^2 + a_z{}^2 = A^2,$$

which is the square of the length of the arrow to point $|A\rangle$. Similarly,

$$\langle B|B \rangle = \begin{pmatrix} b_x & b_y & b_z \end{pmatrix} \begin{pmatrix} b_x \\ b_y \\ b_z \end{pmatrix} = b_x{}^2 + b_y{}^2 + b_z{}^2 = B^2,$$

is the square of the arrow to point $|B\rangle$.

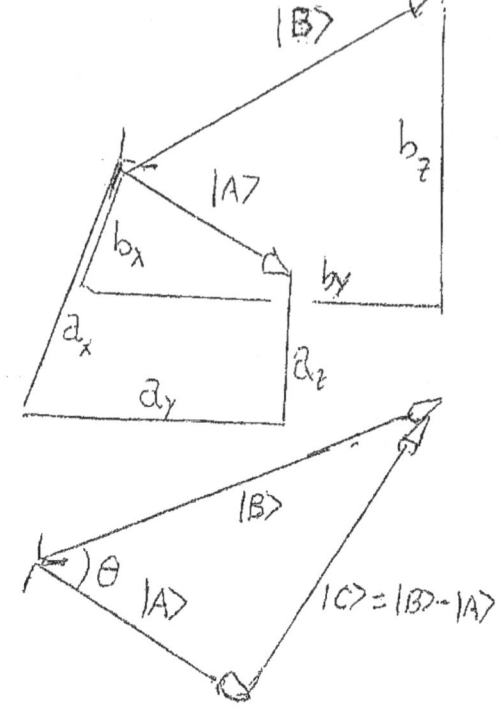

$$|C\rangle = |B\rangle - |A\rangle$$
$$C^2 = \langle C|C \rangle = [\,\langle B| - \langle A|\,][\,|B\rangle - |A\rangle\,]$$
$$C^2 = \langle B|B \rangle + \langle A|A \rangle - \langle A|B \rangle - \langle B|A \rangle$$
$$\langle A|B \rangle = \langle B|A \rangle$$
$$C^2 = B^2 + A^2 - 2\langle A|B \rangle$$

But, the law of cosines states:
$$C^2 = A^2 + B^2 - 2\,A\,B\,\cos\theta$$

So, $$\langle A|B \rangle = \begin{pmatrix} a_x & a_y & a_z \end{pmatrix} \begin{pmatrix} b_x \\ b_y \\ b_z \end{pmatrix} = a_x\,b_x + a_y\,b_y + a_z\,b_z = A\,B\,\cos\theta.$$

Now, $A\cos\theta$ is the projection of $|A\rangle$ onto $|B\rangle$.
Notice that when $\theta = 90°$, $\cos 90° = 0$.

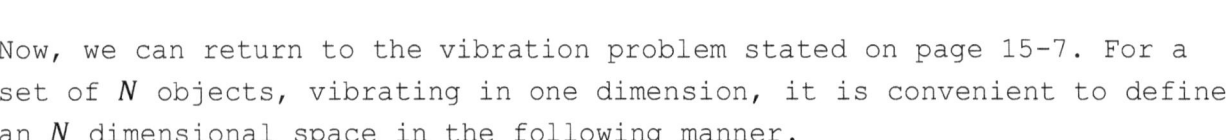

So, when $\langle A|B \rangle = 0$, the arrows, $|A\rangle$ and $|B\rangle$ are perpendicular to each other. This is a very important and useful result.

Now, we can return to the vibration problem stated on page 15-7. For a set of N objects, vibrating in one dimension, it is convenient to define an N dimensional space in the following manner.

At any instant, the set of N objects will have a set of N displacements, y_1, y_2, y_3, y_4, etc which we can write in an ordered column to give a generalized displacement vector $|y\rangle$ describing, at any instant, all the displacements of the N objects. Thus, $|y\rangle = \begin{pmatrix} y_1 \\ y_2 \\ y_3 \\ y_4 \\ \vdots \end{pmatrix}$, which can be written as a sum.

$$|y\rangle = \begin{pmatrix} y_1 \\ y_2 \\ y_3 \\ y_4 \\ \vdots \end{pmatrix} = \begin{pmatrix} y_1 \\ 0 \\ 0 \\ 0 \\ \vdots \end{pmatrix} + \begin{pmatrix} 0 \\ y_2 \\ 0 \\ 0 \\ \vdots \end{pmatrix} + \begin{pmatrix} 0 \\ 0 \\ y_3 \\ 0 \\ \vdots \end{pmatrix} + \begin{pmatrix} 0 \\ 0 \\ 0 \\ y_4 \\ \vdots \end{pmatrix} \cdots$$

We will associate with each object, one of a set of N mutually perpendicular unit length vectors, which form a basis or set of coordinate axes for the N-space. For identification purposes, we will label the unit length vectors by the distance, x_n, locating the object's position on the string. This serves, however, only to relate the object and its associated axis in the N-space. Thus, we write N unit vectors each having N dimensions as follows.

$$|x_1\rangle = \begin{pmatrix} 1 \\ 0 \\ 0 \\ 0 \\ \vdots \end{pmatrix} \quad |x_2\rangle = \begin{pmatrix} 0 \\ 1 \\ 0 \\ 0 \\ \vdots \end{pmatrix} \quad |x_3\rangle = \begin{pmatrix} 0 \\ 0 \\ 1 \\ 0 \\ \vdots \end{pmatrix} \quad |x_4\rangle = \begin{pmatrix} 0 \\ 0 \\ 0 \\ 1 \\ \vdots \end{pmatrix} \quad etc.$$

In the terms of the generalized displacement vector, $|y\rangle$, and the unit vectors denoting the separate objects, we can write the particular displacements by $y_1 = \langle x_1|y\rangle$, $y_2 = \langle x_2|y\rangle$, $y_3 = \langle x_3|y\rangle$, $y_4 = \langle x_4|y\rangle$, etc.

Then, the generalized displacement vector can be expanded into a representation involving components along the vectors, $|x_n\rangle$, taken as axes of the N-space. Hence,

$$|y\rangle = |x_1\rangle\langle x_1|y\rangle + |x_2\rangle\langle x_2|y\rangle + |x_3\rangle\langle x_3|y\rangle + \cdots = \sum_{n=1}^{N}|x_n\rangle\langle x_n|y\rangle$$

As the particles on the string vibrate, the components of $|y\rangle$ will be changing with time, and, hence, the displacement vector, $|y\rangle$, will wander around along some trajectory in the N-space. Note that the unit vectors, $|x_n\rangle$, denoting the particular particles remain fixed, not changing with time.

As $|y\rangle$ wanders through N-space, its projections onto the fixed axes, $|x_n\rangle$ will vary with time variation of the displacements, $y_n = \langle x_n|y\rangle$, of the particular particles. The solution of the vibration of the loaded string, therefore, involves finding the trajectory on which the generalized displacement vector wanders through N-space.

This first derivate of the displacement vector, with respect to time gives a vector of all the particle velocities, and the seconds time derivative gives the vector of all the particle accelerations.

The latter comes out to be:

$$|a\rangle = \frac{d^2}{dt^2}|y\rangle = \frac{d^2}{dt^2}\begin{pmatrix} y_1 \\ y_2 \\ y_3 \\ y_4 \\ \vdots \end{pmatrix} = \begin{pmatrix} \dfrac{d^2 y_1}{dt^2} \\ \dfrac{d^2 y_2}{dt^2} \\ \dfrac{d^2 y_3}{dt^2} \\ \dfrac{d^2 y_4}{dt^2} \\ \vdots \end{pmatrix}$$

Newton's law was $M\dfrac{d^2 y_n}{dt^2} = \dfrac{F}{d}(y_{n-1} - 2y_n + y_{n+1})$. $y_0 = 0$ and $y_{N+1} = 0$ are the end points.

For three objects this is $M\dfrac{d^2 y_1}{dt^2} = \begin{pmatrix} -2\dfrac{F}{d} & \dfrac{F}{d} & 0 \end{pmatrix}\begin{pmatrix} y_1 \\ y_2 \\ y_3 \end{pmatrix}$, $M\dfrac{d^2 y_2}{dt^2} = \begin{pmatrix} \dfrac{F}{d} & -2\dfrac{F}{d} & \dfrac{F}{d} \end{pmatrix}\begin{pmatrix} y_1 \\ y_2 \\ y_3 \end{pmatrix}$,

$M\dfrac{d^2 y_3}{dt^2} = \begin{pmatrix} -2\dfrac{F}{d} & \dfrac{F}{d} & 0 \end{pmatrix}\begin{pmatrix} y_1 \\ y_2 \\ y_3 \end{pmatrix}$. Divide each equation by M and then factor $-\dfrac{F}{d}$ out of the row vectors. Next stack the rows on top of each other.

$$\frac{d^2}{dt^2}\begin{pmatrix} y_1 \\ y_2 \\ y_3 \end{pmatrix} = -\frac{F}{M d}\begin{pmatrix} -2 & -1 & 0 \\ -1 & 2 & -1 \\ 0 & -1 & 2 \end{pmatrix}\begin{pmatrix} y_1 \\ y_2 \\ y_3 \end{pmatrix}$$

The generalized acceleration vector for the system of N objects on a loaded string can be expressed as a matrix operator times the generalized displacement vector for the system. The acceleration of the nth object depends on the displacement of the nth object and on the displacements of its nearest neighbors. These interactions are expressed by the diagonal and just-off-diagonal terms in the interaction matrix in the following equation.

$$\frac{d^2}{dt^2}\begin{pmatrix} y_1 \\ y_2 \\ y_3 \\ y_4 \\ \vdots \end{pmatrix} = -\frac{F}{M d}\begin{pmatrix} 2 & -1 & 0 & 0 & 0 & \cdots \\ -1 & 2 & -1 & 0 & 0 & \cdots \\ 0 & -1 & 2 & -1 & 0 & \cdots \\ 0 & 0 & -1 & 2 & -1 & \cdots \\ \cdot & \cdot & \cdot & \cdot & \cdot & \cdots \\ \vdots & \vdots & \vdots & \vdots & \vdots & \cdots \end{pmatrix}\begin{pmatrix} y_1 \\ y_2 \\ y_3 \\ y_4 \\ \vdots \end{pmatrix}$$

We can symbolize the interaction matrix with the operator, $\hat{G}$.

$$\hat{G} = \frac{F}{M d}\begin{pmatrix} 2 & -1 & 0 & 0 & 0 & \cdots \\ -1 & 2 & -1 & 0 & 0 & \cdots \\ 0 & -1 & 2 & -1 & 0 & \cdots \\ 0 & 0 & -1 & 2 & -1 & \cdots \\ \cdot & \cdot & \cdot & \cdot & \cdot & \cdots \\ \vdots & \vdots & \vdots & \vdots & \vdots & \cdots \end{pmatrix}$$

This allows us to write the differential equation in the shorthand manner:

$$\frac{d^2}{dt^2}|y\rangle = -\hat{G}\,|y\rangle$$

Next, let all particles be vibrating with the same frequency, $\omega = 2\pi f$.

$$y_1 = A_1 \sin(\omega t) \qquad y_2 = A_2 \sin(\omega t) \qquad y_n = A_n \sin(\omega t)$$

Then, in general, $\frac{d^2 y_n}{dt^2} = -\omega^2 A_n \sin(\omega t)$. This gives:

$$-\omega^2 \begin{pmatrix} A_1 \\ A_2 \\ A_3 \\ A_4 \\ \vdots \end{pmatrix} \sin(\omega t) = -\frac{F}{M\,d} \begin{pmatrix} 2 & -1 & 0 & 0 & 0 & \cdots \\ -1 & 2 & -1 & 0 & 0 & \cdots \\ 0 & -1 & 2 & -1 & 0 & \cdots \\ 0 & 0 & -1 & 2 & -1 & \cdots \\ \vdots & \vdots & \vdots & \vdots & \vdots & \cdots \end{pmatrix} \begin{pmatrix} A_1 \\ A_2 \\ A_3 \\ A_4 \\ \vdots \end{pmatrix} \sin(\omega t)$$

Divide both sides by $-\sin(\omega t)$ and let $|A\rangle = \begin{pmatrix} A_1 \\ A_2 \\ A_3 \\ A_4 \\ \vdots \end{pmatrix}$. Then, $\hat{G}\,|A\rangle = \omega^2 |y\rangle$.

This is a unique result. If there are N particles on the string, there will be only N vectors of the sort A each with its unique ω. Using a half-translated German word, the vectors are called eigenvectors and the numbers eigenvalues. It so happens that we now have the primary language of Quantum Mechanics.

Let's explore what this means. The amplitudes of vibration of N balls on a string, vibrating with some single angular frequency, ω, satisfied the equations: $\omega^2 a_n = \frac{F}{M\,d}(-a_{n+1} + 2\,a_n - a_{n-1})$, $a_o = a_{N+1} = 0$.

For $N = 2$, $\omega^2 a_1 = \frac{F}{M\,d}(-a_2 + 2\,a_1)$ and $\omega^2 a_2 = \frac{F}{M\,d}(2\,a_2 - a_1)$

or $\omega^2 \begin{pmatrix} a_1 \\ a_2 \end{pmatrix} = \frac{F}{M\,d} \begin{pmatrix} 2 & -1 \\ -1 & 2 \end{pmatrix} \begin{pmatrix} a_1 \\ a_2 \end{pmatrix}$.

For $N = 3$, $\omega^2 a_1 = \frac{F}{M\,d}(-a_2 + 2\,a_1)$ and $\omega^2 a_2 = \frac{F}{M\,d}(-a_3 + 2\,a_2 - a_1)$ and

$\omega^2 a_3 = \frac{F}{M\,d}(= 2\,a_3 - a_2)$ or $\omega^2 \begin{pmatrix} a_1 \\ a_2 \\ a_3 \end{pmatrix} = \frac{F}{M\,d} \begin{pmatrix} 2 & -1 & 0 \\ -1 & 2 & -1 \\ 0 & -1 & 2 \end{pmatrix} \begin{pmatrix} a_1 \\ a_2 \\ a_3 \end{pmatrix}$.

Let´s see what we can do with the three
particle case. To simplify things put

all the constants together as $k = \dfrac{M\,d\,\omega^2}{F}$.

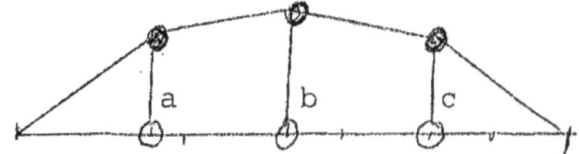

As each shape of vibration has its own frequency, f vibrations per second,
and angular frequency, $\omega = 2\pi f$, each shape has its own k. So that we don't
have to mess with subscripts in the three-particle equations on the previous
page, let $a_1 = a$, $a_2 = b$, and $a_3 = c$. The three particle equations are:
$k\,a = 2\,a - b$; $k\,b = -a + 2\,b - c$; $k\,c = -b + 2\,c$; or we can write:
$b = (2-k)\,a$; $a + c = (2-k)\,b$; $b = (2-k)\,c$.

Multiply the second equation by $(2-k)$. $(2-k)\,a + (2-k)\,c = (2-k)^2\,b$.
Putting the last two lines together gives $2 = (2-k)^2$ or $2 - k = \pm\sqrt{2}$.
This gives $k = 2 - \sqrt{2}$ or $k = 2 + \sqrt{2}$.

Returning to the equations for a, b, and c, it's possible that b could be
zero. Then, $a + c = 0$ and $c = -a$. But, if $b = 0$, then $(2-k)\,a = 0$ and
$(2-k)\,c = 0$. Therefore, if a and c are not 0, then $k = 2$.

The vibration of this shape is…

and $\omega^2 = \dfrac{2F}{M\,d}$.

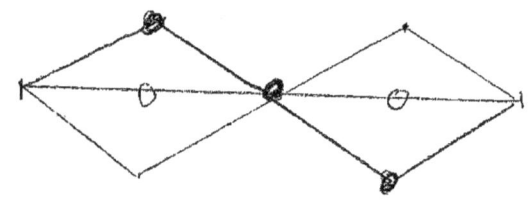

Now, go back and let $k = 2 - \sqrt{2}$, giving
$\omega^2 = \left(2 - \sqrt{2}\right)\dfrac{F}{M\,d}$. $b = \sqrt{2}\,a$ and $b = \sqrt{2}\,c$ requiring $c = a$.

From this shape we get an idea.
It looks as if the particles lie on
a half cycle of a sine wave whose
wavelength is twice the length of the
string. If L is the length of the
string, then at a distance, x, from
the left end, the displacement would

be $y(x) = \sin\left(\dfrac{2\pi x}{2L}\right)$.

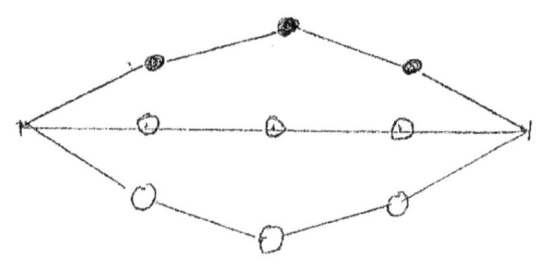

At $x = L/4$, $a = \sin(\pi/4) = 0.707$. At
$x = L/2$, $b = \sin(\pi/2) = 1.000$. At
$x = 3L/4$, $c = \sin(3\pi/4) = 0.707$. Notice
that $c = a$ and $b = \sqrt{2}\,a$ and $b = \sqrt{2}\,c$.

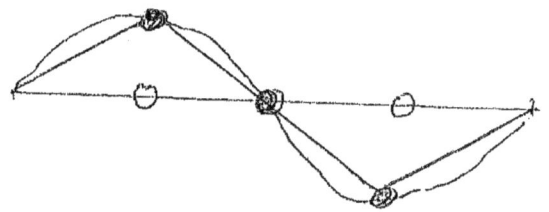

Obviously, when $k = 2$, the particles lie on a sine wave whose wavelength is the length, L, of the string.

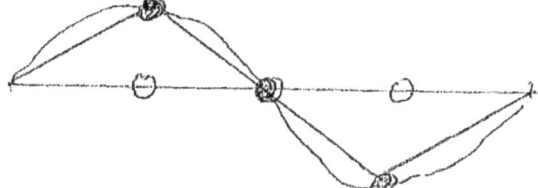

For $k = 2 + \sqrt{2}$, $b = -\sqrt{2}\,a$ and $b = -\sqrt{2}\,c$, giving the vibrating shape as shown on the right.

Now, take a sine wave with wavelength, 2/3 the length of the string.
$y = sin\left(\frac{2\pi x}{2L/3}\right)$ or $y = sin\left(\frac{3\pi x}{L}\right)$.

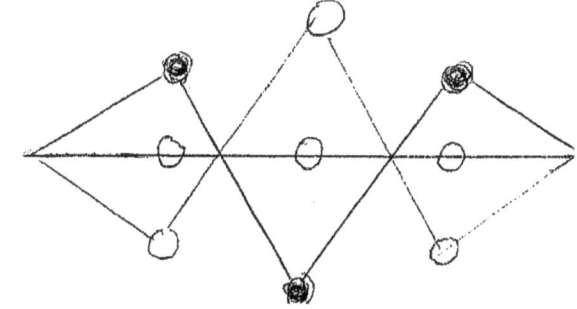

At $x = L/4$, $a = sin(3\pi/4) = 0.707$. At $x = L/2$, $b = sin(3\pi/2) = -1.000$. At $x = 3L/4$, $c = sin(9\pi/4) = 0.707$. $b = -\sqrt{2}\,a$ and $b = -\sqrt{2}\,c$.

Return to $\hat{G}\,|A\rangle = \omega^2|y\rangle$ on 15-11. For each set of amplitudes, let $a_1 = a = 1$.

$$\begin{pmatrix} 2 & -1 & 0 \\ -1 & 2 & -1 \\ 0 & -1 & 2 \end{pmatrix}\begin{pmatrix} 1 \\ \sqrt{2} \\ 1 \end{pmatrix} = (2 - \sqrt{2})\begin{pmatrix} 1 \\ \sqrt{2} \\ 1 \end{pmatrix}$$

$$\begin{pmatrix} 2 & -1 & 0 \\ -1 & 2 & -1 \\ 0 & -1 & 2 \end{pmatrix}\begin{pmatrix} 1 \\ 0 \\ -1 \end{pmatrix} = (2)\begin{pmatrix} 1 \\ 0 \\ -1 \end{pmatrix}$$

$$\begin{pmatrix} 2 & -1 & 0 \\ -1 & 2 & -1 \\ 0 & -1 & 2 \end{pmatrix}\begin{pmatrix} 1 \\ -\sqrt{2} \\ 1 \end{pmatrix} = (2 + \sqrt{2})\begin{pmatrix} 1 \\ -\sqrt{2} \\ 1 \end{pmatrix}$$

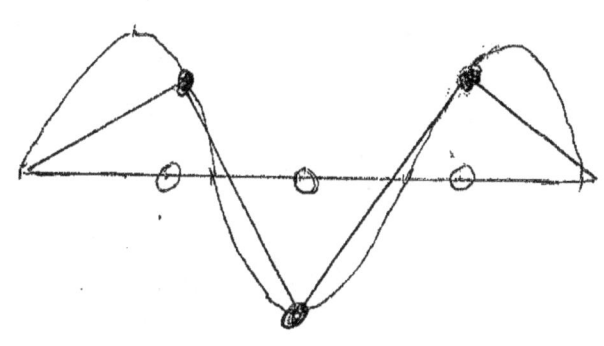

Now take the inner product of any two different vectors.

$$(1 \quad 0 \quad -1)\begin{pmatrix} 1 \\ \sqrt{2} \\ 1 \end{pmatrix} = 0 \qquad (1 \quad 0 \quad -1)\begin{pmatrix} 1 \\ -\sqrt{2} \\ 1 \end{pmatrix} = 0 \qquad (1 \quad \sqrt{2} \quad 1)\begin{pmatrix} 1 \\ -\sqrt{2} \\ 1 \end{pmatrix} = 0$$

The eigenvectors are all perpendicular to each other.

Using $|x_1\rangle = \begin{pmatrix} 1 \\ 0 \\ 0 \end{pmatrix}$, $|x_2\rangle = \begin{pmatrix} 0 \\ 1 \\ 0 \end{pmatrix}$, $|x_3\rangle = \begin{pmatrix} 0 \\ 0 \\ 1 \end{pmatrix}$, as coordinate axes, the eigenvectors

giving the shapes of the vibrating objects on the strings look as shown.

Example:

$$\begin{pmatrix} 1 \\ \sqrt{2} \\ 1 \end{pmatrix} = \begin{pmatrix} 1 \\ 0 \\ 0 \end{pmatrix} \cdot 1 + \begin{pmatrix} 0 \\ 1 \\ 0 \end{pmatrix} \cdot \sqrt{2} + \begin{pmatrix} 0 \\ 0 \\ 1 \end{pmatrix} \cdot 1$$

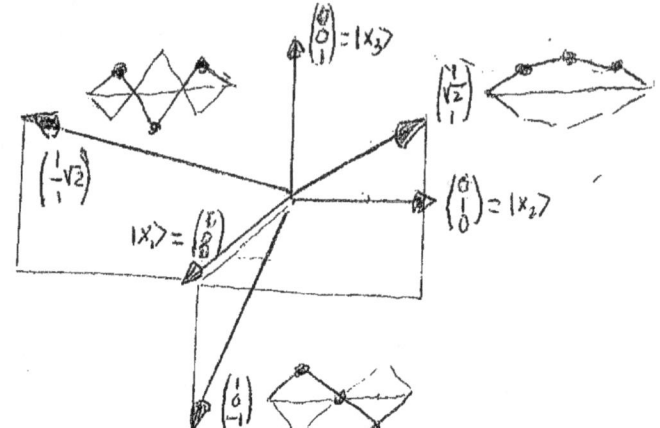

If $|x_1\rangle$, $|x_2\rangle$, and $|x_3\rangle$ isolate things that happen at x_1, x_2, and x_3, any other vector gives a shape of the string joining those points.

If $A(x)$ is the amplitude of vibration at distance, x, from the left end of the string, the displacement at any time, t, is $y(x,t) = A(x)\,sin\,(\omega\,t)$. Let us now have five objects on the string located at $x = L/6$, $x = L/3$, $x = L/2$, $x = 2L/3$, and $x = 5L/6$. At $x = L$, $A(L) = 0$, as that is the end of the string.

With one loop, $A(x) = sin\left(\frac{\pi x}{L}\right)$, giving an eigenvector, with $k = \frac{m\,d\,\omega^2}{F}$, :

$$|1\ loop\rangle = \begin{pmatrix} 0.5 \\ 0.866 \\ 1.000 \\ 0.866 \\ 0.5 \end{pmatrix} \text{ with } k = 0.268.$$

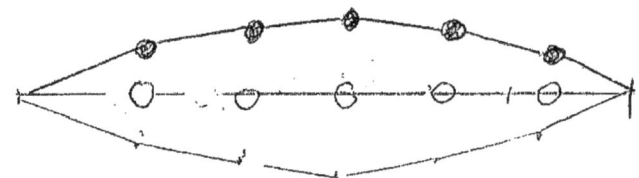

With two loops, $A(x) = sin\left(\frac{2\pi x}{L}\right)$:

$$|2\ loops\rangle = \begin{pmatrix} 0.866 \\ 0.866 \\ 0 \\ -0.866 \\ -0.866 \end{pmatrix} \text{ with } k = 1.000.$$

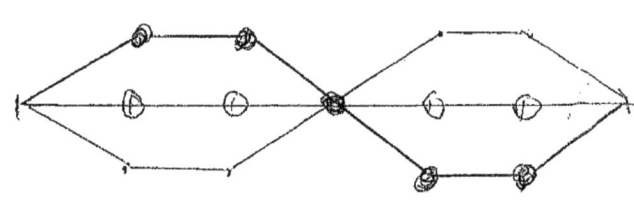

With three loops, $A(x) = sin\left(\frac{3\pi x}{L}\right)$:

$$|3\ loops\rangle = \begin{pmatrix} 1 \\ 0 \\ -1 \\ 0 \\ 1 \end{pmatrix} \text{ with } k = 2.000.$$

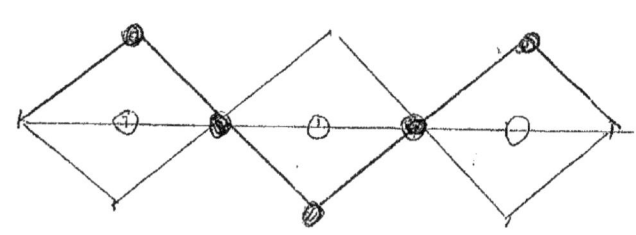

With four loops, $A(x) = sin\left(\frac{4\pi x}{L}\right)$:

$$|4\ loops\rangle = \begin{pmatrix} 0.866 \\ -0.866 \\ 0 \\ 0.866 \\ -0.866 \end{pmatrix} \text{ with } k = 3.000.$$

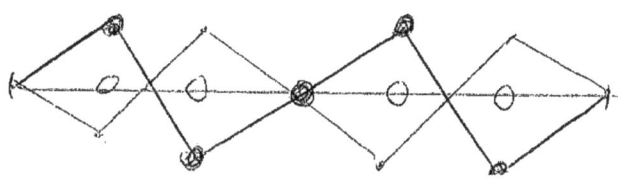

With five loops, $A(x) = sin\left(\frac{5\pi x}{L}\right)$:

$$|5\ loop\rangle = \begin{pmatrix} 0.5 \\ -0.866 \\ 1 \\ -0.866 \\ 0.5 \end{pmatrix} \text{ with } k = 3.732.$$

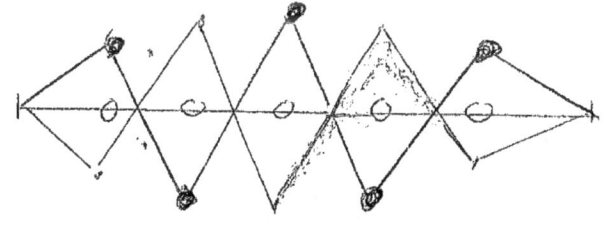

We cannot draw a picture of a five dimension space. However, any one of these vectors gives the amplitudes of the vibrating string. Again, each vector gives a shape. You will notice that the inner product of any two different vectors is zero, such as with $\langle 2\ loops\,|\,4\ loops\rangle = 0$.

It appears that we can write a generalized eigenvector. Let N be the total number of vibrating particles in a length, L. This gives $N+1$ gaps. The length of any gap is $L/(N+1)$. The nth particle will be located at $x_n = nL/(N+1)$.

Let α (alpha) be the number of loops in the sine wave passing through the particle displacements. $A_\alpha(x) = A\ sin(\alpha\pi x/L)$ will be the sine wave with α loops. At location, x, the displacement is $y(x,t) = A_\alpha(x)\ sin(\omega t)$. At the nth location, $A_\alpha(x_n) = A\ sin\left[\frac{\alpha\pi n L}{(N+1)L}\right] = A\ sin\left[\frac{\alpha\pi n}{N+1}\right]$.

In the following discussion we need to refer to an identity involving sine and cosine functions. Theta, θ, and phi, ϕ, are two numbers.

$sin(\theta + \phi) = sin\,\theta\,cos\,\phi + cos\,\theta\,sin\,\phi$ and $sin(\theta - \phi) = sin\,\theta\,cos\,\phi - cos\,\theta\,sin\,\phi$.

Thus, $sin(\theta + \phi) + sin(\theta - \phi) = 2\,sin\,\theta\,cos\,\phi$.

Also, $cos(2\theta) = cos^2\theta - sin^2\theta$ or $cos(2\theta) = 1 - 2\,sin^2\theta$ giving $1 - cos(2\theta) = 2\,sin^2\theta$.

Having gotten a general formula for the eigenvector giving the shape of particles vibrating on a string with some frequency, f, which is in vibrations per second, what is the formula for the frequency that goes with a particular shape? We defined an angular frequency which was $2\pi f$ and then lumped together a bunch of letters in $k = \dfrac{M\omega^2 d}{F}$, with

$$k\,A_n = -A_{n-1} + 2\,A_n - A_{n+1}, \text{ but } A_n = A\,\sin\left(\frac{\alpha\pi n}{N+1}\right).$$

α (alpha) is the number of half-wavelengths in the length, L.

$$(k-2)\,A_n = -[A_{n+1} + A_{n-1}] \qquad\qquad A_{n+1} + A_{n-1} = A\,\sin\left(\frac{\alpha\pi(n+1)}{N+1}\right) + A\,\sin\left(\frac{\alpha\pi(n-1)}{N+1}\right)$$

But, $\sin\left[\dfrac{\alpha\pi n}{N+1} + \dfrac{\alpha\pi}{N+1}\right] + \sin\left[\dfrac{\alpha\pi n}{N+1} - \dfrac{\alpha\pi}{N+1}\right] = 2\,\sin\left[\dfrac{\alpha\pi n}{N+1}\right]\cos\left[\dfrac{\alpha\pi}{N+1}\right].$

But, $A\,\sin\left(\dfrac{\alpha\pi n}{N+1}\right) = A_n$. Hence, $(k-2)\,A_n = -2\,\cos\left[\dfrac{\alpha\pi}{N+1}\right] A_n.$

Finally, $k = 2\left[1 - \cos\left(\dfrac{\alpha\pi}{N+1}\right)\right].$

Then, using an identity above, $k = 2\left[2\,\sin^2\left(\dfrac{\alpha\pi}{2(N+1)}\right)\right].$ But, $\omega_\alpha{}^2 = \dfrac{kF}{Md}$ giving

$$\omega_\alpha = 2\frac{F}{Md}\,\sin\left[\frac{\alpha\pi}{2(N+1)}\right].$$

The angular frequency for vibrations with α half wavelength in the length, L. Check these values for k against the calculations for $n = 5$ on the previous page.

Next, let us look at $A_n = A\,\sin\left(\dfrac{\alpha\pi n}{N+1}\right)$ noticing that d is the distance between particles, $(N+1)\,d = L$, the length of the string and $n\,d = x_n$, the location of the nth particle. Multiplying and dividing "the angle" function by d gives the displacement at location x_n to be $A(x_n) = A\,\sin\left(\dfrac{\alpha\pi x_n}{L}\right).$

This leads into the next discussion. If the number N of particles becomes very large, each particle becoming very small, we approach the case of standing waves in a continuous string (guitar or piano).

As we go from countable to un-countable number of particles, matrix algebra merges into calculus; the two forms of mathematics being merely two aspects of linear operator theory, the mathematics of quantum mechanics.

At this point, we take a momentary break from the matrix formulation and go back to the start with Newton's Law.

If there are many-many particles with decreasing mass per particle the string approaches being a continuum and matrix algebra becomes calculus. Later we shall find that both forms of mathematics are parts of something known as linear operator theory.

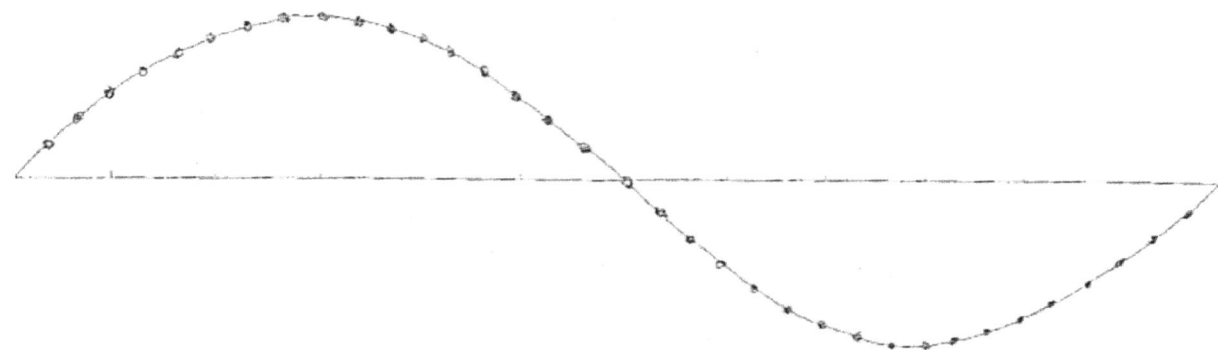

For very many, very small balls the problem approaches that for a continuous string. Recall that d is the jump in x from one ball to the next one; hence in the previous notation d is dx. $y_{n+1} - y_n$ is the change of y on the right side of the nth ball. $\frac{y_{n+1} - y_n}{d} = \left(\frac{dy}{dx}\right)_R$ which is the slope of the string to the right of the nth ball. $\frac{y_n - y_{n-1}}{d} = \left(\frac{dy}{dx}\right)_L$ is the slope of the string to the left of the nth ball. $\left(\frac{dy}{dx}\right)_R - \left(\frac{dy}{dx}\right)_L = d\left(\frac{dy}{dx}\right)$ is the change of the slope from the left of the nth ball to the right of it. Putting this into the general differential equation for a vibrating string and then dividing both sides of the equality by $d = dx$ gives $\frac{M}{d}\left(\frac{d^2 y_n}{dt^2}\right) = F\frac{d\left(\frac{dy}{dx}\right)}{dx}$ or $\frac{M}{d}\left(\frac{d^2 y_n}{dt^2}\right) = F\left(\frac{d^2 y}{dx^2}\right)$. As M is the mass in a length, d, of the string, $\frac{M}{d}$ is a mass per unit length, or linear density, to which we shall assign the symbol, μ. As we now have a displacement, y, at almost every x, we will drop the subscript, n. Then one slight change in notation is necessary when we have a function of several independent variables. $\frac{dy}{dx}$ is the slope of y with a change of x at a fixed moment in time. $\frac{dy}{dt}$ is the velocity of y with changing time at a fixed point of x.

To indicate that the function is changing with respect to one variable, while other variables are constant, we round off the d's to ∂'s.

This gives us, $\mu \dfrac{\partial^2 y}{\partial t^2} = F \dfrac{\partial^2 y}{\partial x^2}$ as Newton's second law of motion for a vibrating continuous string. An α-loop standing wave would be described by

$$y_\alpha(x,t) = A\, sin\left(\frac{\alpha \pi x}{L}\right) cos\left(2\pi f_\alpha\, t\right).$$

The partial derivative is just a normal derivative with respect to the variable specified while the other variable is treated as being constant.

Hence, $\dfrac{\partial y}{\partial x} = \dfrac{\alpha \pi}{L} A\, cos\left(\dfrac{\alpha \pi x}{L}\right) cos\left(2\pi f_\alpha\, t\right)$ and $\dfrac{\partial y}{\partial t} = -2\pi f_\alpha A\, sin\left(\dfrac{\alpha \pi x}{L}\right) sin\left(2\pi f_\alpha\, t\right)$.

One more derivate gives $\dfrac{\partial^2 y}{\partial x^2} = -\left(\dfrac{\alpha \pi}{L}\right)^2 y$ and $\dfrac{\partial^2 y}{\partial t^2} = -\left(2\pi f_\alpha\right)^2 y$. Putting these

results into Newton's second law gives $f_\alpha = \dfrac{\alpha}{2L}\sqrt{\dfrac{F}{\mu}}$, where α is an integer

giving the number of loops.

Returning to pages 13-14 and 13-15, notice that for m and n being integers,

$$\langle\, m\ loops \mid n\ loops\,\rangle = \begin{array}{l} 0 \quad for\ m \neq n \\ 3 \quad for\ m = n \end{array}$$

For $m \neq n$, the $\mid m\ loops\rangle$ vectors and the $\mid n\ loops\rangle$ vectors are perpendicular

to each other. We introduce the Kronecker Delta, $\delta_{mn} = \begin{array}{l} 0 \quad if\ m \neq n \\ 1 \quad if\ m = n \end{array}$.

This will be used often in the following discussion of Fourier Analysis.

The serious study of mathematics starts with axiomatic set theory. A good start might be N. McCoy Introduction to Modern Algebra. Mathematical treatments of this chapter might be:

Paul R. Halmos: Finite Dimensional Vector Spaces

Paul R. Halmos: Introduction to Hilbert Space and the Theory of Spectral Multiplicity

R. V. Churchill: Fourier Series and Boundary Value Problems

Jordan: Linear Operators for Quantum Mechanics

Dennery and Krzywicki: Mathematics for Physicists

If we add two sine waves as shown in the picture, we get a new wave.

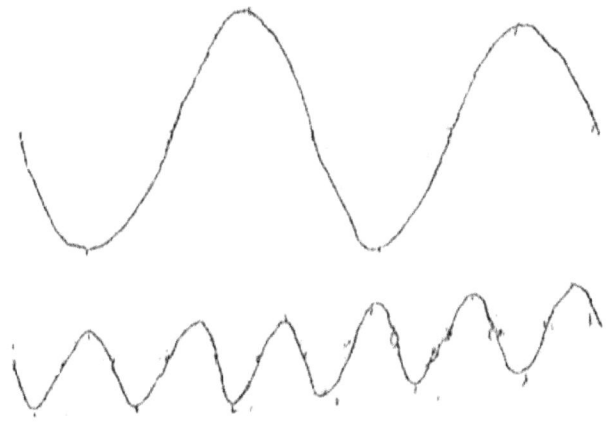

The new wave gotten here is periodic but has a new shape.

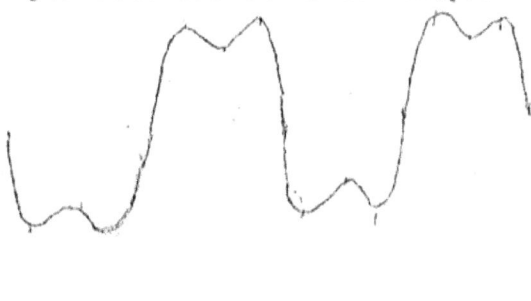

Obviously, if we keep adding waves of higher frequencies, the result would approach what is called a square wave.

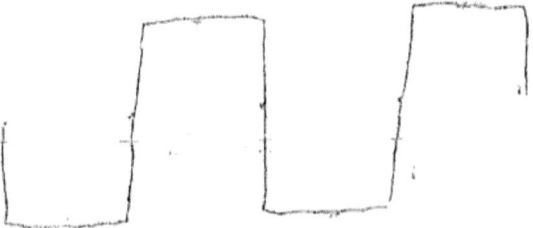

The mathematical method of describing a wave of arbitrary shape by adding sine and cosine functions was invented by the French mathematician Joseph Fourier and is called Fourier analysis.

Before see how this method works, there are facts of sine and cosine functions which must be proved.

Recall that $\cos(\theta + \varphi) = \cos\theta\cos\varphi - \sin\theta\sin\varphi$

and that $\cos(\theta - \varphi) = \cos\theta\cos\varphi + \sin\theta\sin\varphi$.

then $\sin\theta\sin\varphi = \frac{1}{2}\cos(\theta-\varphi) - \frac{1}{2}\cos(\theta+\varphi)$ and $\cos\theta\cos\varphi = \frac{1}{2}\cos(\theta-\varphi) + \frac{1}{2}\cos(\theta+\varphi)$

$\sin(\theta+2\pi) = \sin\theta\cos2\pi + \cos\theta\sin2\pi$ but $\cos2\pi = 1$ $\sin2\pi = 0$
$\sin(\theta+2\pi) = \sin\theta$

$\cos(\theta+2\pi) = \cos\theta\cos2\pi - \sin\theta\sin2\pi$ gives $\cos(\theta+2\pi) = \cos\theta$

Likewise, if n is an integer $\sin(\theta+n2\pi) = \sin\theta$ $\cos(\theta+n2\pi) = \cos\theta$

Now look at $\sin(\frac{n\pi x}{L})$ and notice that $\sin[\frac{n\pi(x+2L)}{L}] = \sin[\frac{n\pi x}{L} + n2\pi]$

$\sin(\frac{n\pi x}{L})$ or $\cos(\frac{n\pi x}{L})$ is periodic with a period $2L$ which suggests a wavelength $\lambda_n = \frac{2L}{n}$

Putting a number of these together $f(x) = \sum_{n=0}^{\infty}\left[A_n\cos(\frac{n\pi x}{L}) + B_n\sin(\frac{n\pi x}{L})\right]$

is a generalized function where $f(x+2L) = f(x)$

The question, then, is how do you get a set of A_n s and B_n s so that such a function can be described this way?

The way that we do the mathematics will be rigorous but without making any intuitive sense to you. We will test the final results with things that do make sense.

First, recall that $\cos\Theta = \frac{e^{i\Theta} + e^{-i\Theta}}{2}$ and $\sin\Theta = \frac{e^{i\Theta} - e^{-i\Theta}}{2i}$

Then $A\cos\Theta + B\sin\Theta = (\frac{A}{2} + \frac{B}{2i})e^{i\Theta} + (\frac{A}{2} - \frac{B}{2i})e^{-i\Theta}$ but $\frac{1}{i} = -i$

THen, let $C = \frac{A-iB}{2}$ and $C^* = \frac{A-iB}{2}$ its complex conjugate

NOw $A\cos\Theta + B\sin\Theta = Ce^{i\Theta} + C^*e^{-i\Theta}$

The periodic function is now $f(x) = \sum_{n=0}^{\infty}\left[C_n e^{in\pi x/L} + C_n^* e^{-in\pi x/L}\right]$

We can write $C_n^* = C_{-n}$

Notice that when n goes from 0 to $+\infty$, $-n$ goes from 0 to $-\infty$

Now $f(x) = F'(x+2L)$ when $f(x) = \sum_{n=-\infty}^{+\infty} C_n e^{in\pi x/L}$

For a given function $f(x)$ how do we find the coefficients C_n?

This will involve integrals of the form, $\int^{L} e^{ikx}dx$

$d(e^{ikx}) = ik\,e^{ikx}dx$, $e^{ikx}dx = d(\frac{e^{ikx}}{ik})$ $x=-L$

$\int_{x=-L}^{+L} e^{ikx}dx = \frac{e^{ikL} - e^{-ikL}}{ik}$ or $\int_{x=-L}^{+L} e^{ikx}dx = \frac{2i\sin(kL)}{ik}$

If m and n are two integers $\int_{x=-L}^{+L} e^{in\pi x/L}e^{-im\pi x/L}dx = \frac{2\sin(n-m)\pi}{(n-m)\pi/L}$

If m and n are integers, $\sin(n-m)\pi = 0$

If m and n are integers and $m \neq n$ $\int_{x=-L}^{+L} e^{in\pi x/L} e^{-im\pi x/L} dx = \delta_{nm}$

If $m = n$, $e^0 = 1$ and $\int_{x=-L}^{+L} e^0 dx = 2L$

We introduce the Kronecker delta $_{mn}$ is zero for $m \neq n$ and one when $m = n$

We end up with $\int_{x=-L}^{+L} e^{in\pi x/L} e^{-im\pi x/L} dx = 2L \delta_{mn}$

Return now to the function $f(x) = \sum_{n=-\infty}^{+\infty} C_n e^{in\pi x/L}$

Pick one of the ns in the summation and call it m.

Multiply each side by $e^{-im\pi x/L} dx$ and integrate from $-L$ to $+L$

$$\int_{x=-L}^{+L} f(x) e^{-im\pi x/L} dx = \sum_{n=-\infty}^{\infty} C_n \int_{x=-L}^{+L} e^{in\pi x/L} e^{-im\pi x/L} dx = \sum_{n=-\infty}^{\infty} 2L C_n \delta_{mn}$$

Every term is multiplied by zero except C_m

Thus $C_m = \frac{1}{2L} \int_{x=-L}^{+L} f(x) e^{-im\pi x/L} dx$

The evaluation of these integrals is not often easy. Several of them will be demonstrated a bit later in this chapter.

Here we take a break to discuss a notation we have seen before. Quantum physics is done in an abstract space called a Hilbert Space. It is not really necessary that we know what this looks like. We will say here a few words about what our Dirac notation looks like when referring to vectors (arrows) in normal space,

The numbers locating a point in three dimensions can be written $\begin{pmatrix} x \\ y \\ z \end{pmatrix}$

or (x,y,z). Then $(x,y,z) \begin{pmatrix} x \\ y \\ z \end{pmatrix} = x^2 + y^2 + z^2$

It is obvious how you do the arithmetic.

In the rest of this chapter, we will be exploring our evolving understanding of quantum physics. It is only right that we do it in the language of 21st century physics.

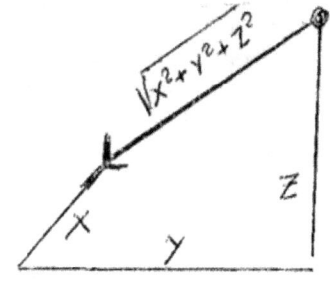

When we write the set of numbers in a row, we have a bra vector*
$\langle x| = (x_1\ x_2\ x_3)$. When we write the numbers in a column, we have a ket
vector $\qquad |x\rangle = \begin{pmatrix} x_1 \\ x_2 \\ x_3 \end{pmatrix}$ Take the vectors $|a\rangle = \begin{pmatrix} 5 \\ 7 \\ 9 \end{pmatrix}$ and $|b\rangle = \begin{pmatrix} 8 \\ 2 \\ 3 \end{pmatrix}$

We can add them $\qquad |a\rangle + |b\rangle = \begin{pmatrix} 13 \\ 9 \\ 12 \end{pmatrix}$ and subtract them $|a\rangle - |b\rangle = \begin{pmatrix} -3 \\ 5 \\ 6 \end{pmatrix}$

by adding and subtracting corresponding terms.

We can multiply a bra vector times a ket vector to give an inner product of
bra(c)ket, which product is a number. For example

$$\langle a|b\rangle = (5,7,9)\begin{pmatrix} 8 \\ 2 \\ 3 \end{pmatrix} = 5X8 + 7X2 + 9X3 = 81$$

we will define three mutually perpendicular vectors of unit length.

$$|u_1\rangle = \begin{pmatrix} 1 \\ 0 \\ 0 \end{pmatrix} \qquad |u_2\rangle = \begin{pmatrix} 0 \\ 1 \\ 0 \end{pmatrix} \qquad |u_3\rangle = \begin{pmatrix} 0 \\ 0 \\ 1 \end{pmatrix}$$

where $\langle u_i | u_j \rangle = \delta_{ij}$, the Kronecker delta which is 1 if $i = j$ and 0 if $i \neq j$.

If we take some vector $|\Psi\rangle = \begin{pmatrix} \Psi_1 \\ \Psi_2 \\ \Psi_3 \end{pmatrix}$ then $\langle u_1|\Psi\rangle = \Psi_1$, $\langle u_2|\Psi\rangle = \Psi_2$, $\langle u_3|\Psi\rangle = \Psi_3$

and $|\Psi\rangle = \begin{pmatrix} \Psi_1 \\ 0 \\ 0 \end{pmatrix} + \begin{pmatrix} 0 \\ \Psi_2 \\ 0 \end{pmatrix} + \begin{pmatrix} 0 \\ 0 \\ \Psi_3 \end{pmatrix} = |u_1\rangle\langle u_1|\Psi\rangle + |u_2\rangle\langle u_2|\Psi\rangle + |u_3\rangle\langle u_3|\Psi\rangle$

which can be written in the shorter form $|\Psi\rangle = \sum_{i=1}^{3} |u_i\rangle\langle u_i|\Psi\rangle$.

The vectors $|u_1\rangle$, $|u_2\rangle$, $|u_3\rangle$ define a set of three mutually perpendicular
axes and are called a BASIS for our space. $\sum_{i=1}^{3} |u_i\rangle\langle u_i|\Psi\rangle$ is called the

representation of the vector $|\Psi\rangle$ in the base vectors $|u_i\rangle$. It follows that
an n-dimensional space id defined by a set of n base vectors $|u_i\rangle$ where
$\langle u_i|u_j\rangle = \delta_{ij}$, in which any vector could be represented by $\sum_{i=1}^{n} |u_i\rangle\langle u_i|\Psi\rangle$

What does this look like in three dimensions?
The base vectors give a set of three axes
that define the directions in a three dimensioned space. Any vector in that space is
gotten by three displacements in the directions of the axes. In the following mathematics, spaces will have more than three dimensions and we cannot draw a picture of what
is going on, but you get the idea.

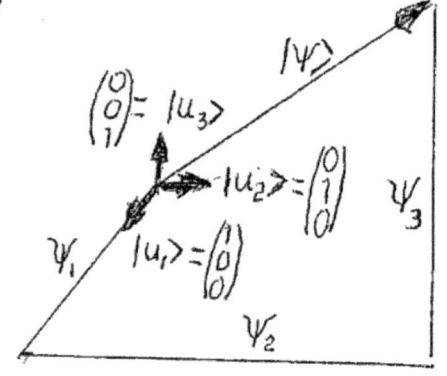

For a moment let us return to the representation of a vector $|\Psi\rangle$ in the base vectors $|u_i\rangle$, namely $|\Psi\rangle = \sum_{i=1}^{3} |u_i\rangle\langle u_i|\Psi\rangle$. This would suggest that $\sum_{i=1}^{3} |u_i\rangle\langle u_i|$ is some sort of an identity or identity operator.

Recall that $|u_1\rangle = \begin{pmatrix} 1 \\ 0 \\ 0 \end{pmatrix}$ $|u_2\rangle = \begin{pmatrix} 0 \\ 1 \\ 0 \end{pmatrix}$ and $|u_3\rangle = \begin{pmatrix} 0 \\ 0 \\ 1 \end{pmatrix}$. From these we get

$$\sum_{i=1}^{3} |u_i\rangle\langle u_i| = \begin{pmatrix} 1 \\ 0 \\ 0 \end{pmatrix}(1,0,0) + \begin{pmatrix} 0 \\ 1 \\ 0 \end{pmatrix}(0,1,0) + \begin{pmatrix} 0 \\ 0 \\ 1 \end{pmatrix}(0,0,1)$$

If these products are treated as matrix products, then taking rows in the left hand terms and columns in the right hand terms, we obtain the products

$$\begin{pmatrix} 1 \\ 0 \\ 0 \end{pmatrix}(1,0,0) = \begin{pmatrix} 1 & 0 & 0 \\ 0 & 0 & 0 \\ 0 & 0 & 0 \end{pmatrix} \quad \begin{pmatrix} 0 \\ 1 \\ 0 \end{pmatrix}(0,1,0) = \begin{pmatrix} 0 & 0 & 0 \\ 0 & 1 & 0 \\ 0 & 0 & 0 \end{pmatrix} \quad \begin{pmatrix} 0 \\ 0 \\ 1 \end{pmatrix}(0,0,1) = \begin{pmatrix} 0 & 0 & 0 \\ 0 & 0 & 0 \\ 0 & 0 & 1 \end{pmatrix}$$

which gives $\sum_{i=1}^{3} |u_i\rangle\langle u_i| = \begin{pmatrix} 1 & 0 & 0 \\ 0 & 1 & 0 \\ 0 & 0 & 1 \end{pmatrix}$ which is the unit or identity matrix.

In the following developments, the notation will stand for very abstract spaces but you need not bother about what they look like. For every number x of a variable x there will be a unit vector $|x\rangle$ which will serve as a base vector for one of the "axes" in that space a function will be written $f(x) = \langle x|f\rangle$

Continuing the waveform analysis, instead of integers m and n we will use alpha (α) and beta (β) to stress that they are integers.

For a function of x we write $f(x) = \langle x|f\rangle$ example being $\langle x|\alpha\rangle = e^{i\alpha\pi x/L}$ for which the complex conjugate is $\langle\alpha|x\rangle = e^{-i\alpha\pi x/L}$

The integral of the product is $\langle\beta|\alpha\rangle = \int_{x=-L}^{+L} \langle\beta|x\rangle\langle x|\alpha\rangle\, dx$

Since it is x that is varying for a fixed β we factor the $\langle\beta|$ from the left side. leaving factoring from the right $|\alpha\rangle = \int_{x=-L}^{+L} |x\rangle\langle x|\alpha\rangle\, dx$

$$\int_{x=-L}^{+L} |x\rangle\langle x|\, dx = \hat{1}$$ which is some sort of an identity

Another change in notation. Going back to chapter 15, we will use α (alph) and β (beta) as integers. Also recall the wave number for a wave of wavelength λ . $k = \frac{2\pi}{\lambda}$ and if $\lambda_\alpha = \frac{2L}{\alpha}$, $k_\alpha = \frac{2\pi\alpha}{2L} = \frac{\pi\alpha}{L}$

Then we can write $e^{i\alpha\pi x/L} = e^{ik_\alpha x}$

Calling things $|\Psi\rangle$ vectors, any vector in the space thusly spanned has the representation

$$|\Psi\rangle = \int_{x=-L}^{+L} |x\rangle \langle x|\Psi\rangle \, dx$$

In this space, define the set of vectors $|k_\alpha\rangle$ such that $\langle x|k_\alpha\rangle = \sqrt{\frac{1}{2L}}\, e^{i\alpha\pi x/L}$ with α taking on all integer values on $-\infty \le \alpha \le +\infty$. In the column or ket representation of $|k_\alpha\rangle$ are now complex numbers. In order that the square of the length of the vector have its usual interpretation as a real positive number — $\langle k_\alpha|K_\alpha\rangle$= positive real number — we state for such vectors the following rule. WHEN A VECTOR HAS A COLUMN REPRESENTATION IN WHICE THE NUMBERS ARE COMPLEX, THE ASSOCIATED ROW REPRESENTATION WILL HAVE THE COMPLEX CONJUGATES OF THE NUMBERS IN THE COLUMN. Hence $\langle k_\alpha|x\rangle = \sqrt{\frac{1}{2L}}\, e^{-i\alpha\pi x/L}$.

<u>Now, for</u> the α th and β th members of this set of vectors we form the inner product which is easily integrated to give

$$\langle k_\alpha|k_\beta\rangle = \int_{x=-L}^{+L} \frac{1}{2L} e^{-i\alpha\pi x/L}\, e^{i\beta\pi x/L}\, dx = \delta_{\alpha\beta}$$

This is interpreted to say that the vectors $|k_\alpha\rangle$ form a set of infinitely many unit length mutually perpendicular vectors. When vectors behaving this way have complex numbers in their representation, we say that they are UNITARY rather than orthogonal and normalized. So the vectors $|k_\alpha\rangle$ form a unitary set. Such a set of vectors may be used as a basis for the Hilbert space, defined on $-L \le x \le +L$. There are certain problems concerning completeness when we are in a Hilbert space. These will be considered later.

On the domain $-L \le x \le +L$, let $\Psi(x)$ be any complex function of x, with its complex conjugate $\Psi^*(x)$ satisfying the condition that $\int_{x=-L}^{+L} \Psi^*(x)\Psi(x)dx$ converges

that is, is a finite number, here positive and real.

In the Hilbert space on $-L \le x \le +L$, let there be a vector $|\Psi\rangle$ corresponding to the function $\Psi(x)$ such that $\Psi(x) = \langle x|\Psi\rangle$, and $|\Psi\rangle = \int_{x=-L}^{+L} |x\rangle \langle x|\Psi\rangle \, dx$

or $|\Psi\rangle = \int_{x=-L}^{+L} |x\rangle \, \Psi(x) \, dx$

This vector could also be represented in the base vectors $|k_\alpha\rangle$ to be written
$$|\Psi\rangle = \sum_{\alpha=-\infty}^{+\infty} |k_\alpha\rangle\langle k_\alpha|\Psi\rangle .$$ Let us denote the projection $\langle k_\alpha|\Psi\rangle$ by the
coefficient C_α giving $|\Psi\rangle = \sum_{\alpha=-\infty}^{+\infty} |k_\alpha\rangle C_\alpha$.

The coefficients C_α are easily obtained from the function $\Psi(x)$ by

$$C_\alpha = \langle k_\alpha|\Psi\rangle = \int_{x=-L}^{+L}\langle k_\alpha|x\rangle\langle x|\Psi\rangle\,dx = \int_{x=-L}^{+L}\sqrt{\frac{1}{2L}}\,e^{-i\alpha\pi x/L}\,\Psi(x)dx$$

Then in terms of the coefficients C_α the function $\Psi(x)$ can be written

$$\Psi(x) = \langle x|\Psi\rangle = \sum_{\alpha=-\infty}^{+\infty}\langle x|k_\alpha\rangle\langle k_\alpha|\Psi\rangle = \sum_{\alpha=-\infty}^{+\infty}\sqrt{\frac{1}{2L}}\,e^{i\alpha\pi x/L}\,C_\alpha$$

which brings us right back to where we started. You will wonder what use
there is for us to be going 'round and 'round in Hilbert space like this.
There are many physical uses for the transformation of axes that we have just
gotten. One of the most important of them is the Fourier analysis of wave
forms, or functions which periodically repeat the same basic shape over and
over again on the entire real axis. Let us digress a moment to see how that
works.

On the domain $-L \le x \le +L$, let x be replaced by ξ as in a previous problem.
In terms of $\Psi(\xi)$ on $-L \le \xi \le +L$, C_α is
$$C_\alpha = \int_{\xi=-L}^{+L}\sqrt{\frac{1}{2L}}\,e^{-i\alpha\pi\xi/L}\,\Psi(\xi)d\xi$$

and then
$$\Psi(x) = \sum_{\alpha=-\infty}^{+\infty}\sqrt{\frac{1}{2L}}\,e^{i\alpha\pi x/L}\,C_\alpha$$

This result was gotten for $\Psi(x)$ on $-L \le x \le +L$, however, using the C_α thusly
gotten, the representation for $\Psi(x)$ can now be extended for all values of x
on the real domain. It is sufficient to observe the behavior of $e^{i\alpha\pi x/L}$,
for any integer α, when x is increased — or decreased — by 2L. Hence
$e^{i\alpha\pi(x+2L)/L} = e^{i\alpha\pi x/L}e^{i\alpha 2\pi} = e^{i\alpha\pi x/L}$. Thus $\Psi(x) = \sum_{\alpha=-\infty}^{+\infty}\sqrt{\frac{1}{2L}}\,C_\alpha\,e^{i\alpha\pi x/L}$
has the property that $\Psi(x+2L) = \Psi(x)$. We say that $\Psi(x)$ is periodic with a
period of 2L . That is, whatever $\Psi(x)$ does on $-L \le x \le +L$, that particular
curve or shape is repeated over and over again everytime x changes by 2L .
The series $\sum_{\alpha=-\infty}^{+\infty}\sqrt{\frac{1}{2L}}\,C_\alpha\,e^{i\alpha\pi x/L}$ is called a FOURIER SERIES or the Fourier
representation of a periodic function, where the shape of one period is given
by $\Psi(\xi)$ on $-L \le \xi \le +L$, and is then repeated in $\Psi(x)$ everytime x increases or
decreases by 2L.

For an example, let us take the straight line $\Psi(\xi) = \xi$ on $-L \le \xi \le +L$. C_α is obtained from

$$C_\alpha = \sqrt{\frac{1}{2L}} \int_{\xi=-L}^{+L} \xi \, e^{-i\alpha\pi\xi/L} \, d\xi.$$ If we notice that

$$e^{-i\alpha\pi\xi/L} = \frac{d}{d\xi}\left[-\frac{L}{i\alpha\pi} e^{-i\alpha\pi\xi/L} \right]$$ we can integrate by parts to get

$$C_\alpha = -\frac{L}{i\alpha\pi}\sqrt{\frac{1}{2L}} \left[L e^{-i\alpha\pi} - (-L) e^{+i\alpha\pi} - \int_{\xi=-L}^{+L} e^{-i\alpha\pi\xi/L} \, d\xi \right]$$

The last integral is zero and you will recognize $e^{i\alpha\pi} = e^{-i\alpha\pi} = (-)^\alpha$. Then
$C_\alpha = \frac{(-)^{\alpha+1}}{i\alpha\pi} L \sqrt{2L}$ for all positive or negative integers α, but with some difficulties for $\alpha=0$. That can be taken care of by direct integration.

$$C_0 = \int_{\xi=-L}^{+L} \sqrt{\frac{1}{2L}} \, e^0 \, \xi \, d\xi = 0$$

From these we get the Fourier series

$$\Psi(x) = \sum_{\alpha=-\infty}^{-1} \frac{(-)^{\alpha+1} L\sqrt{2L}}{i\alpha\pi} \sqrt{\frac{1}{2L}} \, e^{i\alpha\pi x/L} + \sum_{\alpha=1}^{+\infty} \frac{(-)^{\alpha+1} L\sqrt{2L}}{i\alpha\pi} \sqrt{\frac{1}{2L}} \, e^{i\alpha\pi x/L}$$

This is very messy looking, and can be made quite simple by replacing α by $-\alpha$ everywhere in the first series over the negative integers, with the proper changes of signs for the limits of summation. Note that $(-)^{-(\alpha+1)} = (-)^{\alpha+1}$.

$$\Psi(x) = \sum_{\alpha=\infty}^{1} \frac{(-)^{-(\alpha+1)} L}{-i\alpha\pi} e^{-i\alpha\pi x/L} + \sum_{\alpha=1}^{+\infty} \frac{(-)^{\alpha+1} L}{i\alpha\pi} e^{i\alpha\pi x/L}$$

$$= \frac{L}{i\pi} \sum_{\alpha=1}^{+\infty} \frac{(-)^{\alpha+1}}{\alpha} \left[e^{i\alpha\pi x/L} - e^{-i\alpha\pi x/L} \right]$$ but $e^{i\alpha\pi x/L} - e^{-i\alpha\pi x/L} = 2i\sin\frac{\alpha\pi x}{L}$

giving $$\Psi(x) = \frac{2L}{\pi} \sum_{\alpha=1}^{+\infty} \frac{(-)^{\alpha+1}}{\alpha} \sin\frac{\alpha\pi x}{L} = \frac{2L}{\pi}\left[\sin\frac{\pi x}{L} - \frac{1}{2}\sin\frac{2\pi x}{L} + \frac{1}{3}\sin\frac{3\pi x}{L} - \frac{1}{4}\sin\frac{4\pi x}{L} + \dots \right]$$

As more and more terms are added, this representation approaches the 'saw-tooth' wave shown below.

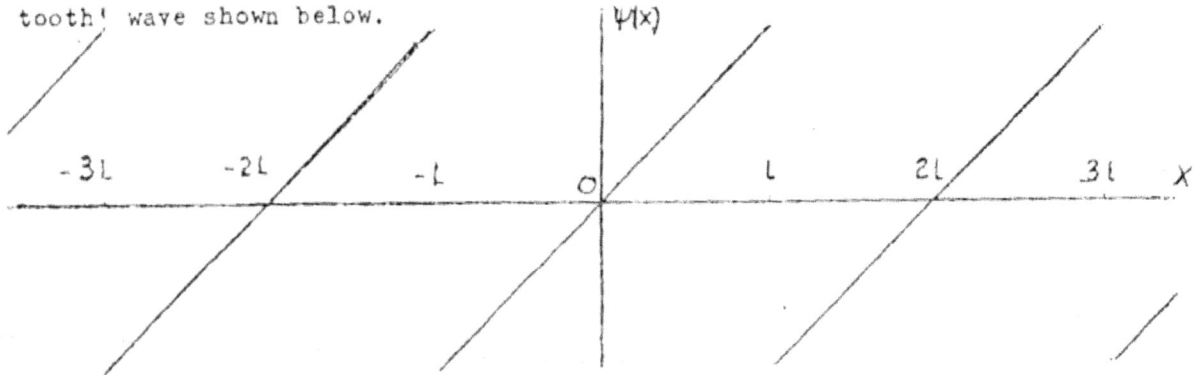

In the same manner the function $\Psi(\xi) = 0$, $-L < \xi < 0$, $\Psi(\xi) = 1$, $0 < \xi \leq +L$, gives rise to the Fourier representation

$$\Psi(x) = \frac{1}{2} + \frac{2}{\pi}\left[\sin\frac{\pi x}{L} + \frac{1}{3}\sin\frac{3\pi x}{L} + \frac{1}{5}\sin\frac{5\pi x}{L} + \frac{1}{7}\sin\frac{7\pi x}{L} + \ldots\right]$$

which repeats the box-like form over and over to give the well known square wave shown below.

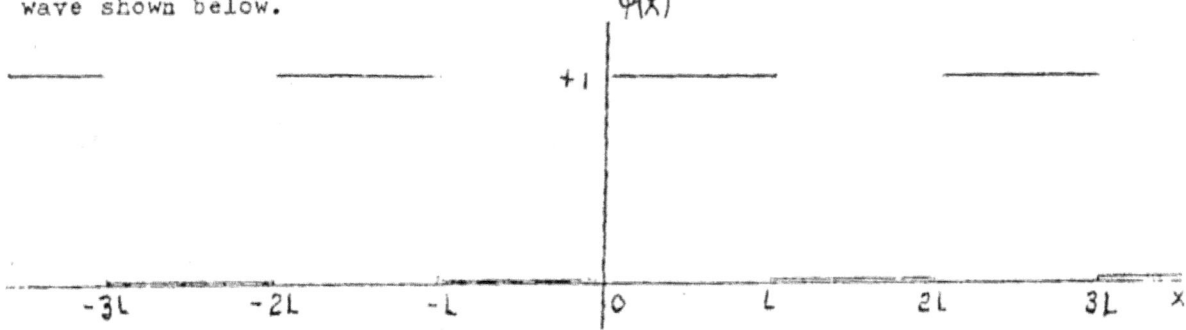

A non-periodic function $\Psi(x)$ on $-\infty \leq x \leq +\infty$ for which $\int_{-\infty}^{+\infty}\Psi^*(x)\Psi(x)\,dx$

converges can be represented as a sum of sine – or cosine – waves. Let us for a moment return to a $\Psi(x)$ that is square integrable on $-L \leq x \leq +L$. We write

$$\Psi(x) = \sum_{\alpha=-\infty}^{+\infty}\sqrt{\frac{1}{2L}}\,e^{i\alpha\pi x/L}\int_{\xi=-L}^{+L}\sqrt{\frac{1}{2L}}\,e^{-i\alpha\pi\xi/L}\Psi(\xi)\,d\xi$$

We desire the limit of this expression as $L \to \infty$. But first, let us rearrange the summations, and, just for kicks, multiply and divide the whole thing by π, which doesn't change the value of anything. Then we write

$$\Psi(x) = \frac{1}{2\pi}\sum_{\alpha=-\infty}^{+\infty}\int_{\xi=-L}^{+L}e^{i\alpha\pi x/L}\,e^{-i\alpha\pi\xi/L}\Psi(\xi)\,d\xi\,\frac{\pi}{L}$$

Now introduce the quantity $k_\alpha = \frac{\alpha\pi}{L}$. As α is an integer, it changes by one at a time, thus the difference between consecutive values of k_α is $\Delta k = \frac{\pi}{L}$. Hence,

$$\Psi(x) = \frac{1}{2\pi}\sum_{\alpha=-\infty}^{+\infty}\int_{\xi=-L}^{+L}e^{ik_\alpha x}\,e^{-ik_\alpha\xi}\Psi(\xi)\,d\xi\,\Delta k$$

Then, for $\Psi(x)$ square integrable on $-\infty \leq x \leq +\infty$, let $L \to \infty$ in the above summation. In the limit, k becomes a continuous variable and so we drop the index α, the summation over α becoming an integral over k. Then

$$\Psi(x) = \frac{1}{2\pi}\int_{k=-\infty}^{+\infty}\int_{\xi=-\infty}^{+\infty}e^{ikx}\,e^{-ik\xi}\Psi(\xi)\,d\xi\,dk$$

This is unscrambled in the following manner. We define a space on the real domain, so that for every x on $-\infty \leq x \leq +\infty$ we define a vector $|x\rangle$, as in the previous discussion. To the square integrable function $\Psi(x)$ we assign a vector $|\Psi\rangle$ in that space with the interpretation that $\Psi(x) = \langle x|\Psi\rangle$. Now, to each real number k, we assign a member of a set of vectors $|k\rangle$. These are defined so that $\langle x|k\rangle = \sqrt{\frac{1}{2\pi}}e^{ikx}$ and, according to the rule for vectors with complex components, $\langle k|x\rangle = \sqrt{\frac{1}{2\pi}}e^{-ikx}$. In terms of the vectors $|k\rangle$ the vector $|\Psi\rangle$ will have a representation such that

$$\langle k|\Psi\rangle = \int_{x=-\infty}^{+\infty} \langle k|x\rangle\langle x|\Psi\rangle \, dx$$

which is $\langle k|\Psi\rangle = \sqrt{\frac{1}{2\pi}} \int_{x=-\infty}^{+\infty} e^{-ikx}\Psi(x)dx$

This is called the FOURIER TRANSFORM of $\Psi(x)$. In terms of it we can write the inverse Fourier transform of $\Psi(x)$ as

$$\langle x|\Psi\rangle = \int_{k=-\infty}^{+\infty} \langle x|k\rangle\langle k|\Psi\rangle \, dk$$

or $\Psi(x) = \sqrt{\frac{1}{2\pi}} \int_{k=-\infty}^{+\infty} e^{ikx}\langle k|\Psi\rangle \, dk$

In this result it is seen that a non-periodic, but square integrable, function on $-\infty \leq x \leq +\infty$ does have a Fourier representation as the sum of sine, or cosine, waves, and $k\Psi$, the Fourier transform of $\Psi(x)$ is nothing more than the amplitude of the wave e^{ikx} in that representation. Obviously, k is a frequency sort of quantity. A function on a finite domain $-L \leq x \leq +L$ was represented in terms of a denumerable set of frequencies. A function on the entire real domain is represented by a summation over a continuous set of frequencies.

The Fourier analysis, that has just been developed, can be applied directly to quantum mechanics. The states of an object moving in one dimension, subject to an harmonic oscillator potential, was studied in the third lecture. There, for the position operator, Q the eigenvalues x and eigenvectors $|x\rangle$ satisfied $Q|x\rangle = x|x\rangle$. In terms of the position eigenvectors, the momentum operator P had a representation as the derivative with respect to x such that $\langle x|P|\Psi\rangle = \frac{\hbar}{i}\frac{\partial}{\partial x}\langle x|\Psi\rangle$. Now, the momentum operator will have eigenvalues p and eigenvectors $|p\rangle$ such that $P|p\rangle = p|p\rangle$. You will recall that $\langle x|p\rangle$ is the probability amplitude for an object with momentum p to be found in the neighborhood of the position x. The probability of finding the particle between x and $x+\Delta x$ is $|\langle x|p\rangle|^2\Delta x$. Projecting $P|p\rangle = p|p\rangle$ onto the bra $\langle x|$ gives $\frac{\hbar}{i}\frac{\partial}{\partial x}\langle x|p\rangle = p\langle x|p\rangle$. This gives $\langle x|p\rangle = C\,e^{ipx/\hbar}$.

The probability amplitude for a particle in the momentum state p is, hence, a wave. You will recall that a wave, with wavelength λ, has the form $\Psi(x) = C\,e^{i2\pi x/\lambda}$. Thus, we see that a particle with a momentum p has an associated probability wave with wavelength $\lambda = \frac{2\pi\hbar}{p} = \frac{h}{p}$, which is exactly the De Broglie prediction of 1924.*

In quantum mechanics, it is convenient to write, instead of p, the quantity $k = p/\hbar$. This has the units of reciprocal length. In classical physics it is called 'wave number', but in quantum physics we just call it momentum. We can write $|p\rangle$ or $|k\rangle$ interchangeably for the momentum eigenket, and $\frac{\partial}{\partial x}\langle x|k\rangle = ik\langle x|k\rangle$ giving $\langle x|k\rangle = C\,e^{ikx}$.

This leads to another experimental interpretation. Consider a pulsed particle accelerator, emitting a bunch of particles with momentum k_0 where the particles are confined to a pulse of length $2L$. The state vector for a particle in the pulse is $|\Psi\rangle$ such that $\langle x|\Psi\rangle = 0$ for points x not in the domain of the pulse, and $\langle x|\Psi\rangle = \sqrt{\frac{1}{2L}}\,e^{ik_0x}$ for points inside the pulse. We have normalized the wave function so that

$$\int_{x=-\infty}^{+\infty} \langle\Psi|x\rangle\langle x|\Psi\rangle\,dx = 1$$

Let the coordinate origin, for specifying the position of a particle, be placed at the center of the pulse. Now the wave function is

$$\langle x|\Psi\rangle = \begin{cases} 0 & \text{for } x < -L \\ \sqrt{\frac{1}{2L}}\,e^{ik_0x} & \text{for } -L \leq x \leq +L \\ 0 & \text{for } x > +L \end{cases}$$

As this function is square integrable on $-\infty \leq x \leq +\infty$, it is instructive to take its Fourier transform. This will show that the momentum of the particle is not as well defined as we thought it was.

The Fourier transform of $\langle x|\Psi\rangle = \sqrt{\frac{1}{2L}}\,e^{ik_0x}$ on $-L \leq x \leq +L$ with $\langle x|\Psi\rangle = 0$ elsewhere is

$$\langle k|\Psi\rangle = \sqrt{\frac{1}{4\pi L}}\int_{x=-\infty}^{+\infty} e^{-ikx}\langle x|\Psi\rangle\,dx = \sqrt{\frac{1}{4\pi L}}\int_{x=-L}^{+L} e^{i(k_0-k)x}\,dx = \sqrt{\frac{L}{\pi}}\,\frac{\sin(k_0-k)L}{(k_0-k)L}$$

This, in return, gives

$$\langle x|\Psi\rangle = \sqrt{\frac{1}{2\hbar}}\int_{k=-\infty}^{+\infty}\langle k|\Psi\rangle\,e^{ikx}$$

Now, as $\langle k|\Psi\rangle$ has considerable value for $k \neq k_0$, $\langle x|\Psi\rangle$ is represented by a sum over many waves with k other than k_0. In the quantum mechanical interpretation, each of the values of k is a momentum, and, hence, $\langle k|\Psi\rangle$ is an amplitude for the particle in the pulsed accelerator beam to have a momentum other than k_0. We interpret $|\langle k|\Psi\rangle|^2 \Delta k$ as the probability that the particle will have a momentum between k and $k+\Delta k$. We plot $|\langle k|\Psi\rangle|^2$ as a function of k in the figure above.

The maximum probability is for $k = k_o$. However, the central region of probabilities is spread between $k = k_o - \pi/L$ and $k = k_o + \pi/L$. The particle's momentum is likely to be anything between these values giving an uncertainty in the momentum $\Delta k = (k_o + \frac{\pi}{L}) - (k_o - \frac{\pi}{L}) = \frac{2\pi}{L}$ or $\Delta p = \frac{2\pi\hbar}{L}$.

The particle is known to be somewhere in a pulse of length $2L$ so we can say that the uncertainty of the particle's position is $\Delta x = 2L$ giving

$$\Delta x \, \Delta p = 4\pi\hbar$$

which, you will recognize, is Heisenberg's uncertainty principle.

The shorter the pulse, the less we can specify the momentum precisely. For an accelerator with a continuous beam, $L \rightarrow \infty$, $\Delta p \rightarrow 0$ and the particle's momentum could be known as well as it could possibly be measured,

The following definitions and theorems are important for the axiomatic development of quantum mechanics. You can read through this section now, or you can wait until later in these lectures when they are needed and then refer back to this section.

Let $|i\rangle$, $|j\rangle$ and $|k\rangle$ be members of a complete set of base vectors that are mutually perpendicular, $\langle i|j\rangle = 0$ for $i \neq j$, and unit length, $\langle i|i\rangle = 1$. Assume that the indices i, j, and k can be denoted by positive integers. When we are dealing with an infinite set of base vectors, mutually perpendicular and unit length, with the indices ranging over all of the positive integers, we shall call the space so described a HILBERT SPACE. If a vector $|\Psi\rangle$ has projections $\langle i|\Psi\rangle$ onto the base vectors in a set, then we can write $|\Psi\rangle = \sum_i |i\rangle\langle i|\Psi\rangle$ in which $\sum_i |i\rangle\langle i| = 1$ is an identity operator.

When a set of vectors are mutually perpendicular we call them ORTHOGONAL, and when they are of unit length we call them NORMALIZED, and when they are both we call them ORTHONORMAL, or UNITARY when their components are complex numbers in the manner discussed below. The condition for orthonormality is described by the KRONECKER DELTA, $\delta_{ij} = \begin{matrix} 0 \text{ if } i \neq j \\ 1 \text{ if } i = j \end{matrix}$, giving $\langle i|j\rangle = \delta_{ij}$ for two vectors from an orthonormal set.

If $z = x + iy$ is a <u>complex</u> <u>number</u> then we indicate by an * its <u>complex</u> <u>conjugate</u>, $z^* = x - iy$. Of course here $i = \sqrt{-1}$. Later * will have another use when used on an <u>operator</u> .

An operator Q will have a matrix form when represented in the base vectors $|i\rangle$, $|j\rangle$, etc. and $\langle i|Q|j\rangle$ will denote the term at the intersection of the i th row and the j th column of that matrix. In mathematics books this is written Q_{ij} but in theoretical physics the Dirac notation is more useful. The matrix terms $\langle i|Q|j\rangle$ may be complex numbers, and their complex conjugates would be $\langle i|Q|j\rangle^*$.

DEFINITION: If Q is an operator with a matrix representation, we get another operator the TRANSPOSE OF Q by interchanging the rows and columns in the matrix. Hence if $\tilde{Q}$ is the transpose of Q, then $\langle i|\tilde{Q}|j\rangle = \langle j|Q|i\rangle$. For example if $Q = \begin{bmatrix} a & b \\ c & d \end{bmatrix}$ then $\tilde{Q} = \begin{bmatrix} a & c \\ b & d \end{bmatrix}$.

DEFINITION: If Q is an operator with a matrix representation, we get the HERMITIAN ADJOINT of Q by interchanging rows and columns and taking the complex conjugate of everything. Hence if Q^* is the hermitian adjoint of Q then $\langle i|Q^*|j\rangle = \langle j|Q|i\rangle^*$. For example, if $Q = \begin{bmatrix} 3+2i & 5-7i \\ 4+5i & 2+9i \end{bmatrix}$ then $Q^* = \begin{bmatrix} 3-2i & 4-5i \\ 5+7i & 2-9i \end{bmatrix}$.

DEFINITION: An operator is orthogonal if its transpose is its inverse, or $\tilde{R} = R^{-1}$ in which $RR^{-1} = R^{-1}R = 1$ for square matrices. Example could be $R = \begin{bmatrix} \cos\theta & -\sin\theta \\ \sin\theta & \cos\theta \end{bmatrix}$.

DEFINITION: An operator U is UNITARY if its hermitian adjoint is its in-
verse, $U^* = U^{-1}$. Example $U = \begin{bmatrix} \cos\theta & i\sin\theta \\ i\sin\theta & \cos\theta \end{bmatrix}$.

DEFINITION: An operator is HERMITIAN if it equals its hermitian adjoint.
$H^* = H$. Example, $H = \begin{bmatrix} 7 & 4+5i \\ 4-5i & -3 \end{bmatrix}$

THEOREM: If P, Q. and R are three matrices such that $R = PQ$, then $R^* = Q^*P^*$.
Let the matrices be representations of the operators in terms of the base
states $|i\rangle$, $|j\rangle$, $|k\rangle$ where $\sum_j |j\rangle\langle j| = 1$. Then as $R = PQ = P1Q$ we write

$\langle i|R|k\rangle = \langle i|PQ|k\rangle = \sum_j \langle i|P|j\rangle\langle j|Q|k\rangle$ and using the definition of the

hermitian adjoint,

$$\langle i|R^*|k\rangle = \langle k|R|i\rangle^* = \sum_j \langle k|P|j\rangle^*\langle j|Q|k\rangle = \sum_j \langle j|P^*|k\rangle\langle i|Q^*|j\rangle$$
$$= \sum_j \langle i|Q^*|j\rangle\langle j|P^*|k\rangle$$
$$= \langle i|Q^*P^*|k\rangle$$

which is the i-k th matrix element of $R^* = Q^*P^*$ which proves the theorem.

THEOREM: If H is hermitian and U is unitary, then $H' = UHU^{-1}$ is also
hermitian.

By definition $\langle j|H|k\rangle = \langle j|H^*|k\rangle = \langle k|H|j\rangle^*$ and also by definition
$\langle k|U^{-1}|m\rangle = \langle k|U^*|m\rangle = \langle m|U|k\rangle^*$ in which we can take the complex conjugate
of the first and last terms and change the indices to get $\langle i|U|j\rangle = \langle j|U^{-1}|i\rangle^*$. Using these identities when needed we prove the theorem.
Let the hermitian adjoint of H' be H'*. Then

$$\langle i|H'^*|m\rangle = \langle m|H'|i\rangle^* = \sum_k \sum_j \langle m|U|k\rangle^*\langle k|H|j\rangle^*\langle j|U^{-1}|i\rangle^*$$
$$= \sum_k \sum_j \langle k|U^{-1}|m\rangle\langle j|H|k\rangle\langle i|U|j\rangle$$
$$= \sum_j \sum_k \langle i|U|j\rangle\langle j|H|k\rangle\langle k|U^{-1}|m\rangle$$
$$= \langle i|UHU^{-1}|m\rangle = \langle i|H'|m\rangle$$

proving that if $H' = UHU^{-1}$ then $H'^* = H'$.

DEFINITION: The transformation $H' = UHU^{-1}$ is called a SIMILARITY TRANSFOR-
MATION, and is very important in modern theoretical physics, especially
when group theory is being emphasized.

DEFINITION: If for an operator H, there exists a number h and a vector
$|h\rangle$ labeled by that number such that $H|h\rangle = h|h\rangle$, then h is called an
EIGENVALUE of H and $|h\rangle$ is called an EIGENVECTOR of H. Almost always there
will be more than one eigenvalue and eigenvector for an operator, and so
we use an index, say n, to distinguish between them. Thus for the n th
eigenvalue and eigenvector we get $H|h_n\rangle = h_n|h_n\rangle$.

DEFINITION: If for an operator there exist several different eigenvectors
having exactly the same eigenvalue, the operator is called DEGENERATE.

In modern physics and chemistry the next theorem is of sufficient importance that it might be called the existence theorem for quantum mechanics. But first a rule about the inner product of two vectors when it gives a complex number. If $|a\rangle$ and $|b\rangle$ are two vectors so that the bra of the first times the ket of the second is $\langle a|b\rangle = c$, a complex number, then what happens if we take $\langle b|a\rangle$? It turns out that it is the complex conjugate of $\langle a|b\rangle$. Hence $\langle b|a\rangle = \langle a|b\rangle^*$. This means that if the terms in the column $|\Psi\rangle = \sum_i |i\rangle\langle i|\Psi\rangle$ are complex numbers $\langle i|\Psi\rangle$, then in the row vector $\langle \Psi| = \sum_i \langle\Psi|i\rangle\langle i|$ the terms $\langle\Psi|i\rangle$ are the complex conjugates of $\langle i|\Psi\rangle$. Then the square of the length of the vector is $\langle\Psi|\Psi\rangle = \sum_i \langle\Psi|i\rangle\langle i|\Psi\rangle = \sum_i |\langle i|\Psi\rangle|^2$ which is a real number as $|\langle i|\Psi\rangle|^2 = \langle i|\Psi\rangle^* \langle i|\Psi\rangle$ is a real number.

THEOREM: If H is a hermitian operator, then its eigenvalues are real numbers and its eigenvectors are mutually perpendicular.

Suppose we start with the special case where H is non-degenerate. This means that if $|h_m\rangle$ and $|h_n\rangle$ are different eigenvectors, their eigenvalues will be different numbers, $h_m \neq h_n$. Then

$H|h_n\rangle = h_n|h_n\rangle$	(1)	which we multiply by $\langle h_m|$
$\langle h_m|H|h_n\rangle = h_n\langle h_m|h_n\rangle$	(2).	Likewise
$H|h_m\rangle = h_m|h_m\rangle$	(3)	can be multiplied by $\langle h_n|$
$\langle h_n|H|h_m\rangle = h_m\langle h_n|h_m\rangle$	(4)	in which we can take the complex con-
$\langle h_n|H|h_m\rangle^* = h_m^*\langle h_n|h_m\rangle^*$	(5)	jugate of everything to give

But H is hermitian, so $\langle h_n|H|h_m\rangle^* = \langle h_m|H^*|h_n\rangle = \langle h_m|H|h_n\rangle$; and by the above rule $\langle h_n|h_m\rangle^* = \langle h_m|h_n\rangle$. Hence step (5) can be rewritten as

$$\langle h_m|H|h_n\rangle = h_m^*\langle h_m|h_n\rangle \qquad (6)$$

If we subtract step (6) from step (2) we get

$$(h_n - h_m^*)\langle h_m|h_n\rangle = 0 \qquad (7)$$

In step (7), first take the case where $m = n$, and $(h_n - h_n^*)\langle h_n|h_n\rangle = 0$. Now $\langle h_n|h_n\rangle$ is the square of the length of $|h_n\rangle$ and, hence, not zero. Thus $h_n - h_n^* = 0$, and the only way that a number can equal its complex conjugate is for that number to be real. Hence the eigenvalues of a hermitian operator are real numbers. Thus, in (7) we can drop the $*$ on h_m^*. Then for $m \neq n$, and $h_m \neq h_n$ as was assumed for the case being considered. Then

$$(h_n - h_m)\langle h_m|h_n\rangle = 0 \qquad (8).$$

As $h_n - h_m \neq 0$ for $n \neq m$, $\langle h_m|h_n\rangle = 0$ for $m \neq n$, and $|h_m\rangle$ is perpendicular to $|h_n\rangle$ and for a nondegenerate hermitian operator, the eigenvalues form a mutually perpendicular set.

As a general rule we arbitrarily require the eigenvalues of an operator to be of unit length. Hence for a nondegenerate hermitian operator the above results can be written $\langle h_m|h_n\rangle = \delta_{mn}$.

Now, suppose that the operator H is a degenerate hermitian operator. Then there will be at least two different eigenvectors having the same eigenvalue. Let $|h_{ni}\rangle$ and $|h_{nj}\rangle$ be non-parallel eigenvectors having the same eigenvalue h_n. Then $H|h_{ni}\rangle = h_n|h_{ni}\rangle$ and $H|h_{nj}\rangle = h_n|h_{nj}\rangle$.

Now, all the steps in the preceeding proof will hold in this case up to step (8). Hence h_n is a real number and $(h_n - h_n)\langle h_{ni}|h_{nj}\rangle = 0$. As $(h_n - h_n)$ is automatically zero, the equation is satisfied without the requirement that $\langle h_{ni}|h_{nj}\rangle = 0$, and thus the eigenvectors having the same eigenvalue do not have to be orthogonal. Never-the-less we can form a set of orthogonal eigenvectors having the eigenvalue h_n by the following method, known as the Gram-Schmidt orthogonalization. It is worth paying close attention to how this is done as the method is very useful in theoretical physics. Let us form a new vector $|h_{nk}\rangle = a|h_{ni}\rangle + b|h_{nj}\rangle$ as a linear combination of the two eigenvectors with a and b being two numbers. Obviously, $H|h_{nk}\rangle = h_n|h_{nk}\rangle$ making $|h_{nk}\rangle$ also to be an eigenvector of H with an eigenvalue h_n. This says that if two eigenvectors have the same eigenvalue, then all vectors lying in the plane of those two vectors are also eigenvectors with the same eigenvalue. Hence, if we want mutually perpendicular eigenvectors with that eigenvalue, just pick any two mutually perpendicular vectors on that plane and they will serve quite all right. The Gram-Schmidt method works in the following way. Let $|h_{ni}\rangle$ and $|h_{nj}\rangle$ be known to be degenerate eigenvectors of H. Suppose that they are of unit length but not mutually perpendicular. To form a new set with the desired orthogonality property we can start by including one member of the old set. Thus, let $|h_{n1}\rangle = |h_{ni}\rangle$. Next form $|h_{n2}\rangle = a|h_{ni}\rangle + b|h_{nj}\rangle$ which is to be unit length and perpendicular to $|h_{n1}\rangle$. The original vectors have unit length, hence $\langle h_{ni}|h_{ni}\rangle = \langle h_{nj}|h_{nj}\rangle = 1$. Thus $\langle h_{n1}|h_{n1}\rangle = 1$ and for $\langle h_{n2}|h_{n2}\rangle = 1$, it is necessary that $a^*a + b^*b + a^*b\langle h_{ni}|h_{nj}\rangle + ab^*\langle h_{nj}|h_{ni}\rangle = 1$ and to satisfy the orthogonality condition, $\langle h_{n1}|h_{n2}\rangle = 0$, $a + b\langle h_{ni}|h_{nj}\rangle = 0$ should hold. If all the components of $|h_{ni}\rangle$ and $|h_{nj}\rangle$ are real numbers, $\langle h_{ni}|h_{nj}\rangle = \langle h_{nj}|h_{ni}\rangle$ and we can get by with a and b being real numbers without any imaginary parts. Then $a^2 + b^2 + 2ab\langle h_{ni}|h_{nj}\rangle^2 = 1$ and $a + b\langle h_{ni}|h_{nj}\rangle = 0$ which are easily solved for a and b.

This method can be extended, one additional eigenvector at a time to cover any dimension of degenerate subspace of the Hilbert space of the operator H with a set of orthogonal eigenvectors.

Hence, if H is a hermitian operator, its eigenvalues are real numbers and its eigenvectors either form or can be made to form a mutually perpendicular set.

Several important observations can be gotten as 'spin off' from the previous theorem. In step three we have $H|h_m\rangle = h_m|h_m\rangle$. Then in going from step (5) to step (6) gives $\langle h_m|H^*|h_n\rangle = \langle h_m|h_n\rangle h_m^*$ in which we can factor out the $|h_n\rangle$ to give $\langle h_m|H^* = \langle h_m|h_m^*$. Thus if an operator operates to the <u>right</u> on an eigenket to give the eigenvalue h_m, its hermitian adjoint H^* will then operate to the <u>left</u> on the eigenbra $\langle h_m|$ to give the eigenvalue h_m^*, the complex conjugate of h_m. For the hermitian operator $H = H^*$ with $h_m^* = h_m$, this becomes $\langle h_m|H = \langle h_m|h_m$.

Next, look at the orthogonality condition $\langle h_m|h_n\rangle = \delta_{mn}$. The Kronecker delta, $\delta_{mn} = \begin{array}{l} 1 \text{ if } m=n \\ 0 \text{ if } m \neq n \end{array}$ is obviously the term at the intersection of the unit matrix $1 = \begin{pmatrix} 1 & 0 & \cdot \\ 0 & 1 & \cdot \\ \cdot & \cdot & \cdot \end{pmatrix}$. Now, suppose that the index n for the

eigenvectors can be counted by the positive integers, $n = 1,2,3,4,\ldots$ Then the set of orthogonal vectors $|h_n\rangle$ can be expanded in terms of another set of orthogonal vectors $|i\rangle$ with $i = 1,2,3,4,\ldots$ and $\langle i|j\rangle = \delta_{ij}$. When this is done $|h_n\rangle = \sum_i |i\rangle\langle i|h_n\rangle$ in which $\langle i|h_n\rangle$ is the projection of any member of one set onto a member of the other set. Now $\sum_i |i\rangle\langle i| = 1$, the unit operator. This permits us to write $\sum_i \langle h_m|i\rangle\langle i|h_n\rangle = \langle h_m|1|h_n\rangle = \delta_{mn}$.

The quantity $\langle i|h_n\rangle$ has two indices i and n which are positive integers, suggesting that it can be thought of as the term at the intersection of i th row and n th column of some matrix which we shall call U. Then $\langle i|h_n\rangle = \langle i|U|n\rangle$. Likewise we can invent a matrix V with $\langle h_m|i\rangle = \langle m|V|i\rangle$, and $\langle n|V|i\rangle = \langle h_n|i\rangle = \langle i|h_n\rangle^* = \langle i|U|n\rangle^*$. The matrix elements of V are gotten from the matrix elements of U by interchanging rows and columns and taking the complex conjugate of everything. This makes V the hermitian adjoint of U, Hence $V = U^*$, and we can write $\langle h_m|i\rangle = \langle m|U^*|i\rangle$. Now the above expression for the orthogonality condition for the $|h_n\rangle$ can be written $\sum_i \langle m|U^*|i\rangle\langle i|U|n\rangle = \langle m|1|n\rangle$ which is the m-n th term in the matrix representation of the product $U^*U = 1$. Hence $U^* = U^{-1}$ and, as its hermitian adjoint is its inverse, U is a unitary matrix.

As the transformation from the orthonormal set $|i\rangle$ to the orthonormal set $|h_n\rangle$ is accomplished by $|h_n\rangle = \sum_i |i\rangle\langle i|h_n\rangle$ in which the projection $\langle i|h_n\rangle$ is the i-n th element of a unitary matrix. Because of this, the transformation is called a UNITARY TRANSFORMATION.

<u>DEFINITION</u>: A matrix is DIAGONAL if it has non-zero elements only along the principal diagonal running from the upper left corner to the lower right corner. An example is $D = \begin{pmatrix} d_1 & 0 & 0 \\ 0 & d_2 & 0 \\ 0 & 0 & d_3 \end{pmatrix}$ in which $\langle i|D|j\rangle = d_i \delta_{ij}$ is the ij th term.

THEOREM: A hermitian operator gives a diagonal matrix when it is represented in terms of its eigenvectors, the elements along its principal diagonal being its eigenvalues.

Proof: If $H|h_n\rangle = h_n|h_n\rangle$, then $\langle h_m|H|h_n\rangle = h_n\langle h_m|h_n\rangle = h_n\delta_{mn}$.

THEOREM: If two matrices, A and B, are diagonal, their product is commutative, i.e. AB = BA. The proof is easy, and left for you to do yourself.

THEOREM: If the set of vectors $|h_n\rangle$ are simultaneously eigenvectors of the two hermitian operators G and H, then the matrices for G and H commute when represented in any orthonormal basis.

If the vectors $|h_n\rangle$ are eigenkets of both G and H, then $H|h_n\rangle = h_n|h_n\rangle$ and $G|h_n\rangle = g_n|h_n\rangle$, and if they are a complete orthonormal set, then they give the identity operator $\sum_n |h_n\rangle\langle h_n| = 1$. Let the vectors denoted by $|i\rangle$, $|j\rangle$, or $|k\rangle$ form another complete orthonormal set in a space of the same dimension, having the identity $\sum_j |j\rangle\langle j| = 1$. In this second basis, G has the matrix elements $\langle i|G|j\rangle$ and H has the matrix elements $\langle j|H|k\rangle$, and the product GH has the matrix element $\langle i|GH|k\rangle = \sum_j \langle i|G|j\rangle\langle j|H|k\rangle$. Now, inserting and removing the above identity operators when convenient we can prove the theorem.

$$\sum_j \langle i|G|j\rangle\langle j|H|k\rangle = \langle i|GH|k\rangle = \sum_n \langle i|G|h_n\rangle\langle h_n|H|k\rangle = \sum_n \langle i|g_n|h_n\rangle\langle h_n|h_n|k\rangle$$

$$= \sum_n \langle i|h_n\rangle g_n h_n \langle h_n|k\rangle = \sum_n \langle i|h_n\rangle h_n g_n \langle h_n|k\rangle$$

$$= \sum_n \langle i|h_n|h_n\rangle\langle h_n|g_n|k\rangle = \sum_n \langle i|H|h_n\rangle\langle h_n|G|k\rangle$$

$$= \langle i|HG|k\rangle = \sum_j \langle i|H|j\rangle\langle j|G|k\rangle$$

proving that if the eigenvectors for H are also eigenvectors for G, then GH = HG when represented in any orthonormal basis. This and the following theorem are very important in the quantum theory of measurements.

THEOREM: If G and H are nondegenerate hermitian operators that do not commute, GH $\neq$ HG, then the same set of vectors can not serve simultaneously as eigenvectors of both G and H.

Let the set $|h_n\rangle$ be eigenkets of H, so $H|h_n\rangle = h_n|h_n\rangle$. When represented in terms of the basis of $|h_n\rangle$ the inequality GH $\neq$ HG says that for at least some values of the indices m and n, $\langle h_m|GH|h_n\rangle \neq \langle h_m|HG|h_n\rangle$. Consider such matrix elements. If this inequality holds, then $\langle h_m|G|h_n\rangle h_n \neq h_m\langle h_m|G|h_n\rangle$ and for these values of m and n — obviously excluding m = n — we get $(h_m - h_n)\langle h_m|G|h_n\rangle \neq 0$. If H is nondegenerate, then $(h_m - h_n) \neq 0$ for $m \neq n$, and the last result requires $\langle h_m|G|h_n\rangle \neq 0$ for the same $m \neq n$. If this is so then the matrix elements for G represented in the basis of $|h_n\rangle$ are nonzero for $m \neq n$, telling us that the matrix is not diagonal. But, if the $|h_n\rangle$ were eigenvectors of G, the matrix would be diagonal. Hence the eigenvectors of H are not also eigenvectors of G when GH $\neq$ HG.

DEFINITION: If M is an operator with square matrix representations such that the powers of M, M^2, M^3, etc exist, and e is the base of the natural logarithms, then we <u>define</u> $e^M = 1 + M + \frac{1}{2!}M^2 + \frac{1}{3!}M^3 + \frac{1}{4!}M^4 + \ldots$ in which the 1 stands for the identity operator.

THEOREM: If M and N are operators such that M^2, MN, N^2, etc. exist, then $e^M e^N = e^N e^M$ or $e^{M+N} = e^{N+M}$ if and only if MN = NM.

Represent each exponential operator by its power series expansion. If MN = NM, then any power of M commutes with any power of N and, hence the series commute and $e^M e^N = e^N e^M$. If MN $\neq$ NM, then the series do <u>not</u> commute and $e^M e^N \neq e^N e^M$.

The next theorem is very important in the application of the abstract algebraic theory of Lie groups to quantum mechanics, especially in the theory of high energy physics and strongly interacting particles.

THEOREM: If H is a <u>hermitian</u> operator, λ a real number and $i = \sqrt{-1}$, then the operator $U = e^{i\lambda H}$ is a <u>unitary</u> operator.

If $U = e^{i\lambda H} = \sum_{n=0}^{\infty} \frac{(i\lambda)^n}{n!} H^n$ is unitary, then, with $H^* = H$, its hermitian adjoint $U^* = \sum_{n=0}^{\infty} \frac{(-i\lambda)^n}{n!} (H^*)^n = \sum_{n=0}^{\infty} \frac{(-i\lambda)^n}{n!} H^n = e^{-i\lambda H}$ should be its inverse. As H commutes with itself, $i\lambda H$ should commute with $-i\lambda H$ and hence $U^*U = UU^* = e^{i\lambda H - i\lambda H} = e^0 = 1$ which is the requirement for $U^* = U^{-1}$

Another, longer, proof is the following.

$$UU^* = \sum_{m=0}^{\infty} \frac{(i\lambda)^m}{m!} H^m \sum_{n=0}^{\infty} \frac{(-i\lambda)^n}{n!} H^n = \sum_{m=0}^{\infty}\sum_{n=0}^{\infty} \frac{(-)^n (i\lambda)^{m+n}}{m!\,n!} H^{m+n}$$

in which we replace N = m+n and reorder the summations so that $0 \leq n \leq N$ and $0 \leq N < \infty$. Then

$$UU^* = \sum_{N=0}^{\infty}\sum_{n=0}^{N} \frac{(-)^n (i\lambda)^N}{n!(N-n)!} H^N = \sum_{N=0}^{\infty} \frac{(i\lambda)^N}{N!} \left[\sum_{n=0}^{N} \frac{N!}{n!(N-n)!} (1)^{N-n}(-)^n \right] H^N$$

$$= \sum_{N=0}^{\infty} \frac{(i\lambda)^N}{N!} (1-1)^N H^N = e^{i\lambda 0 H} = e^0 = 1$$

THE PROOFS OF THE FOLLOWING THEOREMS ARE LEFT FOR THE STUDENT TO DO.

THEOREM: If $|h_n\rangle$ is an eigenket of H with the eigenvalue h_n, then it is an eigenket of $e^{i\lambda H}$ with the eigenvalue $e^{i\lambda h_n}$.

THEOREM: Let P and Q be two hermitian operators for which $QP - PQ = i\hbar$ where $i = \sqrt{-1}$ and $\hbar$ is a number. By mathematical induction show that for any integer n, $Q^n P - P Q^n = i\hbar n Q^{n-1}$ and $Q P^n - P^n Q = i\hbar n P^{n-1}$.

THEOREM: Let F( ) be some function whose coefficients c_0, c_1, c_2 etc are numbers with $F(Q) = c_0 + c_1 Q + c_2 Q^2 + c_3 Q^3 + \ldots$ and $F(P) = c_0 + c_1 P + c_2 P^2 + \ldots$ being operator functions of the operators P and Q. Use the result of the previous theorem to show that if $QP - PQ = i\hbar$, then $F(Q)P - PF(Q) = i\hbar \frac{dF}{dQ}$ and $QF(P) - F(P)Q = i\hbar \frac{dF}{dP}$.

EXERCISE: Let $|\Psi\rangle = |\Psi(t)\rangle$ be a family of vectors whose components are differentiable functions of t. For the i th component, $d\Psi_i/dt = D_t\Psi_i(t)$ and by $D_t^n\Psi_i(t_o)$ we mean the n th derivative of the i th component evaluated at $t = t_o$. When the operator D_t operates on the vector $|\Psi(t)\rangle$ it gives a new vector $D_t|\Psi(t)\rangle$ whose components are the derivatives of the components of $|\Psi(t)\rangle$ determined at the indicated value of t. When t changes from t_o to $t_o+\Delta t$, the change in $|\Psi(t)\rangle$ is given by the INFINITESSIMAL TRANSFORMATION
$$|\Psi(t_o+\Delta t)\rangle = |\Psi(t_o)\rangle + \Delta t D_t|\Psi(t_o)\rangle = \left[1 + \Delta t D_t\right]|\Psi(t_o)\rangle$$
Hence, show that the Taylor's series deriving $|\Psi(t)\rangle$ from $|\Psi(t_o)\rangle$ can be written as $|\Psi(t)\rangle = e^{(t-t_o)D_t}|\Psi(t_o)\rangle$ with an inverse transformation
$$|\Psi(t_o)\rangle = e^{-(t-t_o)D_t}|\Psi(t)\rangle$$

EXERCISE: This exercise depends on the results of the previous exercise. Let $Q(t)$ be an operator whose matrix elements are differentiable functions of the variable t. Let $|\Psi(t)\rangle$ and $|\Phi(t)\rangle$ be two families of vectors related by $Q(t)$ so that $|\Phi(t)\rangle = Q(t)|\Psi(t)\rangle$, the vectors both varying with t accor-ding to the t-development operator derived in the previous exercise. When $t = t_o$, the vectors are related by $|\Phi(t_o)\rangle = Q(t_o)|\Psi(t_o)\rangle$. Using the t-development operator, and its inverse, show that the form of this rela-tion is unchanged at any other value of t if the operator varies according to $Q(t) = e^{(t-t_o)D_t}Q(t_o)e^{-(t-t_o)D_t}$

EXERCISE: In terms of the x- and y-axes a point on a plane can be described by a vector $|\Psi\rangle = \begin{pmatrix} x \\ y \end{pmatrix}$ where if the vector makes an angle α with the positive x-axis, $x = r\cos\alpha$ and $y = r\sin\alpha$, with r being the length of the vector. If the axes are rotated counterclockwise through an angle φ to give an x' and a y' axis the vector, remaining where it was on the plane will be transformed into $|\Psi'\rangle = \begin{pmatrix} x' \\ y' \end{pmatrix}$ where $x' = r\cos(\alpha-\varphi)$ and $y' = r\sin(\alpha-\varphi)$. Show that this transformation can be written $|\Psi'\rangle = R(\varphi)|\Psi\rangle$ where $R(\varphi) = \begin{pmatrix} \cos\varphi & \sin\varphi \\ -\sin\varphi & \cos\varphi \end{pmatrix}$

Hence show that $R(\varphi)$ can be written $R(\varphi) = e^{i\sigma\varphi}$ where $\sigma = \begin{pmatrix} 0 & -i \\ i & 0 \end{pmatrix}$.

15-38

CHAPTER SIXTEEN: INTRODUCTION TO FORMAL QUANTUM MECHANICS

On the scale of the universe with which we are familiar, the numbers describing things run together as continuous variables. But for the things on the very small scale of the universe, their numbers come with a number here and a number there with no continuous variables connecting them. Physical quantities whose values have only certain numbers and no others are said to be QUANTIZED.

Just in case you are starting quantum mechanics here, without reading previous chapters, we will repeat the discussion of the hydrogen atom given in previous chapters. Then we will develop the formal mathematics of the subject and rigorously apply them to the quantization of a particle in a box and the quantization of the harmonic oscillator. There are so many phenomena that can be analyzed with the tools. We will stop here and get along with thinking about physical problems. The first application will be to the quantization of the energy of a vibrating string. This will give us some particles called phonons, and by analogy particles of electromagnetic energy called photons.

A hydrogen atom consists of an electron with charge, $-e$, existing around a proton with charge, $+e$, where $e = 1.6 \times 10^{-19}$ coulomb. In the chapter on electricity, you learned that a charge, q, at a distance, r, from a charge, Q, has an energy, $E = \frac{1}{2} M v^2 + \frac{e Q}{4 \pi \epsilon r}$.

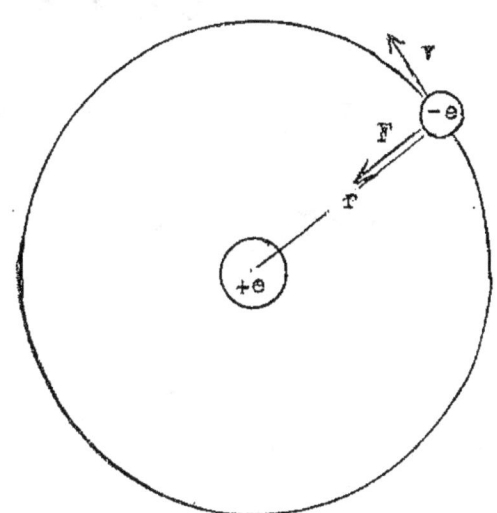

In that development, we could not avoid using the letter E for both electric force field and for total energy. The context must determine the sense in which the letter is used. The resulting energy formula is good for motion with velocity, v, being in any direction. In the following development we will simplify the formulas by using,

$\frac{1}{4 \pi \epsilon} = k = 9 \times 10^9$ Newton-meter²/coulomb², in a vacuum.

We will now consider an atom of the simplest element, hydrogen, consisting of one electron going in a circle about a nucleus consisting of a single nuclear particle, or proton. Each particle carries one unit of charge, being $e = 1.6 \times 10^{-19}$ coulomb. The electron carries, $q = -e$, and the proton carries, $q = +e$. The force attracting the electron to the proton, $F = k \frac{(-e)(+e)}{r^2} = -\frac{k e^2}{r^2}$, is a centripetal force in Newton's second law, $F = M a$, with the centripetal acceleration, $a = -r\,\omega^2$, first seen in chapter 4, where the velocity was, $v = r\,\omega$; hence, $a = -\dfrac{v^2}{r}$ also.

From $M\,a = F$, we get $-M\dfrac{v^2}{r} = -\dfrac{k\,e^2}{r^2}$, giving $M\,v^2 = \dfrac{k\,e^2}{r}$.

If energy is $E = \dfrac{1}{2}M\,v^2 - \dfrac{k\,e^2}{r}$, then $E = -\dfrac{k\,e^2}{2\,r}$.

You may ask how is it that we can determine the electron's energy or orbital radius. This is done with light which involves waves of electric and magnetic force fields. On being emitted and absorbed and on passing through matter, the vibrating fields of a light wave interact with the electrical structure of the atoms of matter. Hot solid surfaces, when emitting light give off all colors of the spectrum. Excited gases and vapors give off only certain colors and no others. Each element has a unique set of colors which its excited vapor gives off. If the light from a particular source is passed through a diffraction grating in a spectrometer, the wavelengths of the emitted colors can be determined. Below are some examples of emission spectra for various elements.

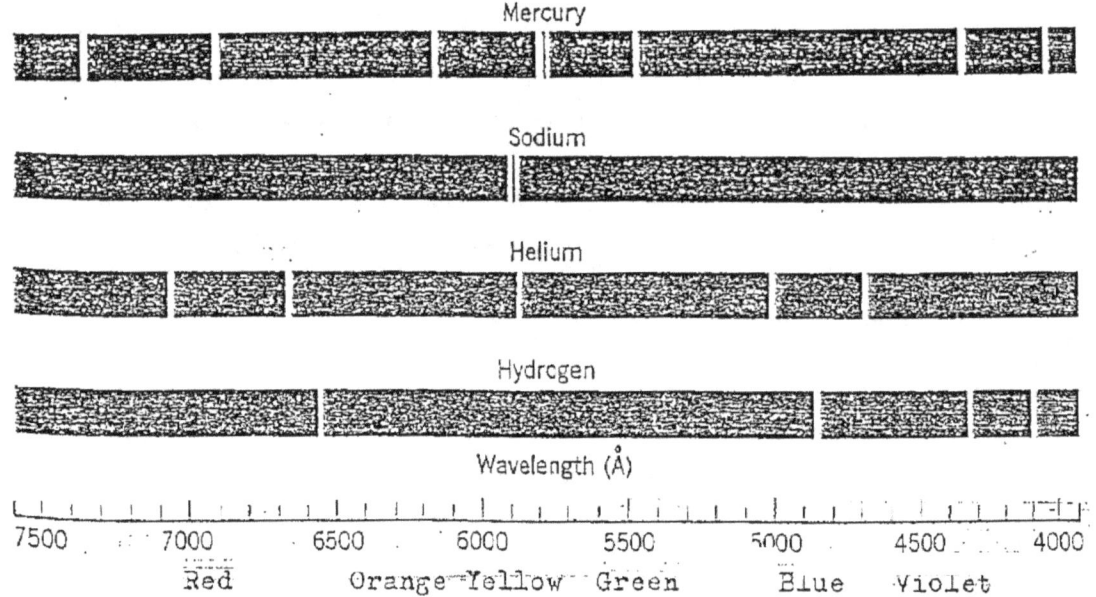

For all of these sources, there are other emissions with wavelengths longer than red, called infra-red, and with wavelengths shorter than violet, called ultraviolet. For the visible emissions from excited hydrogen, the wavelengths are: $\lambda_{red} = 6563\,\text{Å}$, $\lambda_{blue-green} = 4861\,\text{Å}$, $\lambda_{violet} = 4340\,\text{Å}$, in which one Angstrom is $1.0\,\text{Å} = 10^{-10}$ meter. Expressing the emitted wavelengths from hydrogen in meters, the wavelengths can be described by the formula, $\dfrac{1}{\lambda} = R\left[\dfrac{1}{2^2} - \dfrac{1}{n^2}\right]$, where $n = 3$ for red, $n = 4$ for blue-green, and $n = 5$ for violet. R is a constant called the Rydberg constant.

Using the wavelengths given on the previous page in meters, we can calculate a number for the Rydberg constant, $R = 1.097 \times 10^7$ meters^{-1}. Including ultraviolet and infrared emissions, all components of the hydrogen spectrum can be described by this same formula if the first term in the bracket is also $\frac{1}{1^2}$ or $\frac{1}{3^2}$ or $\frac{1}{4^2}$ and so forth.

Other spectra do not lend themselves to such simple description. The spectral data, however, suggests that there is something very specific and limited about the structure of various atoms. Only certain values of atomic properties are allowed by Nature to exist. As a result, when atoms are excited, they give off only certain colors of light.

Before we go further with light from atoms, we must consider light given off by hot surfaces. A hot glowing coal gives off red light. A very hot wire in a light bulb gives off white light. Using a prism or a grating, we know that white light is made up of all colors from red to violet. Why is there a relation between the quality of emitted light and the temperature of the emitting surface? The brightest emitted color associates increasing temperature with increasing frequency. One Hertz is one vibration per second.

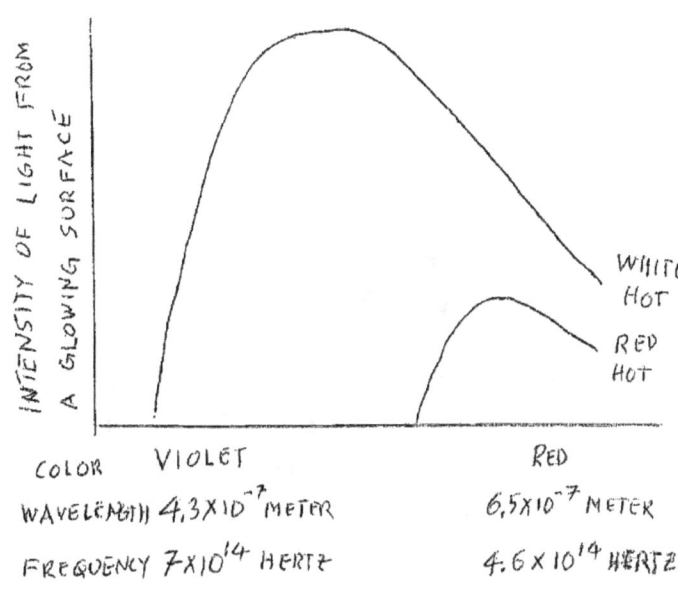

In the year 1900, Max Planck came up with the right idea. Although light is mostly thought of as involving waves, its energy is carried in definite sized bundles of energy, which we now call photons. Planck reasoned that the higher the frequency of the light wave, the greater the size of the bundle of its energy. Coming off a hot surface, a photon is emitted locally from a small piece of the surface area. Hence, the greater the energy the photon takes, the greater the energy that must be in that piece of area. This is measured by the concentration of energy per unit area on the surface.

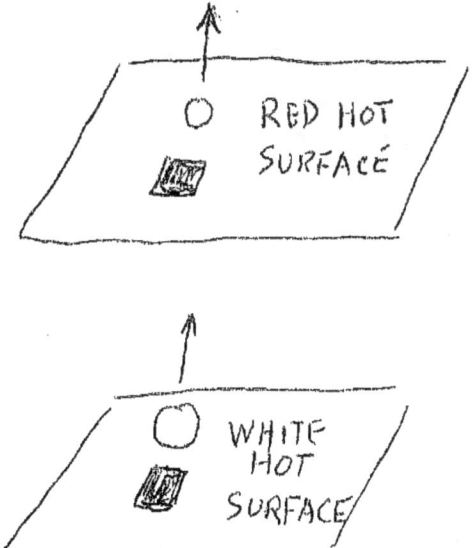

The thermal energy per unit area of a hot surface increases with the temperature. Hence, to get photons of higher energy, and light of increasing frequency, it is necessary to increase the temperature of the surface. As the temperature of the surface increases, the surface can emit light of frequency increasing from red toward the violet, which is what we observe. Obviously, the energy of a photon is proportional to the frequency, and we write $dE = hf$, where h is Planck's constant.

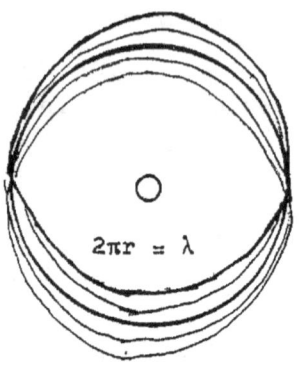

Measurements give $h = 6.625 \times 10^{-34}$ Joule-seconds. For a light wave, $f = c$, where the velocity of light is $c = 3 \times 10^8$ m/sec. Thus, a photon has energy, $dE = \dfrac{hc}{\lambda}$. Now recall the Rydberg formula for the hydrogen specta. A photon of one of these has

$dE = hcR \left[\dfrac{1}{2^2} - \dfrac{1}{\lambda^2} \right]$. A hydrogen electron in an orbit of

radius, r, has energy, $E = -k \dfrac{e^2}{2} r$. Maybe the orbit is limited to having only certain radii.

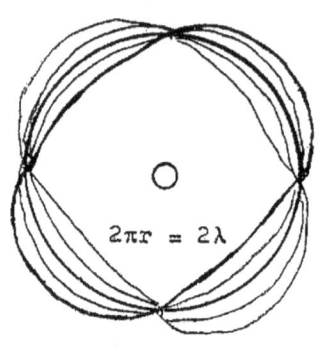

The photon energy equals the difference between the energy of two orbits. Therefore, if E_n is the energy of the nth orbit and E_2 is the energy of the 2nd orbit, then the difference is $dE = E_n - E_2$. Noting that E is negative, we get $E_n = -\dfrac{hcR}{n^2}$,

$hcR = 2.18 \times 10^{-18}$ Joules.

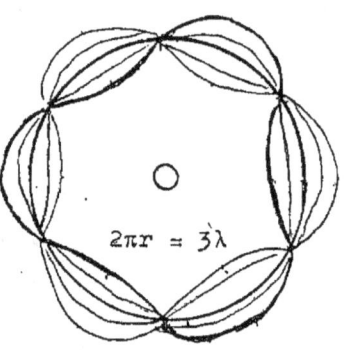

Dividing this by $e = 1.6 \times 10^{-19}$, gives $hcR = 13.6$ eV, energy in electron volts. Starting from the orbit, $n = 1$, this equals the measured ionization potential of hydrogen.

Equating E_n to $E_n = -\dfrac{ke^2}{2r}$, gives $r_n = \dfrac{ke^2}{hcR} n^2$. This gives $r_n = (5.28 \times 10^{-11}) n^2$

meter. Let $a = \dfrac{ke^2}{hcR}$. Then, $r_n = a n^2$.

From a standard (before 1900) physics point of view, the limitation of radii to only certain numbers had no reason and did not make sense. In classical physics, the only thing that was so limited was the frequencies of standing waves. Now if light waves had particles, photons, could particles have waves? Let us assume that they do, and for an electron, its orbit was an integer number of wavelengths. $2 \pi a n^2 = n \lambda_n$ or $\lambda_n = 2 \pi a n$. This then, gives $r_n = \frac{a^2 n^2}{a} = \frac{\lambda_n^2}{4 \pi^2 a}$. Now $M v^2 = \frac{k e^2}{r_n}$, where for the electron, $M = 9.1 \times 10^{-31}$ kilogram. Focusing on momentum, $M^2 v^2 = \frac{M k e^2 4 \pi^2 a}{\lambda^2}$, or $M^2 v^2 \lambda^2 = M k e^2 4 \pi^2 a$, giving $M v \lambda = 6.6 \times 10^{-34}$. This looks a lot like, $(M v)(\lambda) = h$.

Louis de Broglie came up with the idea in 1924. If light waves have particles called photons, which carry their energy, maybe by some symmetry in physics, electron particles have some sort of waves associated with them. Then the electron would exist in orbits whose circumference was associated with a standing wave of that wave. That would require the circumference to equal an integer number of wavelengths, as shown in the pictures on the previous page.

De Broglie suggested that any particle with mass, M, and velocity, v, would have a wave associated with it, with $\lambda = \frac{h}{M v}$, as its wavelength. These waves, for a beam of electrons, were observed directly by G. P. Thomson, C. J. Davisson and L. H. Germer in 1927.

INTRODUCTION TO QUANTUM MECHANICS

"Now are they undone, the ancient laws" Aeschylus <u>The Eumenides</u>

The unforeseen may sometimes end
Conditions which are perfect

> Shih Nai-An
> <u>All Men are Brothers</u> or <u>Water Margin</u>

It is the goal of theoretical physics that we may be able to construct a consistent mathematical scheme which provides a unified description of the measurable appearance of the universe. To accomplish this aim, we attempt to formulate axioms from which we may derive mathematically as conclusions both the results of past measurements and predictions of future measurements to be done in the laboratory. The development of a theory begins with a body of experimental evidence – in this case the observation of the nature of light emitted from hot solids and excited gases. After much enlightened guesswork, some bright people stumble onto a set of statements from which a description of the experiments may be derived. As a bonus for good mathematics, we may expect predictions of phenomena which had not yet been observed – e.g. antimatter, mesons, semiconductors, lasers, etc. Unfortunately, in going from known physics to unknown physics, our mathematics may carry no physical intuition along with it. Quantum mechanics is an extreme case of this; it tells you nothing physical about what it is doing until the point when it says that if you go into the laboratory and do such-and-such then this and that will be the results. As a result, we are totally dependent on the rigor of our mathematics to hold everything together until that point where it suddenly predicts the result of an experiment.

Max Planck (1901) postulated that a light wave with a frequency, f, had its energy carried in packets (quanta) or particles (photons) of energy, $E = hf$. Robert A. Millikan (1916) measured Planck's constant to be 6.624×10^{-34} Joule second. Planck's postulate was directly observed by Arthur H. Compton (1923) when he scattered x-rays off of electrons. Louis de Broglie (1923) postulated that particles with momentum, $p = mv$, would have associated waves whose wavelength was $\lambda = \dfrac{h}{mv}$. These waves for electrons were directly observed by C. J. Davisson, L. H. Germer and G. P. Thomson in 1927.

Assume now that a free particle, with no interactions, experiences the Planck and de Broglie wave-particle relationships. In chapters 9 and 13, we discussed waves of frequency, f, wavelength, λ, and wave velocity, $u = f\lambda$.

We can write for them a wave function, $\psi = A \, \dfrac{sin}{cos} \, 2\pi f\left(\dfrac{x}{u} - t\right)$. At a moving point located at $x = x_o + ut$, ψ has a constant value of $\psi = A \, \dfrac{sin}{cos} \left(\dfrac{2\pi f x_o}{u}\right)$.

If either a sine function or a cosine can describe our wave, we can use $e^{i\theta} = cos\,\theta + i\,sin\,\theta$ to write a generalized wave, $\psi = A\,e^{i\,2\pi f\left(\frac{x}{u} - t\right)}$, but $u = f\lambda$ gives $\psi = A\,e^{i\,2\pi\left(\frac{x}{\lambda} - f t\right)}$. But suppose that $f = \dfrac{E}{h}$ and $\dfrac{1}{\lambda} = \dfrac{p}{h}$ with $p = mv$, then $\psi = A\,e^{i\frac{2\pi}{h}(p\,x - E\,t)}$. In all the discussions that follow, Planck's constant will always be divided by 2π, so we will follow a suggestion by P. A. M. Dirac and define an haitch-bar as $\hbar = \dfrac{h}{2\pi}$. Then our wavefunction is $\psi = A\,e^{i\left(\frac{p\,x - E\,t}{\hbar}\right)}$.

If $\psi = A\,e^{i\left(\frac{p\,x}{\hbar}\right)} e^{-i\left(\frac{E\,t}{\hbar}\right)}$, then $\dfrac{\partial\psi}{\partial x} = i\dfrac{p}{\hbar}\psi$ and $\dfrac{\partial^2\psi}{\partial x^2} = -\dfrac{p^2}{\hbar^2}\psi$.

Similarly, $\dfrac{\partial\psi}{\partial t} = -i\dfrac{E}{\hbar}\psi$ and $\dfrac{\partial^2\psi}{\partial t^2} = -\dfrac{E^2}{\hbar^2}\psi$.

But $E = \dfrac{p^2}{2M}$ for a particle with no interactions where $p = Mv$. Multiplying this by ψ gives $E\psi = \dfrac{p^2}{2M}\psi$, and substitution gives $i\hbar\dfrac{\partial\psi}{\partial t} = \dfrac{\hbar^2}{2M}\dfrac{\partial^2\psi}{\partial x^2}$. For a particle whose interaction has a potential energy, $V(x)$, $E = \dfrac{p^2}{2M} + V(x)$. If we multiply this by ψ and substitute, we get $i\hbar\dfrac{\partial\psi}{\partial t} = \dfrac{\hbar^2}{2M}\dfrac{\partial^2\psi}{\partial x^2} + V(x)\,\psi$. This result is the time-dependent Schrödinger equation.

Returning to $\psi = A\,e^{i\left(\frac{p\,x - E\,t}{\hbar}\right)}$, $\dfrac{\hbar}{i}\dfrac{\partial\psi}{\partial x} = p\,\psi$ and $i\hbar\dfrac{\partial\psi}{\partial t} = E\,\psi$. Substituting $i\hbar\dfrac{\partial\psi}{\partial t} = E\,\psi$ into the time-dependent Schrödinger equation gives the time-independent Schrödinger equation, $-\dfrac{\hbar^2}{2M}\dfrac{\partial^2\psi}{\partial x^2} + V(x)\,\psi = E\,\psi$.

Yes, this is more of an excuse for the Schrödinger equation than a derivation, but a rigorous derivation might confuse things rather than clarify them. Given the Schrödinger equation, we will justify it by the answers we get in using it.

To start, consider an electron wave in a confined one-dimensional box. Such a confined particle exists in a standing wave with nodes (zero points) at the end walls. The length of the box will be taken to be an integer number of wavelengths.

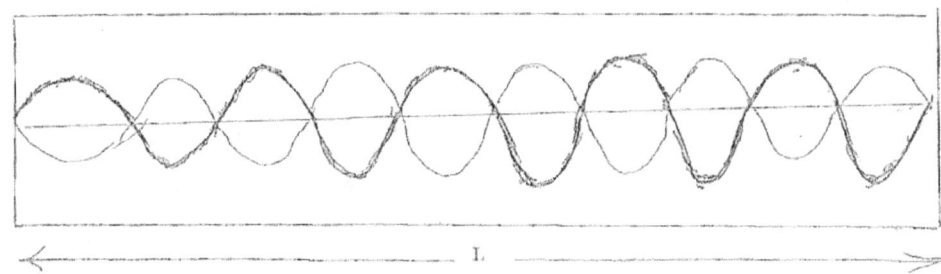

In the previous picture, the length, L, of the box contains five wavelengths of a standing wave. For the "particle" related to the wave, its momentum, $p = m v$, is $p = \frac{h}{\lambda}$ and its energy is $E = \frac{1}{2} m v^2 = \frac{p^2}{2 m}$. If there are n wavelengths in the length, L, then $\lambda = \frac{L}{n}$ and $p = \frac{n h}{L}$. The particle's energy is $E = \frac{n^2 h^2}{2 m L^2}$.

Notice that the more nodes there are in the wave function, the higher the energy value. This is a general principle. With the momentum, $p = M v$, the difference between any two quantum values is $dp = h/L$. The actual location of the particle is anywhere in the length of the box and is unknown to an extent, $dx = L$. Then, $dx\, dp = h$. The better you know the location of a particle the less you know its momentum and vice versa. This is the Heisenberg uncertainty principle.

QUANTIZATION OF THE HARMONIC OSCILLATOR

Throughout physics, many phenomena will be able to be stated in the form of a harmonic oscillator. Hence, the quantization of an oscillator will give us a result which can be applied to these cases without any more mathematics. For an object with mass, M, on a spring of stiffness, k, Newton's law was $M \frac{dv}{dt} = -k x$ when multiplied by $v = \frac{dx}{dt}$ gives $M v \frac{dv}{dt} = -k x \frac{dx}{dt}$ which gives

$$\frac{d}{dt}\left(\frac{1}{2} M v^2\right) = -\frac{d}{dt}\left(\frac{1}{2} k x^2\right).$$

A total energy, $\frac{1}{2} M v^2 + \frac{1}{2} k x^2$ is constant. The object will vibrate as

$x = A \sin(\omega t)$ with $\omega = 2 \pi f t$. Then, $\frac{dv}{dt} = \frac{d^2 x}{dt^2} = -\omega^2 x$ in Newton's law gives

$-M \omega^2 x = -k x$ or $\omega^2 = \frac{k}{M}$.

Then, $k = M \omega^2$ and the energy is $E = \frac{1}{2} M v^2 + \frac{1}{2} M \omega^2 x^2$. The potential energy is

$V(x) = \frac{1}{2} M \omega^2 x^2$. The Schrödinger equation is now $-\frac{\hbar^2}{2 M} \frac{\partial^2 \psi}{\partial x^2} + \frac{1}{2} M \omega^2 x^2 \psi = E \psi$.

We especially want the quantized energies, E, but we also want the functions, ψ, which are called wave functions even if they don't look anything like waves.

This may appear to be an impossible task, but it can be done, and we will use an algebra of things called operators to do it. First, we will turn calculus into an algebra.

The derivative of some function, call it $f(x)$, will give another function call it $g(x)$. In other words, $g(x) = df/dx$. Now, let the symbol, $\hat{D}$, stand for whatever you do to $f(x)$ to get $g(x)$. $g = \hat{D} f$. We say that $\hat{D}$ OPERATES on f to get g. $\hat{D}$ is called an OPERATOR. We can also operate on $f(x)$ by merely multiplying it by x giving $x f$.

Now, $\frac{d}{dx}(x f) = f \frac{dx}{dx} + x \frac{df}{dx}$ or $\hat{D}(x f) = f + x \hat{D} f$.

We write $\hat{D}(x f) = \hat{D} x f$, understanding that $\hat{D}$ takes the derivative of everything in the product to the right of it.

This gives us $\hat{D} x f - x \hat{D} f = f$, written $[\hat{D} x - x \hat{D}] f = f$.

If there is an understanding that these are operators and are to operate on anything to the right of them, we get an algebra, $\widehat{D}x - x\widehat{D} = 1$.

The Schrödinger equation is now $-\dfrac{\hbar^2}{2M}\widehat{D}^2\psi + \dfrac{1}{2}M\omega^2 x^2\psi = E\psi$.

This could be written, $\dfrac{1}{2}\left(\dfrac{M\omega}{\hbar}\right)^2 x^2\psi - \dfrac{1}{2}\widehat{D}^2\psi = \dfrac{ME}{\hbar^2}\psi$.

As they involve measured constants, set $\dfrac{M\omega}{\hbar} = 1$. Then, $\dfrac{M}{\hbar} = \dfrac{1}{\omega}$.

Then, let $\dfrac{E}{\hbar\omega} = e$. We get, $\left[\dfrac{1}{2}x^2 - \dfrac{1}{2}\widehat{D}^2\right] = e$.

The most unlikely method for solving this equation was first developed by Paul A. M. Dirac, and is outlined in his book, <u>The Principles of Quantum Mechanics</u>. They were extended to this type of differential equation by L. L. Infeld and T. E. Hull in their monumental paper, "The Factorization Method", in the January 1951 issue of the Reviews of Modern Physics. We will follow the Infield-Hull approach.

Looking at the operator, $\dfrac{1}{2}\left[x^2 - \widehat{D}^2\right]$, invent two factor operators, $\widehat{L} = \dfrac{1}{\sqrt{2}}\left[x + \widehat{D}\right]$ and $\widehat{R} = \dfrac{1}{\sqrt{2}}\left[x - \widehat{D}\right]$, recalling $\widehat{D}x - x\widehat{D} = 1$.

Then, $\widehat{R}\,\widehat{L} = \dfrac{1}{2}\left[x - \widehat{D}\right]\left[x + \widehat{D}\right] = \dfrac{1}{2}\left[x^2 - \widehat{D}^2\right] + \dfrac{1}{2}\left[x\widehat{D} - \widehat{D}x\right]$. This gives, $\widehat{R}\,\widehat{L} = e - \dfrac{1}{2}$.

Next, $\widehat{L}\,\widehat{R} = \dfrac{1}{2}\left[x + \widehat{D}\right]\left[x - \widehat{D}\right] = \dfrac{1}{2}\left[x^2 - \widehat{D}^2\right] + \dfrac{1}{2}\left[\widehat{D}x - x\widehat{D}\right] = e + \dfrac{1}{2}$.

This gives the algebra, $\widehat{L}\,\widehat{R} - \widehat{R}\,\widehat{L} = 1$. Also notice, $e = \widehat{R}\,\widehat{L} + \dfrac{1}{2}$.

In this case, e and $\widehat{R}$ and $\widehat{L}$ are operators. On the right side of the equation, $1/2$ is just a number. But $\widehat{R}\,\widehat{L}$ is an operator. We will call $\widehat{N} = \widehat{R}\,\widehat{L}$, a number giving operator. Going back to the psi functions, ψ, let them be called f. The functions we want are those satisfying, $\widehat{N}f_n = nf_n$, where n are pure numbers, called characteristic numbers (or eigenvalues in formal language). The functions, f_n, are characteristic functions (or eigenfunctions in formal language).

For each n, there is a number, $e_n = n + \dfrac{1}{2}$. Recall that $e = \dfrac{E}{\hbar\omega}$.

This gives energy values, $E_n = \left(n + \dfrac{1}{2}\right)\hbar\omega$. But how do we find the values of n?

If this doesn't make any sense to you, continue on and it will. It is the sort of person who thinks up clever tricks like this who wins Nobel Prizes. With $\hat{N} = \hat{R}\,\hat{L}$, let the function, f_n, have a number, n, where $\hat{N}\,f_n = n\,f_n$ or $\hat{R}\,\hat{L}\,f_n = n\,f_n$. Now, invent another function, $\hat{L}\,f_n$.

$\hat{N}\,\hat{L}\,f_n = \hat{R}\,\hat{L}\,\hat{L}\,f_n$. But $\hat{R}\,\hat{L} = \hat{L}\,\hat{R} - 1$, $\hat{R}\,\hat{L}\,\hat{L}\,f_n = \left[\hat{L}\,\hat{R} - 1\right]\hat{L}\,f_n$.

Thus, $\hat{N}\,\hat{L}\,f_n = \hat{L}\,\hat{N}\,f_n - \hat{L}\,f_n$.

But $\hat{N}\,f_n = n\,f_n$, and n is just a number. $\hat{N}\,\hat{L}\,f_n = (n-1)\,\hat{L}\,f_n$.

$\hat{L}\,f_n$ is a characteristic function with a characteristic number, $n-1$.

$\hat{L}\,f_n = f_{n-1}$.

Next, form the function, $\hat{R}\,f_n$, and investigate $\hat{N}\,\hat{R}\,f_n = \hat{R}\,\hat{L}\,\hat{R}\,f_n$. In this case, $\hat{L}\,\hat{R} = \hat{R}\,\hat{L} + 1$, and $\hat{N}\,\hat{R}\,f_n = \hat{R}\left[\hat{R}\,\hat{L} + 1\right]f_n = (n+1)\,\hat{R}\,f_n$, $\hat{R}\,f_n = f_{n+1}$.

If n is a characteristic number, $n+1$ and $n+2$ and $n+3$ etc. are also and $n-1$, $n-2$, $n-3$, etc. are characteristic numbers. As $\hat{R}$ and $\hat{L}$ step us up and down a ladder of numbers, they are called LADDER OPERATORS.

Now, $E_n = \left(n + \frac{1}{2}\right)\hbar\,\omega$ is a positive number for an oscillator. Therefore, as n steps downward, it must have a minimum value. Call that m.

Now, $\hat{N} = \hat{R}\,\hat{L}$, is the number giving operator. Then, $\hat{N}\,f_m = m\,f_m$, but $\hat{L}\,f_m = f_{m-1}$. If there is a f_{m-1}, then $\hat{N}\,f_{m-1} = (m-1)\,f_{m-1}$ and $m-1$ is not permitted and $f_{m-1} = 0$. If $\hat{L}\,f_m = 0$ and $\hat{N} = \hat{R}\,\hat{L}$, then $\hat{N}\,f_m = 0$. But, f_m is not zero. If $\hat{N}\,f_m = m\,f_m$, then $m = 0$.

We have broken into the chain of numbers, n. For every n there is a $n+1$ giving the values of n to be 0, 1, 2, 3, and so on up.

We now have quantized energies for the oscillator:
$E_n = \left(n + \frac{1}{2}\right)\hbar\,\omega$, $n = 0$, 1, 2, 3, etc.

Next, what are the functions, $f_n(x)$? We know that $\hat{L}\,f_n = 0$, where $\hat{L} = \frac{1}{\sqrt{2}}\left[x + \hat{D}\right]$ hence, $\left[x + \hat{D}\right]f_0 = 0$. $\hat{D}\,f_0 = -x\,f_0$. If $\frac{df}{dx} = -x\,f$, there is a function $g(x)$ such that $f(x) = e^g$. Then, $\dfrac{df}{dx} = \dfrac{df}{dg}\dfrac{dg}{dx}$. If $f = e^g$, then $\dfrac{df}{dg} = f$ and $\dfrac{dg}{dx} = -x$. This gives $g = -\dfrac{1}{2}x^2$.

We have now broken into the set of functions, f_n. Recall $f_{n+1} = \hat{R}\,f_n$. So $f_1 = \hat{R}\,f_0$, $f_2 = \hat{R}\,f_1$ or $f_2 = \hat{R}^2\,f_0$, so $f_n = \hat{R}^n\,f_0$, where $\hat{R} = \dfrac{1}{\sqrt{2}}\left[x - \hat{D}\right]$. Neglect the $\dfrac{1}{\sqrt{2}}$ and look at the operator, $\left[x - \hat{D}\right]$. Keep $f_0 = e^{-x^2/2}$ in mind. $\left[x - \hat{D}\right]f = xf - \dfrac{df}{dx}$.

How do we raise this to the nth power? Let us pull some tricks.

We have seen that $\dfrac{d}{dx}e^{-x^2/2} = -x\,e^{-x^2/2}$. Multiply this by $e^{+x^2/2}$. This gives $e^{+x^2/2}\left[\dfrac{d}{dx}e^{-\frac{x^2}{2}}\right] = -x$. Then notice $e^{+x^2/2}\,e^{-x^2/2} = 1$.

We get $\left[x - \hat{D}\right]f = -e^{+x^2/2}\left[f\left(\dfrac{d}{dx}e^{-x^2/2}\right) + e^{-x^2/2}\left(\dfrac{df}{dx}\right)\right]$.

But if $u(x)$ and $v(x)$ are two functions of x, then $\dfrac{d}{dx} = (u\,v) = u\dfrac{dv}{dx} + v\dfrac{du}{dx}$.

Then, $\left[x - \hat{D}\right]f = -e^{+x^2/2}\dfrac{d}{dx}\left[f\,e^{-x^2/2}\right]$.

We are not interested in $\sqrt{2}$, so let $\hat{R} = \left[x - \hat{D}\right]$, $f_{n+1} = \hat{R}\,f_n$.
$f_1 = -\,e^{+x^2/2}\dfrac{d}{dx}\left[e^{-x^2/2}\,f_0\right]$, but $f_0 = e^{-x^2/2}$. $f_1 = -\,e^{+x^2/2}\dfrac{d}{dx}\left[e^{-x^2}\right]$ and $f_2 = -\,e^{+x^2/2}\dfrac{d}{dx}\left[e^{-x^2/2}\,f_1\right]$. But, $e^{+x^2/2}\,e^{-x^2/2} = 1$. $f_1 = (-)^2\,e^{+x^2/2}\dfrac{d^2}{dx^2}\left[e^{-x^2}\right]$.

From this you get the idea, $f_n = (-)^n\,e^{+x^2/2}\dfrac{d^n}{dx^n}\left[e^{-x^2}\right]$.

We have now gotten the characteristic energies, $E_n = \left(n + \dfrac{1}{2}\right)\hbar\,\omega$, and their related functions. Instead of f_n these are usually written as a Greek ψ_n and in the early days of quantum mechanics called wave functions.

Remember that $f_0(x) = e^{-x^2/2}$. Then $f_1 = -\,e^{+x^2/2}\dfrac{d}{dx}\left[e^{-x^2}\right]$.

This gives $f_1(x) = x\,e^{-x^2/2}$. Taking more derivatives, you can show that $f_2(x) = (4x^2 - 2)\,e^{-x^2/2}$ and $f_3(x) = (4x^3 - 6x)\,e^{-x^2/2}$.

Notice that e^{-x^2} is being multiplied by polynomials. These are called Hermite polynomials and are written $H_n(x)$.

For reference, remember where these results are located in the text. You will see them often again, even in unexpected places as fringes in the output of certain lasers.

On page 16-10 the Schrödinger equation was $-\frac{\hbar^2}{2M}\hat{D}^2\psi_n + \frac{1}{2}M\omega^2 x^2\psi_n = E_n\psi_n$

The energy operator here is written, $\hat{H} = -\frac{\hbar^2}{2M}\hat{D}^2 + \frac{1}{2}M\omega^2 x^2$, and is called the Hamiltonian after the great 19th century mathematician Sir W. R. Hamilton. If $\hat{N}$ is the number-giving operator, then $\hat{H} = \left(\hat{N} + \frac{1}{2}\right)\hbar\omega$.

Each characteristic value, E_n, and characteristic function define a state in quantum mechanics. (Properly, they are called eigenvalues and eigenfunctions in half-translated German.)

Instead of $\hat{N}\psi_n = n\psi_n$, we introduce what we will call a state symbol, $|n\rangle$, and write, $\hat{N}|n\rangle = n|n\rangle$. Then, $\hat{H}|n\rangle = \left(n + \frac{1}{2}\right)\hbar\omega|n\rangle$.

The nth state has a state function (eigenfunction) ψ_n and this is written in our new notation, $\psi_n(x) = \langle x|n\rangle$ which is interpreted as the function of x associated with the nth state.

- -

Recall the notation from chapter 15. A function, $f(x)$, would have a vector $|f\rangle$ and a location, x, would have a vector, $\langle x|$, where $\langle x|f\rangle = f(x)$. The derivative function is written, $\langle x|\hat{D}|f\rangle = \frac{df}{dx}$.

We have just now developed the quantization of a harmonic oscillator. You probably are saying: "So what? Who is interested in harmonic oscillators." It so happens that all sorts of questions in physics can be stated in such a form that their quantization can be solved by harmonic oscillator quantization. In the next chapter, we will use this tool to get the energy states of protons and neutrons in the atomic nucleus. Later in this chapter, we will discover that the quantized frequencies of a vibrating string gives us the energy states of particles called phonons and the harmonic oscillator quantization gives the number of phonons in the string's energy. At this point, welcome to second quantization in quantum field theory, which is developing into the major tool in theoretical physics of the 21st century.

In the next topic, we find that Hermite-Gaussian functions actually happen in the laboratory with the output of certain solid-state lasers. The treatment here is not to found in any textbook in laser physics. But the physics involved is coming into engineering practice. Please consider harmonic oscillator quantization as a major tool in quantum physics.

The light output of some solid-state lasers exhibits a fringe structure which cannot be identified with a diffraction pattern. Functionally it appears like the square of Hermite-Gaussian functions. We next explore the implications of such an identification.

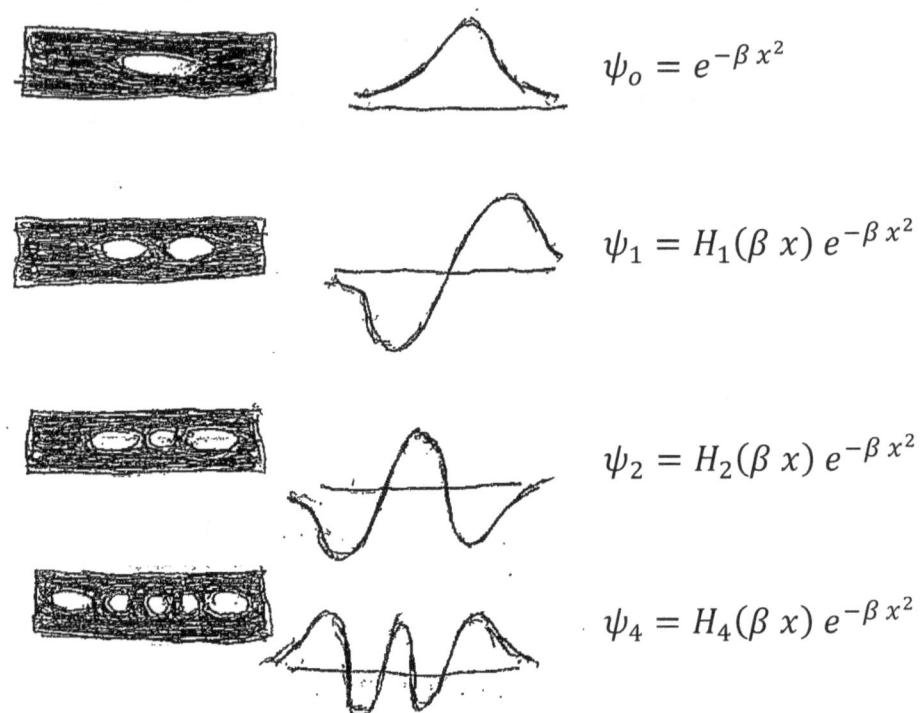

$$\psi_o = e^{-\beta x^2}$$

$$\psi_1 = H_1(\beta x)\, e^{-\beta x^2}$$

$$\psi_2 = H_2(\beta x)\, e^{-\beta x^2}$$

$$\psi_4 = H_4(\beta x)\, e^{-\beta x^2}$$

For a light wave if ψ is either electric or magnetic field the ψ function satisfies, $\dfrac{\partial^2 \psi}{\partial x^2} = \dfrac{1}{u}\dfrac{\partial^2 \psi}{\partial t^2}$. And if $\psi(x,t) = A(x)\, e^{i\omega t}$, then $\dfrac{\partial^2 \psi}{\partial x^2} = -\dfrac{\omega^2}{u^2}\psi$, where u is the wave velocity in the solid material of the laser. In the quantum theory of a harmonic oscillator, Hermite-Gaussian functions satisfy,

$-\dfrac{\hbar^2}{2M}\dfrac{\partial^2 \psi_n}{\partial x^2} + \dfrac{1}{2}M\omega^2 x^2 \psi_n = \left(n + \dfrac{1}{2}\right)\hbar\,\omega\,\psi_n$. This could also be written,

$\dfrac{\partial^2 \psi_n}{\partial x^2} = -\left(\dfrac{M\omega}{\hbar}\right)\left[2n + 1 - \dfrac{M\omega}{\hbar}x^2\right]\psi_n$. n is an integer. Let $k = \dfrac{M\omega}{\hbar}$.

Comparing the two equations would give a wave velocity, u, to obey

$u^2 = \dfrac{\omega^2}{k\,[2n+1-k\,x^2]}$, suggesting an index or refraction to vary by $\sqrt{2n+1-k\,x^2}$, but what does this mean?

This says that the refractive index decreases as we go out from the center, or conversely increases toward the center of the laser cylinder. That means that the wave velocity increases as we go out from the center. The waves on the outside moving faster than the waves at the center produces a self-focusing effect as we will now investigate.

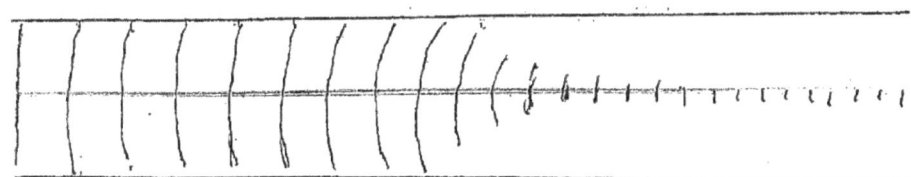

As the wave moves down the solid-state laser cylinder, the light intensity (Pointing vector) increases. Notice, now, that the light intensity and the refractive index increases as we go from the outside toward the center. This is a nonlinear effect on refractive index which increases with light intensity. This occurs in very high-power lasers.

Damage from this self-focusing was early noticed in yttrium-aluminum-garnet, $Y_3Al_5O_{16}$ (YAG) lasers with intense output beams. A core that was melted down the center of the cylinder.

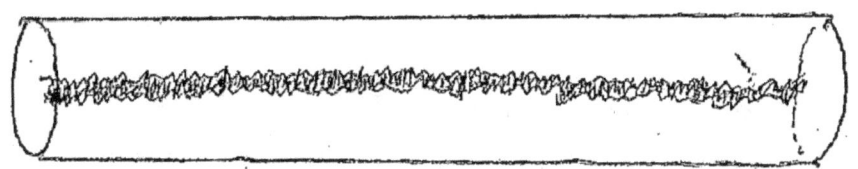

At this point, review the discussion of index of refraction beginning on page 13-20 of chapter 13.

In the case of conventional (i.e., linear) optics, the induced polarization depends linearly on the electric field strength in a manner that can often be described by the relationship

$$\tilde{P}(t) = \epsilon_0 \chi^{(1)} \tilde{E}(t),$$

where the constant of proportionality $\chi^{(1)}$ is known as the linear suscepti-bility and ϵ_0 is the permittivity of free space. In nonlinear optics, the optical response can often be described   by expressing the polarization $\tilde{P}(t)$ as a power series in the field strength $\tilde{E}(t)$ as

$$\tilde{P}(t) = \epsilon_0 \left[\chi^{(1)} \tilde{E}(t) + \chi^{(2)} \tilde{E}^2(t) + \chi^{(3)} \tilde{E}^3(t) + \cdots \right]$$

$$\equiv \tilde{P}^{(1)}(t) + \tilde{P}^{(2)}(t) + \tilde{P}^{(3)}(t) + \cdots.$$

The quantities $\chi^{(2)}$ and $\chi^{(3)}$ are known as the second- and third-order non-linear optical susceptibilities, respectively.

We shall refer to $\tilde{P}^{(2)}(t) = \epsilon_0 \chi^{(2)} \tilde{E}^2(t)$ as the second-order nonlinear po-larization and to $\tilde{P}^{(3)}(t) = \epsilon_0 \chi^{(3)} \tilde{E}^3(t)$ as the third-order nonlinear polariza-tion.

The theory of refractive index starting on page 13-20 of chapter 13 has the electric field of the light wave inducing induced electric dipole moments in the matter through which the light is passing. The phenomenon which we are discussing involves a third order non-linear refractive index. This implies the charge dipoles to be produced in the light itself. By normal intuition this is nonsense.

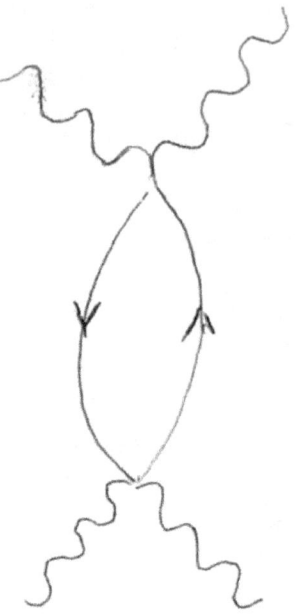

An explanation assumes the photons to be particles that can collide. Suppose in the process of colliding, a short-lived pair of positive and negative electrons are created as with the vacuum polarization described on page 9-17 of chapter 9 and page 17-27 of chapter 17. During the interval of time that they exist the electric field of the light wave could pull them apart creating a dipole, αE, where α is a polarizability.

At the end of chapter 13, we found that the energy per unit volume of a light wave was $\epsilon_o E^2$. Now the energy of a photon is hf giving the photons per unit volume to be $\epsilon_o E^2/hf$.

Multiplying this by the dipole moment, αE, gives a volume polarization, $\frac{\alpha}{hf}\epsilon_o E^2 E$. This has a magnitude, $P = \epsilon_o \frac{\alpha}{hf} E^3$, which gives a third order nonlinear index of refraction which can be derived as in chapter 13.

Question: Although the outgoing photons have the same energy as the incoming photons, they may not have the same frequencies. Is there any spectral line broadening observed with this phenomenon?

Refer to:

T. Ozawa et al: TOPOLOGICAL PHOTONICS in THE REVIEWS OF MODERN PHYSICS January to March 2019

H. Injeyan and G. D. Goodno: HIGH-POWER LASER HANDBOOK, McGraw Hill, 2011

When a photon emerges from the laser, its velocity is
$c = f\lambda$ with a momentum $p = \frac{h}{\lambda}$ or $p = \frac{hf}{c}$. If the energy, hf,
has a mass equivalent, $hf = mc^2$ giving $p = mc$.

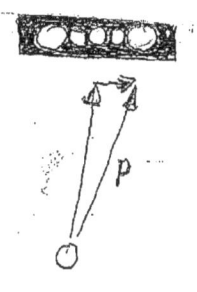

With a Hermite-Gaussian output mode, photons may have an
outward momentum component, causing a broadening of the
laser beam. A cylindrical lens will play games with that
component.

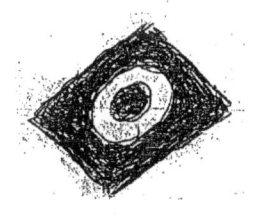

With a laser beam with a Hermite-Gaussian mode coming up
from the bottom of the page (see the figure on the right),
make the axis of the pattern be diagonal to the x and y
axes. Pass the light through a cylindrical lens, as shown.
Points on the y-axis will be focused across the x-axis.
Points on the x-axis are not affected. The pattern is
rotated by **90°**.

The two bright spots on the Hermite-Gaussian mode have
linear momentum. If there is no force with the optical
focusing, the momentum is conserved. The momentum
perpendicular to the axis of the laser beam is directed
outward before it is focused. After being focused across
the x-axis, it will be tangential. This will cause the
spots to rotate and a rotating momentum creates angular
momentum. The laser beam will now have orbital angular
momentum.

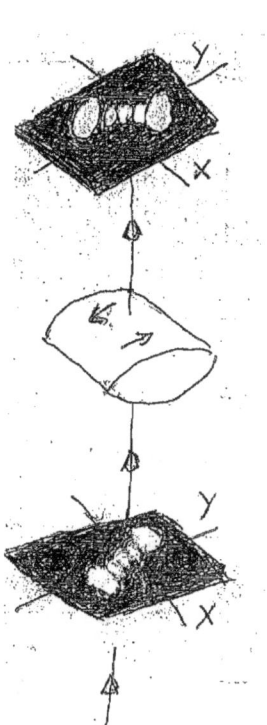

Les Allen and Miles
Padgett at Glasgow
University and
their research
group have done
major work on the
orbital angular
momentum of light.

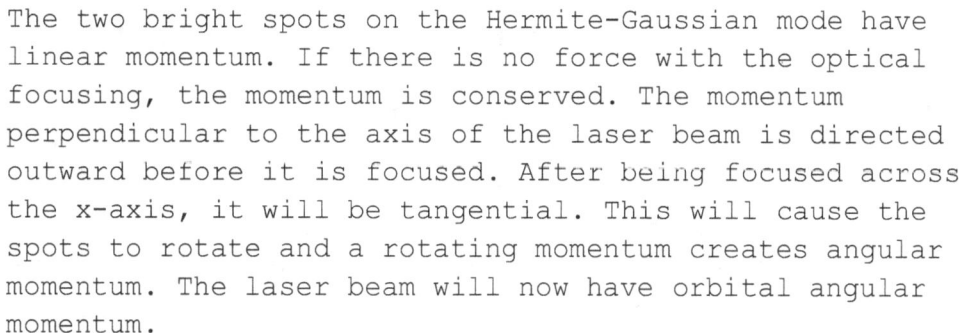

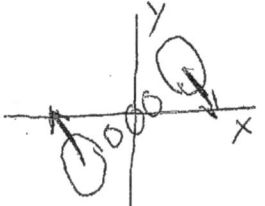

Laser beam after
reflection across
the x-axis, coming
out of the page

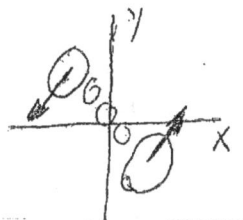

Laser beam after
reflection across
the x-axis, coming
out of the page.

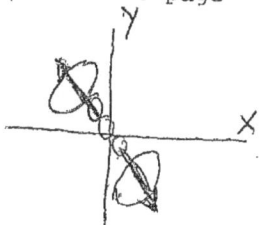

Incoming laser
beam, directed
out of the page

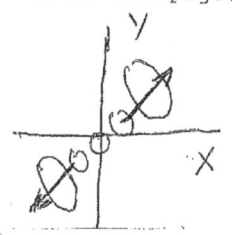

Incoming laser
beam, directed
out of the page.

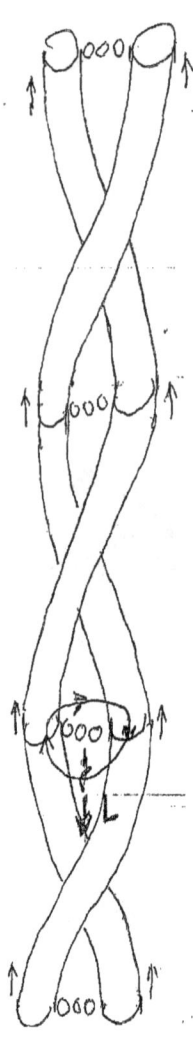

What does a rotating Hermite-Gaussian pattern look like as it advances from the bottom toward the top of the page?

For a photon with energy, $E = hf$, we set $hf = mc^2$ to give a mass, m. If a single photon has a mass, then a lot photons have mass, and mass going 'round and 'round has angular momentum. As the rotating Hermite-Gaussian pattern goes 'round and 'round, the light involved has angular momentum. As the light moves up the page, it carries angular momentum with it.

The pictures show a Hermite-Gaussian pattern advancing up the page. No, this is not a strand of DNA, this is a laser beam. In the beam on the left notice the pattern at a point near the bottom of the picture. As the waves advance up from the bottom of the picture, the bright spot on the right will be moving out of the picture and that on the left will be moving in. This gives an angular momentum down and opposite to the wave's direction of motion. With the waves on the right as it advances, the spot on the left is moving out of the picture and the one on the right is going into the picture. This gives angular momentum in the direction of the wave's motion.

Thus, a laser beam with a rotating Hermite-Gaussian cross section will carry angular momentum. If a particle or very small object absorbs a photon from the beam, the angular momentum is absorbed, and the particle or small object will be made to rotate. In this case it is a very delicate experiment, but the effect has been noticed with very strong laser beams. Actually, a photon carries what is called an intrinsic spin angular momentum, but we need a bit more physics to develop that idea. The angular momentum just discussed is called orbital angular momentum.

The Hermite-Gaussian functions have a many dot pattern with the obvious feature of the brightest dots at the ends of the array. It is these brightest spots that are used to form the helical beams. With increasing order of Hermite-Gaussian functions, these spots are farther from the center and their off-axis momentum increases. When rotated **90°** this gives higher angular momentum decreasing the length of one turn of their helix. Higher order H-G functions give shorter wavelength helices.

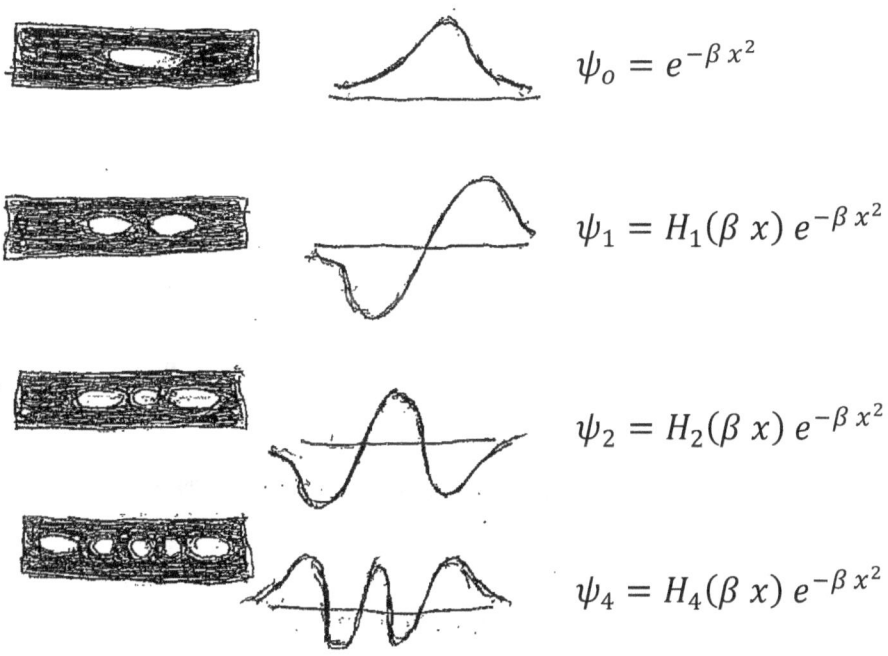

$$\psi_o = e^{-\beta x^2}$$

$$\psi_1 = H_1(\beta x)\, e^{-\beta x^2}$$

$$\psi_2 = H_2(\beta x)\, e^{-\beta x^2}$$

$$\psi_4 = H_4(\beta x)\, e^{-\beta x^2}$$

Much of the basic work investigating the orbital angular momentum (OAM) of a laser beam was done by Miles Padgett and Les Allen at the University of Glasgow. Their discussion of their work was published in the May 2004 issue of PHYSICS TODAY, the magazine of the American Institute of Physics. More recently, a group working under Alan Willner at the University of Southern California sent several channels of data over OAM from different Hermite-Gaussian laser outputs. There was no mixing (interference) of OAM channels. Willner discussed his work in the August 2016 issue of SPECTRUM, the magazine of the Institute of Electrical and Electronic Engineers. Following is Willner's account of his work:

There are a variety of ways to create and transmit helical beams, we chose a conventional approach, built with as many off-the-shelf components as possible. The transmitter generates regular laser beams which are then passed through a spatial light modulator, based on a liquid crystal, in order to impart a twist to the beam. At the receiving end, each OAM beam was converted back into a regular plane wave by passing it through a spatial modulator with the inverse pattern. The data could then be recovered by a conventional optical receiver.

In 2012, we published our first journal article on this approach. Our experiment sent **32** different optical beams of the same frequency, each carrying **80** gigabits per second of data, over a modest distance of just one meter in the laboratory. But the total transmission rate, some **2.5** terabits per second, was actually quite high for free-space communications. And it held out the promise for longer-distance transmissions and, because we used only one frequency, much higher data rates.

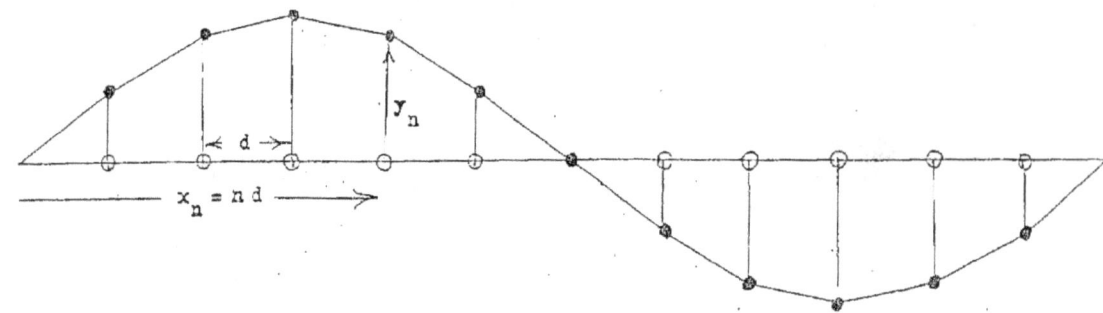

We now consider a two-loop standing wave in a light-weight string loaded with equally spaced balls, each of which has a mass, M. Let the integers, n, denote the balls, where in this case, n goes from 1 to 11. In general, let N be the total number of balls on the string. Let d be the distance between each pair of balls and $x_n = nd$ be the distance from the left end of the string to the nth ball. When the string is vibrating, let y_n be the displacement of the nth ball. Referring to the description of a vibrating standing wave in chapter 6, $y_n = A \sin \frac{2 \pi x_n}{L} \cos (2 \pi f_2 t)$, where L is the length of the string and f_2 is the frequency when vibrating with two loops.

Obviously, each ball, being in vibrating motion, has a force acting on it. Let F stand for the tension force in the string. When the string is horizontal, the string on either side of any ball exerts a tension force, and when the string is horizontal and straight, the tension forces cancel each other.

When the objects are displaced, the strings on either side of any ball are not pulling in opposite directions, and the tension forces cannot cancel each other. Consider the effective force on the nth ball. But first we must describe some geometry. The nth object has a displacement, y_n, while the objects on either side have displacements, y_{n-1} and y_{n+1}. Let θ_n be the angle between the string and the horizontal to the right of the nth ball and θ_{n-1} be the angle to the right of the $n-1$th ball and to the left of the nth ball. If d is the horizontal distance between balls, then $\tan \theta_n = \frac{y_{n+1} - y_n}{d}$ and $\tan \theta_{n-1} = \frac{y_n - y_{n-1}}{d}$.

The tension force to the right of the nth ball has horizontal and vertical components, $F_h = F \cos \theta_n$ and $F_v = F \sin \theta_n$. The string to the left has horizontal and vertical components, $F_h = F \cos \theta_{n-1}$ and $F_v = F \sin \theta_{n-1}$.

Newton's law for a vibrating string can be written as follows:

$$M \frac{\partial^2 y_n}{\partial x^2} = F \left[\frac{y_{n+1} - y_n}{d} - \frac{y_n - y_{n-1}}{d} \right]$$

If there are many-many particles with decreasing mass per particle the string approaches being a continuum, and Newton's law can be stated as calculus.

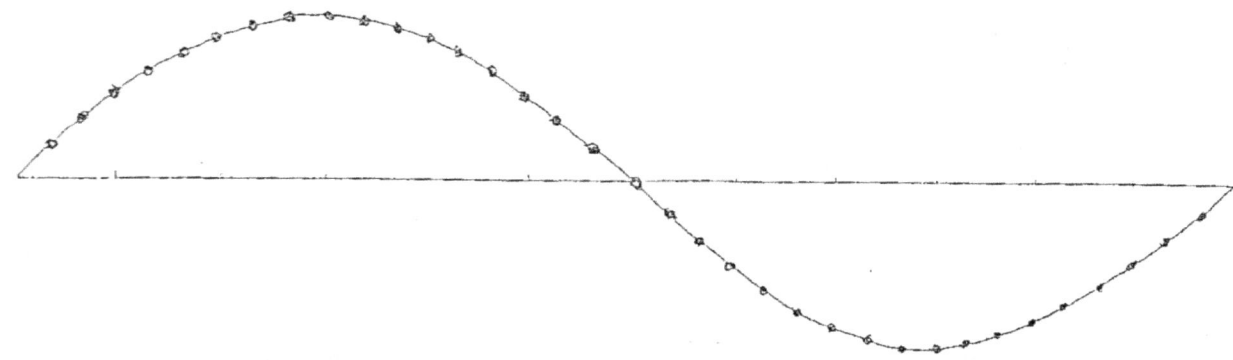

For very many, very small, balls the problem approaches that for a continuous string as treated in chapter 6. Recall that d is the jump in x from one ball to the next one; hence in the previous notation, d is dx. $y_{n+1} - y_n$ is the change of y on the right side of the nth ball. $\frac{y_{n+1} - y_n}{d} = \left(\frac{dy}{dx} \right)_R$, which is the slope of the string to the string to the right of the nth ball. $\frac{y_n - y_{n-1}}{d} = \left(\frac{dy}{dx} \right)_L$ is the slope of the string to the left of the nth ball. $\left(\frac{dy}{dx} \right)_R - \left(\frac{dy}{dx} \right)_L = d \left(\frac{dy}{dx} \right)$ is the change of the slope from the left of the nth ball to the right of it. Putting this into the general differential equation for a vibrating string and the dividing both sides of the equality by $d = dx$ gives $\frac{M}{d} \left(\frac{d^2 y_n}{dt^2} \right) = F \frac{d \left(\frac{dy}{dx} \right)}{dx}$ or $\frac{M}{d} \left(\frac{d^2 y_n}{dt^2} \right) = F \left(\frac{d^2 y}{dx^2} \right)$ and as M is the mass in a length, d, of the string, $\frac{M}{d}$ is a mass per unit length, or linear density, to which we shall assign the symbol mu, μ. As we now have a displacement, y, at almost every x, we will drop the subscript, n. Then one slight change in notation is necessary when we have a function of several independent variables. $\frac{dy}{dx}$ is the slope of y with a change of x at a fixed moment of time. $\frac{dy}{dt}$ is the velocity of y with changing time at a fixed point of x. To indicate that the function is changing with respect to one variable, while other variables are constant, we round off the $d's$ to $\partial's$ giving

$\mu\left(\frac{\partial^2 y}{\partial t^2}\right) = F\left(\frac{\partial^2 y}{\partial x^2}\right)$ as Newton's second law of motion or a vibrating continuous string. An α-loop standing wave would be described by

$$y_\alpha(x,t) = A \sin\frac{\alpha\pi x}{L}\cos(2\pi f_\alpha t).$$

The partial derivative is just a normal derivative with respect to the variable specified while the other variable is treated as being constant.

Hence, $\frac{\partial y}{\partial x} = \frac{\alpha\pi}{L}A\cos\left(\frac{\alpha\pi x}{L}\right)\cos(2\pi f_\alpha t)$ and $\frac{\partial y}{\partial t} = -2\pi f_\alpha A \sin\left(\frac{\alpha\pi x}{L}\right)\sin(2\pi f_\alpha t).$

One more derivative gives $\frac{\partial^2 y}{\partial x^2} = -\left(\frac{\alpha\pi}{L}\right)^2 y$ and $\frac{\partial^2 y}{\partial t^2} = -\left(2\pi f_\alpha\right)^2 y.$

Putting these results into Newton's second law gives $f_\alpha = \frac{\alpha}{2L}\sqrt{\frac{F}{\mu}}$, where α is an integer number of loops.

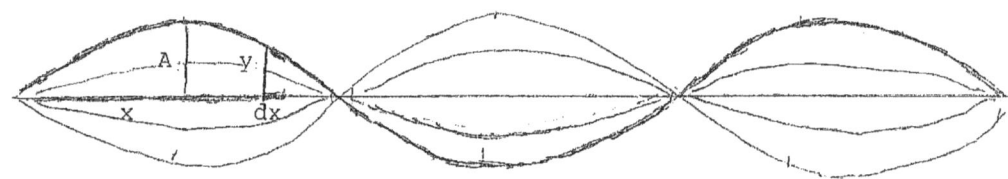

Look at a standing wave, in this case with three loops ($\alpha = 3$) in a string under tension force, F, and with a mass-per-unit length, μ. A section of the string with length, dx, has a mass, $\mu\,dx$.

If the string is vibrating, the kinetic energy of $\mu\,dx$ is $KE = \frac{1}{2}\mu\,dx\left(\frac{\partial^2 y}{\partial t^2}\right)^2.$

This is a time rate of change $\frac{d}{dt}[KE] = \frac{1}{2}\mu\,dx\,2\left(\frac{\partial y}{\partial t}\right)\left(\frac{\partial^2 y}{\partial t^2}\right).$ But, $\mu\left(\frac{\partial^2 y}{\partial t^2}\right) = F\left(\frac{\partial^2 y}{\partial x^2}\right).$
From the equation for a standing wave above,

$\frac{\partial^2 y}{\partial t^2} = -\left(\frac{\alpha\pi}{L}\right)^2 y$. This gives $\frac{d}{dt}[KE] = -\frac{1}{2}dx\,F\left(\frac{\alpha\pi}{L}\right)^2 2\,y\left(\frac{\partial y}{\partial t}\right).$

But $2\,y\left(\frac{\partial y}{\partial t}\right) = \frac{\partial}{\partial t}(y^2)$ and above, $\omega = 2\pi f_\alpha = \frac{\alpha\pi}{L}\sqrt{\frac{F}{\mu}}.$

From this, $\left(\frac{\alpha\pi}{L}\right)^2 F = \omega^2 \mu.$ Combining things, $\frac{d}{dt}[KE] = -\frac{d}{dt}\left[\frac{1}{2}\mu\,dx\,\omega^2\,y^2\right].$

Let $m = \mu\,dx$ be the mass of the piece, dx, of the string. Further combining,

$\frac{d}{dt}\left[\frac{1}{2}m\left(\frac{dy}{dt}\right)^2 + \frac{1}{2}m\,\omega^2\,y^2\right] = 0$ or $E = \frac{1}{2}m\left(\frac{dy}{dt}\right)^2 + \frac{1}{2}m\,\omega^2\,y^2$ is constant.

On page 16-9, using y now instead of x. the energy of a traditional oscillator is $E = \frac{1}{2} m v^2 + \frac{1}{2} m \omega^2 y^2$. This was quantized with values,

$E_n = \left(n + \frac{1}{2}\right) \hbar \omega$, $n = 0$, 1, 2, 3... etc. Now $\hbar = \frac{\hbar}{2\pi}$ and $\omega = 2\pi f$, so $\hbar \omega = hf$.

As the mechanical energy of a vibrating string is $E_n = \left(n + \frac{1}{2}\right) hf$ and $n = 0$, 1, 2, 3... etc, the quantum number changes by $dn = 1$ and the energy changes by $dE = hf$. This packet of energy actually acts as a particle, and we call it a PHONON. The energy of an electromagnetic wave of frequency, f, is quantized in the same way, $dE = hf$, and we call that PHOTON.

But the standing wave frequencies are also quantized $f_\alpha = \frac{\alpha}{2L} \sqrt{\frac{F}{\mu}}$ where $\alpha = 1$, 2, 3... etc.

α becomes a quantum number, quantizing particle energies. The quantizing of particle energies is called FIRST QUANTIZATION. Then with $E_n = \left(n + \frac{1}{2}\right) h f_\alpha$, the quantum number, n, now n_α, gives the number of phonons carrying the total energy. This quantization of the number of particles is called SECOND QUANTIZATION.

In the second quantization, the operator, $\hat{R}$, creates a phonon and the operator, $\hat{L}$, annihilates a phonon. The second quantization of electromagnetic waves giving photons is discussed on page 13-66.

THE QUANTUM HALL EFFECT

With the material developed in this chapter, we can study one of the most
puzzling phenomena in today's physics. First, we think about normal
electrical resistance. Next, we find out how strange that property of matter
can be at low temperature and high magnetic fields.

We shall now develop a simple-minded theory of electrical conduction. An
electric field, E, will accelerate electrons in a direction opposite to
itself, creating an electric current. But the electrons will be bombarded by
phonons knocking electrons out of the current. Assume that the average
effect of a phonon collision is to reduce the electron's velocity on the
current to be zero and so the field has to start all over again to bring the
electron into the current. If t is the average time between collisions, the
electron will move a distance, $x = \frac{1}{2}\left(\frac{e E}{m}\right)t^2$, with an average velocity,
$\overline{v} = \frac{x}{t} = \frac{e E}{2 m}t$. If v is the rate of collisions per second, $t = \frac{1}{v}$, and the average
velocity is $\overline{v} = \frac{e E}{2 m v}$.

Let the current be flowing in a
wire of cross-section area, A. Let
there be N electrons per unit
volume. An average electron goes a
distance, $\overline{v}t$, in a time, t, and
all electrons before it will cross

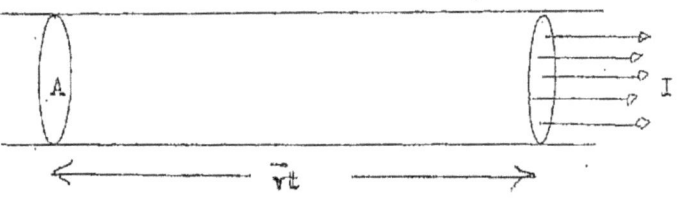

the far cross-section in that time. All electrons in a volume, $\overline{v}A$, will
pass that point in a unit time. This gives an electric current,
$I = N e A \overline{v} = \left(\frac{N e^2}{2 m v}\right)A E.$

$\sigma = \left(\frac{N e^2}{2 m v}\right)$ is the conductivity and the electrical resistance is $R = \frac{L}{\sigma A}$ for a
path of length, L.

NOW TAKE A GAS (TODAY OFTEN CALLED A LIQUID) OF CONDUCTION ELECTRONS

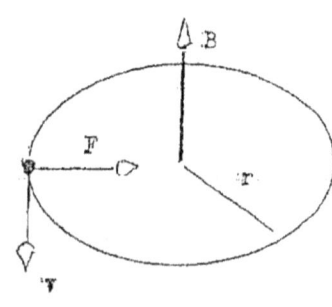

Slap a magnetic field on our electron gas and an
electron has a force, $F = e v B$, making it move in a
circle. Then, $m\frac{v^2}{r} = e v B$ or $m v = E B r$. $T = \frac{2 \pi r}{v}$ is the
time for one revolution giving the cyclotron frequency,
$f = \frac{v}{2 \pi r} = \frac{e B}{2 \pi m}$, so called because it is the principle by
which E. O. Lawrence designed the particle accelerator
called the cyclotron.

Lev. D. Landau, applying oscillator quantization, gave $\frac{1}{2}mv^2 = \left(n + \frac{1}{2}\right)hf$ to be the quantization of electron energy, where $n = 1$, 2, 3, etc. (some authors write i instead of n but others of us think n to look more like a quantum number. The resulting energy levels are called Landau levels.

Now, $\frac{1}{2}mv^2 = \left(n + \frac{1}{2}\right)\frac{heB}{2\pi m}$ and $mv = eBr$.

Combining these equations gives us $\left(n + \frac{1}{2}\right)\frac{heB}{2\pi m} = \frac{B^2 e^2 r^2}{2m}$ or $B\pi r^2 = \left(n + \frac{1}{2}\right)\frac{h}{e}$ is the magnetic flux through the loop. But $\Phi = \pi r^2 B$ is the magnetic flux through the loop.

This makes it appear that the magnetic flux is quantized, and it is physically interpreted that way. The quantum of magnetic flux is $d\Phi = \frac{h}{e}$, or some authors write $\Phi_o = \frac{h}{e}$. The free electrons gas now becomes a gas of quantized vortices.

To get the effects of the flux quantization it is necessary to get all of the electron motion to be in a plane to which B is perpendicular. If the electrons are in a slab of thickness, D, their wavelengths, perpendicular to the slab, are

$\lambda_z = D/n_z$ giving moments, $p_z = \frac{n_z h}{D}$, and energies,

$E_z = \frac{n_z^2 h^2}{2 m_e D^2}$.

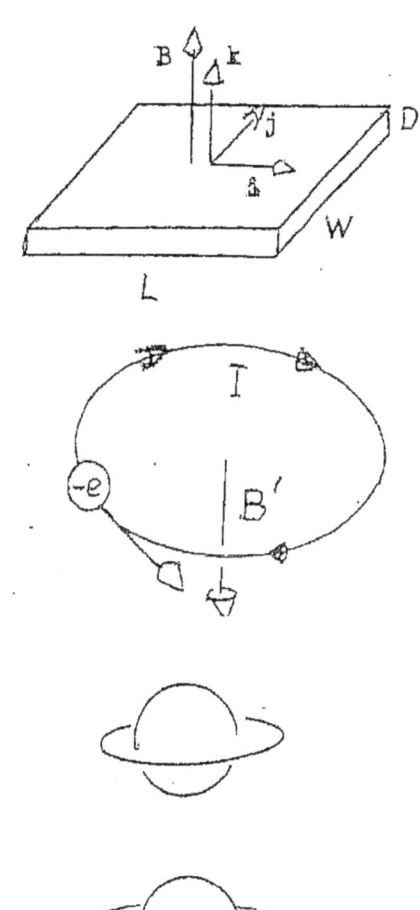

For a thickness of 1.0 nm $= 10^{-9}$ m, the first energy jump is 1.5 eV. At a temperature of 1.0 Kelvin, the energy of one degree of freedom is 4.3×10^{-5} eV.

Now let us look at one quantum of flux. If it is created by an electron going counter-clockwise in a circle, the equivalent positive charge is going clockwise producing a magnetic field correction, B', as shown. Once this is established, the flux quantum acts as a unique particle, independent of the electron which we said produced it. Hence, we have two-dimensional gases.

In the next discussion, the quantized Hall effect, some strange things will happen. So, let us consider some strange things which might happen with the two-dimensional gases. Suppose the flux quanta act as spin-1/2 particles and combine with electrons to form composite particles. An electron with an even number of flux quanta would be a Fermion, but an electron with an odd number of flux quanta would be a Boson. Then, as the magnetic field increases in intensity, the two-dimensional composite gas would go between satisfying Fermi and Bose statistics. Welcome to the two-dimensional world of quantized fluids. If N electrons form N composite particles, each with an integer, i, number of flux quanta, and the N CP's have a total area, A, each electron sees a reduced magnetic field, B^*, where $B^* A = B A - i N \Phi_0$, giving

$$B^* = B - \frac{i N \Phi_0}{A}.$$

If N is the number of electrons per unit volume, $N L W D$ will be the number of electrons in a slab of dimensions, L, W, and D, giving the number of electrons per unit area to be $n_s = \frac{N L W D}{L W} = N D$.

If a magnetic field, B, is perpendicular to an area, $L W$, the total flux through the area is $B L W$, and the total flux quanta in that area is $\frac{B L W}{\Phi_0}$, giving the flux quanta per unit area to be $\frac{B}{\Phi_0} = \frac{e B}{h}$.

These give what is known as the filling factor, the fraction on flux quanta filled by electrons. $\nu = n_s / \left(\frac{e B}{h} \right) = \frac{n_s h}{e B}$ or electrons per quantum.

Thus, $\frac{1}{\nu}$ is the flux quanta per electron. The filling factor will be a significant parameter in discussing the quantum Hall effect and the fractional quantum Hall effect.

THE HALL EFFECT was discovered by Edwin Hall in 1878.

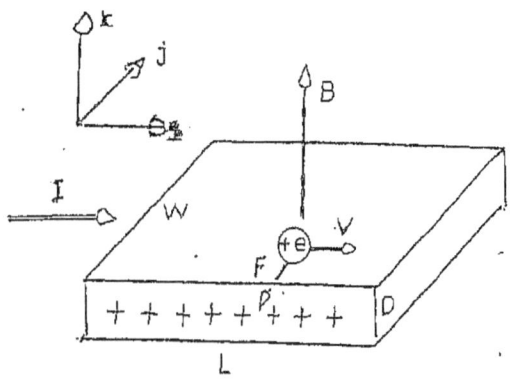

Take a piece of p-type semiconductor (chapter 14) as shown having a magnetic field perpendicular to it in the z-direction. The positive carriers will experience a force in the negative y-direction making the near face positive and the far face negative. The separated charges will produce an electric field in the positive y-direction. A steady state configuration will occur when the electric force balances the magnetic force.

$e\,E_y = e\,v_x\,B_z$ or $E_y = v_x\,B_z$ and a Hall voltage, $V_H = E_y\,W$, across width, W. If there are N carriers per unit volume and each is moving with a velocity, v, we showed that the current across the area, $W\,D$, is

$I_x = e\,N\,W\,D\,v_x$, giving $v_x = \dfrac{I_x}{e\,N\,D\,W}$ and

$E_y = \dfrac{1}{N\,e}B_z\,J_x$. $J_x = \dfrac{I_x}{W\,D}$ is the current density for I_x. For negative carriers the opposite polarization of charge occurs.

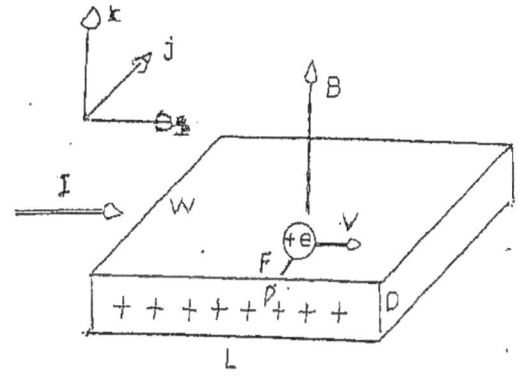

With the presence of a magnetic field, B_z, a current, I_x, produces an electric field, E_y, and thus a voltage, $V_y = E_y\,W$, across the width, W. Using a strip of gold foil, Hall placed a galvanometer across the width, W, and measured the current due to the voltage, V_y. Taking the ratio of that current, call it I_y, to the original current, he plotted the ratio, $\dfrac{I_y}{I_x}$, against the strength of B_z and got the result shown in figure on the right.

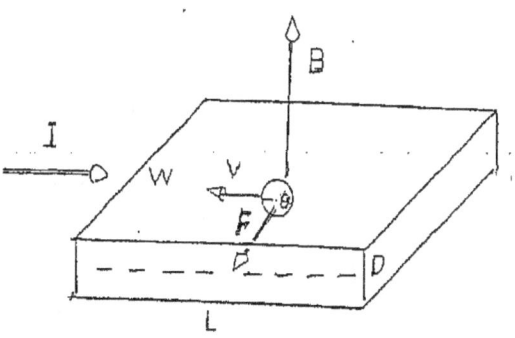

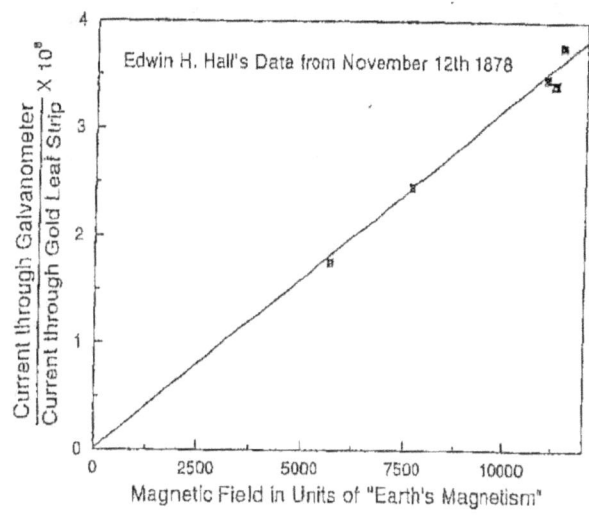

Edwin H. Hall's Data from November 12th 1878

We have seen that a stream of charges moving with a velocity, v, across an area, A, will carry a current, $I = N\,e\,A\,v$. In this case, I_x flows across an area, $D\,W$, giving a current formula, $I_x = N\,e\,W\,D\,v_x$ or $v_x = \dfrac{I_x}{N\,e\,W\,D}$. But, $N\,D = n_s$ electrons per unit surface area.

But, the Hall voltage is $V_y = v_x\,B_z\,W$, giving $V_y = \dfrac{I_x\,B_z}{n_s\,e}$.

The quantity, $R_{xy} = \dfrac{B_z}{n_s\,e}$, is the Hall resistance. Recall that the filling factor (ratio of electrons to flux quanta) was $\nu = \dfrac{n_s\,h}{e\,B}$. The Hall resistance can be written, $R_{xy} = \dfrac{h}{e^2\,\nu}$.

The quantity, $R_{xx} = V_x/I_x$, is the normal resistance of the channel, where V_x is the voltage drop in the length, L, of the channel. If σ_{xx} is the conductivity, then $R_{xx} = \dfrac{L}{\sigma_{xx} D W}$.

If this formula held, we would infer that as the magnetic field pushed the current to one side of the strip, the cross section of the current would decrease, and the resistance of the channel would increase. But the resistance tends to decrease producing the magneto-resistance effect on σ_{xx}.

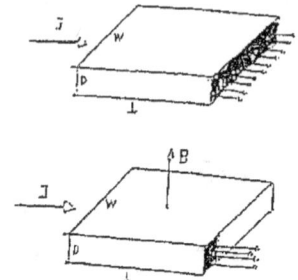

The QUANTUM HAIL EFECT discovered by Klaus von Klitzing and the fractional quantum Hall effect discovered by Horst Störmer, Daniel Tsui and Arthur Gossard require much thinner strips than the gold foil of Hall's experiment. Their strips were deposited by molecular beam epitaxy on a non-conducting substrate to form the channel of a field effect transistor (chapter 14). With the channel current, I_x, V_x gives the magneto-resistance, R_{xx}, and the Hall voltage, V_y, gives the Hall resistance, R_{xy}.

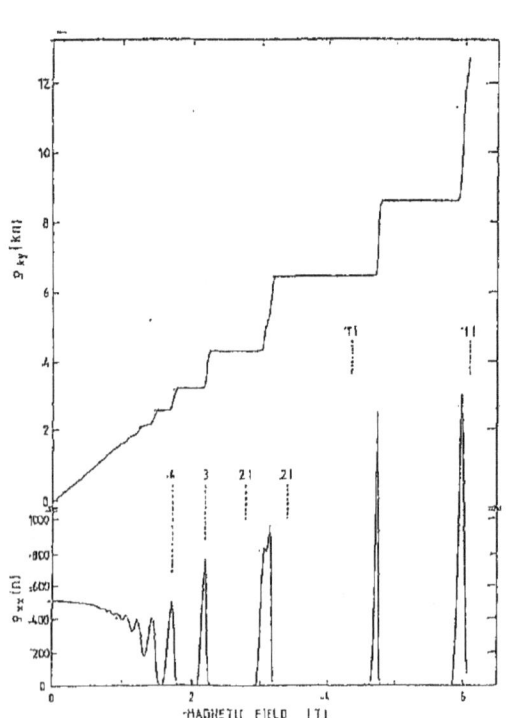

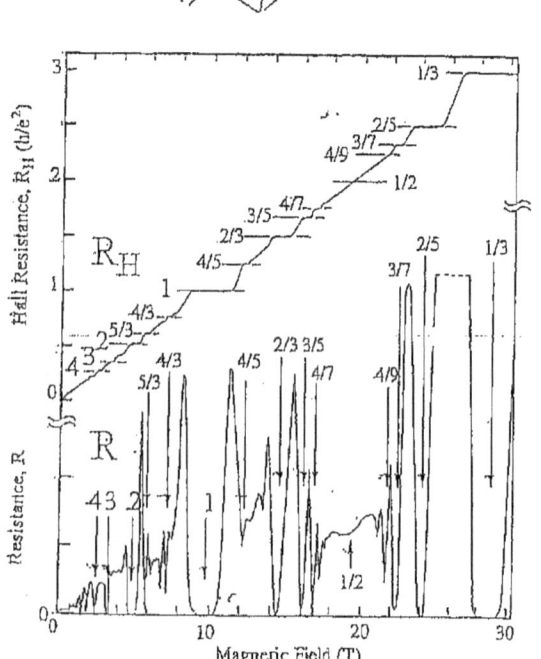

The effect of jumps in the nature of composite particles are shown in the figures on the previous page. The left-hand figure shows data reported by von Klitzing in 1980 and the right-hand figure shows data reported by Tsui, Störmer and Gossard in 1982. In each figure the upper curve is the Hall resistance, R_{xy}, and R_{xx}, the lower curve, is the magnetoresistance of the channel. The horizontal axis is the magnetic field in Teslas. Temperature is in milli-Kelvin.

Instead of a smooth variation of R_{xy} with magnetic field, we now see a series of quantized jumps. R_{xx} displays an exotic jumpiness. What is going on? Let us look at the von Klitzing data in the left-hand graph, now known as the integer quantum Hall effect (IQHE). We later will turn to the fractional quantum Hall effect (FQHE) of Tsui at al in the right-hand graph.

Von Klitzing (1986) included another set of data which may help to interpret the IQHE. Here we recall that the experimental setup on page 16-27 was the channel of an N-channel metal-oxide-field-effect-transistor (MOSFET) of the enhancement type, where a positive voltage at the gate draws more carriers into the channel (see chapter 14). If the carriers in the channel are a free electron fluid, their Fermi energy for N carriers in a volume, V, is

$$E_f = \left(\frac{3}{8\pi}\right)^{2/3} \left(\frac{h}{2m}\right)^3 \left(\frac{N}{V}\right)^{2/3}.$$

Now, $R_{xy} = \frac{h}{e^2 \nu}$, where $\nu = \frac{n_s h}{e B}$. Then if D is the thickness of the channel, $n_s = \frac{ND}{V}$. Von Klitzing noticed that when R_{xy} was plotted as a function of E_f, the jumps in R_{xy} could be accounted for if ν was limited to integer values. If that is so, then taking jumps of ν to be one at a time, we get a value for what is called the von Klitzing resistance, $R_K = \frac{h}{e^2}$ which is approximately $R_K = 25,813$ ohms.

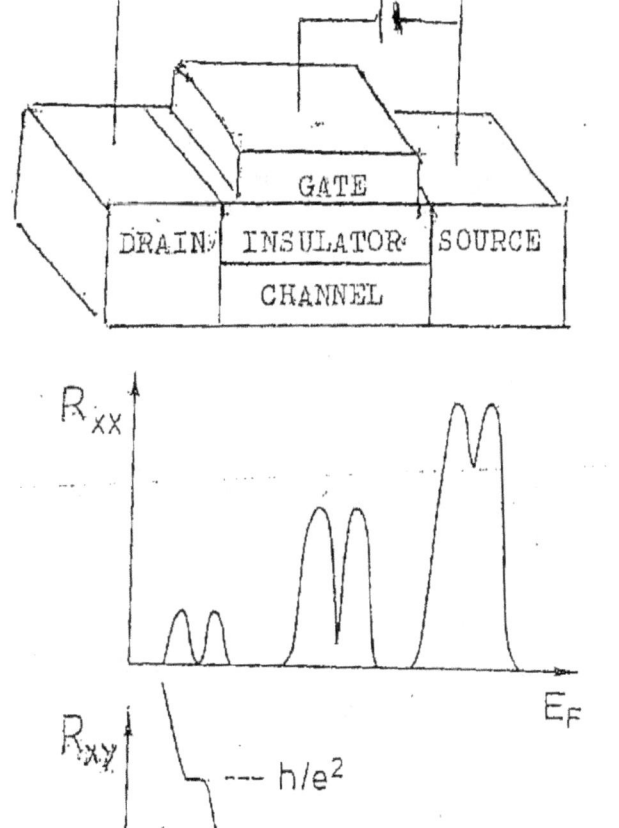

Von Klitzing and succeeding experimenters found that this quantity could be determined to increasing precision, and thus, it is taken to be one of the fundamental measured quantities in physics. For more information on this, refer to Peter J. Mohr and Barry N. Taylor (2005) CODATA RECOMMENDED VALUES OF THE FUNDAMENTAL PHYSICAL CONSTANTS RMP (Reviews of Modern Physics) 77,1.

With the IQHE, the normal resistance to current, R_{xx}, starts off normally and then reduces to zero as if it had become a superconductor. After this, R_{xx} varies between superconducting and conducting phases. While R_{xx} is zero, the Hall resistance, R_{xy}, is not affected by the change in magnetic field, but when R_{xx} is a normal conductor, R_{xy} jumps to catch up to where it should have been on a smooth Hall effect curve.

A significant phenomenon is that an external magnetic field will not penetrate into a superconducting material. It is rejected at the surface. This is the Meissner effect.

Superconductivity (and superfluidity) happen in systems where the particles are bosons – not obeying the Pauli Exclusion Principle. When such a system is cooled to near absolute zero temperature, the particles all condense into their ground state energy (Bose condensation). Fermions condense into the distribution shown on the left-hand figure on page 16-27. Locking in a significant zero-point energy compared with bosons. We found on page 16-25 that electrons could combine with an odd number of flux quanta to form bosons. But with integer filling factors higher than $\nu = 1$, we have more electrons (fermions) than composite particles (quasi-particles) which are bosons, and this particular theory doesn't appear promising.

In discussing his fractional quantum Hall effect, H. L. Störmer (RMP,1999) notices that major plateaus in R_{xy} and near superconducting states in R_{xx} do occur with $\nu = 1$ and $\nu = 1/3$, when the composite particle is a boson, but not $\nu = 1/2$ when it is a fermion. Although Tsui, Störmer, and Gossard (RMP, 1982) repeated von Klitzing's 1980 data for lower magnetic fields, at higher fields they got plateaus in R_{xy} and dips in R_{xx} for unexpected fractional values for ν. Störmer (RMP, 1999) gives the formula, $\nu = \dfrac{p}{2p \pm 1}$, $p = 1, 2, 3...$

Now, let us go back and look at the quantum of flux, $\Phi_o = \dfrac{h}{e}$. Suppose we take a contour on the conducting channel. The magnetic field, B, produces a flux through the contour. Suppose we introduce one quantum of flux. The flux through the contour increases by $d\Phi = \Phi_o$.

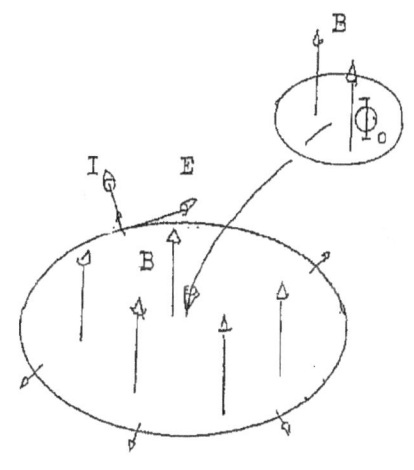

During the insertion, the flux changes at the rate, $\dfrac{\partial\Phi}{\partial t}$, which produces an electric field on the contour, as shown. In one circuit of the contour, there will be a voltage, $V = \dfrac{\partial\Phi}{\partial t}$.

A current, I_x, in a magnetic field experiences a Hall voltage, $V_y = R_{xy} I_x$, perpendicular to it. Suppose now that the electric field, E, produces Hall current perpendicular to it. Then, the induced voltage around the loop produces $I = \dfrac{\frac{\partial\Phi}{\partial t}}{R_{xy}}$.

Then, $I\,dt = \dfrac{1}{R_{xy}}d\Phi$ would be the positive charge moving out across the loop turning out to be $dq = \dfrac{1}{R_{xy}}d\Phi$. But, $R_{xy} = \dfrac{h}{e^2 v}$ and $d\Phi = \Phi_o = \dfrac{h}{e}$, giving $dq = v\,e$. This leaves a negative charge, $-v\,e$, behind. It would appear that when a quantum of flux is introduced it gets a charge, $-v\,e$, and hence all flux quanta would carry such a charge.

Working in the neighborhood of $v = 1/3$ of the Fractional Quantum Hall Effect, we then have quasiparticles carrying charge, $-e/3$. This is a favorite region to work as the flat step in R_{xy} is very wide.

Such particles have been claimed to be observed by Goldman and Su: <u>Science</u> 267, 1010 (1995), de Picciotto, et al: <u>Nature</u> (London) 389, 162 (1997) and Saminadayar, et al: <u>Phys. Rev. Lett.</u> 79, 2526 (1997).

Thus, instead of flux quanta being added to electrons to give quasi-particles that are alternately fermions and bosons (which would give Hall effect steps at $v = 1$ and $v = 1/3$ but not at $v = 1/2$), something even more bizarre happens. Remember that v is the ratio of electrons to flux quanta. At $v = 1/3$, there are three flux quanta for every electron. It would appear that the flux quanta steal the charges from the electrons and divide them among themselves.

What else does the electron have to be stolen? Well, spin is left. If for $\nu = 1/3$, the flux quanta stole spin-1/2 and divided it among themselves, each would have spin-1/6. What does that mean? Let us look at the relation between wave functions and statistics.

In spherical coordinates the wave function has the form,
$\psi(r, \theta, \varphi) = f(r, \theta) \, e^{i m \varphi}$, where m is the z-component of angular momentum. Now, statistics relate to a rotation of φ through 2π.

If m is an integer, $m = p$, where $p = 0, \ 1, \ 2...,$ $\psi(r, \theta, \varphi + 2\pi) = \psi(r, \theta, \varphi)$ obeying Bose-Einstein statistics.

If $m = (2p + 1)/2$, a half odd integer, $\psi(r, \theta, \varphi + 2\pi) = -\psi(r, \theta, \varphi)$ obeying Fermi-Dirac statistics.

But if $m = 1/6$? Then, $\psi(r, \theta, \varphi + 2\pi) = \psi(r, \theta, \varphi) e^{i\pi/3}$

Welcome to the world of fractional statistics. Frank Wilczek has labeled such particles with the name, anyons.

For another 2-dimensional electron fluid, we turn to graphite. This carbon material is composed of one atom thick sheets of carbon atoms in an extension of the benzene ring structure. A single sheet is called graphene. Roll it into a cylinder and you have a carbon nanotube. Graphite involves layers of these sheets held together by van de Waals forces. Because they slide easily over each other, the material is handy for pencils and use as a lubricant.

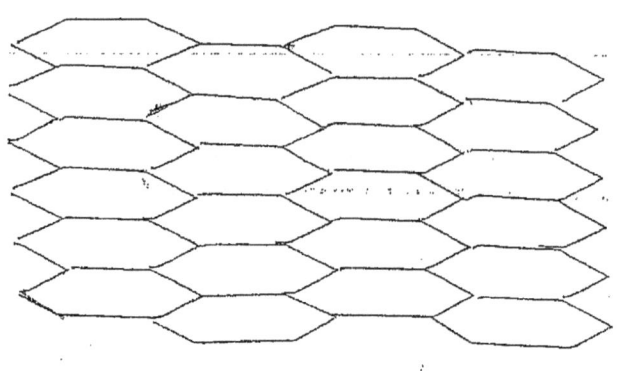

Because of the extreme fragility of the material, it is only recently that research has been able to be done on graphene, although theoretical speculation has been going on for quite some time.

Pioneering work on grapheme has been done by the group under Andre Geim at the University of Manchester (UK). See the articles: K. S. Novoselov et al, Science 306, 666 (2004), Geim and MacDonald Physics Today August 2007 p 35, A. H. Castro Neto et al RMP (2009) 81, 109.

The integer quantum Hall effect shown here was reported by Novoselov, Geim, et al in 2005. The carrier density, n, is for electrons. Negative density of electrons would give density of holes or positive carriers. Observation of the fractional quantum Hall effect in graphene has been more elusive, but Eva Andrei of Rutgers University reported experimental evidence of it in 2009.

For more details on research in this fascinating area, you are referred to the paper by Castro Neto (2009).

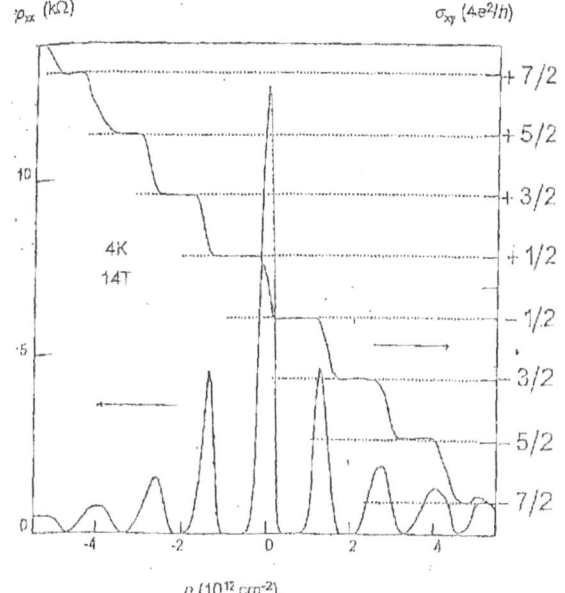

Reference: B. I. Halperin and J. K. Jain *Fractional* **Quantum Hall Effects New Developments** World Scientific

CHAPTER SEVENTEEN: THE ATOMIC NUCLEUS

The atomic nucleus appears to be composed of
positively charged protons and neutral neutrons.
Protons repel each other by electrostatic forces,
but protons and neutrons, alike, are attracted by
the strong interaction under which we shall lump
them together as nucleons. A nucleon in the
interior is equally surrounded by other nucleons,
and being equally attracted to each other, these
nucleons experience no net force. A nucleon on the
surface has nearest neighbors only on the interior
and hence would experience a strong negative force
inward. This surface-tension-type force suggested
to Niels Bohr a liquid-drop model of the nucleus.

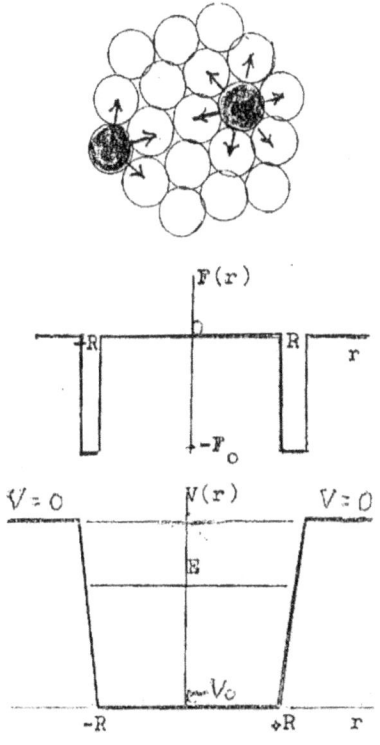

Moving along a radius, consider a nucleon to have
no force on it until it gets to the nuclear
radius, R, where it experiences a strong negative
(inward) force. The potential energy will be such
that $F = -\dfrac{dV}{dr}$. Taking the potential energy to be

$-V_o$ for radii less than R, it will experience a sharp rise, $\dfrac{dV}{dr} = +F_o$, in a

neighborhood of the nuclear radius, R; rising to $V(r) = 0$ for r greater than
R. This picture of the potential energy is a cross section of a three-
dimensional potential well. Think of the nucleus as a potential bubble.

Although the potential energy of a particle in a box was easy to solve, the
quantum mechanics of a particle in a spherical bubble requires things like
spherical Bessel functions for which we don't have time to develop.

An alternative to a three-dimensional
potential well, a bubble with sharp edges
would be a three-dimensional oscillator.
Here think of a ball on the end of a
spring, the other end tied to a single
point in space. In three dimensions, the
ball is free to oscillate in any
direction. This would correspond to a
spring force, $F = -k\,r$, and would give
rise to oscillations with $\omega = 2\pi f$ where
$\omega^2 = \dfrac{k}{M}$.

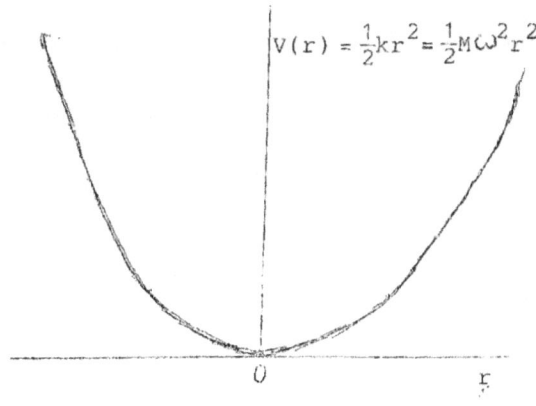

$$V(r) = \tfrac{1}{2}kr^2 = \tfrac{1}{2}M\omega^2 r^2$$

First, look at an object on springs that can
vibrate in three dimensions. Let the object
in the box be at rest at the center of the
box. Let any displacement be described by
coordinates x, y, and z. If r is distance from
the center, then $r^2 = x^2 + y^2 + z^2$ and if v is
velocity, it has components v_x, v_y and v_z with
$v^2 = v_x{}^2 + v_y{}^2 + v_z{}^2$ and if the energy is
$E = \frac{1}{2}M\,v^2 + \frac{1}{2}M\,\omega^2\,r^2$, we can write it

$E = \frac{1}{2}M\left[v_x{}^2 + v_y{}^2 + v_z{}^2\right] + \frac{1}{2}M\,\omega^2[x^2 + y^2 + z^2]$.

This can be broken into three pieces:

$E = \left[\frac{1}{2}M\,v_x{}^2 + \frac{1}{2}M\,\omega^2\,x^2\right] + \left[\frac{1}{2}M\,v_y{}^2 + \frac{1}{2}M\,\omega^2\,y^2\right] +$

$[the\ same\ for\ z]$

The total motion is the sum of three separate motions, each having its own

energy. We get: $E_x = \frac{1}{2}M\,v_x{}^2 + \frac{1}{2}M\,\omega^2\,x^2$, $E_y = \frac{1}{2}M\,v_y{}^2 + \frac{1}{2}M\,\omega^2\,y^2$,

$E_z = \frac{1}{2}M\,v_z{}^2 + \frac{1}{2}M\,\omega^2\,z^2$.

The particles in the nucleus are not really attached to springs. They are
subject to the forces in the nucleus with the particles acting on each
other. This is called the STRONG INTERACTION. We are merely describing it as
if it behaved like a three-dimensional oscillator.

Go back and look at page 16-9 of chapter 16. Each dimension of motion will
have its own Schrödinger equation. Each particle will have its own "wave
function" $\psi_x(x)$, $\psi_y(y)$, $\psi_z(z)$, and the derivatives will be written as partial
derivatives looking forward to combining things into a unified treatment.

Thus, $\hat{D}_x\,\psi(x) = \frac{\partial\psi}{\partial x}$, $\hat{D}_y\,\psi(y) = \frac{\partial\psi}{\partial y}$, $\hat{D}_z\,\psi(z) = \frac{\partial\psi}{\partial z}$.

Now, review page 16-9 and what follows in chapter 16.

Each oscillator will have its own set of quantized energies and its own set

of quantum numbers: $E_x = \left(n_x + \frac{1}{2}\right)\hbar\omega$ $n_x = 0,\ 1,\ 2$, etc

$E_y = \left(n_y + \frac{1}{2}\right)\hbar\omega$ $n_y = 0,\ 1,\ 2$, etc. $\qquad E_z = \left(n_z + \frac{1}{2}\right)\hbar\omega$ $n_z = 0,\ 1,\ 2$, etc.

Each energy state will have its own Hermite-Gaussian "wave function" as
discussed on page 16-10. These will be discussed with regard to nuclear
"wave functions" later in this chapter.

All of this is put together in the following manner. A total "wave function" is put together as follows. Following traditional methods of physics, the total function is the product of functions of the separate variables.

$$\psi(x, y, z) = \psi_{n_x}(x)\, \psi_{n_y}(y)\, \psi_{n_z}(z)$$

$$\hat{D}_x\, \psi(x, y, z) = \left[\hat{D}_x\, \psi_{n_x}(x)\right] \psi_{n_y}(y)\, \psi_{n_z}(z) \qquad \hat{D}_y\, \psi(x, y, z) = \psi_{n_x}(x)\left[\hat{D}_y\, \psi_{n_y}(y)\right] \psi_{n_z}(z)$$

and similarly, for $\hat{D}_z$. Operating on the entire product function, each partial derivative finds the function of its variable, operates on that and leaves functions of other variables alone.

On page 16-10, the energy operator is written, $\hat{H}_x = -\dfrac{\hbar^2}{2M}\hat{D}_x{}^2 + \dfrac{1}{2}M\,\omega^2\,x^2$, and is called the Hamiltonian. This is for x and similarly for y and z. $\hat{H}_x$ operates only on functions of z. Now, on the middle of page 16-13, $\hat{H}_x\, \psi_{n_x}(x) = E_{n_x}\, \psi_{n_x}(x)$. Similarly, for $\hat{H}_y$ and $\hat{H}_z$.

Now, $\hat{H}_x\, \psi(x, y, z) = \left[\hat{H}_x\, \psi_{n_x}\right] \psi_{n_y}\, \psi_{n_z}$ or $\hat{H}_x\, \psi(x, y, z) = \left[E_{n_x}\, \psi_{n_x}\right] \psi_{n_y}\, \psi_{n_z}$.

E_{n_x} is a number, not an operator. Hence, $\hat{H}_x\, \psi(x, y, z) = E_{n_x}\, \psi(x, y, z)$

Likewise, $\hat{H}_y\, \psi(x, y, z) = E_{n_y}\, \psi(x, y, z)$ and $\hat{H}_z\, \psi(x, y, z) = E_{n_z}\, \psi(x, y, z)$.

Each operator finds its function, multiplies it by an energy number which can be made to multiply anything and hence the entire product function. Now,

$$E_{n_x} = \left(n_x + \tfrac{1}{2}\right)\hbar\,\omega \qquad E_{n_y} = \left(n_y + \tfrac{1}{2}\right)\hbar\,\omega \qquad E_{n_z} = \left(n_z + \tfrac{1}{2}\right)\hbar\,\omega$$

If $\hat{H} = \hat{H}_x + \hat{H}_y + \hat{H}_z$ is the energy operator for the three-dimensional oscillator, then $\hat{H}\, \psi(x, y, z) = \left(n_x + n_y + n_z + \tfrac{3}{2}\right)\hbar\,\omega\,\psi(x, y, z)$.

Return to the state symbol notation on page 16-13. Let $\psi_{n_x}(x) = \langle x \mid n_x \rangle$, $\psi_{n_y}(x) = \langle y \mid n_y \rangle$, $\psi_{n_z}(x) = \langle x \mid n_z \rangle$.

Forgetting about variables, we now write:

$$\hat{H}_x \mid n_x \rangle = \left(n_x + \tfrac{1}{2}\right)\hbar\,\omega \mid n_x \rangle$$

$$\hat{H}_y \mid n_y \rangle = \left(n_y + \tfrac{1}{2}\right)\hbar\,\omega \mid n_y \rangle \qquad\qquad \left.\begin{array}{l} n_x \\ n_y \\ n_z \end{array}\right\} = 0,\ 1,\ 2,\ 3,\ 4, \ldots$$

$$\hat{H}_z \mid n_z \rangle = \left(n_z + \tfrac{1}{2}\right)\hbar\,\omega \mid n_z \rangle$$

Now, form $|n_x, n_y, n_z\rangle = |n_x\rangle|n_y\rangle|n_z\rangle$. Each of the operators, $\hat{H}_x$, $\hat{H}_y$ and $\hat{H}_z$, operates only in its separate component symbol.

$$\hat{H}_x|n_x\ n_y\ n_z\rangle = [\hat{H}_x|n_x\rangle]|n_y\rangle|n_z\rangle = \left(n_x + \frac{1}{2}\right)\hbar\omega|n_x\ n_y\ n_z\rangle$$

$$\hat{H}_y|n_x\ n_y\ n_z\rangle = |n_x\rangle[\hat{H}_y|n_y\rangle]|n_z\rangle = \left(n_y + \frac{1}{2}\right)\hbar\omega|n_x\ n_y\ n_z\rangle$$

$$\hat{H}_z|n_x\ n_y\ n_z\rangle = |n_x\rangle|n_y\rangle[\hat{H}_z|n_z\rangle] = \left(n_z + \frac{1}{2}\right)\hbar\omega|n_x\ n_y\ n_z\rangle$$

Thus, each symbol, $|n_x\ n_y\ n_z\rangle$, is an eigenket of $\hat{H} = \hat{H}_x + \hat{H}_y + \hat{H}_z$.

$$\hat{H}|n_x\ n_y\ n_z\rangle = \left(n_x + n_y + n_z + \frac{3}{2}\right)\hbar\omega|n_x\ n_y\ n_z\rangle \qquad \left.\begin{array}{l}n_x\\n_y\\n_z\end{array}\right\} = 0,\ 1,\ 2,\ 3,\ 4,\dots$$

in which the n's are the quantum numbers of the three-dimensional harmonic oscillator.

In the following table we list the first several eigenkets, $|n_x\ n_y\ n_z\rangle$, and their energies.

$|0, 0, 0\rangle$ one state with energy $\frac{3}{2}\hbar\omega$

$|1, 0, 0\rangle$, $|0, 1, 0\rangle$, $|0, 0, 1\rangle$ three states with energy $\frac{5}{2}\hbar\omega$

$|2, 0, 0\rangle$, $|0, 2, 0\rangle$, $|0, 0, 2\rangle$ six states with energy $\frac{7}{2}\hbar\omega$

$|1, 1, 0\rangle$, $|1, 0, 1\rangle$, $|0, 1, 1\rangle$

$|3, 0, 0\rangle$, $|0, 3, 0\rangle$, $|0, 0, 3\rangle$, $|1, 1, 1\rangle$ ten states with energy $\frac{9}{2}\hbar\omega$

$|2, 1, 0\rangle$, $|1, 2, 0\rangle$, $|1, 0, 2\rangle$, $|2, 0, 1\rangle$

$|0, 1, 2\rangle$, $|0, 2, 1\rangle$

Then 15 states with energy, $\frac{11}{2}\hbar\omega$, and so forth.

With the three-dimensional harmonic oscillator, each set of eigenstates having the same energy eigenvalues is called an ENERGY SHELL. As we do not know the exact potential energy function of the interaction holding the nucleons – protons and neutrons – in the nucleus, we proceed to make some reasonable guess about the interaction in order that some calculations can be made on nuclear structure.

The three-dimensional oscillator is a convenient guess as to the average potential interactions seen by a nucleon. In addition to the motion through space, described by the states derived on the previous page, each nucleon has the property of intrinsic spin angular momentum in which it acts something, but not altogether, like a spinning top. In a partial analogy to the planets moving about the sun, we talk of states of orbital motion and states of spin motion. For each of the orbital states, which is what we call the quantum states just derived, there are two states for intrinsic spin giving the direction of the nucleon's spin angular momentum vector with respect to an external magnetic field, the two spin states being called spin-up and spin-down.

Therefore, each orbital state is split into two spin-orbital states, now giving 2 states with energy, $\frac{3}{2}\hbar\omega$, 6 states with energy, $\frac{5}{2}\hbar\omega$, 12 states with energy, $\frac{7}{2}\hbar\omega$, 20 states with energy, $\frac{9}{2}\hbar\omega$, and so forth.

Now the nucleons obey the Pauli Exclusion Principle which, applied to the nucleus, forbids that more than one nucleon of either type can be found in any particular spin-orbital state. Thus, when the nucleons are distributed among the allowed energy quantum states, a particular proton state will be either empty or occupied by one proton and a particular neutron state will be either empty or occupied by one neutron.

Normally a nucleus is found in its lowest energy, or ground, state. Below a certain level, all energy states will be occupied by nucleons and above that level all nucleon states will be empty. If a nucleon finds itself in an excited energy state, it will fall back into a lower energy state emitting a gamma ray photon.

So, when a nucleus contains many protons, only two of them can occupy the lowest energy shell, six in the next energy shell, twelve in the next, twenty in the next and so on. If the three-dimensional oscillator is to describe the behavior of protons in the nucleus, the same numbers would apply to the distribution of neutrons in the nucleus.

This means that, when it is in its ground state, if a nucleus contains either 2, 8, 20, 40, or 90 protons and/or any one of these numbers of neutrons, then any proton and/or neutron energy shell will be either all full or all empty.

In the electron structure of an atom in its ground state, when the atom is such a one that the electron energy shells are either all full or all empty, we then happen to have an atom of one of the noble gases, helium, neon, argon, etc. These structures are characterized by exceptional stability and chemical inertness.

Experimentally, the same thing is observed to happen in nuclear structure, exceptionally stable nuclei having either 2, 8, 20, 28, 50, or 82 protons or neutrons or both. These numbers are called MAGIC NUMBERS. With the three-dimensional oscillator, the states in the first three shells correctly add up to give the first three magic numbers. After this, the theoretical results are not in agreement with experiment. In addition to the harmonic oscillator potential, which is an approximate but not exact picture of the nuclear interaction, Mayer and Jensen included an attractive interaction arising from the coupling of a nucleon's spin and orbital angular momentum states. This resulted in a splitting of the states in the higher energy shells which rearranged the shells so as to give all the observed magic numbers exactly.

The atomic number, Z, is the number of protons in the nucleus. A, the atomic mass number, is the total number of nucleons. $A - Z$ is the number of neutrons.

Oxygen 17 with atomic number 8 is written $_8O^{17}$ having eight protons and nine neutrons.

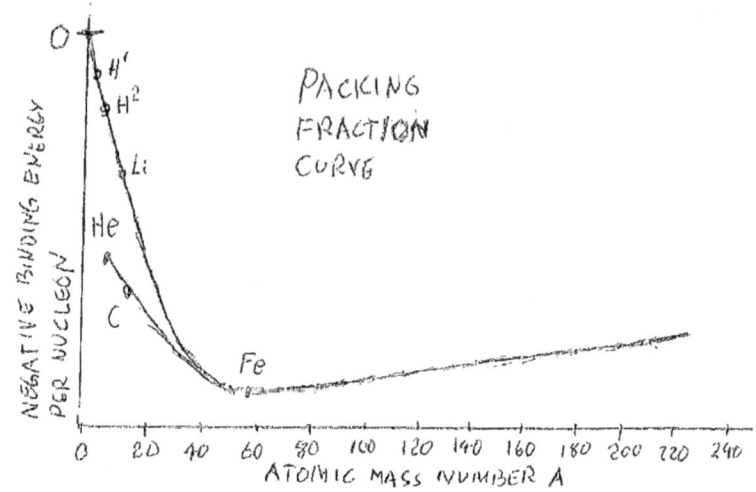

The packing fraction curve represents the loss of mass of the nucleons in nuclei as atomic number increases. Starting with the proton, $_1H^1$, and the neutron, n, the mass per nucleon decreases in the nuclei with increasing atomic number up to Iron (Fe). Following Iron, the mass per nucleon increases. These data are gotten by using a small spectrometer to measure nuclear masses. Recall that mass has energy, $E = mc^2$. Also recall that the hydrogen electron is bound to the nucleus with negative binding energy. The loss of mass of the nucleons is related to their negative binding energy holding the nucleus together.

This has important meaning in the physics of stellar processes. The big bang left us with a universe that was almost all hydrogen. With gravitational attraction this condensed into galaxies and stars. The intense gravitational pressure in the center of stars began to push hydrogen atoms together to form helium. Notice the big loss of mass involved. The loss of mass comes off as heat energy. This energy diffuses out to the surface and comes off as radiation. The star in the core is mostly converted to helium, the star hiccups and expands from the main sequence to be a red giant. Helium synthesis continues in a shell around the core and in the core, helium is pushed together to form heavier elements. In the biggest stars, this process continues until iron starts to accumulate. Synthesis after this needs the adding of mass or decrease of binding energy. To produce the rest of the periodic table, we must wait for a supernova. When a star has produced all the energy it is going to make, it can no longer support itself against its own gravity. As the outer layers of the star collapse, they gather mass before them, forming a shock wave of intense density, pressure and temperature. This shock wave bounces off a core and explodes outward giving a supernova. The kick given to the core drives it inward giving a white dwarf, neutron star (in which electrons are pushed by gravity into protons giving all neutrons) or a black hole. With the intense pressure and temperature of the shock wave, the rest of the periodic table past iron is synthesized. The universe had to wait for supernovas before chemistry (and life) to happen. We are all star dust.

We now will study the nuclei synthesized in the stars. Many of them do not occur in normal matter because of radioactivity; but using neutrons produced in atomic reactors and other methods we have been able to produce them and then study them. We will use the nuclear shell model to explain nuclear processes including radioactivity. With radioactivity, nuclei throw out alpha particles (helium-four nuclei), beta particles (electrons) and gammas (high energy protons).

Separately for both protons and neutrons there were two quantum states with energy, $\frac{3}{2}\hbar\omega$, six quantum states with energy, $\frac{5}{2}\hbar\omega$, and twelve quantum states with energy, $\frac{7}{2}\hbar\omega$, and twenty with $\frac{9}{2}\hbar\omega$.

Next, we notice that there is a principle in the universe (stated by Wolfgang Pauli) that for electrons, protons and neutrons, that only one particle can exist in any quantum state when there are multiple particles in any object. We will start our study with helium-four.

$_2He^4$ has two protons and two neutrons filling the $\frac{3}{2}\hbar\omega$ states for both particles. Let some process produce $_2He^5$. We find $_2He^5 \rightarrow {}_2He^4 + n$. The extra neutron must go into the next shell where it is lonely (not tightly bound) and falls off immediately. Let $_3Li^5$ be produced. Immediately we get $_3Li^5 \rightarrow {}_2He^4 + p$. The extra proton is lonely and falls off.

This establishes the reality of the quantum states in the first shell and the jump to the second shell.

Lithium, $_3Li^6$, has a proton and a neutron in the second shell. $_3Li^7$ has a proton and two neutrons in the second shell. They like each other, form a club and hang around. These nuclei are stable and occur in nature.

Next, $_3Li^8$ contains three protons and five neutrons. A neutron can lose mass and become a proton. The mass loss must be just enough to produce an electron, losing a negative charge in the nucleus's process of gaining a positive charge. $_3Li^8 \rightarrow _4Be^8 + e^-$

Next, $_4Be^8 \rightarrow 2\,_2He^4$. In this case it is possible for the system to go to a lower energy, and more stable state, and it does so.

$_4Be^9$ cannot do this and is stable found in nature.

For reference, the proton mass is 1.673×10^{-27} kg, the neutron mass is 1.675×10^{-27} kg, and for the electron 9.1×10^{-31} kg.

Obviously for any neutrons produced in the big bang, they could go to a lower energy (proton) by giving off an electron. $n \rightarrow p^+ + e^- + \bar{\nu}$

The extra particle (a neutrino) is required by conservation of spin angular momentum. We won't get into that here.

The reality of the jump from the first to the second shell has been established. What about the jump from the second to the third?

Let us look at $_7N^{16}$. We find $_7N^{16} \rightarrow _8O^{16} + e^-$.

$_7N^{16}$ has a lonely neutron in the third shell and can go to a lower energy if the neutron gives a negative electron becoming a proton.

Let us look at $_9F^{16}$ this has a lonely proton in the third shell which could go to a lower energy if it became a neutron. This it does, giving off a positive electron.

$_9F^{16} \rightarrow _8O^{16} + e^+$

The jump from the second to the third shell seems validly predicted.

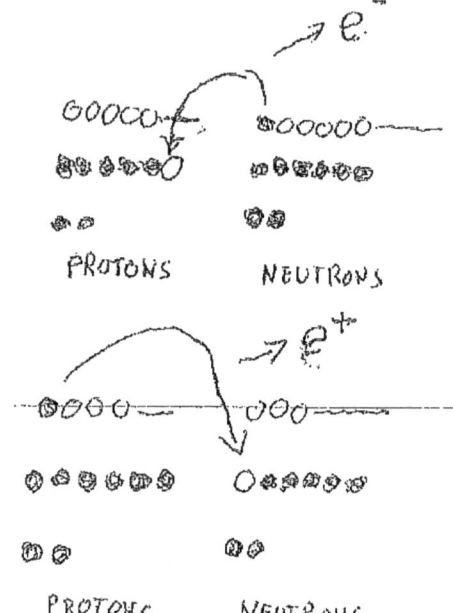

What we have just now discussed shows the validity of the shell picture of the nucleus with nucleons limited to shells of energy quantum states. Although all neutrons produced in the big bang decayed into protons, we see that neutrons are produced again by processes in the nucleus. It should be mentioned here that in beta (electron) emission there is often more energy involved than the mass of the electron. This energy is emitted as a gamma (high energy photon) particle.

Looking at the packing fraction curve we see that past iron, the nucleon mass starts to increase. This means that if very big nuclei break up, what remains goes to a lower energy and a more stable condition. This goes along with the fact that the nucleons in the outermost shell are not as tightly held as those in inner shells. What is significant is when the nucleus sheds mass, it does it with helium nuclei, again showing the stability of the first shell. A helium nucleus tossed out by a big nucleus is called an alpha particle. At the top of the periodic table, uranium is the most stable nucleus and hence, remains in the periodic table.

$_{92}U^{238}$ starts the chain of alpha (and some beta) decay.

$_{92}U^{238} \rightarrow {}_{90}Th^{234} + {}_2He^4$ but $_{90}Th^{234} \rightarrow {}_{91}Pa^{234} + e^-$.

Then, $_{91}Pa^{234} \rightarrow {}_{92}U^{234} + e^-$. Then mass loss by alpha decay again.

$_{92}U^{234} \rightarrow {}_{90}Th^{230} + {}_2He^4$ and $_{90}Th^{230} \rightarrow {}_{88}Ra^{226} + {}_2He^4$.

This process continues by alphas and betas until we reach $_{82}Pb^{206}$ which is extremely stable.

But, starting with $_{92}U^{238}$ if it absorbs a neutron, n, we can go the other way.

$_{92}U^{238} + n \rightarrow {}_{93}Np^{239} + e^-$

But, $_{93}Np^{239} \rightarrow {}_{94}Pu^{239} + e^-$ in which case plutonium is stable.

In this manner, continued, it is possible to create new, but not very stable, elements past the end of the periodic table.

There is another isotope of uranium (not very abundant) which is interesting when it absorbs a neutron.

$_{92}U^{235} + n \rightarrow$ two middle sized nuclei plus a lot of neutrons plus a lot of energy. If the neutrons are contained and absorbed by other $_{92}U^{235}$ nuclei, the process accelerates with energy release and we have nuclear bomb. If the neutron production is controlled, we have the steady state production of energy of a nuclear reactor producing energy for electrical power production.

Such a reactor is also a source of neutrons which can be used in nuclear experiments.

The same thing happens with $_{94}Pu^{239}$. When it absorbs neutrons it also breaks up into middle sized nuclei with more neutrons and a lot of energy. If $_{92}U^{238}$ is in the neutron flux of a reactor, it can be turned into plutonium creating more energy source than we had before. Such a reactor is called a breeder reactor.

In the following table we list the first several eigenkets, $|n_x\ n_y\ n_z\rangle$, and their energies.

$|0, 0, 0\rangle$	one state with energy $\frac{3}{2}\hbar\omega$
$|1, 0, 0\rangle,\ |0, 1, 0\rangle,\ |0, 0, 1\rangle$	three states with energy $\frac{5}{2}\hbar\omega$
$|2, 0, 0\rangle,\ |0, 2, 0\rangle,\ |0, 0, 2\rangle$	six states with energy $\frac{7}{2}\hbar\omega$
$|1, 1, 0\rangle,\ |1, 0, 1\rangle,\ |0, 1, 1\rangle$	
$|3, 0, 0\rangle,\ |0, 3, 0\rangle,\ |0, 0, 3\rangle,\ |1, 1, 1\rangle$	ten states with energy $\frac{9}{2}\hbar\omega$
$|2, 1, 0\rangle,\ |1, 2, 0\rangle,\ |1, 0, 2\rangle,\ |2, 0, 1\rangle$	
$|0, 1, 2\rangle,\ |0, 2, 1\rangle$	

Then 15 states with energy, $\frac{11}{2}\hbar\omega$, and so forth.

The position of the three-dimensional oscillator is described by three numbers, x, y, and z. For the x-axis we construct a Hilbert space where the vector, $|x\rangle$, corresponds to the point, x, on that axis. Likewise, we can form Hilbert spaces for the y- and z- axes, where $|y\rangle$, corresponds to the point, y, and $|z\rangle$, to the point, z, on those axes respectively. For the operator, $\hat{H}_x$, when we project the eigenvectors, $|n_x\rangle$, onto the vectors, $|x\rangle$, we get the EIGENFUNCTIONS of $\hat{H}_x$ in the manner previously described for the one-dimensional oscillator.

Hence, $\langle x\,|\,n_x\rangle = (-)^n\ e^{+x^2/2}\ \dfrac{d^n}{dx^n}\left(e^{-x^2}\right)$. Where $n = 0,\ 1,\ 2,\ 3,\ 4, \ldots$

The first four eigenfunctions are written out on the following page.

The first four eigenfunctions are:

$$\langle x \mid 0 \rangle = e^{-x^2/2} \qquad\qquad \langle x \mid 1 \rangle = x\, e^{-x^2/2}$$

$$\langle x \mid 2 \rangle = (4x^2 - 2)\, e^{-x^2/2} \qquad\qquad \langle x \mid 3 \rangle = (4x^3 - 6x)\, e^{-x^2/2}$$

With identical eigenfunctions for $\langle y \mid n_y \rangle$ and $\langle z \mid n_z \rangle$ except that x is replaced by y or z.

Corresponding to the points (x, y, z) in three physical dimensions, we construct a Hilbert space, $\mid x,\ y,\ z \rangle = \mid x \rangle \mid y \rangle \mid z \rangle$, which is the direct product space of $\mid x \rangle$, $\mid y \rangle$, and $\mid z \rangle$. We write the total eigenfunctions of the three-dimensional oscillator in the form, $\langle x,\ y,\ z \mid n_x\, n_y\, n_z \rangle = \langle x \mid n_x \rangle \langle y \mid n_y \rangle \langle z \mid n_z \rangle$.

The first few eigenfunctions for the lowest energy states are:

$$\langle x,\ y,\ z \mid 0,\ 0,\ 0 \rangle = e^{-r^2/2} \qquad \text{where} \quad r^2 = x^2 + y^2 + z^2.$$

$$\langle x,\ y,\ z \mid 1,\ 0,\ 0 \rangle = x\, e^{-r^2/2} \qquad\qquad \langle x,\ y,\ z \mid 0,\ 1,\ 0 \rangle = y\, e^{-r^2/2}$$

$$\langle x,\ y,\ z \mid 0,\ 0,\ 1 \rangle = z\, e^{-r^2/2} \qquad\qquad \langle x,\ y,\ z \mid 1,\ 1,\ 0 \rangle = x\, y\, e^{-r^2/2}$$

$$\langle x,\ y,\ z \mid 1,\ 0,\ 1 \rangle = x\, z\, e^{-r^2/2} \qquad\qquad \langle x,\ y,\ z \mid 0,\ 1,\ 1 \rangle = y\, z\, e^{-r^2/2}$$

$$\langle xyz \mid 200 \rangle = (x^2 - \tfrac{1}{2})\, e^{-r^2/2} \qquad\qquad \langle xyz \mid 020 \rangle = (y^2 - \tfrac{1}{2})\, e^{-r^2/2}$$

Let us rewrite these functions in spherical coordinates. We re-write them as,

$\psi_{n_x n_y n_z} = \langle x\, y\, z \mid n_x\, n_y\, n_z \rangle$, and notice that a linear combination of eigenvectors with the same eigenvalue is also an eigenvector with that eigenvalue, that is if, $\mid \psi \rangle = a \mid 1,\ 0,\ 0 \rangle + b \mid 0,\ 1,\ 0 \rangle + c \mid 0,\ 0,\ 1 \rangle$, then $\hat{H} \mid \psi \rangle = \frac{5}{2} \hbar \omega \mid \psi \rangle$.

In each of the following functions on the following page, we let C be whatever the appropriate constants may be.

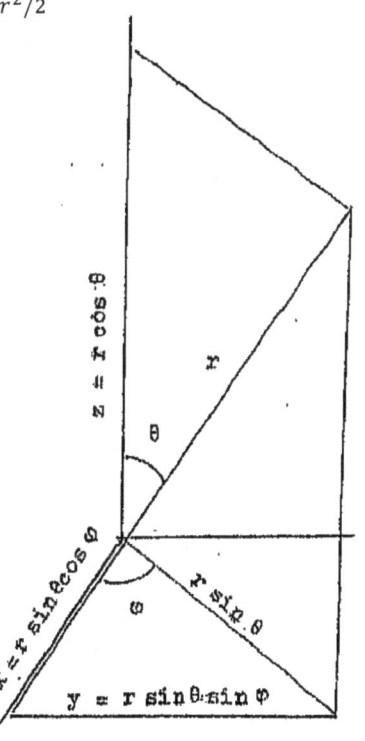

$\Psi_{000} = Ce^{-k^2r^2/2}$

A spherically symmetric
orbital is an s-orbital.

$\Psi_{001} = Ce^{-k^2r^2/2} \, r\cos\theta$

$\Psi_{100} = Ce^{-k^2r^2/2} \, r\sin\theta\cos\varphi$

$\Psi_{010} = Ce^{-k^2r^2/2} \, r\sin\theta\sin\varphi$

$\Psi_{101} = Ce^{-k^2r^2/2} \, r^2\sin\theta\cos\theta\cos\varphi$

$\Psi_{010} = Ce^{-k^2r^2/2} \, r^2\sin\theta\cos\theta\sin\varphi$

$\Psi_{110} = Ce^{-k^2r^2/2} \, r^2\sin^2\theta\sin\varphi\cos\varphi$

or $\Psi_{110} = \dfrac{C}{2}e^{-k^2r^2/2} \, r^2\sin^2\theta\sin(2\varphi)$

If ψ^2 can be transformed into itself by a rotation of 90° about a coordinate axis, it is called a d-orbital.

The linear combination, $\psi_{200} + \psi_{020} + \psi_{002} = C\,e^{-k^2 r^2/2}\left(k^2 r^2 - \dfrac{3}{2}\right)$, is an s-orbital with and energy, $\dfrac{7}{2}\hbar\,\omega$.

Notice that $|200\rangle - |020\rangle$ is orthogonal to $|200\rangle + |020\rangle + |002\rangle$ giving,

$\psi_{x^2-y^2} = C\,e^{-k^2 r^2/2} \, r^2 \sin^2\theta(\cos^2\phi - \sin^2\phi) = C\,e^{-k^2 r^2/2} \, r^2 \sin^2\theta \cos(2\phi)$,

which is another d-orbital.

There is yet another linear combination orthogonal to the last two. Let us write it, $|\psi\rangle = a\,|\,200\,\rangle + a\,|\,020\,\rangle + b\,|\,002\,\rangle$, orthogonal to $|\,200\,\rangle - |\,020\,\rangle$. Then

$[a\,\langle\,200\,| + a\,\langle\,020\,| + b\,\langle\,002\,|\,][\,|\,200\,\rangle + |\,020\,\rangle + |\,002\,\rangle] = 2\,a + b = 0$ gives $a = -\frac{1}{2}$.

In rectilinear coordinates, $\psi = C\,e^{-k^2\,r^2/2}\left[-\frac{1}{2}x^2 - \frac{1}{2}y^2 + z^2\right]$.

Add $\frac{1}{2}z^2 - \frac{1}{2}z^2 = 0$ inside the brackets and $\psi = C\,e^{-k^2\,r^2/2}\left[\frac{3z^2 - r^2}{2}\right]$, which is,

$\psi = C\,e^{-k^2\,r^2/2}\,r^2\left[\frac{3\cos^2\theta - 1}{2}\right]$, in polar coordinates. Considering its rotation about the z-axis, we shall consider this as another d-orbital.

The question arises: what in the Schrödinger equation,

$$\left[\frac{1}{2M}\left(P_x{}^2 + P_y{}^2 + P_z{}^2\right) + V(x,y,z)\right]|\,E_n\,\rangle = E_n|\,E_n\,\rangle,$$

is responsible for breaking the spherical symmetry of these orbitals?

Certainly not the potential energy, $V(r) = \frac{1}{2}M\,\omega^2\,r^2$, nor the constant, E_n.

Hence, $\frac{1}{2M}\left(P_x{}^2 + P_y{}^2 + P_z{}^2\right)$, the kinetic energy operator, is responsible for breaking the spherical symmetry. This is the same for all non-relativistic problems. Thus, the directional shapes which we have derived are the same for all problems where the potential energy is spherically symmetric, being a function only of the length of the radius, $V(x,y,z) = V(r)$. This would include a single electron on an atom of atomic number, Z, $V(r) = -\frac{Z\,e^2}{r}$.

Hence, we have derived the directional shapes of the atomic electron eigenfunctions, which you have seen in your chemistry book. The radial part of those functions will rarely be used in practice.

Further linear combinations of $cos(m\,\phi) + i\,sin(m\,\phi) = e^{i\,m\,\phi}$ and $cos(m\,\phi) - i\,sin(m\,\phi) = e^{-i\,m\,\phi}$ give the SPHERICAL HARMONICS, $Y_{l,m}(\theta,\phi)$, which are one of the most important sets of functions in mathematical physics. The index, l, denotes a set of these functions, and m denotes a member of that set. In atomic physics, $l = 0$ gives an s-orbital, $l = 1$ gives a p-orbital, $l = 2$ gives a d-orbital, and so forth as described in your chemistry book. l and m are quantum numbers for angular momentum.

In the following tabulation, C, stands for the appropriate normalization constant.

$Y_{0,0}(\theta,\phi) = C,$ $\quad Y_{1,1}(\theta,\phi) = C \sin\theta \, e^{i\phi},$ $\quad Y_{1,0}(\theta,\phi) = C \cos\theta,$ $\quad Y_{1,-1}(\theta,\phi) = C \sin\theta \, e^{-i\phi}$

$Y_{2,2}(\theta,\phi) = C \sin^2\theta \, e^{i\,2\,\phi},$ $\quad Y_{2,1}(\theta,\phi) = C \sin\theta \cos\theta \, e^{i\phi},$ $\quad Y_{2,0}(\theta,\phi) = C \frac{1}{2}(3\cos^2\theta - 1)$

$Y_{2,-1}(\theta,\phi) = C \sin\theta \cos\theta \, e^{-i\phi},$ $\quad Y_{2,-2}(\theta,\phi) = C \sin^2\theta \, e^{-i\,2\,\phi}$

From these you might guess, correctly, that $Y_{l,l}(\theta,\phi) = C \sin^l\theta \, e^{il\phi}$

From this, the functions, $Y_{l,m}(\theta,\phi)$, can be derived using the lowering operator for m in the quantization of angular momentum.

Although derivations of raising and lowering operators for spherical harmonics involve mathematics beyond the level of this book, we will give the operators for your information.

Raising $Y_{l,m}$ to $Y_{l,m+1}$: $\qquad e^{i\phi}\left[m \cot\theta - \frac{d}{d\theta}\right] Y_{l,m} = \sqrt{l(l+1) - m(m+1)}\, Y_{l,m+1}$

Lowering $Y_{l,m}$ to $Y_{l,m-1}$: $\qquad e^{-i\phi}\left[m \cot\theta + \frac{d}{d\theta}\right] Y_{l,m} = \sqrt{l(l+1) - m(m-1)}\, Y_{l,m-1}$

Exercise 1: Starting with $Y_{0,0} = 1$, $Y_{1,1} = C \sin\theta \, e^{i\phi}$, and $Y_{2,2}(\theta,\phi) = C \sin^2\theta \, e^{i\,2\,\phi}$, derive the spherical harmonics at the top of this page.

Exercise 2: Starting with $Y_{3,3}(\theta,\phi) = C \sin^3\theta \, e^{i\,3\,\phi}$, letting C be whatever constants multiply each function, show that $Y_{3,2}(\theta,\phi) = C \sin^2\theta \cos\theta \, e^{i\,2\,\phi}$.

$Y_{3,3}(\theta,\phi) = C \sin^3\theta \, e^{i\,3\,\phi}$

$Y_{3,2}(\theta,\phi) = C \sin^2\theta \cos\theta \, e^{i\,2\,\phi}$

$Y_{3,1}(\theta,\phi) = C \left[5\sin\theta \cos^2\theta - 3\sin\theta\right] e^{i\phi}$

$Y_{3,0}(\theta,\phi) = C \left[5\cos^3\theta - 3\cos\theta\right]$

$Y_{3,-1}(\theta,\phi) = C \left[5\sin\theta \cos^2\theta - \sin\theta\right] e^{-i\phi}$

$Y_{3,-2}(\theta,\phi) = C \sin^2\theta \cos\theta \, e^{-2\,i\,\phi}$

$Y_{3,-3}(\theta,\phi) = C \sin^3\theta \, e^{-3\,i\,\phi}$

INTRODUCTION TO THE PARTICLES OF PHYSICS

The existence of sub-atomic particles was known early in the twentieth century. The electron and the proton were the first to be known and in 1932 James Chadwick discovered the neutron and showed it not to be a combination of an electron and a proton. Also, in 1932, using cosmic rays, Carl Anderson used a cloud chamber invented by C. R. T. Wilson to discover a positive electron or positron that had already been predicted by P. A. M. Dirac.

The next question was what was the interaction that held the protons and neutrons together in the atomic nucleus against the repulsion of proton positive charge? Atoms in molecules and crystal lattices are held together by sharing electrons. This is rather like it was when you were in freshman camp. In order to get the newly arrived freshmen to have a feeling of togetherness, they had you play volleyball. Tossing a ball back forth was supposed to give you a feeling of togetherness. Starting with molecular bond, it was found that the basic interactions of nature appeared to involve one vast volleyball game.

In 1935, Hideki Yukawa proposed that the strong interaction holding the protons and neutrons (nucleons) together involved the exchange of an intermediate mass particle that came to be called the meson. We won't go through all of Yukawa's reasoning but rather approach the problem using Heisenberg's uncertainly principle. A fluctuation, ΔE, of an object's energy is un-detectable if it happens in a time, Δt, or less, where $\Delta E \, \Delta t = \hbar$. It had been established that the radius of a nucleus atomic mass number, A, was $r = r_0 A^{1/3}$, where $r_0 = 1.5 \times 10^{-15}$ meter. Hence, r_0 was the radius of a nucleon. The time for a particle moving with the velocity of light to go a distance, r_0, was $\Delta t = r_0/c$. In that time the rest mass of a nucleon could have an un-detectable fluctuation of $\Delta E = c^2 \Delta M$ if $\left(c^2 \Delta M\right)(r_0/c) = \hbar$ or $\Delta M = \dfrac{\hbar}{r_0 c}$ in kilograms. But in particle physics, energies are given in the voltage through which a particle of charge, e, would have to be accelerated to give it that energy (electron volts). In this case it would be $c^2 \Delta M/e = \hbar c/r_0 e$, now simply written, $\Delta M = \dfrac{\hbar c}{r_0 e}$.

If $\hbar = 1.055 \times 10^{-34}$ joule-sec, $c = 3 \times 10^8$ m/sec and $e = 1.6 \times 10^{-19}$ Coulomb, we get $\Delta M = 1.32 \times 10^8$ eV or $\Delta M = 132$ MeV (million electron volts).

Yukawa predicted that the strong interaction could arise from nucleons exchanging a particle of mass, $M = 132$ MeV (often written MeV/c^2).

PROBLEM: The electron, proton and neutron have masses in kilograms of

$M_e = 9.0195 \times 10^{-31}$, $M_p = 1.6726 \times 10^{-27}$, $M_n = 1.6750 \times 10^{-27}$. Letting $M c^2 = e V$, express these masses in million electron volts (MeV) or (MeV/c²).

In 1936, again using a cloud chamber and cosmic rays, Carl Anderson and S. H. Nedermeyer discovered a particle of mass 106 MeV. But Yukawa's meson was supposed to engage in the strong interaction and the new particle was later shown to be able to pass through solid matter without interacting with anything. Thus, the newly discovered particle, now called the muon, was to be in a group of particles related to electrons. These particles, called leptons, do not engage in the strong interaction. Yukawa's particles, called pions, were discovered by Powell, Occhialini and Lattes in 1947, working with cosmic rays at a high elevation in the Andes.

In Wilson's cloud chambers, a charged particle created a trail of condensation in a supersaturated vapor. In 1952, Donald Glaser invented a bubble chamber in which a charged particle created a vapor trail in a volatile liquid. With both of these detectors, the masses of particles were determined by placing the detecting chamber in a strong magnetic field. While this work with cosmic rays was going on, machines producing dependable beams of high energy particles, mostly protons by successive voltage boosts were being invented. The first accelerators producing protons with more than a billion volts of energy came into operation in 1952.

Suppose we wish to have a high energy proton to hit a proton in matter and convert energy of the incoming proton into a pion by the reaction:

p + p $\rightarrow$ p + n + π^+ or   p + p $\rightarrow$ p + p + π^-

If the rest mass of a proton is M_p and the rest mass of a pion is M_π, what must be the voltage through which the fast proton must be accelerated? The energy of the proton is $E = M_p c^2 + e V$ and its momentum is given from

$E^2 = p^2 c^2 + M_p{}^2 c^4$ as derived on page 9 of chapter 9.

Combining these two equations gives $p^2 c^2 = 2 M_p c^2 e V + e^2 V^2$.

Combining the energy of the fast proton with the energy of the target proton gives the total initial energy to be $E = 2 M_p c^2 + e V$.

In the collision both energy and momentum are conserved. The final total energy is $E = p^2 c^2 + (2 M_p + M_\pi)^2 c^4$. Letting $E = E$ and $p = p$ gives,

$2 M_p c^2 e V = \left(4 M_p M_\pi + M_\pi{}^2 \right) c^4$ or $e V = M_\pi \left(2 + M_\pi/2 M_p \right) c^2$.

PROBLEM: Fill in the missing steps of the previous derivation.

PROBLEM: If $M_p = 938$ MeV and $M_n = 140$ MeV, what is the minimum accelerating voltage required to produce a pion?

Once pions could be produced, if they had enough energy they could hit protons in matter and produce particles. It then happened that un-expected new particles related to pions but heavier and related to protons but heavier began to show up. These particles were called strange particles.

There was now a whole set of medium mass strongly interacting particles (mesons) and strongly interacting heavy particles called baryons. The pion is a meson and the proton and neutron are baryons. All strongly interacting particles are called hadrons.

It was noticed that if sufficiently energetic pions hit protons of matter, two strange particles could be produced. A meson to be called a K-on (K) and a baryon to be called a lambda (Λ) were produced by a strong interaction.

$$\pi^- + p \rightarrow \Lambda^\circ + K^\circ$$

Each of these decayed by a weak interaction (as with radioactive decay) as for example: $\Lambda^\circ \rightarrow p + \pi^-$ and $K^\circ \rightarrow \pi^+ + \pi^-$

$M_\Lambda = 1116$ MeV and $M_K = 497$ MeV.

After 1952, when the high energy particle accelerators came into operation, more and more strange particles began to show up. Order was brought into this apparent chaos of particles by the work of Murray Gell-Mann and others.

First, we must consider the angular momentum of these particles and an idea that Heisenberg derived from it. The quantization of angular momentum is discussed on page 37 of chapter 12. The intrinsic angular momentum (spin of a proton) has a value of $\frac{\sqrt{3}}{2}\hbar$ and gives the charge a magnetic moment which will precess around a magnetic field. The angular momentum parallel to the magnetic field is quantized to have only two values. These are $L_3 = +\frac{1}{2}\hbar$ or $L_3 = -\frac{1}{2}\hbar$, called spin-up and spin-down.

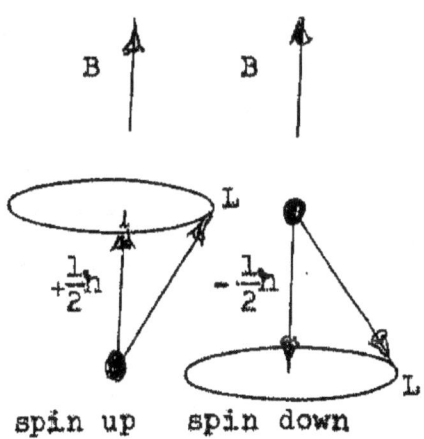

spin up   spin down

A proton and a neutron can combine to give a heavy hydrogen nucleus (deuteron). If their spins add, the angular momentum is $\sqrt{2}\,\hbar$ and its component along the magnetic field will be $+\hbar$, 0, or $-\hbar$. For a proton or neutron, L_3 has quantum numbers, $+\frac{1}{2}$ and $-\frac{1}{2}$. For the deuteron with spins adding, L_3 has quantum numbers, $+1$, 0, -1.

Heisenberg considered the proton and neutron to be two states of the same particle (Nucleon). In addition to two spin states, it has two charge states. So, he invented a name of isospin (no angular momentum involved). The proton has isospin, $+\frac{1}{2}$ and the neutron has isospin, $-\frac{1}{2}$, there turned out to be three pions with charges, $+e$, 0, and $-e$. So, the pion has isospin states of $+1$, 0, and -1.

Return, now to the interaction, $\pi^- + p \rightarrow \Lambda^\circ + K^\circ$, a strange baryon and a strange meson. Then a new strange particle showed up (Σ) which was heavier that the Λ°. $\pi^- + p \rightarrow \Sigma^\circ + K^+$, and it turned out that there were three Σ particles, Σ^-, Σ°, and Σ^*. Gell-Mann invented a strangeness number which he surmised was conserved in the strong interaction but not in the weak interaction.

He assigned $S = 0$ to the nucleons and pions, and $S = -1$ to the Λ° and Σ particles.

He gave $S = +1$ to the K° and K^+. Later, two more K-ons showed up requiring $S = -1$, and two even heavier baryons (Ξ) requiring $S = -2$.

The particles are displayed according to isospin and strangeness with masses given in MeV. A heavy η meson has been omitted. Gell-Mann's quarks, u, d, and s will be discussed later.

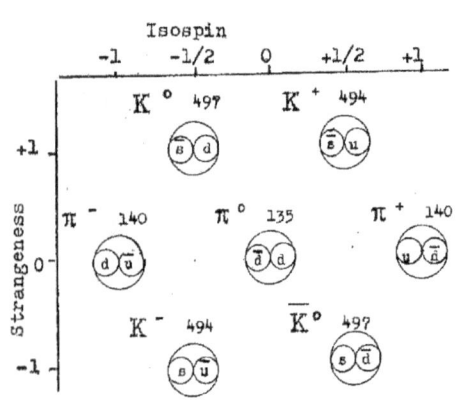

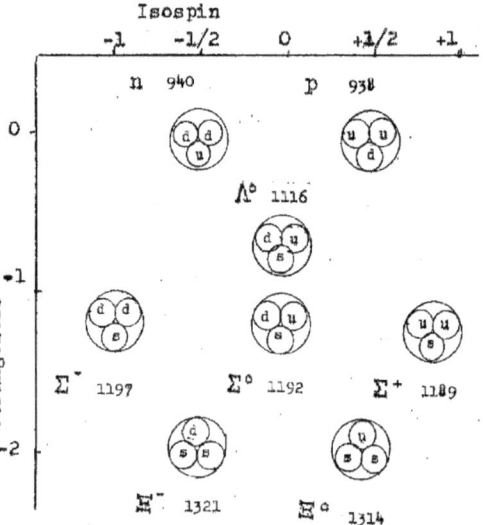

With more energy to play with, a new set of particles showed up, called deltas (Δ) with masses of about 1238 MeV. Examples were:

$$\pi^- + p \rightarrow \Delta^o \rightarrow \Lambda^o + K^o \qquad\qquad \pi^+ + p \rightarrow \Delta^{++} \rightarrow \pi^+ + p + \gamma$$

$$\text{or } \Delta^o \rightarrow \pi^- + p + \pi^o$$

These appeared to be excited states (resonances) of the baryons we have already discussed. The deltas appeared to have strangeness zero, thus appearing to be excited states of nucleons, except that there were four of them, Δ^-, Δ^o, Δ^+, Δ^{++}. Next to show up were Y^* resonances (1385 MeV).

$$K^- + p \rightarrow Y^* \rightarrow \Lambda^o + \pi^+ + \pi^-$$

$$\text{or } Y^* \rightarrow p + K^o + \pi^-$$

There were three of these, Y^{*-}, Y^{*o}, Y^{*+}, having strangeness, $S = -1$.

In the figure to the right we see a new event of interest. From the left of the picture, a charged particle known to be K^- enters the field of view. It hits a proton and two uncharged particles are produced. These then decay into two pairs of particles, identified as a proton/π^- pair and π^+/π^- pair. What happened at the first collision?

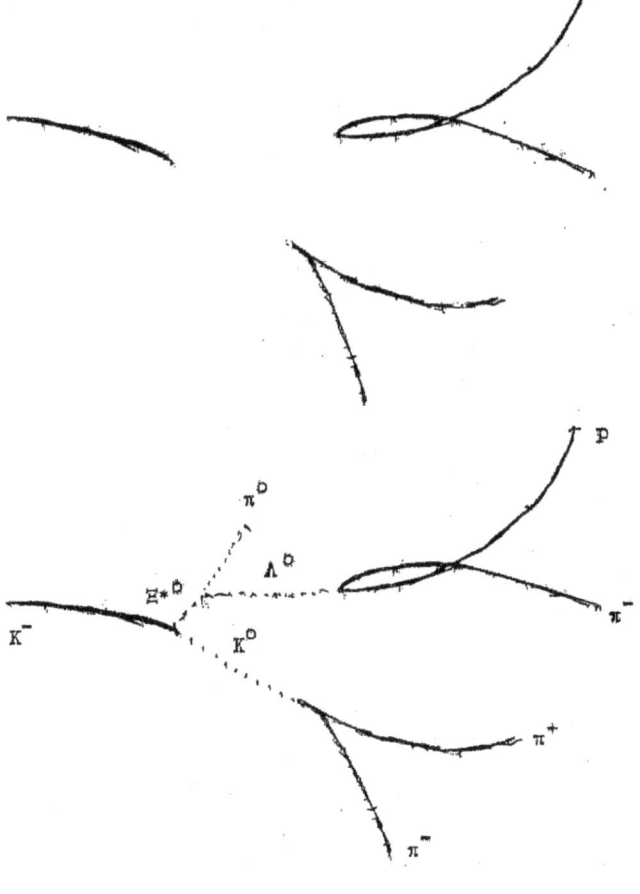

It is interpreted that a K^o and a new resonance, Ξ^{*o}, are produced strongly. The K^o decays into a pion pair but the resonance decays into a π^o and a Λ^o, the latter decaying into a p and π^-.

The Ξ^{*o}, thus, has strangeness, $S = -2$, and is an excited state of the Ξ^o of the first baryon set. Because of the decay from this particle, it is called a cascade particle. Its mass turns out to be about 1530 MeV.

When these new hyperons (particles heavier than the nucleons) were arranged according to isospin and strangeness, the pattern suggested to Gell-Mann the existence of strangeness,

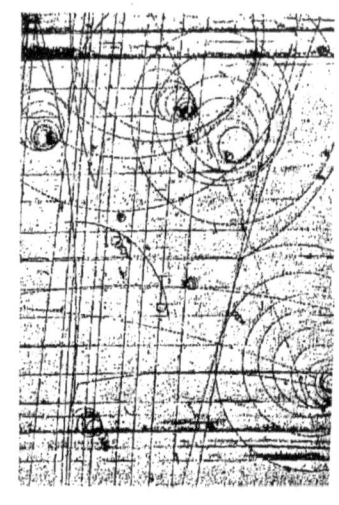

Δ⁻   Δ⁰   Δ⁺   Δ⁺⁺   1238 Mev
  Y*⁻   Y*⁰   Y*⁺   1385 Mev
      Ξ*⁻   Ξ*⁰   1530 Mev
            ?   1672 Mev
          O

$S = -3$. He predicted its mass and weak decay schemes. Exactly the particle that he predicted was seen at Brookhaven National Laboratory in 1964.

The new hyperon resonance was called the omega minus. The bubble chamber picture from Brookhaven National Laboratory is shown here with the interpretation of the event. It decayed according to the following scheme.

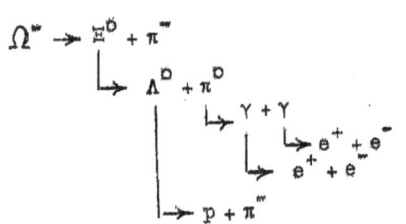

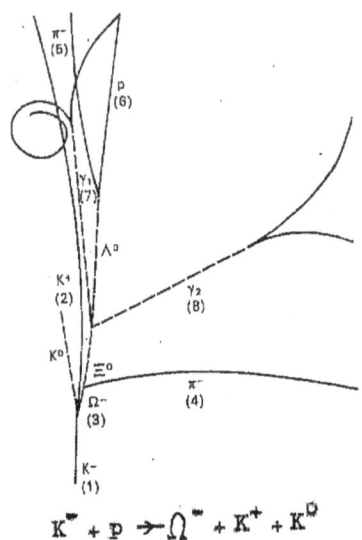

$$K^- + p \rightarrow \Omega^- + K^+ + K^0$$

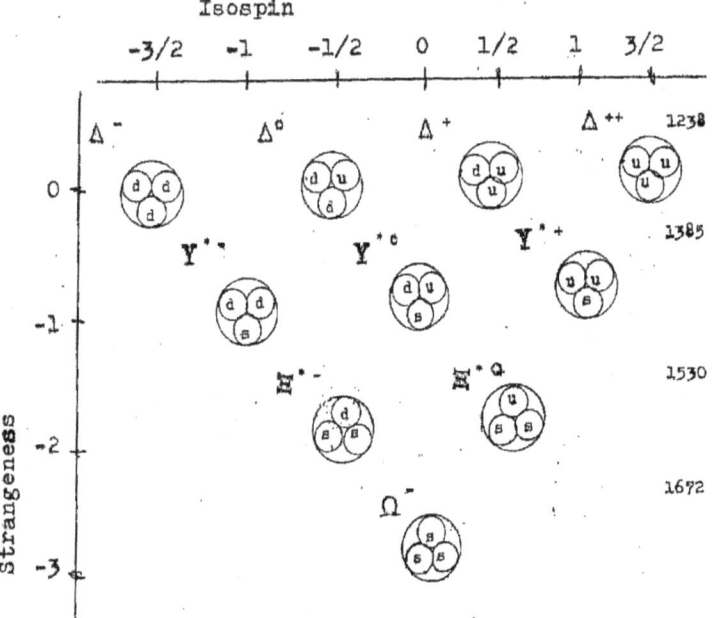

Here, we show the hyperon resonances, displayed according to isospin (I_3) and strangeness (S). To get a formula for the charges on the mesons and baryons, we introduce a baryon number (B) where $B = 0$ for mesons and $B = 1$ for baryons. The charge, in multiples of e is given by $Q = I_3 + \frac{1}{2}(B + S)$.

PROBLEM: Check if this gives the proper charge for each of the hadrons we have discussed. The numbers by each type of particle are the mass in MeV.

The pion said to the mu:
"Don't go with those hyperons. You
    have a weak interaction.
        So form no strong attraction.
Besides, they are strange particles too."

17-20

We called the hyperon resonances to be excited states of less massive particles. But, if something has an excited state it must have a structure. Murray Gell-Mann next proposed that the hadrons were composed of more elementary particles which he called quarks (citing the passage "Three quarks for Muster Mark" in James Joyce's book, Finnegan's Wake) because he proposed the baryons to be composed of three quarks. He proposed an up quark (u), a down quark (d) and a strange quark (s), with antiquarks, $\bar{u}$, $\bar{d}$, and $\bar{s}$. More recently seen particles require what are called charmed quark, bottom quark and top quark.

If there are to be quarks in each baryon, each quark must have a baryon number of $1/3$, and also a charge (Q) that is a fraction of e. Below is a table of quark properties.

	B	I_3	S	Q
u	1/3	+ 1/2	0	+ 2/3
d	1/3	− 1/2	0	− 1/3
s	1/3	0	− 1	− 1/3
$\bar{u}$	− 1/3	− 1/2	0	− 2/3
$\bar{d}$	− 1/3	+ 1/2	0	+ 1/3
$\bar{s}$	− 1/3	0	+ 1	+ 1/3

PROBLEM: In the display of each set of particles, each particle is labeled with its component quarks. Each baryon has three quarks and each meson is a quark-antiquark pair. For each particle, combine the values of S and Q to see if they add up to give the strangeness and charge for that particle.

We have been dividing things up into pieces. Matter is composed of atoms. The atom is composed of a nucleus and electrons. The nucleus is composed of protons and neutrons. Protons and neutrons are composed of quarks. This begins to sound like a poem:

Every dog should have some fleas
To flit on 'em bite 'em.
And every flea should have some fleas,
And so ad infinitum.

We now look at the quantum mechanical operators involved in the particle transmutations. In second quantum field theory, $\hat{a}^*$ creates a particle and $\hat{a}$ annihilates a particle. Let $\hat{a}_u{}^*$, $\hat{a}_d{}^*$ and $\hat{a}_s{}^*$ create up, down and strange quarks, and $\hat{a}_u$, $\hat{a}_d$ and $\hat{a}_s$, annihilate up, down and strange quarks.

Then, the weak interaction, $\Omega^- \rightarrow \Xi^o + \pi^-$, would involve the operator, $\hat{V}^+ = \hat{a}_u{}^* \hat{a}_s$, and the decay,

$\Xi^o \rightarrow \Lambda^o + \pi^o$, would involve $\hat{U}^+ = \hat{a}_d{}^* \hat{a}_s$.

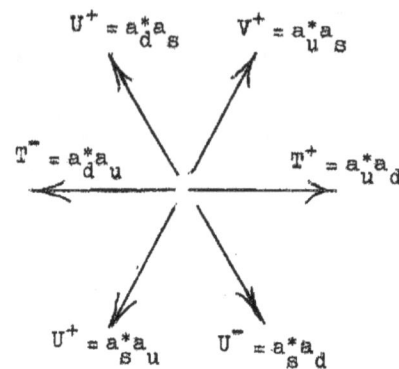

Gradually, a theory of quark-quark interactions was developed. Recall Yukawa's model involving fluctuations of nucleon mass which involved emission and absorption of pions in small intervals of time. Exchanging pions between adjacent nucleons produced their cohesion. The quarks are said to exchange gluons which provide the glue to hold them together. But things become complicated. The three quarks form a core of the baryon, but they are emitting short lived gluons along with the emission of short lived pions, all of which form a cloud of attracting particles around the core of three quarks.

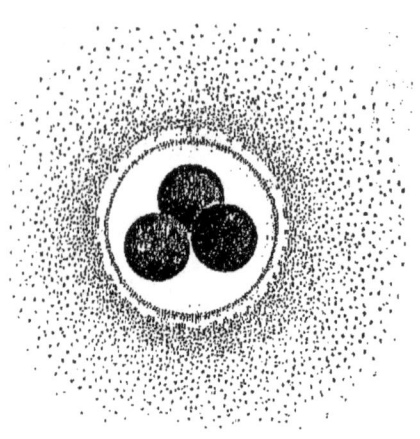

So far, we have discussed the finding and identification of elementary particles and their decay schemes. But if we want to study the structure of a nucleon, our tool must not engage in the strong interaction. This requires photons (γ) or electrons. Obviously, this involves a variation of a microscope, optical or electron. Let us consider an electron microscope.

Detail in a specimen cannot be resolved at intervals less than the wavelength of the radiation. The size of a nucleon is on the order of 10^{-15} meter. Suppose we want $\lambda = 10^{-16}$ meter for our electrons. Through what voltage must they be accelerated? Now, $p \lambda = h$. Hence, $p = h/\lambda$.

Now, $E^2 = p^2 c^2 + m_o{}^2 c^4$ and $p c = 2 \times 10^{-9}$ Joules but $m_o c^2 = 8 \times 10^{-14}$ Joules.

Hence, $E = p c$ for these electrons. But, $E = e V + m_o c^2$. Again, neglect $m_o c^2$.

Thus, $V = \frac{p c}{e} = 12 \times 10^9$ volts.

Hence, the construction of the two mile long, 30 billion volts (30Gev), Stanford Linear Accelerator (SLAC). From simple mechanics, if a particle is scattered off a target much heavier than itself, the scattering object gains negligible momentum and the energy of the scattered particle is conserved (elastic scattering). If the scattering object's mass is on the order of that of the scattered particle, there will be considerable momentum transfer and the scattered particle loses energy (in-elastic scattering). E. D. Bloom and others at SLAC in 1969 reported electrons into nuclear targets being scattered off of particles of much less mass than the nucleon itself. Not jumping to a hasty conclusion, Feynman called the lighter scattering particles "Partons", but physicists considered that most partons were quarks. However, no matter how much one banged away at a nucleon, nothing ever came out that could be identified as a quark.

To get a start on thinking about nucleon structure, A. Chodos and others at MIT proposed in 1974 a simple but intuitive model in which the cloud of strongly interacting particles shown around the three core quarks in the figure on the previous page act as an elastic membrane forming a stretchable "bag". In the ground (relaxed) state the core quarks act as if they were free un-attached particles (asymptotic freedom). In an excited state the quarks would stretch the elastic bag with the result that they would behave as if they were attached to the center by springs. This would allow the use of a harmonic oscillator potential as a basis for description.

In deep in-elastic scattering at SLAC an incoming electron transferring momentum and energy comes off with reduced energy. The quark takes off, stretches the "spring" (bag) and bounces back. But why can't the quark be hit strongly enough to knock it out of the nucleon?

Suppose the incoming electron hits the quark with enough energy transferred to the quark that the spring is stretched to where the energy in the spring is greater than the mass of a quark-antiquark pair. Something new happens, the spring breaks, its energy producing a quark and antiquark. The quark falls back into the nucleon and the antiquark goes off with the original quark, producing a meson.

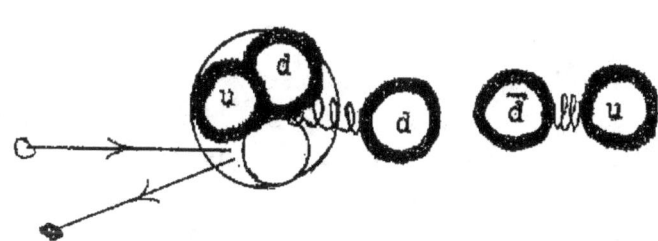

In the picture:

$$e + p \rightarrow n + \pi^+ + e$$

During the process, a short lived Δ^+ resonance may occur.

In this chapter we found that a particle with rest mass, m_o, and momentum, p, had energy, $E = p^2 c^2 + m_o{}^2 c^4$. What quantum mechanics does this give? If we replace momentum by its operator, $\hat{P}$, and energy by the Hamiltonian operator, $\hat{H}$, we get the operator relation, $\hat{H}^2 = \hat{P}^2 c^2 + m_o{}^2 c^4$, where m_o and c are numbers.

Paul A. M. Dirac (1902-1984) created factors for $\hat{H}^2 = \hat{P}^2 c^2 + m_o{}^2 c^4$ by introducing four new operators: α_x, α_y, α_z, and β.

Instead of, $\hat{H}^2 = \hat{P}_x{}^2 c^2 + \hat{P}_y{}^2 c^2 + \hat{P}_z{}^2 c^2 + m_o{}^2 c^4$, he wrote,

$$\hat{H} = \alpha_x \hat{P}_x c + \alpha_y \hat{P}_y c + \alpha_z \hat{P}_z c + \beta m_o c^2 .$$

Thus, we get an energy operator which doesn't look like any H we have seen before.

If this applies to electrons. Suppose the electron is at rest. $P_x = P_y = P_z = 0$.

Then,

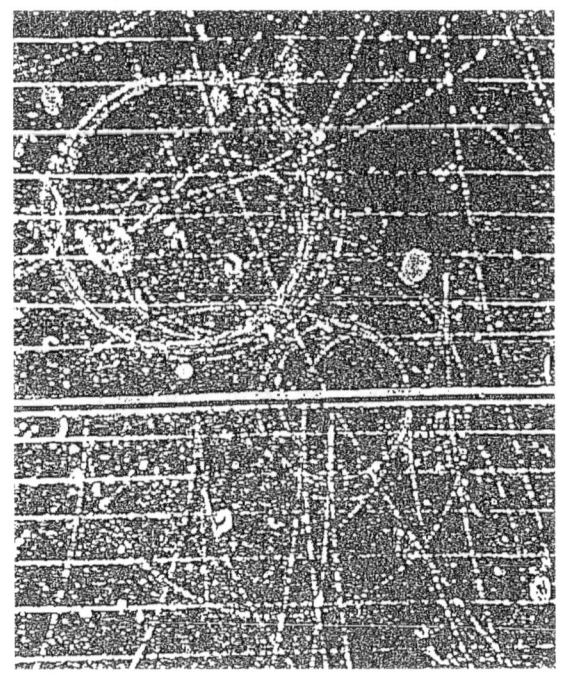

$$H = \beta m_o c^2 = \begin{pmatrix} 1 & 0 & 0 & 0 \\ 0 & 1 & 0 & 0 \\ 0 & 0 & -1 & 0 \\ 0 & 0 & 0 & -1 \end{pmatrix} \times m_o c^2$$

An electron at rest has energies $E = + m_o c^2$ and $E = - m_o c^2$. What are we to make of negative rest mass?

Dirac used this picture to account for both positive and negative rest-mass energy states. Assume the universe to be filled with a potentially infinite number of both positive and negative rest-mass energy states for electrons and protons. Assume that in a vacuum, all of the negative states are filled. In the presence of an electric field nothing happens, just as if there were no particles at all. If a photon with sufficient energy entered the vacuum region, an electron could be knocked from a negative to a positive rest-mass energy state, where it would act as normal electron, but newly created. The vacant energy state would produce a deficiency of electrons and hence, a surplus of protons in negative energy states.

The vacancy would act as a positive electron, newly created. The event predicted by Dirac was discovered in 1932 by Carl Anderson (1905-1991). He used a cloud chamber where super-saturated water vapor would condense on the trails of charged particles moving through the region. In the presence of a magnetic field, charged particles would move in circular paths, positive and negative particles curving in opposite directions.

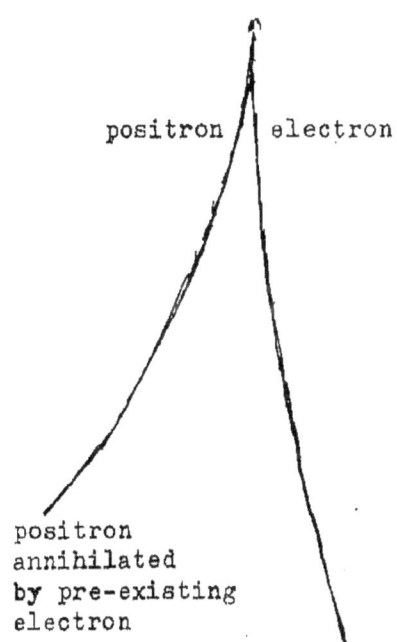

Events such as what Anderson saw appear in the dark image on the previous page. Photons with approximately **200** MeV energy enter from the top right. "Catalyzed" by the nuclear fields in a strip of lead across the center of the picture, they produce the positive and negative electrons as predicted by Dirac.

The positive electrons are called positrons. On the right, an electron-positron pair is produced just above the lead strip. With the magnetic field directed out of the picture, the positron curves to the left and at a point below the lead strip it disappears. At this point, a pre-existing electron disappears by falling down into the vacant negative mass-energy state and filling it. Two photons carry off the $2\, m_o\, c^2$ energy released.

It was evident that protons also could be knocked out of their negative mass-energy states producing negatively charged mass-energy proton states. To do this, the incident photons must have around **2000** times the energy of those producing positrons. On the completion of the Bevatron at U. C. Berkeley, Owen Chamberlain and Emilio Segré predicted this event in 1955. Negative protons and positive electrons are called anti-particles, being particles of anti-matter.

Being bothered by the existence of an infinite sea of virtual electrons and protons in negative mass-energy states, Richard P. Feynman came up with an alternate model of how to interpret the negative energies. Thinking of world-lines in space-time, time is plotted on a vertical axis and some space dimension is plotted horizontally. The positron-electron events are depicted at the top of the following page. A photon and a pre-existing electron progress in time from the bottom of the diagram. The photon decays into a positive and negative electron. The negative electron continues to exist, but the positive electron is annihilated by the pre-existing electron producing two photons. You will notice that charge and energy are conserved at all times.

Richard Feynman (1918-1988) devised an alternate interpretation. Momentarily forget about the time axis. The pre-existing electron progresses to a point where it emits two photons each with its own rest-mass energy. Conserving energy, it comes off with a negative rest-mass energy. Going backwards in time, it absorbs a photon with twice its rest-mass energy at which point it has positive rest-mass energy and continues moving forward in time. Here, a negative electron with negative rest-mass energy will be going backwards in time, but will appear as a positive electron with positive rest-mass energy going forward in time. Notice that in the Feynman model, charge and energy are conserved at each vertex as the same particle goes both forward and backwards in time. Feynman, considering only one particle has no need for infinitely many virtual particles in negative energy states.

Where do physicists go from here?

At this point we touch on what may becoming the 21st century´s fundamental picture of the universe, that is, QUANTUM FIELD THEORY. We won´t be able to take you very far into this theory as the mathematics is some of the most difficult ever put together. (The philosopher Plato said that God was a mathematician, and was he ever right.) But, let us look at one consequence of the previous discussion.

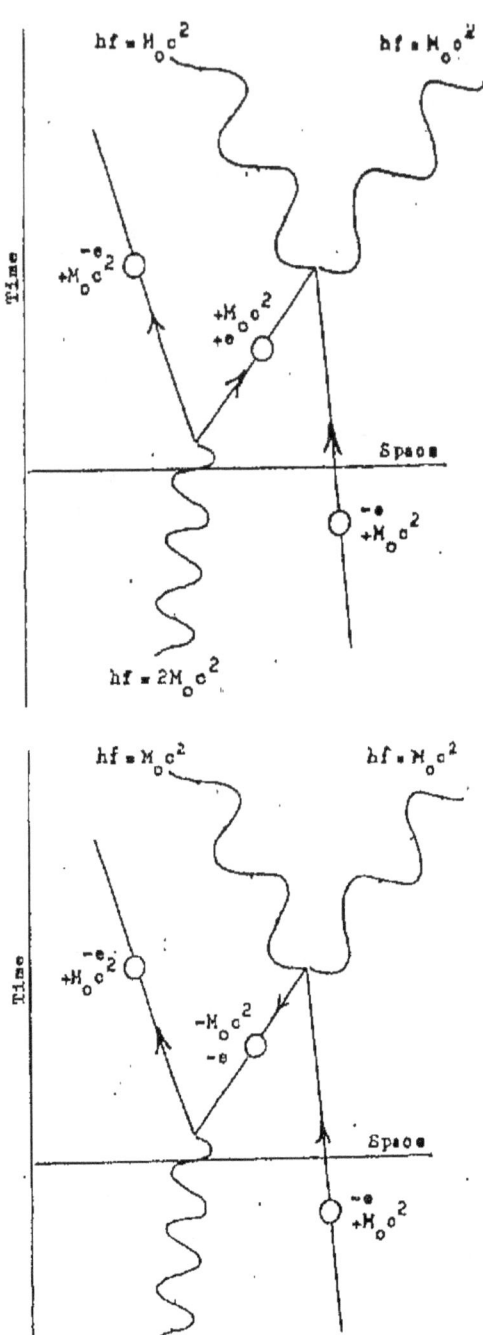

When we talked about a particle in a box, we found that we cannot simultaneously know the position and momentum of a particle. If we knew exactly where a particle was located, we couldn´t know how fast it was going, and vice versa. By the Heisenberg uncertainty principle, if Δx was how much we don´t about position, and Δp was the error in our knowing its momentum, then $\Delta x \, \Delta p = h$.

Now, $E = \dfrac{p^2}{2M}$, then $dE = \dfrac{2\,p\,dp}{2\,M}$, but $p = M\dfrac{dx}{dt}$, which gives $dE = \dfrac{dx\,dp}{dt}$, or

$dE\,dt = dp\,dx$. This suggests that if Δt is the error in knowing when a particle's energy was measured, and ΔE was the resulting error in knowing its energy, then $\Delta E\,\Delta t = h$.

This means that if in an interval, Δt, the energy of something varies by ΔE, where $\Delta E\,\Delta t = h$, the variation is not observable.

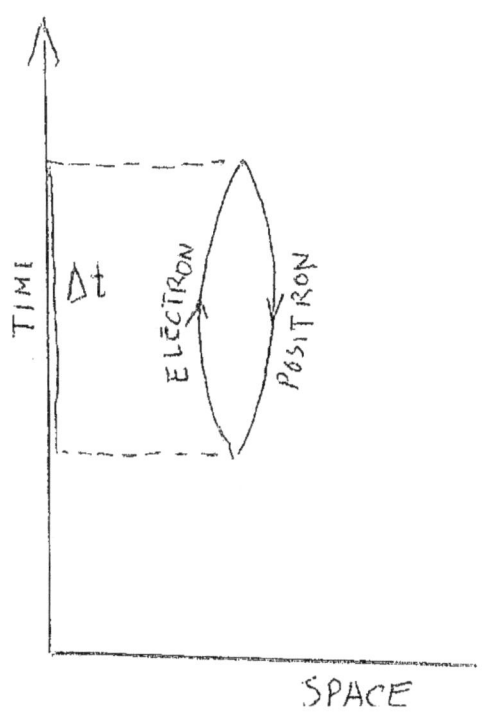

This is interpreted that if there is an energy surge in the vacuum of more than twice the electron's rest mass energy, an electron-positron pair can appear out of nothing and disappear within the time interval, Δt, we could not directly observe it. By quantum field theory, the vacuum is boiling with unobservable electron-positron pairs. Never-the-less, is there a net effect on the vacuum related to dark matter or the repulsive gravity recently observed in the vacuum? In spite of the advances of quantum field theory, we do not understand the vacuum.

You may wish to study atomic, nuclear, and particle physics past what is in this book. Some titles of other books are listed on the next page and you may find more books online. Before buying a book, check its listing on the web to see if it is a book you may use.

Basic physics in the 21st century is moving past quantum mechanics into quantum field theory and even topological quantum field theory (to bring in some mathematics). No one ever said that quantum field theory was an easy study, but the following titles might be places to start.

T. Lancaster & S. J. Blundell: Quantum Field Theory for the Gifted Amateur, Oxford (2014)

D. McMahon: Quantum Field Theory Demystified, McGraw Hill (2008)

M. Maggiore: A Modern Introduction to Quantum Field Theory, Oxford (2005)

A. M. Tsvelik: Quantum Field Theory in Condensed Matter Physics, Cambridge (2003)

P. Labelle: Supersymmetry Demystified, McGraw Hill (2010)

RENORMALIZATION IN QUANTUM FIELD THEORY

Modern physics recognizes that what we call particles aren't the simple little blobs of things we had thought, but complicated combinations of basic things with all sorts of virtual stuff hanging on to them. The process of renormalization involves taking a bare particle and determining its properties when all sorts of things are hanging on to it.

An idea of this can be gained by taking a copper ion in water. The water molecule (because of the 2p orbital functions of oxygen) has its hydrogen attached to one side of the oxygen with the result that the oxygen has a negative charge and the hydrogen end is positive. In liquid water, the negative ends of some water molecules attach themselves to the copper ion which

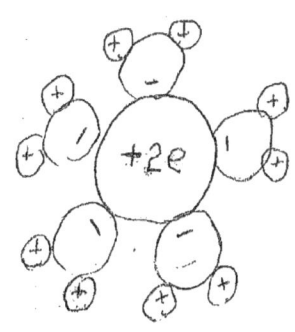

results in the complex object of copper and water. What is moving through the water, is not a simple copper particle, but a complex of copper and water. It still has a charge of $+2e$ but a greater mass and drag coefficient. It can be called a "dressed particle".

In quantum field theory a charged particle is dressed with virtual particles, mostly photons, but also other things. Recall that energy (and mass) are not conserved if in a time interval, Δt, an energy variation, ΔE, is such that $\Delta E \, \Delta t$ is less than h.

With conditions obeyed, a charged particle can be moving through time, emitting and absorbing virtual photons, which we cannot observe as photons. However, these virtual photons do have a physical effect. The cloud of virtual photons surrounding a charged particle constitutes its electric field. Another charged particle in the neighborhood could absorb these virtual photons and this gives the electric interaction.

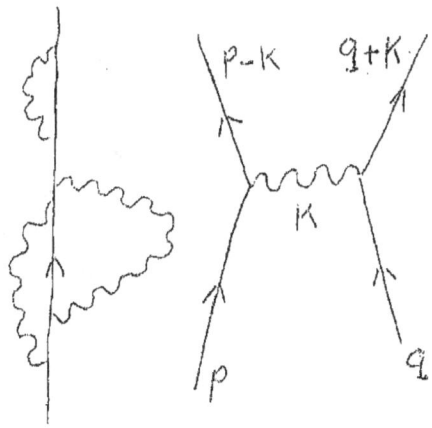

In the graph above, two charged particles, one with momentum, p, and the other with momentum, q, exchange a photon. The photon carries energy, $E = hf$, and momentum, $k = E/c$. This graph represents coulomb repulsion. An example of this was when Lord Rutherford bounced $_2\text{He}^4$ or alpha particles off of gold foil, discovering that it was not hard-shell scattering, but rather, coulomb repulsion of positive alphas by a positive nucleus.

How does the coulomb interaction come out of this? If a virtual photon (moving with the velocity of light) goes out a radius, r, and falls back, it goes a distance, $2r$, taking a time, $\Delta t = 2r/c$. This results in an energy fluctuation, $E = \dfrac{hc}{2r}$, where $\dfrac{hc}{2} = 1.0 \times 10^{-25}$.

Two particles with charge, e, a distance, r, apart, have $E = \dfrac{ke^2}{r}$ potential energy. $ke^2 = 2.3 \times 10^{-28}$, which is less that the photon's energy.

That photons carry momentum was demonstrated by A.H. Compton in 1923 when he bounced x-ray photons off hydrogen atoms in paraffin. The scattered photons coming off had longer wavelengths and less momentum than the incoming photons showing that the photons had lost momentum, and thus, that they carried momentum.

The renormalized charged particle can have even stranger things happening around it. The emitted photon can give rise to a short-electron-position pair which decays back in to a photon which is absorbed. All in $\Delta E\, \Delta t$ less than h.

The novelist Douglas Adams (1952-2001) commented:

"There is a theory which states that if ever anybody discovers exactly what the Universe is for and why it is here, it will instantly disappear and be replaced by something even more bizarre and inexplicable. There is another theory which states that this has already happened."

Further reading:

J. C. Foot: Atomic Physics, Oxford (2005)

Mark Fox: A Student's Guide to Atomic Physics, Cambridge (2018)

Gerhard Herzberg: Atomic Spectra and Atomic Structure, Dover Press (2010)

H. Haken, H. E. Wolf: Physics of Atoms and Quanta, Springer (2005)

J. Hamilton, F. Yang: Modern Atomic and Nuclear Physics, World Scientific (2010)

J-L. Basdevant, J. Rich, M. Spiro: Fundamentals of Nuclear Physics, Springer (2010)

Samuel Wong: Introductory Nuclear Physics, Prentice-Hall (1999)

Richard Dunlap: The Physics of Nuclei and Particles, Brooks/Cole (2004)

Frank Close, M. Martin, C. Sutton: The Particle Odyssey, A Journey to the Heart of Matter, Oxford (2002)

Ben Still: Particle Physics, Brick by Brick (2018)

Martinus Veltman: Facts and Mysteries in Elementary Particle Physics, World Scientific (2003)

Andrew Sessler, Edmund Wilson: Engines of Discovery: A Century of Particle Accelerators, World Scientific (2007)

Klaus Wille: The Physics of Partilce Accelerators, An Introduction, Oxford (2000)

Mario Conte, William W. MacKay: An Introduction to the Physics of Particle Accelerators, World Scientific (2008)

R. L. Sproull, W. A. Phillips: Modern Physics: The Quantum Physics of Atoms, Solids,and Nuclei, Dover (1980

CHAPTER EIGHTEEN: QUANTUM OPTICS AND THE THEORY OF LASERS

In chapter 13 on semiconductors, we have already discussed Light
Amplification by Stimulated Emission of Radiation with solid state diode
lasers. In this chapter we look more closely into the process of stimulated
emission, now considering gas phase lasers.

A typical gas phase laser is a modification of the spectrum tube used as a
source in measuring wavelengths of visible light from atomic sources.

The tube now has a fully
reflecting mirror at one end
and a partially surfaced mirror
at the other end to capture
light in standing waves, while
letting some light out through
the partial surfaced mirror.

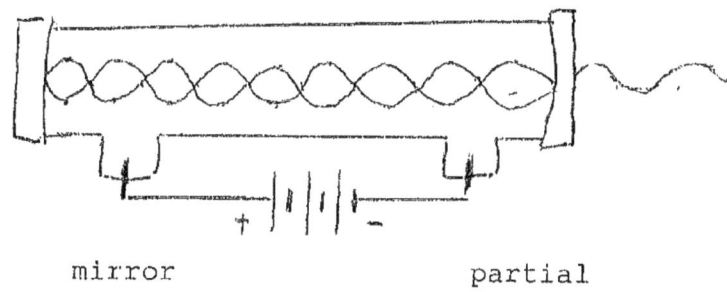

mirror   partial
                                          mirror

Light is produced in the following manner. Electrons produced at a negative
cathode are accelerated in an electric field and collide with atoms. The
collisions can result in three effects. Nothing can happen to an atom being
called an elastic collision. An atomic electron can be bumped into a higher
energy state (starting from its normal lowest energy state) this being an
in-elastic collision. Or the electron can be knocked totally out of the atom
leaving a free electron and an ion (charged atom).

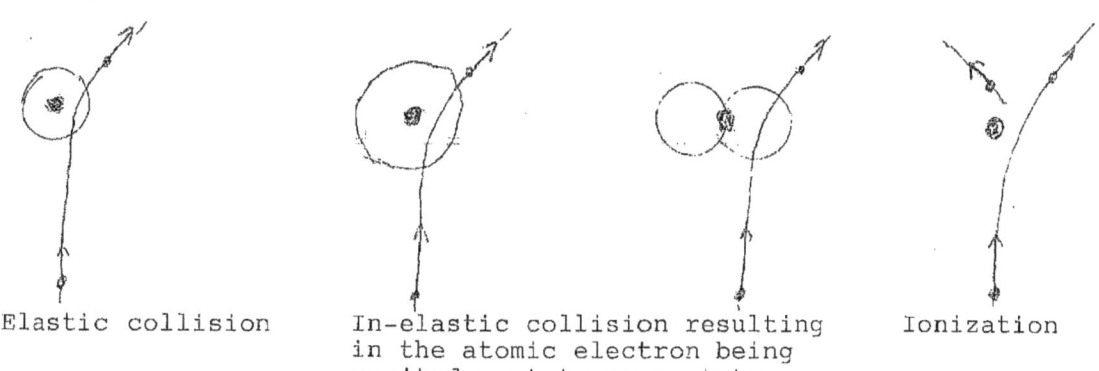

    Elastic collision   In-elastic collision resulting   Ionization
                             in the atomic electron being
                             excited s-state or p-state

To keep the picture simple, you can think of the gas being hydrogen. The gas
in the tube may be partially evacuated so that with a low density the
distance between atoms is greater. The greater distances between atoms allow
the free electrons to accelerate to higher velocities, giving a bigger bump
when they collide with atomic electrons. When there is considerable

ionization, additional free electrons are produced, and the tube contains a gas of charged ions which is called a plasma.

When the atomic electrons fall back to their lowest energy state, they emit photons of light. There are no restrictions on the excited atomic electron states that can result from a collision.

To keep a steady-state current through the tube, a voltage between **400** and **600** volts may be adequate. But to get a current started, a quick surge of up to **40,000** volts may be necessary. This concentrates so many electrons on the negative cathode that their repulsive forces push them off the electrode producing free electrons to start the current flowing.

TIME DEPENDENT QUANTUM MECHANICS

Light amplification by stimulated emission of radiation requires a detailed study of the interaction of atomic electrons with vibrating electric fields, and that requires that we employ time dependent quantum mechanics. So far in these chapters we have used quantum mechanics to describe things that did not change with time. Now bring time into the discussion. Let us review what we have said about quantum mechanics, but let us bring time variation into the discussion.

Assume now that a free particle, with no interactions, experiences the Planck and de Broglie wave-particle relationships. In chapters 9 and 13, we discussed waves of frequency, f, wavelength, λ, and wave velocity, $u = f\lambda$.

We can write for them a wave function, $\psi = A \dfrac{sin}{cos} 2\pi f \left(\dfrac{x}{u} - t\right)$.

If either a sine function or a cosine can describe our wave, we can use $e^{i\theta} = \cos\theta + i\sin\theta$ to write a generalized wave, $\psi = A e^{i 2\pi f\left(\frac{x}{u} - t\right)}$, but $u = f\lambda$ gives $\psi = A e^{i 2\pi\left(\frac{x}{\lambda} - ft\right)}$. But suppose that $f = \dfrac{E}{h}$ and $\dfrac{1}{\lambda} = \dfrac{p}{h}$ with $p = mv$, then $\psi = A e^{i\frac{2\pi}{h}(px - Et)}$. In all the discussions that follow, Planck's constant will always be divided by 2π, so we will follow a suggestion by P. A. M. Dirac and define an haitch-bar as $\hbar = \dfrac{h}{2\pi}$. Then our wavefunction is $\psi = A e^{i\left(\frac{px - Et}{\hbar}\right)}$.

If $\psi = A e^{i\left(\frac{px}{\hbar}\right)} e^{-i\left(\frac{Et}{\hbar}\right)}$, then $\dfrac{\partial\psi}{\partial x} = i\dfrac{p}{\hbar}\psi$ and $\dfrac{\partial^2\psi}{\partial x^2} = -\dfrac{p^2}{\hbar^2}\psi$.
Similarly, $\dfrac{\partial\psi}{\partial t} = -i\dfrac{E}{\hbar}\psi$ and $\dfrac{\partial^2\psi}{\partial t^2} = -\dfrac{E^2}{\hbar^2}\psi$.

But $E = \dfrac{p^2}{2M}$ for a particle with no interactions where $p = Mv$. Multiplying this by ψ gives $E\psi = \dfrac{p^2}{2M}\psi$, and substitution gives $i\hbar\dfrac{\partial\psi}{\partial t} = \dfrac{\hbar^2}{2M}\dfrac{\partial^2\psi}{\partial x^2}$. For a particle whose interaction has a potential energy, $V(x)$, $E = \dfrac{p^2}{2M} + V(x)$.

If we multiply this by ψ and substitute, we get $i\hbar\dfrac{\partial\psi}{\partial t} = \dfrac{\hbar^2}{2M}\dfrac{\partial^2\psi}{\partial x^2} + V(x)\psi$. This result is the time-dependent Schrödinger equation.

At this point we start to talk about the atoms in the laser shown on the first page of this chapter. The directional shapes of the nuclear wave functions developed in chapter 17 are identical to the directional shapes of atomic electron orbitals. The radial functions differ in the two cases, but they fall off with negative exponentials in both cases. Plus and minus are algebraic signs on the functions and are not charge which for electrons is negative everywhere.

For a discussion of atomic orbital functions, we can start with the eigenfunctions of the three-dimensional oscillator, given in chapter 17. The reason the set of functions are labeled s, p, d, etc, comes from aspects of atomic spectroscopy, and should not concern us here. For each function, the letter C will stand for the appropriate normalizing constant.

$\langle x, y, z \,|\, 0, 0, 0 \rangle = C\,e^{-k^2 r^2/2}$ is called an s-orbital.

It is the only orbital occupied in the group state of hydrogen or helium. Notice that the s-orbital is transformed into its self by any rotation about any one of three mutually perpendicular axes.

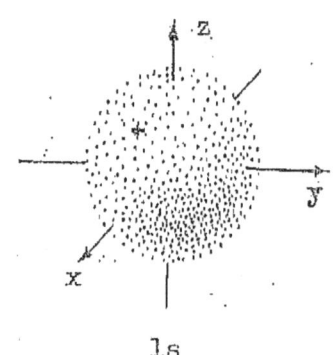

1s

$\langle x, y, z \,|\, 1, 0, 0 \rangle = C\,x\,e^{-k^2 r^2/2}$, $\langle x, y, z \,|\, 0, 1, 0 \rangle = C\,y\,e^{-k^2 r^2/2}$, and

$\langle x, y, z \,|\, 0, 0, 1 \rangle = C\,z\,e^{-k^2 r^2/2}$ are first occupied by electrons in the elements in the second row of the periodic table. They are called p-orbitals and are shown below. In the order given they are p-x, p-y, and p-z. Notice each has two mutually perpendicular axes of symmetry, transforming into its self by a rotation of **180°** about either axis.

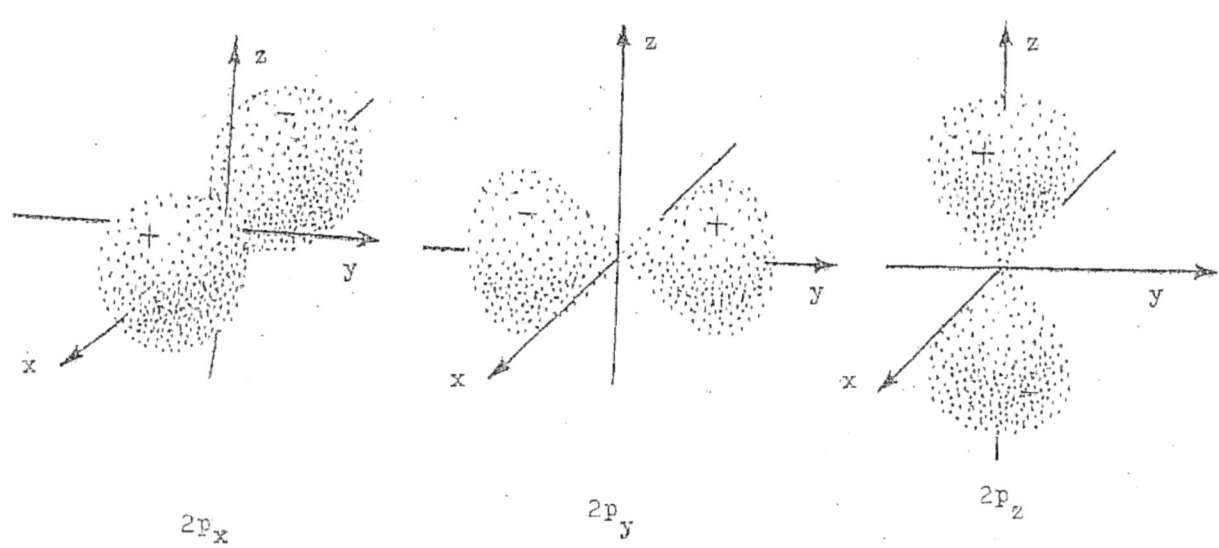

2p$_x$ 2p$_y$ 2p$_z$

The next set of functions are called d-orbitals. They first show up in the third row elements.

For example, $\langle x, y, z \, | \, 1, 1, 0 \rangle = C \, x \, y \, e^{-k^2 r^2/2}$, called d-xy, is shown to the right. Notice that it has one axis of symmetry and transforms into itself with a rotation of 90° about that axis. For the rest of the three-dimensional oscillator orbitals related to d-xy:

$3d_{xy}$

$$\langle x, y, z \, | \, 1, 0, 1 \rangle = C \, x \, z \, e^{-k^2 r^2/2}$$

$$\langle x, y, z \, | \, 0, 1, 1 \rangle = C \, y \, z \, e^{-k^2 r^2/2}$$

$$\langle x, y, z \, | \, 2, 0, 0 \rangle = C \left(k^2 x^2 - \frac{1}{2} \right) e^{-k^2 r^2/2}$$

$$\langle x, y, z \, | \, 0, 2, 0 \rangle = C \left(k^2 y^2 - \frac{1}{2} \right) e^{-k^2 r^2/2}, \quad \langle x, y, z \, | \, 0, 0, 2 \rangle = C \left(k^2 z^2 - \frac{1}{2} \right) e^{-k^2 r^2/2}$$

As these are eigenfunctions of the three-dimensional oscillator, all with the energy eigenvalue, $E = \frac{7}{2} \hbar \omega$, any linear combination of them will be an eigenfunction with the same energy. Notice that,

$$\langle x, y, z \, | \, 2, 0, 0 \rangle + \langle x, y, z \, | \, 0, 2, 0 \rangle + \langle x, y, z \, | \, 0, 0, 2 \rangle = C \left(k^2 r^2 - \frac{3}{2} \right) e^{-k^2 r^2/2}, \text{ is}$$

not a d-orbital, but rather a 3s orbital.

$$\langle x, y, z \, | \, 2, 0, 0 \rangle - \langle x, y, z \, | \, 0, 2, 0 \rangle = C \, (x^2 - y^2) \, e^{-k^2 r^2/2} \text{ is called } 3d_{x^2-y^2}.$$

A third eigenvector can be formed that is normal to the last two, then giving the function,

$$\langle x, y, z \, | \, [2 \, | \, 0, 0, 2 \rangle - | \, 2, 0, 0 \rangle - | \, 0, 2, 0 \rangle] = C \, (3z^2 - r^2) \, e^{-k^2 r^2/2}, \text{ called } 3d_{z^2}.$$

We now look at how electrons fill these orbitals in atomic structure. The Pauli exclusion principle allows only one electron to occupy any quantum state. In an atom's ground state, electrons fill lower energy quantum states before occupying higher energy states. The chart on the following page displays the relative energies of the atomic orbital shapes. The spherical s-orbital has only one shape, the dumbbell shape p-orbital has three shapes, the d-orbitals have five shapes, and f-orbitals have seven shapes. For each orbital there are two quantum states of spin angular momentum, shown with arrows on the next page. For atomic orbitals, the number prefix is for the corresponding Bohr orbit. Hydrogen and helium fill the 1-s orbits. The next eight elements fill the 2-s and 2-p orbits. These elements are of the most interest to the organic chemist. The 3-d orbits are metals starting with scandium. The chemists interested in these are inorganic chemists.

$\uparrow\downarrow$ $\uparrow\downarrow$ $\uparrow\downarrow$ $\uparrow\downarrow$ $\uparrow\downarrow$ $\uparrow\downarrow$ $\uparrow\downarrow$ 5F

$\uparrow\downarrow$ $\uparrow\downarrow$ $\uparrow\downarrow$ $\uparrow\downarrow$ $\uparrow\downarrow$ $\uparrow\downarrow$ $\uparrow\downarrow$ 4F $\uparrow\downarrow$ $\uparrow\downarrow$ $\uparrow\downarrow$ $\uparrow\downarrow$ $\uparrow\downarrow$ 5d

ENERGY

$\uparrow\downarrow$ 7S

The 4-F (lanthanides) and 5-F (actinides) are rare earths and of great interest to physicists.

$\uparrow\downarrow$ 6S

$\uparrow\downarrow$ $\uparrow\downarrow$ $\uparrow\downarrow$ $\uparrow\downarrow$ $\uparrow\downarrow$ 4d $\uparrow\downarrow$ $\uparrow\downarrow$ $\uparrow\downarrow$ 5p

$\uparrow\downarrow$ 5S

$\uparrow\downarrow$ $\uparrow\downarrow$ $\uparrow\downarrow$ 4P

A theoretical explanation of the periodic table is one of the great achievements of quantum mechanics.

$\uparrow\downarrow$ $\uparrow\downarrow$ $\uparrow\downarrow$ $\uparrow\downarrow$ $\uparrow\downarrow$ 3d $\uparrow\downarrow$ 4S

$\uparrow\downarrow$ $\uparrow\downarrow$ $\uparrow\downarrow$ 3P $\uparrow\downarrow$ 3S

Note: at the same Bohr level, the more nodes in the orbital, the higher the energy, as you might expect.

$\uparrow\downarrow$ $\uparrow\downarrow$ $\uparrow\downarrow$ 2P $\uparrow\downarrow$ 2S

$\uparrow\downarrow$ 1S

PERIODIC TABLE OF THE ELEMENTS

The question comes up, what is the relation of these functions to the old-fashioned idea of a point-like particle? With the particle in a box, the wave function extended throughout the whole box and the particle was somewhere in the box. Maybe we can say that the particle is somewhere in the region where the wave is not zero.

When all this first came together as the new physics, physicists wondered if there was any physical meaning to $\psi_n(x)$. The idea of PROBABILITY popped up. A probability of one (1) for an event means that it certainly will happen. A probability of zero (0) means that it definitely will not happen. Probability between zero and one means that it may or may not happen.

18-6

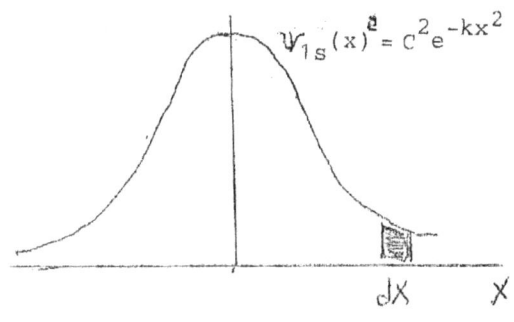

$\psi_{1s}(x)^2 = c^2 e^{-kx^2}$

dX X

The sum of the probabilities for anything happening is unity (1).

It was decided that for a particle in quantum state, E_n, $\psi_n(x)^2 dx$ would be the probability that the particle would be in the dx neighborhood of x.

With that interpretation, ψ^2 might be called a probability density.

To give this a bit of physical meaning, if the particle has charge say $-e$ for an electron, the charge would be distributed over the space occupied by the de-localized particle's wavefunction. So, we might call $-e\,\psi^2$, a charge density.

A bit of new words: in $\hat{H}\psi_n = E_n\psi_n$, E_n is called an EIGENVALUE which means a unique value. ψ_n is called an EIGENFUNCTION. Such things are also called wavefunctions although most of them don't look like waves.

We are demonstrating these ideas in one dimension. For actual atomic physics, all this is done in three dimensions. If $\psi_n(x)^2 dx$ is the probability of the particle being in a dx neighborhood, it has a total probability of unity (1) of being somewhere. This requires the sum:

$$\int \psi_n(x)^2 \, dx = 1$$

letting the sum be over the entire real axis.

Obviously, this sum gives a value for the constant, C.

We also need sums of the type:

$$\int \psi_m \psi_n \, dx = \langle m \mid n \rangle$$

(notice the new notation)

Using pictures rather than formal integration, we can save a lot of time and effort. On the following page, observe the symmetry and anti-symmetry of the orbital functions. The s-function is symmetric with reflection about all axes. Each p-function is anti-symmetric with one particular axis.

The d-functions are either symmetric or anti-symmetric depending on the reflection. Let us cut each function through in this plane of symmetry and multiply the results.

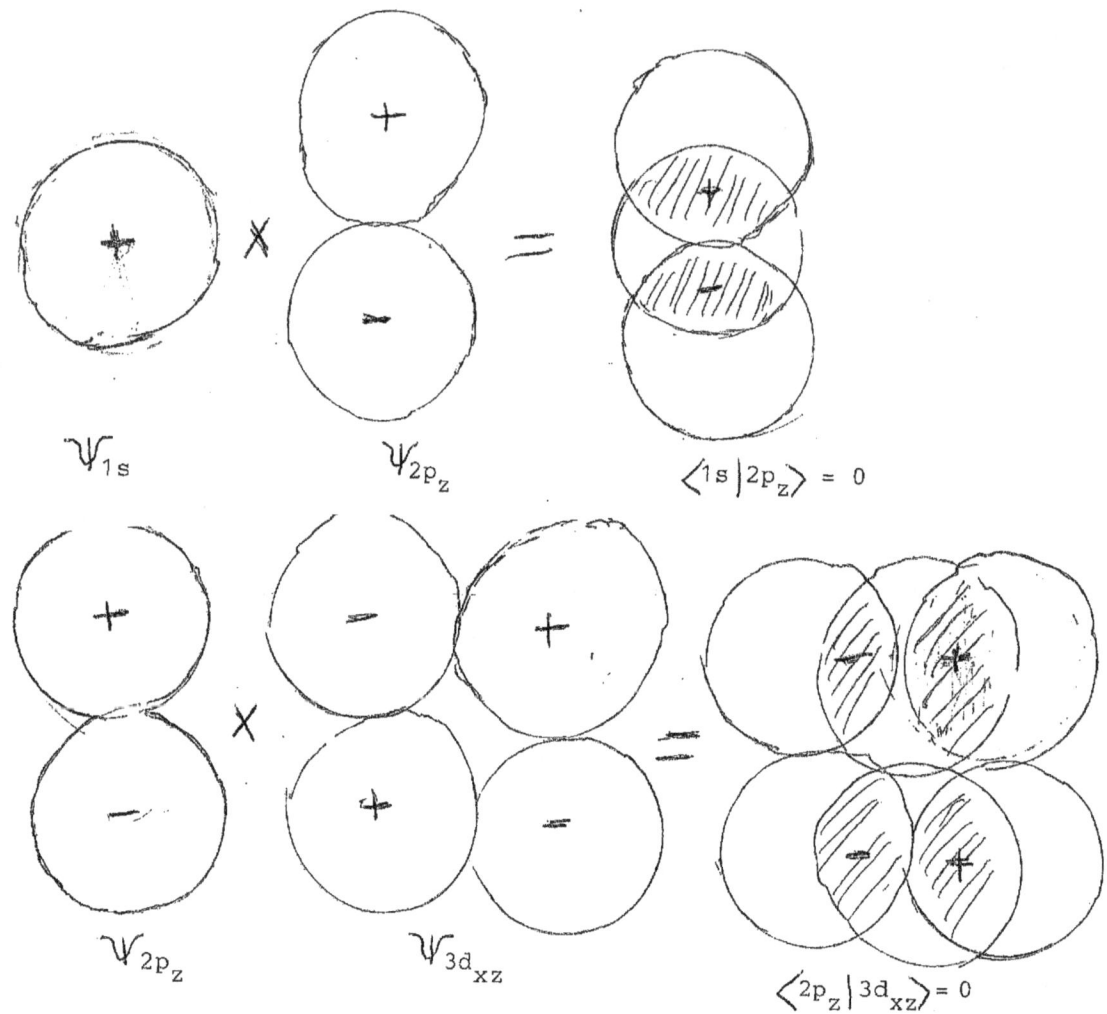

This suggest a general result. m and n are two different orbital functions.

If $E_m \neq E_n$, then $\langle m \,|\, n \rangle = 0$. When $E_m = E_n$, then $\langle n \,|\, n \rangle = 0$ is not zero.

In quantum mechanics we define ψ_n so that $\langle n \,|\, n \rangle = 1$.

Thus, $\langle m \,|\, n \rangle = \begin{matrix} 0 & if \ m \neq n \\ 1 & if \ m = n \end{matrix}$.

There is a symbol called the Kronecker delta in mathematics:

$\delta_{mn} = \begin{matrix} 0 & if \ m \neq n \\ 1 & if \ m = n \end{matrix}$.   So, $\langle m \,|\, n \rangle = \delta_{mn}$.

Before we develop time-dependent quantum mechanics for stimulated emission
and absorption of radiation, we should say more about the theory of atoms
and systems of atoms. We shall slightly change our notation. Let ϕ_n (phi) be
the wavefunction for an atomic quantum state. Let $\hat{H}^o$ be the energy operator
(Hamiltonian) for the internal electron's interactions $\hat{H}^o \phi_n = E_n{}^o \phi_n$.
Actually, there are a bunch of quantum numbers, so n is a representative
quantum number. Also, the electron exists in three dimensions, x, y, and z,
but since we don't have to calculate anything, we will do everything with
one representative position variable. Remember that $\langle n | n \rangle = 1$.

Now introduce the notation:

$$\langle n | \hat{H}^o | n \rangle = \int \phi_n \, \hat{H}^o \, \phi_n \, dx = E_n{}^o \langle n | n \rangle = E_n{}^o$$

Now, let the atom (say hydrogen in its 1s state) be
acted on by some external effect. In the picture,
there is an electric force field which pulls plus
charges in its direction and minus charges in the
opposite direction, distorting the 1s orbital as
shown. Let the normal (classical) energy for charges
in an electric field be $\hat{H}'$. Then, $\hat{H} = \hat{H}^o + \hat{H}'$ is the
total energy. Then, $\langle 1 | \hat{H} | 1 \rangle = \langle 1 | \hat{H}^o | 1 \rangle + \langle 1 | \hat{H}' | 1 \rangle$.

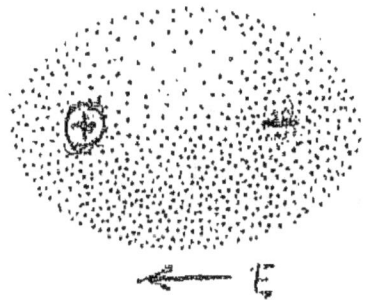

The new energy is $E = E_1{}^o + \langle 1 | \hat{H}' | 1 \rangle$. The integral, $\langle 1 | \hat{H}' | 1 \rangle$, is the energy
of the electric field acting on the electron's 1s orbital.

Next, take a gas of atoms, many in excited states. A general wave function
for an atom in the gas must take in to account that the atom might be in any
quantum state. We write the general function, ψ, to include possibilities
from each and all states. $\psi = \sum_n c_n \phi_n$, where c_n is constant.

Just as $\langle \phi_n | \phi_n \rangle = 1$, we want $\langle \psi | \psi \rangle = 1$.

When a summed function, ψ, appears twice in an integral, each time it
appears it must be summed separately.

$$\langle \psi | \psi \rangle = \int \left[\sum_m c_m \phi_m \right] \left[\sum_n c_n \phi_n \right] dx = 1$$

$$\langle \psi | \psi \rangle = \sum_m \sum_n c_m c_n \int \phi_m \phi_n \, dx = 1$$

Now, $\int \phi_m \phi_n \, dx = \langle m \mid n \rangle = \delta_{mn} = \begin{array}{ll} 0 & if \ m \neq n \\ 1 & if \ m = n \end{array}$. $\langle \psi \mid \psi \rangle = \sum_m \sum_n c_m \, c_n \, \delta_{mn}$.

Taking the summation over m first, δ_{mn} eliminates all terms except $m = n$. This leaves $\sum_n c_n^2 = 1$.

Thinking now of probabilities, unity (1) is the probability of something definitely happening. As we are talking about a generalized treatment of the states of a gas of molecules, there is unit probability of a molecule being in some allowed state. Then the probability of any molecule being in the nth state is given by c_n^2, interpreted as the fraction of molecules in the nth quantum state.

Next, consider $\langle \psi \mid \hat{H}^o \mid \psi \rangle = \int \psi \, \hat{H}^o \, \psi \, dx$, with $\hat{H}^o \phi_n = E_n^{\ o} \, \phi_n$.

As the c_n's are constant, $\langle \psi \mid \hat{H}^o \mid \psi \rangle = \sum_m \sum_n c_m \, c_n \int \phi_m \, \hat{H}^o \, \phi_n \, dx$.

$\langle \psi \mid \hat{H}^o \mid \psi \rangle = \sum_m \sum_n c_m \, c_n \, E_n^{\ o} \int \phi_m \, \phi_n \, dx$ or $\langle \psi \mid \hat{H}^o \mid \psi \rangle = \sum_n c_n^2 \, E_n^{\ o}$.

Multiplying each quantized energy by the probability that a molecule would have that energy and summing over all quantum states gives the average energy of the molecules in the gas. This times the number of molecules in the system would give the energy of the system. This total energy is called the internal energy.

Recall, if $\psi(t)$ is a time varying state function, $i\hbar\frac{d\psi}{dt} = \hat{H}\psi$.

First, what is the time variation of an atomic state function, $\phi_n(t)$.

Now, $\hat{H}^o\phi_n = E_n{}^o\phi_n$, so $i\hbar\frac{d}{dt}\phi_n = \hat{H}^o\phi_n$, or $\frac{d}{dt}\phi_n = \frac{E_n{}^o}{i\hbar}\phi_n$.

This gives $\phi_n(t) = \phi_n(0)\,e^{-i\,E_n{}^o\,t/\hbar}$. $\int\phi_n(0)^2\,dx = 1$, recall $\frac{1}{i} = -i$.

If $\phi_n(0)$ is a real valued function, how do we preserve the real value of its squared valued? Let us look at complex numbers in general.

Let $z = x + iy = R\,e^{i\theta}$ locate a point on a circle.
$z^* = x - iy$ is the complex conjugate of z.

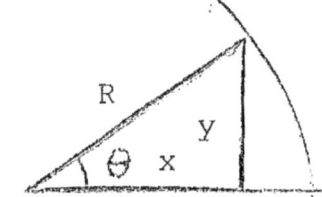

Now $z^*z = (x - iy)(x + iy) = x^2 + y^2 = R^2$. This is constant, regardless of the phase angle, θ.

The complex conjugate of $\phi_n(t)$ is $\phi_n^*(t) = \phi_n(0)\,e^{+i\,E_n{}^o\,t/\hbar}$.

Now, $\int\phi_n^*(t)\,\phi_n(t)\,dx = \int\phi_n(0)^2\,dx = 1$.

An atom not affected by an outside influence is said to be un-perturbed. Any outside influence that might change the state of the atom is called a perturbation. Let $\hat{H}'$ be such an outside effect that is time dependent.

The Hamiltonian is now $\hat{H} = \hat{H}^o + \hat{H}'$.

As we don't know the result of $\hat{H}'$, let the wavefunction be some $\psi(t)$.

Now, $i\hbar\frac{d}{dt}\psi = \hat{H}\psi(t)$.

Let $\psi(t)$ be expressed in terms of the probabilities that the resulting atomic state will be some one of the states, $\phi_n(t)$.

$\psi(t) = \sum_n a_n(t)\,\phi_n(t)$ and unlike previous c_n, $a_n(t)$ varies with time.

Without being disturbed from the outside, $\hat{H}^o\phi_n = E_n{}^o\phi_n$ gives the atomic electron's energy states and wavefunctions. Now let $\hat{H}'(t)$ be the energy of the atom's external interaction. $\hat{H} = \hat{H}^o + \hat{H}'(t)$ is now the Hamiltonian. If ψ is the state function, $i\hbar\frac{d}{dt}\psi = \hat{H}\,\psi(t)$.

Now, let $\psi(t) = \sum_n a_n(t)\,\phi_n(t)$.

In the atom's undisturbed state, $i\hbar\frac{d}{dt}\phi_n = \hat{H}^o\phi_n = E_n{}^o\,\phi_n$.

In quantum optics, $\hat{H}'(t)$ will be the interaction of the atom with the electromagnetic wave of light. Next, we put all of this together.

$$\sum_n a_n\, i\hbar\frac{d}{dt}\phi_n + \sum_n \phi_n\, i\hbar\frac{d}{dt}a_n = \sum_n a_n\,\hat{H}^o\,\phi_n + \sum_n a_n\,\hat{H}'\,\phi_n$$

From what we have just now said about the atom's undisturbed state, the first term to the left of the equal sign is identical to the first term to the right of the equal sign, hence they cancel each other. We are left with:

$$\sum_n \phi_n\, i\hbar\frac{d}{dt}a_n = \sum_n a_n\,\hat{H}'\,\phi_n$$

Now, pick one value of the n's and call it m. multiply our result by ϕ_m^* and integrate:

$$\sum_n \int \phi_m^*\,\phi_n\,dx\; i\hbar\frac{d}{dt}a_n = \sum_n a_n \int \phi_m^*\,\hat{H}'\,\phi_n\,dx$$

But, $\int \phi_m^*\,\phi_n\,dx$ is zero for $n \neq m$ and 1 only when $n = m$. This eliminates all of the terms in the left-hand summation except the nth term leaving,

$$i\hbar\frac{d}{dt}a_m = \sum_n a_n \int \phi_m^*\,\hat{H}'\,\phi_n\,dx$$

This is not the only time in quantum physics that we must know all of the a_n's in order to find any one of them. To a first order approximation, we will assume that when n is not the initial state, a_n, for $n \neq i$ is not zero, but much less than and that a_i for the initial state is approximately unity.

Thus, in the right-hand summation we drop all terms except for $n = i$.

We now expect $\hat{H}'(t)$ to start happening at some time, $t = -T$, and cease at some time, $t = +T$. At $t = -T$, $a_i = 1$ and all other a's are zero. At any time later than $t = -T$, we want the amplitudes for states other than the initial one. In what follows we let $m = k$ for $k \neq i$.

All this gives, $i\hbar\frac{d}{dt}a_k = \int \phi_k^* \hat{H}' \phi_i \, dx$, where $\phi_k^*(t) = \phi_k(0) \, e^{+i E_k^o t/\hbar}$ and $\phi_i(t) = \phi_i(0) \, e^{-i E_i^o t/\hbar}$. This gives us:

$$i\hbar\frac{d}{dt}a_k = \int \phi_k(0) \hat{H}' \phi_i(0) \, dx \, e^{i(E_k^o - E_i^o) t/\hbar}$$

We will label the symbol for the integral by the initial and final energies.

$$\langle E_k^o \, | \, \hat{H}' \, | \, E_i^o \rangle = \int \phi_k(0) \hat{H}' \phi_i(0) \, dx$$

We have considered an atomic electron in some initial quantum state and subject it to some external disturbance described by $\hat{H}'(t)$ from $t = -T$ to $t = +T$. The amplitude for any quantum states evolves in time according to $i\hbar\frac{d}{dt}a_k = \langle E_k^o \, | \, \hat{H}^1(t) \, | \, E_i^o \rangle \, e^{i(E_k^o - E_i^o) t/\hbar}$ for $k \neq i$.

One of the simplest problems is that of an object or system that is subject to no external perturbation for $t < -T$. At $t = -T$ we turn on a constant external field that gives $\hat{H}^1 = $ a constant until $t = +T$, at which time we turn off the perturbation and no external perturbation exists for $t > +T$. We describe this by $\hat{H}^1 = 0$ for $t < -T$; $\hat{H}^1 = V_o$, a constant, for $-T < t < +T$; and $\hat{H}^1 = 0$ for $t > +T$.

Now for $t < -T$, the object or system is in some initial state, ψ_i, that is assumed to be known. The problem is to find the final state, $\psi(t) = \phi_i a_i + \sum_{k \neq i} \phi_k a_k(+T)$, for $t > +T$, that is after the external field has been removed. The differential equation for $a_k(t)$ becomes:

$$\frac{da_k}{dt} = \begin{cases} 0 & \text{for } t < -T \\ -\frac{i}{\hbar}\langle E_k^o \, | \, V_o \, | \, E_i^o \rangle \, e^{i(E_k^o - E_i^o) t/\hbar} & \text{for } -T < t < +T \\ 0 & \text{for } t > +T \end{cases}$$

This gives $a_k(+T) = -\frac{i}{\hbar} \langle E_k{}^o | V_o | E_i{}^o \rangle \int_{t=-T}^{+T} e^{i(E_k{}^o - E_i{}^o)t/\hbar} dt$

$$= -\frac{i}{\hbar} \langle E_k{}^o | V_o | E_i{}^o \rangle \frac{e^{i(E_k{}^o - E_i{}^o)T/\hbar} - e^{-i(E_k{}^o - E_i{}^o)T/\hbar}}{i(E_k{}^o - E_i{}^o)/\hbar}$$

$$= -\frac{2i}{\hbar} \langle E_k{}^o | V_o | E_i{}^o \rangle \frac{sin\left[(E_k{}^o - E_i{}^o)T/\hbar\right]}{(E_k{}^o - E_i{}^o)/\hbar}$$

in which we have recalled that $sin\,\theta = \frac{e^{i\theta} - e^{-i\theta}}{2i}$.

You will notice that $(E_k{}^o - E_i{}^o)/\hbar$ is an angular frequency sort of quantity which according to the Bohr theory would be the angular frequency of radiation that would be emitted or absorbed by the system going from energy, $E_i{}^o$, to a final $E_k{}^o$. We therefore define, $\omega_{ik} = (E_k{}^o - E_i{}^o)/\hbar$. The expression for $a_k(+T)$ is made a bit simpler if we make this substitution, and also multiply and divide the whole thing by T. Then,

$$a_k(+T) = -\frac{2i}{\hbar} \langle E_k{}^o | V_o | E_i{}^o \rangle T \frac{sin(\omega_{ik}T)}{\omega_{ik}T}.$$

Now $a_k(+T)$ is the probability amplitude that, after having applied the constant perturbation, $\hat{H}^1 = V_o$, for a length of time of $2T$, the system will have made a transition from the state, $|\phi_i\rangle$, to the state $|\phi_k\rangle$. The probability of the transition is $|a_k(+T)|^2$. Now it is useful to think of the result of this problem as being expressed in terms of the transition probability per unit time. We define $\omega_{i\to k} = \frac{|a_k(+T)|^2}{2T}$.

Using the absolute value, or modulus, of the complex amplitude given above, we get the interaction being discussed.

$$\omega_{i\to k} = \frac{2}{\hbar^2} |\langle E_k{}^o | V_o | E_i{}^o \rangle|^2 T \left[\frac{sin(\omega_{ik}T)}{\omega_{ik}T}\right]^2$$

The function $\frac{sin^2\theta}{\theta^2}$ has the curve given to the right when plotted against θ. You will recall the limit, $\lim\limits_{\theta \to 0} \frac{sin\,\theta}{\theta} = 1$.

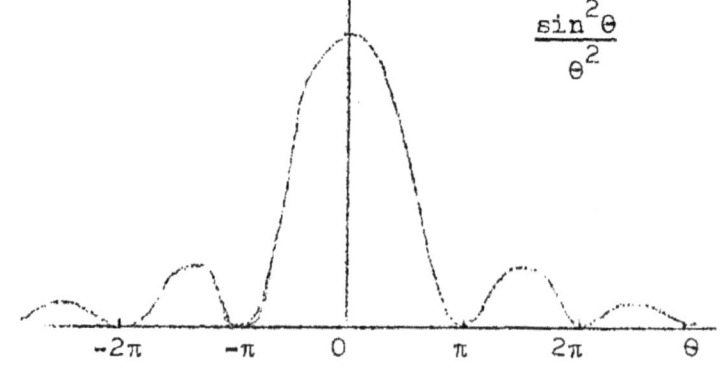

18-14

Likewise, the function, $\left(\dfrac{\sin(\omega T)}{\omega T}\right)^2$ can be plotted against ω (see next page).

When $\omega = 0$, its value is unity, from which it drops to zero for $\omega = \pm\dfrac{\pi}{T}$.

Now, $\omega_{i\,k} = (E_k{}^o - E_i{}^o)/\hbar$ and, if we start with the energy, $E_i{}^o$, $\omega_{i\,k}$ is the transition probability per unit time of the system going to a final state with energy, $E_k{}^o$.

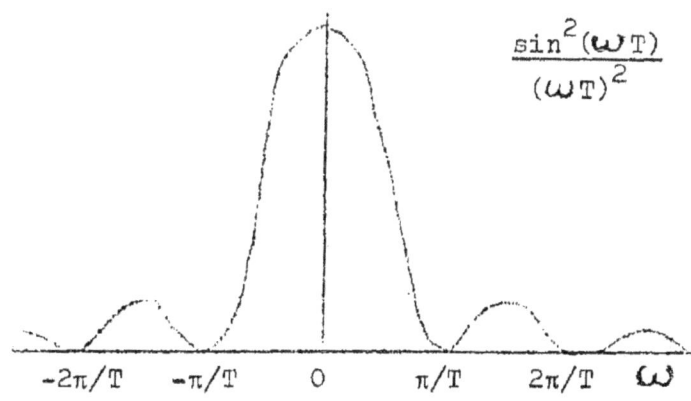

Hence, if the transition is permitted by the matrix element, $\langle E_k{}^o| V_o| E_i{}^o \rangle$, being not zero, there are relatively large probabilities for transitions to final states having energies between $E_{k'}{}^o$, for $\omega_{i\,k'} = -\dfrac{\pi}{2T}$, and $E_{k''}{}^o$ for $\omega_{i\,k''} = +\dfrac{\pi}{2T}$. For the energy limits we have arbitrarily picked the values of ω for which the probability curve has $(2/\pi)^2 \approx 0.4$ of its maximum height. This gives the higher energy limit, $E_{k''}{}^o = E_i{}^o + \dfrac{\pi\hbar}{2T}$, and the lower energy limit, $E_{k'}{}^o = E_i{}^o - \dfrac{\pi\hbar}{2T}$.

The difference between these limits is $\Delta E = E_{k''}{}^o - E_{k'}{}^o = \dfrac{2\pi\hbar}{2T}$.

If we take $\Delta t = 2T$ to be the time during which the energy operator was perturbed, yielding an uncertainty, ΔE, in the final energy of the system, then $\Delta E\,\Delta t = 2\pi\hbar$. This is, of course, nothing more that Heisenberg's uncertainly principle.

When an electromagnetic wave of light passes any point, it produces a vibrating electric field, say in the x-direction, $E = E_o \cos(\omega t)$. On the charge, $-e$, this exerts a force, $F = -e\,E_o \cos(\omega t)$.

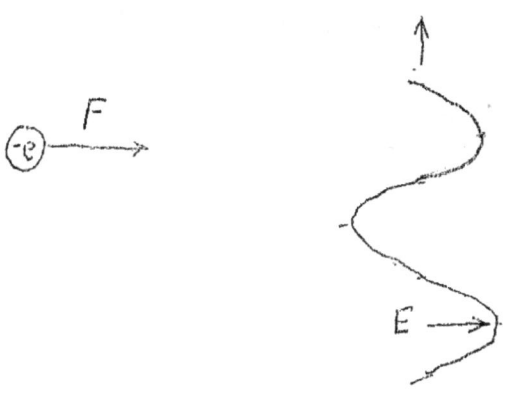

But $F = -\dfrac{dV}{dx}$, giving a potential energy, $V = e\,x\,E_o \cos(\omega t)$.

Letting $V_0 = e R_0 x$, this gives an operator, $\hat{H}^1 = V_0 \cos(\omega t)$.

Let $\hat{H}^1 = V_0 \cos(\omega t)$ be turned on at some time, say $t = -T$, and turned off at some time, say $t = +T$, where the time, $2T$, during which the system is perturbed is much longer than the period of vibration of $V_0 \cos(\omega t)$. Note that T stands for the time limits of the interaction and not the period of vibration of the interaction.

In treating this interaction, it is well to recall, $\cos(\omega t) = \dfrac{e^{i\omega t} + e^{-i\omega t}}{2}$, and treat $e^{i\omega t}$ and $e^{-i\omega t}$ separately. In these problems, V_0 may be a function of the position coordinates or other internal variables of the system under study. All that we say about it here is that it is not a function of time.

For weak perturbations, the general problem is stated on the middle of page 18-13.

$$\frac{da_k}{dt} = -\frac{i}{\hbar} \left\langle E_k{}^o \left| \hat{H}^1(t) \right| E_i{}^o \right\rangle e^{i\omega_{ik} t} \text{ for } k \neq i \text{ where } \omega_{ik} = \frac{(E_k{}^o - E_i{}^o)}{\hbar}.$$

Let $\hat{H}^1(t) = (V_0/2) e^{i\omega t}$ for $-T < t < +T$ and zero for $t < -T$ and $t > +T$. Then,

$$\frac{da_k}{dt} = \begin{cases} 0 & \text{for } t < -T \\ -\dfrac{i}{2\hbar} \left\langle E_k{}^o \left| V_0 \right| E_i{}^o \right\rangle e^{i(\omega_{ik} - \omega)t} & \text{for } -T < t < +T \\ 0 & \text{for } t > +T \end{cases}$$

By the same manner of integration performed on 8-13,

$$a_k(+T) = -\frac{i}{\hbar} \left\langle E_k{}^o \left| V_0 \right| E_i{}^o \right\rangle \frac{\sin((\omega_{ik} + \omega)T)}{(\omega_{ik} + \omega)}.$$

In the same manner the perturbation, $\hat{H}^1(t) = (V_0/2) e^{-i\omega t}$, acting during the same time interval gives,

$$a_k(+T) = -\frac{i}{\hbar} \left\langle E_k{}^o \left| V_0 \right| E_i{}^o \right\rangle \frac{\sin((\omega_{ik} - \omega)T)}{(\omega_{ik} - \omega)}.$$

For these time dependent Hamiltonians, we can write a general transition probability per unit time.

For $\hat{H}^1(t) = (V_0/2) e^{\pm i\omega t}$, we get $\omega_{i \to k} = \dfrac{T}{2\hbar^2} |\langle E_k{}^o | V_0 | E_i{}^o \rangle|^2 \left[\dfrac{\sin((\omega_{ik} \pm \omega)T)}{(\omega_{ik} \pm \omega)T} \right]^2.$

For $\hat{H}^1(t) = V_0 \cos(\omega t) = (V_0/2) \left(e^{i\omega t} + e^{-i\omega t} \right)$, we get,

$$a_k(+T) = -\frac{i}{\hbar} \langle E_k{}^o \,|\, V_0 \,|\, E_i{}^o \rangle\, T \left[\frac{\sin\left((\omega_{ik} + \omega)\, T\right)}{(\omega_{ik} + \omega)\, T} + \frac{\sin\left((\omega_{ik} - \omega)\, T\right)}{(\omega_{ik} - \omega)\, T} \right].$$

In finding the transition probability per unit time, it is necessary to take the square of the magnitude or modulus of $a_k(+T)$. You will notice that this gives a term of the sort, $\left[\dfrac{\sin\left((\omega_{ik} + \omega)\, T\right)}{(\omega_{ik} + \omega)\, T} \right]\left[\dfrac{\sin\left((\omega_{ik} - \omega)\, T\right)}{(\omega_{ik} - \omega)\, T} \right]$, which is a function of ω_{ik} for a particular ω, which is the angular frequency of the perturbing Hamiltonian. In this product, each term is negligible at the value of ω_{ik} for which the other term has a large amplitude. Thus, each term cancels the other term's contribution to the transition probability and the product can be dropped altogether from the final equation. Hence for $\hat{H}^1(t) = V_0 \cos(\omega t)$, we get for $\omega_{i \to k} = \dfrac{|a_k(+T)|^2}{2\, T}$, the expression:

$$\omega_{i \to k} = \frac{1}{2\,\hbar^2}\ |\langle E_k{}^o \,|\, V_0 \,|\, E_i{}^o \rangle|^2\ T \left[\left(\frac{\sin\left((\omega_{ik} + \omega)\, T\right)}{(\omega_{ik} + \omega)\, T} \right)^2 + \left(\frac{\sin\left((\omega_{ik} - \omega)\, T\right)}{(\omega_{ik} - \omega)\, T} \right)^2 \right]$$

This involves the sum of two terms, each of the type, $\left(\dfrac{\sin(\omega T)}{\omega T} \right)^2$. One of them has a maximum at $\omega_{i \to k} = \omega$, and the other has a maximum at $\omega_{i \to k} = -\omega$. When $\omega_{i \to k}$ is plotted as a function of ω_{ik} the curve looks like the following figure.

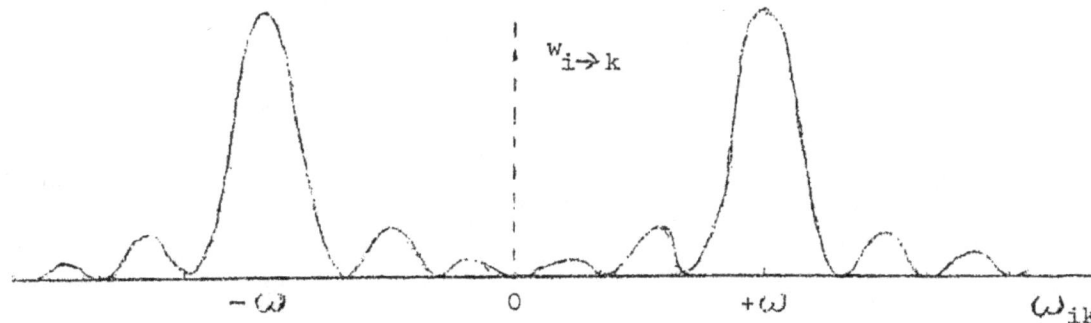

You will recall that $\omega_{ik} = \dfrac{(E_k{}^o - E_i{}^o)}{\hbar}$, where $E_i{}^o$ is the initial energy of the system and $E_k{}^o$ is the energy of a possible final state of the system. The left-hand peak gives a relatively large probability of the final energy to be $E_k{}^o = E_i{}^o - \hbar\omega$ and for the right-hand peak, $E_k{}^o = E_i{}^o + \hbar\omega$.

If $\hat{H}^1(t)$ arises from the interaction of the system with an electromagnetic field of frequency, ω, then a final energy, $E_i{}^o - \hbar\omega$, indicates that the system has emitted a photon of energy, $\hbar\omega$, to the radiation and a final energy, $E_i^o + \hbar\omega$, indicates that the system has absorbed a photon from the radiation. This is, of course, the nature of the interaction with radiation.

Plotting $\left(\dfrac{sin\,(\omega T)}{\omega T}\right)^2$ against ω, it is unit height when $\omega = 0$ and has its first zeros when $\omega T = \pm\pi$ or $\omega = \pm\pi/T$. Hence, the longer the excitation lasts, the narrower the range of frequency response.

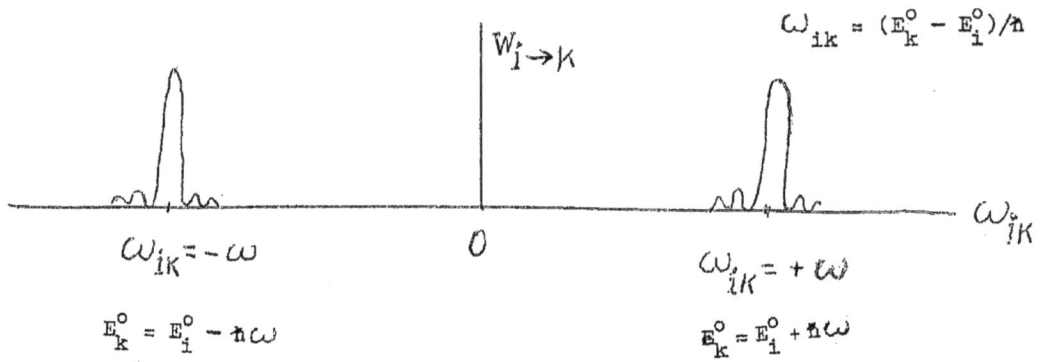

With this interaction the final atomic energy state is less than the initial state. This is STIMULATED EMMISSION OF RADIATION.	With the interaction the final atomic energy state is greater than the initial state. This is STIMULATED ABSORPTION OF RADIATION.

Returning to $\omega_{i\rightarrow k} = \dfrac{T}{2\,\hbar^2}|\langle\,E_k^o\,|\,V_o|\,E_i^o\,\rangle|^2\left[\dfrac{sin\,((\omega_{i\,k}\pm\omega)\,T)}{(\omega_{i\,k}\pm\omega)\,T}\right]^2$, which is the probability per unit time for the atomic transition from an initial state of energy, E_i^o, to a final energy state, E_k^o, the remaining discussion is on the matrix, $\langle\,E_k^o\,|\,V_o|\,E_i^o\,\rangle$, where $V_o = e\,E_o\,X$ is the energy of a charge, e, in an electric field, E_o. Note here that E_o is electric field and not energy.

The matrix element is now $\langle\,E_k^o\,|\,V_o\,|\,E_i^o\,\rangle = \langle\,E_k^o\,|\,e\,E_o\,X\,|\,E_i^o\,\rangle$, and $\langle\,E_k^o\,|\,V_o|\,E_i^o\,\rangle^2 = e^2\,E_o^2\,\langle\,E_k^o\,|\,X\,|\,E_i^o\,\rangle^2$, in which we have the square of the electric field strength.

We now see the reason for mirrors trapping radiation in standing waves in the tube. By the result of collisions, the atomic electrons are excited into higher energy states. Normally in the standard spectrum emission tube they would give off radiation be spontaneous emission. But by trapping radiation in the tube, we build up the energy density of radiation, thereby greatly increasing the probability per unit time for what is now stimulated emission. That still increases the radiation in the tube which further increases the probability. Although there is stimulated absorption going on, the limiting factor is the rate by which collisions can keep atomic electrons excited into higher energy states. The voltage across the tube times the current through the tube is the rate the source delivers energy to the process. Volts times amperes are watts or joules per second.

Lastly, we discuss the matrix, $\langle E_k^o | X | E_i^o \rangle$. What is the X in this?

Consider an atom of helium, neon or any atom whose electrons are symmetrically distributed about the positive nucleus. Because of its spherical symmetry, it has no evidence of charge.

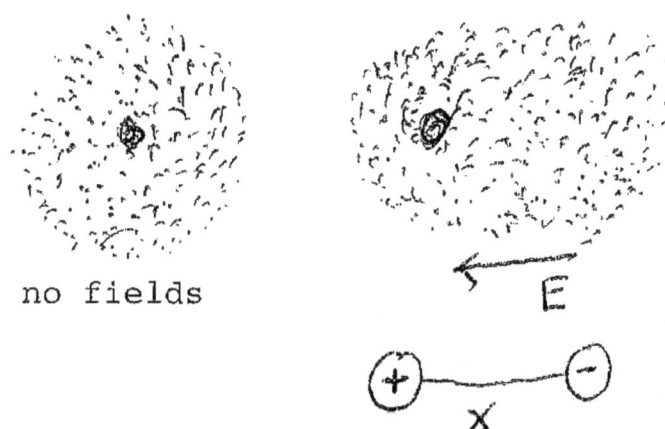

no fields

Put the atom in an electric field and the resulting force on the electrons pulls them in the opposite direction from the field direction. This creates a charge dipole where any two opposite charges are separated by a distance, X; this is the X in the matrix.

The primary distinction between spontaneous
emission and stimulated emission of radiation
is covered in the paragraph at the bottom of
the previous page. The main use of the matrix,
$\langle E_k{}^o | X | E_i{}^o \rangle$, will be to say which state
transitions will happen and which are
forbidden.

Because the electrons are diffusely
distributed, evaluating the matrix will
require a summation of wavefunction products
over the volume of the atom. For this purpose,
an element of volume can be either *dx dy dz* or
$(dr)\,(rd\theta)\,(r \sin \theta\, d\varphi)$. For atomic calculations,
the second is most convenient.

In summing over space with spherical
coordinates, the angle, θ, goes from zero (0)
to π. The angle, φ, goes all the way around
the z-axis, going from the zero (0) to 2π.

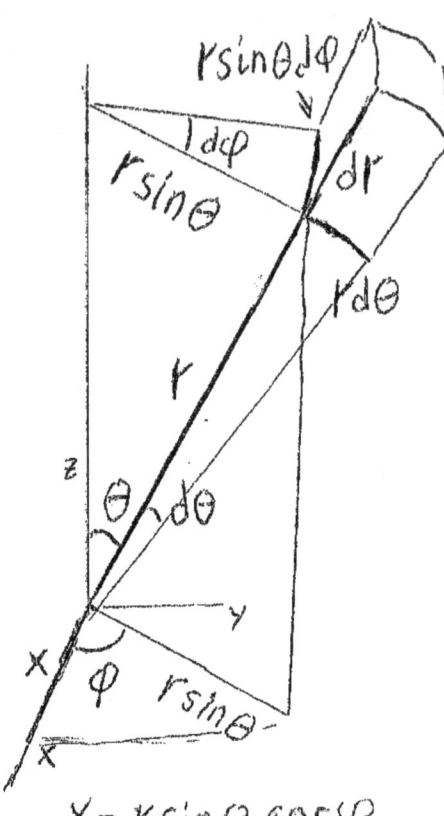

$$X = r \sin \theta \cos \varphi$$

Evaluation of the matrix will need atomic wave
functions.

$\psi_i(r,\ \theta,\ \varphi) = \langle r,\ \theta,\ \varphi | E_i{}^o \rangle$, is the initial wavefunction and

$\psi_k(r,\ \theta,\ \varphi) = \langle E_k{}^o | r,\ \theta,\ \varphi \rangle$ is the final state.

For reference, we repeat the orbital functions derived in chapter 17.

In each of the following functions we let C be whatever the appropriate
constants may be.

$$\Psi_{000} = Ce^{-k^2 r^2/2}$$

A spherical symmetric
orbital is an s-orbital.

$$\Psi_{001} = Ce^{-k^2 r^2/2}\, r \cos \theta$$

$$\Psi_{100} = Ce^{-k^2 r^2/2}\, r \sin \theta \cos \varphi$$

$$\Psi_{010} = Ce^{-k^2 r^2/2}\, r \sin \theta \sin \varphi$$

If ψ^2 can be transformed into itself by a rotation of 180° about a
coordinate axis, it is called a p-orbital.

$$\Psi_{101} = Ce^{-k^2 r^2/2} r^2 \sin\theta \cos\theta \cos\varphi$$

$$\Psi_{011} = Ce^{-k^2 r^2/2} r^2 \sin\theta \cos\theta \sin\varphi$$

$$\Psi_{110} = Ce^{-k^2 r^2/2} r^2 \sin^2\theta \sin\varphi\cos\varphi$$

$$\text{or} \quad \Psi_{110} = \frac{C}{2} e^{-k^2 r^2/2} r^2 \sin^2\theta \sin(2\varphi)$$

If ψ^2 can be transformed into itself by a rotation of 90° about a coordinate
axis, it is called a d-orbital.

The linear combination, $\Psi_{200} + \Psi_{020} + \Psi_{002} = Ce^{-k^2 r^2/2}(k^2 r^2 - \frac{3}{2})$, is an
s-orbital with an energy, $\frac{7}{2}\hbar\omega$.

Notice that $|200\rangle - |020\rangle$ is orthogonal to $|200\rangle + |020\rangle + |002\rangle$,
giving another d-orbital:

$$\Psi_{x^2-y^2} = Ce^{-k^2 r^2/2} r^2\sin^2\theta (\cos^2\varphi - \sin^2\varphi) = Ce^{-k^2 r^2/2} r^2 \sin^2\theta \cos(2\varphi)$$

There is another linear combination to the last two. Let us
write it $|\psi\rangle = a|200\rangle + a|020\rangle + b|002\rangle$, orthogonal to $|200\rangle - |020\rangle$, then
$[a\langle 200| + a\langle 020| + b\langle 002|][|200\rangle + |020\rangle + |002\rangle] = 2a + b = 0$ gives $a = -\frac{1}{2}b$.

In rectilinear coordinates, $\psi = C e^{-k^2 r^2/2}\left[-\frac{1}{2}x^2 - \frac{1}{2}y^2 + z^2\right]$.

Add $\frac{1}{2}z^2 - \frac{1}{2}z^2 = 0$ inside the brackets and $\psi = C e^{-k^2 r^2/2}\left[\frac{3z^2 - r^2}{2}\right]$,

which is, in polar coordinates, $\psi = C e^{-k^2 r^2/2} r^2 \left[\frac{3\cos^2\theta - 1}{2}\right]$.

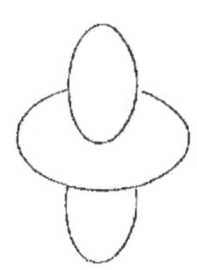

Considering its rotation about the z-axis, we shall consider
this as another d-orbital.

$$\langle E_k^0 |x| E_i^0\rangle = \int_r \int_\theta \int_\varphi \psi_k(r,\theta,\varphi) \, r\sin\theta\cos\varphi \, \psi_i(r,\theta,\varphi) \, r^2 dr \sin\theta \, d\theta \, d\varphi$$

18-21

This will not be as bad as it looks. Let us try it out on two s-orbitals, say decay from an excited 3s state to a 1s state.

$$\int_r \psi_{1s}(r)\,\psi_{3s}(r)\,r^3\,dr \int_{\theta=0}^{\pi} \sin^2\theta\,d\theta \int_{\varphi=0}^{2\pi} \cos\varphi\,d\varphi$$

S-orbitals are functions only of the radius and go to zero rapidly with increasing radius. Because of this rapid fall off, the radial summation need not bother us. We can start with the sum over φ. Now, $\cos\varphi\,d\varphi = d(\sin\varphi)$ gives:

$$\int_{\varphi=0}^{2\pi} \cos\varphi\,d\varphi = \sin(2\pi) - \sin(0) = 0$$

Here we are talking about hydrogen or helium lasers. Excitation by collisions from 1s to 3s is allowed, but $\langle 1s\,|\,X\,|\,3s \rangle = 0$ decay from 3s to 1s by emitting electromagnetic radiation (light) will not happen. That color (actually in the ultra-violet) will not appear in the spectrum.

Now try the decay from a 3s to a $2p = \psi_{100}$ from page 18-20.

As in most of these questions, only the integral over φ is needed.

$$\int_{\varphi=0}^{2\pi} \cos^2\varphi\,d\varphi = \frac{1}{2}\int_{\varphi=0}^{2\pi} d\varphi + \frac{1}{2}\int_{\varphi=0}^{2\pi} \cos(2\varphi)\,d\varphi = \pi$$

$\langle 2p\,|\,X\,|\,3s \rangle$ is not zero and that line (red) will appear in the hydrogen spectrum. Any radiative emission or absorption between s-states and p-states can happen.

Starting with its electron in the 1s or
ground state, the hydrogen electron can
land in any excited state as a result of a
collision. In decaying by emitting light,
only certain decays are allowed. These
limitations are called SELECTION RULES.

Integrating over φ, you can investigate
other selection rules using the functions
on the previous pages.

Gerhard Herzberg's, ATOMIC SPECTRA AND
ATOMIC STRUCTURE, is still the standard
reference on this matter.

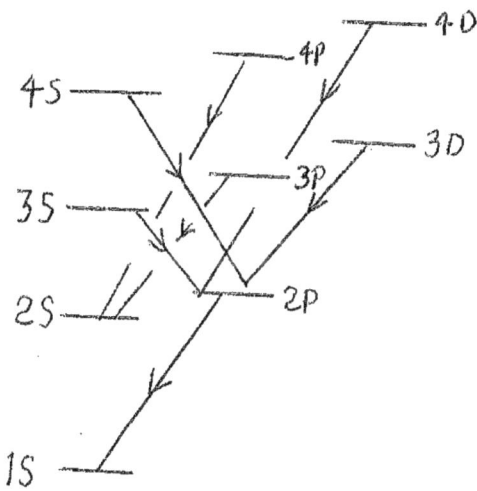

Hydrogen spectra

CHAPTER NINETEEN: ELECTRON BONDING IN MOLECULES AND SOLID MATTER

In previous chapters we have derived a set of functions that describe the behavior of electrons in atoms. Thinking of the circular orbits of electrons in the simple model of atoms, we call these functions orbitals. Bonding of molecules and solids involves electrons spreading out to exist on many atoms. These are called molecular orbitals. In describing these functions, we assume that the part of a molecular orbital in the neighborhood of any atom can be described by using the atomic orbitals of that atom. This is called LINEAR COMBINATION OF ATOMIC ORBITALS (LCAO)

For a discussion of atomic orbital functions, we can start with the eigenfunctions of the three-dimensional oscillator, given on page 21-15. Why it is that the sets of functions are labeled s, p, d, etc. comes from aspects of atomic spectroscopy, and should not concern us here. For each function, the letter C will stand for whatever the appropriate normalizing constant.

$\langle x,y,z|0,0,0\rangle = C\,e^{-k^2r^2/2}$ is called an s-orbital. It is the only orbital occupied in the ground state of hydrogen or helium. Notice that the s-orbital is transformed into its self by any rotation about any one of three mutually perpendicular axes.

1s

$\langle x,y,z|1,0,0\rangle = Cxe^{-k^2r^2/2}$, $\langle x,y,z|0,1,0\rangle = Cye^{-k^2r^2/2}$

and $\langle x,y,z|0,0,1\rangle = Cze^{-k^2r^2/2}$ are first occupied by electrons in the elements in the second row of the periodic table. They are called p-orbitals and are shown below. In the order given they are p-x, p-y, and p-z. Notice each has two mutually perpendicular axes of symmetry, transforming into its self by a rotation of $180°$ about either axis.

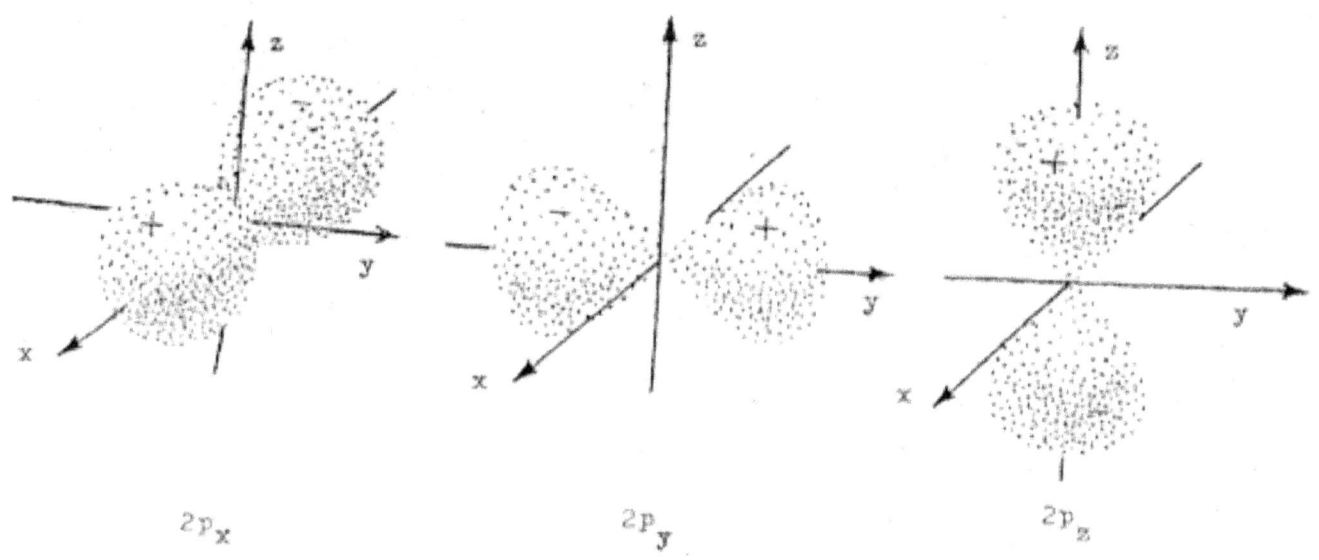

$2p_x$ $2p_y$ $2p_z$

The next set of functions are called d-orbitals. They first show up in the third row elements.

For example, $\langle x,y,z|1,1,0\rangle = C\,xy\,e^{-k^2 r^2/2}$, called d-xy, is shown to the right. Notice that it has one axis of symmetry and transforms into itself with a rotation of 90° about that axis. For the rest of the three dimensional oscillator orbitals related to d-xy,

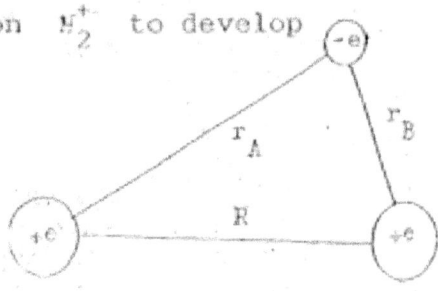

$3d_{xy}$

$$\langle x,y,z|1,0,1\rangle = C\,xz\,e^{-k^2 r^2/2}$$

$$\langle x,y,z|0,1,1\rangle = C\,yz\,e^{-k^2 r^2/2}$$

$$\langle x,y,z|2,0,0\rangle = C(k^2 x^2 - \tfrac{1}{2})e^{-k^2 r^2/2}, \quad \langle x,y,z|0,2,0\rangle = C(k^2 y^2 - \tfrac{1}{2})e^{-k^2 r^2/2}$$

$$\langle x,y,z|0,0,2\rangle = C(k^2 z^2 - \tfrac{1}{2})e^{-k^2 r^2/2}$$

As these are eigenfunctions of the three dimensional oscillator, all with the energy eigenvalue $E = \tfrac{7}{2}\hbar\omega$, any linear combination of them will be an eigenfunction with the same energy. Notice that

$$\langle x,y,z|2,0,0\rangle + \langle x,y,z|0,2,0\rangle + \langle x,y,z|0,0,2\rangle = C(k^2 r^2 - \tfrac{3}{2})e^{-k^2 r^2/2}$$

is not a d-orbital, but rather a 3s orbital.

$\langle x,y,z|2,0,0\rangle - \langle x,y,z|0,2,0\rangle = C(x^2 - y^2)e^{-k^2 r^2/2}$ is called $3d_{x^2-y^2}$

A third eigenvector can be formed that is normal to the last two, then giving the function

$$\langle x,y,z|\left[2|0,0,2\rangle - |2,0,0\rangle - |0,2,0\rangle\right] = C(3z^2 - r^2)e^{-k^2 r^2/2} \text{ called } 3d_{z^2}$$

We next will use the hydrogen molecular ion H_2^+ to develop detail of molecular orbital calculations. The hydrogen molecule ion consists of two protons with a distance R between them, and an electron at distances r_A and r_B from the two nuclei. A reasonable assumption is that when in the neighborhood of either nucleus, the electron is governed by its wave functions for that nucleus. In the ground state those will be the 1s functions.

Let the molecular orbital be a linear combination of 1s atomic orbitals assuming that the behavior of the electron in the neighborhood of any nucleus is governed by the orbitals characteristic of that atom.

Let $|A\rangle$ be the state vector for a 1S orbital on atom A giving an atomic wave function $\varphi_A(x,y,z) = \langle x,y,z,|A\rangle$ φ is small Greek phi

$\frac{p^2}{2M} - \frac{e^2}{r_A}$ would be its energy, neglecting

atom B   Then $[\frac{p^2}{2M} - \frac{e^2}{r_A}]|A\rangle = E^0|A\rangle$

$E^0 = -13.6$ electron volts.

We neglect writing $\frac{1}{4\pi\epsilon_0}$

$\langle A|[\frac{p^2}{2M} - \frac{e^2}{r_A}]|A\rangle = \int \varphi_A[\frac{p^2}{2M} - \frac{e^2}{r_A}]\varphi_A \, dv = E^0\langle A|A\rangle$

$dv = dxdydz$ is an element of volume.   Let state vectors be unit length

Let $\varphi_B = \langle x,y,z,|B\rangle$ be a 1S orbital on atom B. $\langle A|A\rangle = 1, \langle B|B\rangle = 1$ etc

Then $\langle B|[\frac{p^2}{2M} - \frac{e^2}{r_B}]|B\rangle = \int \varphi_B[\frac{p^2}{2M} - \frac{e^2}{r_B}]\varphi_B \, dv = E^0$

$\varphi^2 dv$ is the probability that an electron in the orbital   will be in the volume dv=dxdydz

$-e\varphi^2 dv$ is the amount of electron charge in the volume dv. If the nucleus (here a proton)

has a charge $+e$, $\frac{(+e)(-e\varphi^2 dv)}{r}$ will be the potential energy of the electron in volume dv a distance r from the nucleus.

In the following discussion $\langle A|B\rangle = \int \varphi_A \varphi_B dv$ may occur.
This involves what is called an overlap region being non-zero only in the shaded region. These terms can be ignored in our treatment.

Putting the molecule together, we define a state vector as a linear combination $|M\rangle = C_A|A\rangle + C_B|B\rangle$ the Cs being constants. If $\langle M|M\rangle = 1$

then $[C_A\langle A| + C_B\langle B|][C_A|A\rangle + C_B|B\rangle] = C_A^2\langle A|A\rangle + C_B^2\langle B|B\rangle + C_A C_B\langle A|B\rangle + C_B C_A\langle B|A\rangle = 1$

Assuming unit length vectors and neglecting overlap $C_A^2 + C_B^2 = 1$

It will be assumed that the distance between nuclei, R, has a fixed value. Motion of the nuclei can be considered after we have worked out a molecular orbital. In addition to the kinetic energy operator, the electron has a Coulomb attraction to each nucleus and there is a Coulomb repulsion between the two nuclei.

$$H = \frac{p^2}{2M} - \frac{e^2}{r_A} - \frac{e^2}{r_B} + \frac{e^2}{R}$$

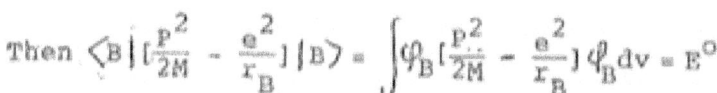

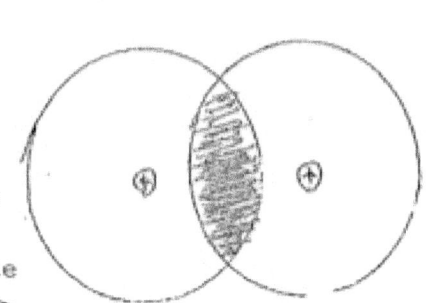

For the moment, we neglect the nuclear repulsion energy $+ e^2/R$

The electron energy is $[C_A \langle A| + C_B \langle B|][\frac{p^2}{2M} - \frac{e^2}{r_A} - \frac{e^2}{r_B}][C_A |A\rangle + C_B |B\rangle]$

Neglecting overlap, this gives $C_A^2 E^O + C_B^2 E^O - C_A^2 \langle A|\frac{e^2}{r_B}|A\rangle - C_B^2 \langle B|\frac{e^2}{r_A}|B\rangle$

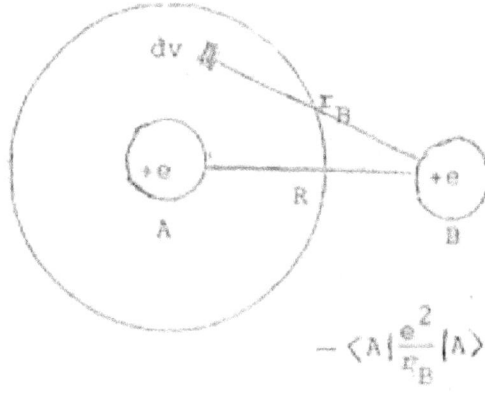

The two new
terms are the
energy of the
electron on   A
and nucleus   B
and the energy
of the electron
on   B   with
nucleus   A.

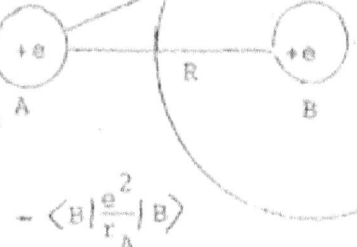

$- \langle A|\frac{e^2}{r_B}|A\rangle$ Symmetry here $- \langle B|\frac{e^2}{r_A}|B\rangle$
makes them the
same, call them
both   V.

Notice that $E^O \approx -13.6$ ev   and $V = -\langle A|\frac{e^2}{r_B}|A\rangle = -\langle B|\frac{e^2}{r_A}|B\rangle$
are all negative.

This gives a total electron energy $E = C_A^2 E^O + C_B^2 E^O + C_A^2 V + C_B^2 V$

This leads to a new game. Multiply   E   by $C_A^2 + C_B^2 = 1$

Then $(C_A^2 + C_B^2)E = (E^O + V)C_A^2 + (E^O + V)C_B^2$ giving $E = E^O + V$

To this we must add back the nuclear repulsion term $+ e^2/R$ and the
total energy is $E = E^O + V + e^2/R$.

At large separation,   V   and e^2/R are
zero, the electron is on one nucleus or

the other and $E = E^O$

With decreasing   R,   V   becomes more
negative and the energy is less than

E^O until positive $+ e^2/R$ becomes
stronger and the energy is greater than

E^O resulting in the energy plotted
against nuclear distance   R   as shown.

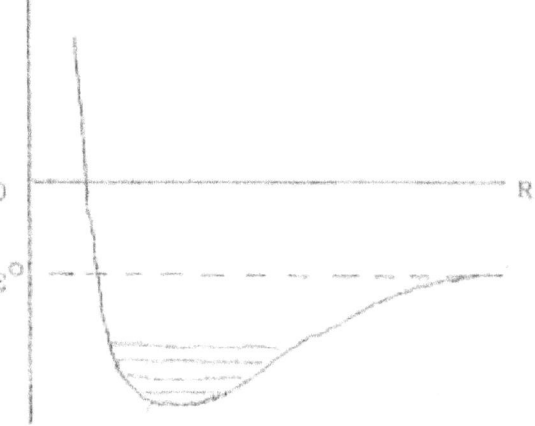

The resulting curve is parabolic about
the point of minimum energy giving
rise to oscillations about that   R,
shown by the quantized oscillator
energy levels as shown. Molecular bond acts as a spring for molecular
vibrations. Experimentally, these are studied with radiation in the
infrared region of the spectrum.

With decreasing nuclear separation, the overlap region becomes larger
and energy terms of the type $\langle A|U|B\rangle$ must be considered. The integrals
involved are called exchange integrals and are also negative. We could
call them $U_{AB} = \langle A|U|B\rangle$ but we will just call them   U   in this case.

It is evident that as the distance between nucleii is decreased the overlap cannot be ignored. Now terms with $C_A C_B$ and $C_B C_A$ cannot be ignored in the electron energy at the top of page 19-5. Now, energy terms of the type $U = \langle A | \left[\frac{p^2}{2M} - \frac{e^2}{r_A} - \frac{e^2}{r_B} \right] | B \rangle$ must be included, giving them the symbol U.

The electron energy now is $(C_A^2 + C_B^2) E = C_A^2 (E^O + V) C_B^2 (E^O + V) + C_A C_B U + C_B C_A U$

This suggests
$$[C_A, C_B] \begin{bmatrix} E^O + V & U \\ U & E^O + V \end{bmatrix} \begin{bmatrix} C_A \\ C_B \end{bmatrix} = [C_A, C_B] \begin{bmatrix} E C_A \\ E C_B \end{bmatrix}$$

giving
$$\begin{bmatrix} E^O + V & U \\ U & E^O + V \end{bmatrix} \begin{bmatrix} C_A \\ C_B \end{bmatrix} = E \begin{bmatrix} C_A \\ C_B \end{bmatrix}$$
or
$$\left\{ \begin{array}{l} (E^O + V) C_A + U C_B = E C_A \\ U C_A + (E^O + V) C_B = E C_B \end{array} \right\}$$

Combining $U^2 C_A^2 = (E - E^O - V)^2 C_A^2$

$(E - E^O - V) = +U$ or $(E - E^O - V) = -U$
$\left\{ \begin{array}{l} E^+ = E^O + V + U \\ E^- = E^O + V - U \end{array} \right\}$ and $\left\{ \begin{array}{l} C_B^+ = C_A^+ \\ C_B^- = -C_A^- \end{array} \right\}$

If U is negative E^- is higher energy than E^+. $C_B^+ = C_A^+$ is a bonding orbital

E | $C_B = -C_A$ antibonding orbital

$C_A = C_B$
bonding orbital

bonding   anti-bonding

Hydrogen electrons will be in their 1s orbitals. Putting two hydrogen atoms close enough together that their orbitals overlap, the electrons can lose energy because of the exchange integral, and their orbitals fuse into molecular orbitals. The 1s orbitals are symmetric under any rotation about any axis. The molecular orbital is symmetric under any rotation about the bond axis. Because the s-orbital symmetry is reflected about the bond axis, we call this a sigma (σ) orbital.

It is evident that the methods of linear algebra in chapter 15 will be necessary in the following discussion, especially pages 15-7 to 15-19. Later in this chapter, we will let $E^O + V = \alpha$ and $U = \beta$.

The energy equation above now will be $\begin{bmatrix} \alpha & \beta \\ \beta & \alpha \end{bmatrix} \begin{bmatrix} C_A \\ C_B \end{bmatrix} = E \begin{bmatrix} C_A \\ C_B \end{bmatrix}$

Ethylene, C_2H_4 is shown in the pictures below. It has four hydrogen nuclei each with charge $+e$, and two carbon nuclei, each shielded by two 1s electrons. Because of this shielding, the carbon nuclei appear to have charges $+4e$. Counting all valence electrons, we have twelve electrons to play with. Each hydrogen contributes one 1s orbital, and each carbon contributes a 2s and three 2p orbitals. Linear combinations of twelve atomic orbitals should give us twelve orthogonal molecular orbitals. But for each molecular orbital there are two spin states, giving twenty four spin-orbitals. The twelve electrons will occupy the lowest energy spin-orbitals which are called bonding orbitals. The twelve highest energy spin-orbitals will be empty when the system is in its ground state. These are called anti-bonding states, Sums are only over occupied states.

Hydrogen 1s orbitals and 2s, $2p_x$, and $2p_y$ orbitals from the carbons form a two dimensional molecular orbital.

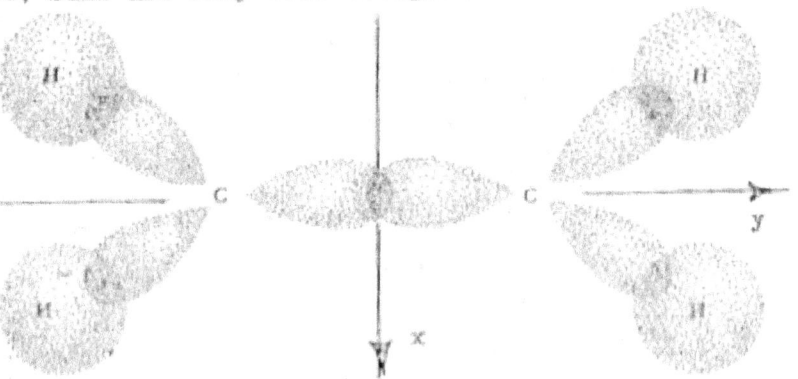

The overlap of $2p_y$ orbitals from the carbons contributes to the carbon-carbon bond. Notice that a linear combination of the $2p_x$ with the $2p_y$ and the 2s breaks the symmetry of the $2p_x$ and gives the direction of the C—H bond. This linear combination of orbits from one atom to give an orbit of a new shape is called hybridization, and the new orbital is called a hybrid orbit.

Looking at any particular C—C or C—H bond, this molecular orbital has total rotational symmetry about that bond axis. For that reason, although it involves p orbitals, this is a sigma molecular orbital.

Sticking up and below the x-y plane, the carbon $2p_z$s overlap, and can then fuse, as shown below. About the C—C axis, a $180°$ rotation trans-forms this orbital into itself. Because it retains the $2p_z$ symmetry, we call this a pi (π) orbital.

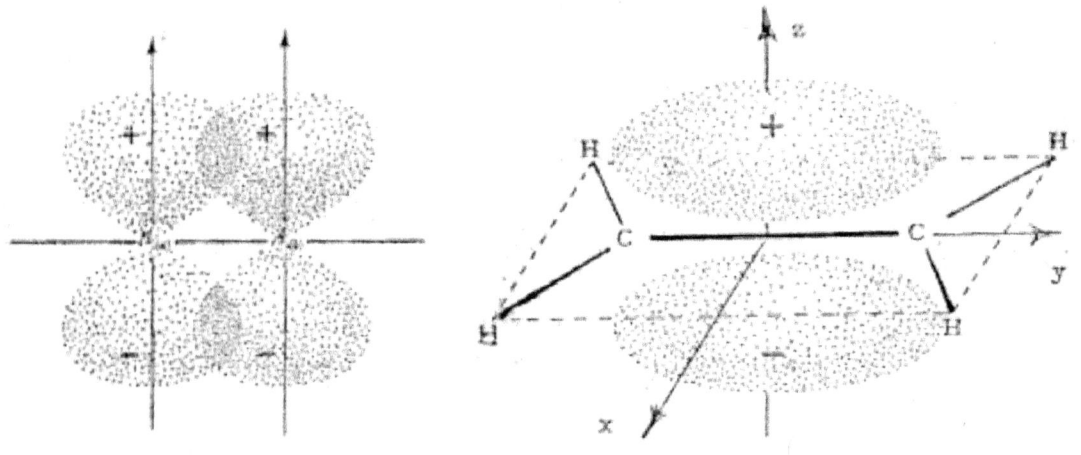

Please remember that the algebraic signs
shown here are on the orbital functions.
They are _not_ charge. The charge on the
electron is -e and the charge density

is $\rho = -e\varphi^2$.
If the sign on the right hand orbital is
reversed, the overlap is cancelled and
we get an anti-bonding orbital.

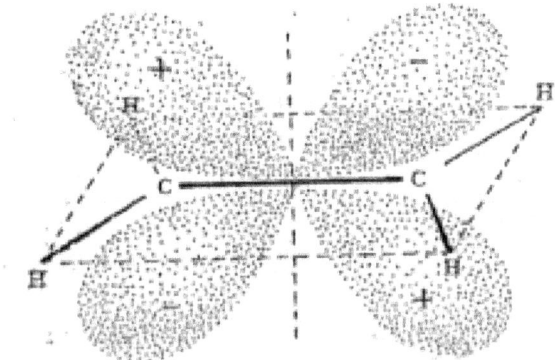

The importance of molecular orbital calculations is that they let us
study properties of molecules that we can not determine experimentally.
An important property is the distribution of electric charge. This is
important in the study of enzymes and drugs.

To start, look at an electron wave function.
Take an electron in a 2px state. Something
tells us that it exists most strongly
in the x-direction. Its algebraic sign is
both positive and negative, making it in
itself not suitable for giving probability,
but if we square the wavefunction, we get a
function of the same shape, but having a
positive value everywhere. (The algebraic
sign is on the function. It does not imply
charge.)

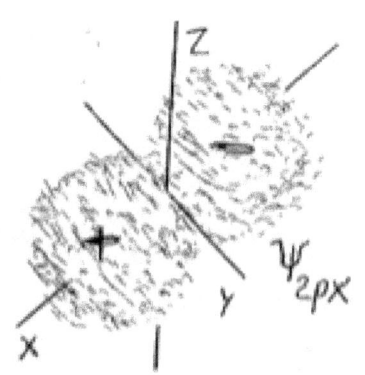

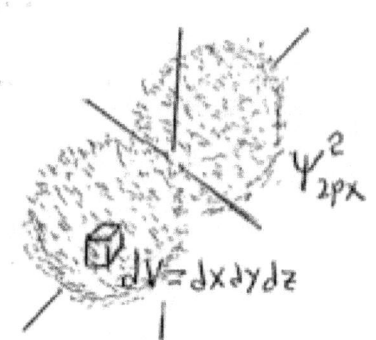

This is interpreted in the following manner
The square of the wave function is taken to
be a density of probability, so that the
probability of the electron being in any
element of volume $dv = dxdydz$ is
$\psi^2 dv$ requiring $\int \psi^2 dv = 1$

The electron's charge, -e , is interpreted
as being distributed as the probability thus
$-e\psi^2$ is the electron's charge density.

We may take the positive charge at an atomic site to be the nuclear charge
shielded by closed inner shells of electrons. The wave function Ψ_μ of the

μ th bonding orbital has a charge density $\rho_\mu = -e\Psi_\mu^2$

As there are two electrons in each bonding orbital (with opposite spins)
the charge density of bonding electrons is $\rho = \sum_\mu -2e\Psi_\mu^2$ summed over

occupied orbitals.

HÜCKEL MOLECULAR ORBITAL THEORY

By looking at a molecular orbital, we can say something about its energy. Recall the energies of wave functions in a box. The more spread out (de-localized) the wave, the lower its energy. The more compact (localized) the wave function, the higher its energy. The more nodal planes (boundaries across which the function's sign changes), the higher the energy. The fewer nodal planes, the lower the energy.

To allow you to get a feeling for what all this means in practice, we introduce a method developed by Erich Hückel in the days before computers. This applies to an extensive and important class of molecules developed from ethylene. Looking at our pictures of ethylene, the hydrogens are connected to the carbons by sigma orbitals which the chemists consider to add up to a single bond. The carbons are also connected by electrons in pi orbitals which the chemists consider to add up to another single bond. Thus the carbons are connected by a double bond. The electrons in the difuse sigma bonds will be at the lowest bonding energies, and therefore will be relatively stable. The electrons in the localized pi bonds will be more unstable and will respond the most to changes in molecules of this sort. The varying proprties of different molecules of this sort will be described most by the behavior of pi orbitals.

The chemist's symbol for ethylene with its double bond is

If you take two ethylenes; knock off two hydrogens and join the bared carbons with sigma bonds, you get butadiene.

Adding another ethylene in the same manner we get hexatriene.

If with hexatriene, we rotate C_2 and C_3 about their sigma bond , and C_4 and C_5 about their sigma bond, knock off two more hydrogens and join the now adjacent carbon with a double bond; we get cyclohexatriene which we recognize as benzene.

If we keep adding more ethylenes to hexatriene, as shown, we get a class of molecules known as linear polyenes. Of course, if we keep adding things to benzene we get a vast class of interesting and important molecules

Applied to molecules gotten by attaching ethylenes to each other in the manner shown, Hückel adapted the semi-empirical approach at the top of page 37-21 in the following manner. The α s will all be the same, namely α. The β s will all be the same, namely β, for adjacent carbons and zero for non-adjacent carbons. For ethylene this gives

$$\begin{pmatrix} \alpha & \beta \\ \beta & \alpha \end{pmatrix} \begin{pmatrix} c_{\mu 1} \\ c_{\mu 2} \end{pmatrix} = \varepsilon_\mu \begin{pmatrix} c_{\mu 1} \\ c_{\mu 2} \end{pmatrix}$$

This obviously gives eigenvalues and eigenvectors

$$\varepsilon_1 = \alpha + \beta \qquad |\Psi_1\rangle = \frac{1}{\sqrt{2}} \begin{pmatrix} 1 \\ 1 \end{pmatrix} \quad \text{and} \quad \varepsilon_2 = \alpha - \beta \qquad |\Psi_2\rangle = \frac{1}{\sqrt{2}} \begin{pmatrix} 1 \\ -1 \end{pmatrix}$$

Remember that both α and β are negative, with β having a smaller magnitude than α. Both energies are negative, but ε_1 is more negative than ε_2. Thus $\mu = 1$ is a bonding orbital and $\mu = 2$ an anti-bonding orbital.

Let φ_i be the $2p_z$ orbital on the i th carbon.

For ethylene $\Psi_\mu = c_{\mu 1}\varphi_1 + c_{\mu 2}\varphi_2$ which gives

$$\Psi_1 = \frac{1}{\sqrt{2}}\left[\varphi_1 + \varphi_2 \right] \text{ for the bonding orbital}$$

and $\Psi_2 = \frac{1}{\sqrt{2}}\left[\varphi_1 - \varphi_2 \right]$ for the anti-bonding orbital.

Notice the orbitals are exactly what we showed in the figures on page 37-19. For the bonding orbital φ_1 and φ_2 add to give an overlap. For the anti-bonding orbital, the opposite signs cancel the overlap.

For a linear polyene with N carbon atoms, the Hückel equation is

$$\begin{pmatrix} \alpha & \beta & 0 & 0 & \cdots & & & & 0 \\ \beta & \alpha & \beta & 0 & \cdots & & & & 0 \\ 0 & \beta & \alpha & \beta & \cdots & & & & 0 \\ \cdot & \cdot & \cdot & \cdot & \cdots & & & & 0 \\ \cdot & \cdot & \cdot & \cdot & \cdots & & \beta & \alpha & \beta \\ 0 & 0 & 0 & 0 & 0 & 0 & 0 & \beta & \alpha \end{pmatrix} \begin{pmatrix} c_{\mu 1} \\ c_{\mu 2} \\ c_{\mu 3} \\ \cdot \\ \cdot \\ c_{\mu N} \end{pmatrix} = \varepsilon_\mu \begin{pmatrix} c_{\mu 1} \\ c_{\mu 2} \\ c_{\mu 3} \\ \cdot \\ \cdot \\ c_{\mu N} \end{pmatrix}$$

Comparing this with the matrix on page 19-16 we see that the eigenvectors on page 19-17 can be adapted to give

$$c_{\mu i} = \sqrt{\frac{2}{N+1}} \sin\left(\frac{\mu \pi i}{N+1}\right) \text{ conveniently giving } c_{\mu 0} = c_{\mu, N+1} = 0$$

Any row of the matrix times this eigenvector gives

$$\sqrt{\frac{2}{N+1}}\left[\alpha \sin\left(\frac{\mu \pi i}{N+1}\right) + \beta \sin\left(\frac{\mu \pi(i-1)}{N+1}\right) + \beta \sin\left(\frac{\mu \pi(i+1)}{N+1}\right) \right] = \varepsilon_\mu \sqrt{\frac{2}{N+1}} \sin\left(\frac{\mu \pi i}{N+1}\right)$$

This gives $\qquad \varepsilon_\mu = \alpha + \beta \cos\left(\frac{\mu \pi}{N+1}\right)$

For butadiene $\quad c_{\mu i} = \sqrt{\frac{2}{5}} \sin\left(\frac{\mu \pi i}{5}\right)$ and $\varepsilon_\mu = \alpha + 2\beta \cos\left(\frac{\mu \pi}{5}\right)$

19-9

For butadiene, the results are

μ	$c_{\mu 1}$	$c_{\mu 2}$	$c_{\mu 3}$	$c_{\mu 4}$	ε_μ
1	0.372	0.602	0.602	0.372	$\alpha + 1.62\,\beta$
2	0.602	0.372	-0.372	-0.602	$\alpha + 0.62\,\beta$
3	0.602	-0.372	-0.372	0.602	$\alpha - 0.62\,\beta$
4	0.372	-0.602	0.602	-0.372	$\alpha - 1.62\,\beta$

For a single carbon atom, if the $2p_z$ orbital were described in this manner, it would have a negative energy $\varepsilon = \alpha$. For an electron to have zero energy, it would be free of the atom. The positive energy $-\alpha$ would be the energy (ionization potential) required to remove the $2p_z$ from the atom. If k is a positive number, molecular electrons with energies $\varepsilon = \alpha + k\beta$ are more tightly bound than on a single carbon atom. Those with energies $\varepsilon = \alpha - k\beta$ are less tightly bound.

The energy levels given by $\varepsilon_\mu = \alpha + 2\beta \cos\left(\frac{\mu\pi}{5}\right)$ can be located on a circle of radius 2β with its center at the energy level $\varepsilon = \alpha$. $\frac{\pi}{5}$ radians $= 36°$.

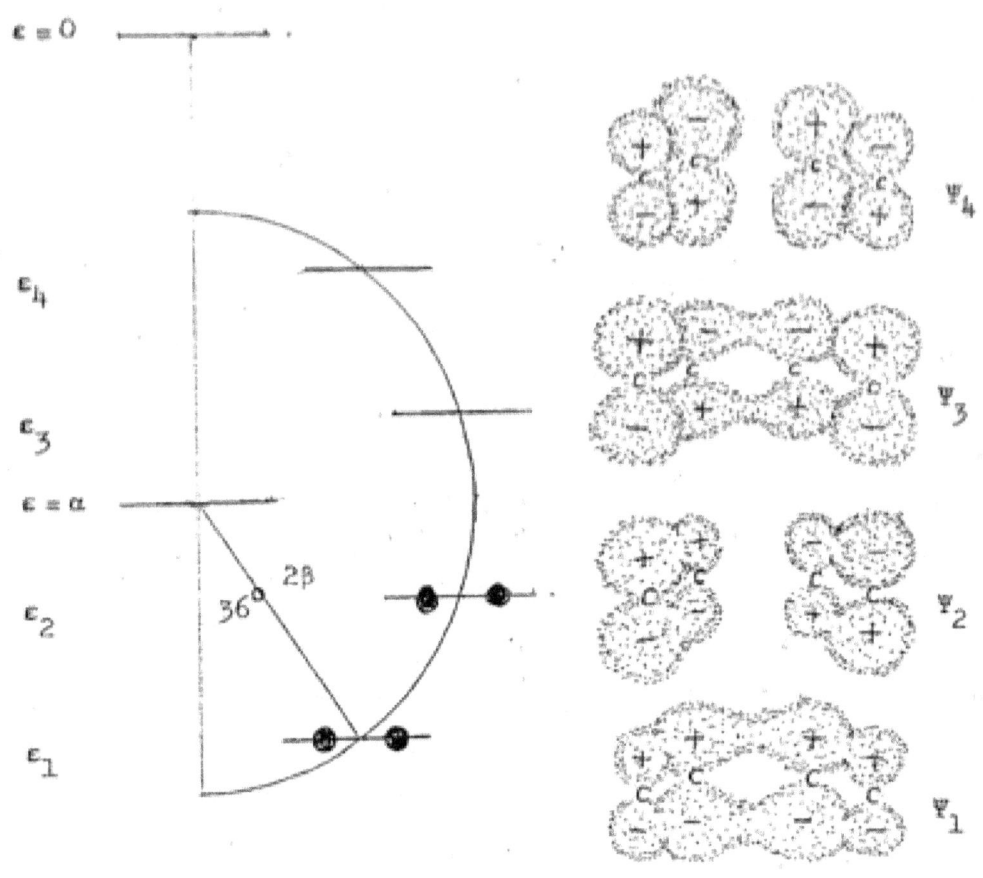

we introduced the summation over <u>occupied</u> orbitals $P_{ij} = \sum\limits_\mu 2c_{\mu i}c_{\mu j}$. If the i th and j th carbons are adjacent atoms, this is called the bond order. C. A. Coulson used this in a semi-empirical formula for the radius (in angstroms) between those carbons. $R_{ij} = 1.517 - 0.180\,P_{ij}$

For ethylene, the bonding orbital had coefficients $c_{11} = \frac{1}{\sqrt{2}}$, $c_{12} = \frac{1}{\sqrt{2}}$ giving $P_{12} = 1$ and $R_{12} = 1.337$ Å

For butadiene, the bonding orbitals have $\mu = 1$ and $\mu = 2$. Between carbons 1 and 2, $P_{12} = 0.896$ and $R_{12} = 1.356$ Å. The same for R_{34}. Between carbons 2 and 3 $P_{23} = 0.448$ and $R_{23} = 1.44$ Å. Compounds containing carbon-carbon double bonds are called alkenes, and where as in the case of butadiene there are alternating double and single bonds in the standard picture, the pi bonds are said to be conjugated. In the case of butadiene, the pi bond is delocalized; between carbons 1 and 2 the bond is less than double and between carbons 2 and 3 the bond is stronger than single. Conjugated pi bonds will always be de-localized to some extent.

With conjugated cyclic polyenes such as benzene it will be convenient for our purpose to label the carbons with i going from 0 to $N-1$, rather than from 1 to N. Now there will be a β for the $C_{N-1} - C_0$ bond.

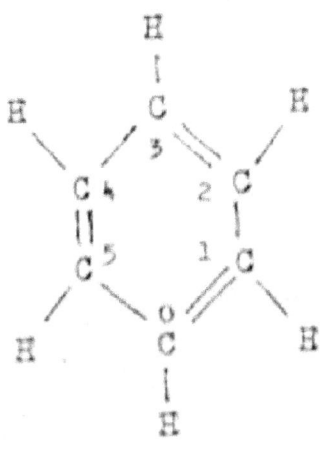

The Huckel equation becomes

$$\begin{pmatrix} \alpha & \beta & 0 & 0 & 0 & \beta \\ \beta & \alpha & \beta & 0 & 0 & 0 \\ 0 & \beta & \alpha & \beta & 0 & 0 \\ 0 & 0 & \beta & \alpha & \beta & 0 \\ 0 & 0 & 0 & \beta & \alpha & \beta \\ \beta & 0 & 0 & 0 & \beta & \alpha \end{pmatrix} \begin{pmatrix} c_{\mu 0} \\ c_{\mu 1} \\ c_{\mu 2} \\ c_{\mu 3} \\ c_{\mu 4} \\ c_{\mu 5} \end{pmatrix} = \varepsilon_\mu \begin{pmatrix} c_{\mu 0} \\ c_{\mu 1} \\ c_{\mu 2} \\ c_{\mu 3} \\ c_{\mu 4} \\ c_{\mu 5} \end{pmatrix}$$

At the i th carbon, $\beta c_{\mu,i-1} + \alpha c_{\mu,i} + \beta c_{\mu,i+1} = \varepsilon_\mu c_{\mu,i}$

In general for a ring of N carbons, the Nth carbon is the 0th carbon. Therefore $c_{\mu N} = c_{\mu 0}$.

Using normalizing constants from page 19-8, these conditions are satisfied by $c_{\mu i} = \sqrt{\frac{2}{N}}\cos\left(\frac{2\mu i}{N}\right)$ or $c_{\mu i} = \sqrt{\frac{2}{N}}\sin\left(\frac{2\mu i}{N}\right)$ with $\varepsilon_\mu = \alpha + 2\beta\cos\left(\frac{2\mu}{N}\right)$ exceptions are, for the cosines $c_{0i} = \sqrt{\frac{1}{N}}$ $c_{N/2,i} = \sqrt{\frac{1}{N}}(-)^i$

Benzene has the results:

For the cosines,

μ	$c_{\mu 0}$	$c_{\mu 1}$	$c_{\mu 2}$	$c_{\mu 3}$	$c_{\mu 4}$	$c_{\mu 5}$	ε_μ
0	0.408	0.408	0.408	0.408	0.408	0.408	$\alpha + 2\beta$
1	0.577	0.289	-0.289	-0.577	-0.289	0.289	$\alpha + \beta$
2	0.577	-0.289	-0.289	0.577	-0.289	-0.289	$\alpha - \beta$
3	0.408	-0.408	0.408	-0.408	0.408	-0.408	$\alpha - 2\beta$

For the sines, $c_{0i} = c_{3i} = 0$

μ	$c_{\mu 0}$	$c_{\mu 1}$	$c_{\mu 2}$	$c_{\mu 3}$	$c_{\mu 4}$	$c_{\mu 5}$	For both sets of orbitals
1	0.0	0.5	0.5	0.0	-0.5	-0.5	$c_{4i} = c_{2i}$
2	0.0	0.5	-0.5	0.0	0.5	-0.5	$c_{5i} = c_{1i}$

The energy levels for benzene are spaced at $60°$ angles apart on a circle of radius 2β with the ethylene energy α at its center. The lowest energy states are doubly occupied by electrons with opposite spins. Looking down onto the plane

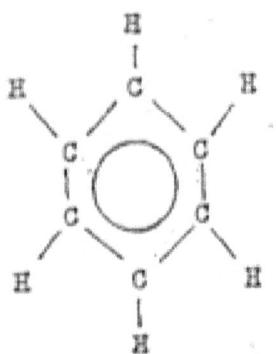

of the molecule, the bonding orbitals are shown below. All bond orders are $P_{ij} = 0.666$ giving all bond lengths $R_{ij} = 1.40 \text{ Å}$

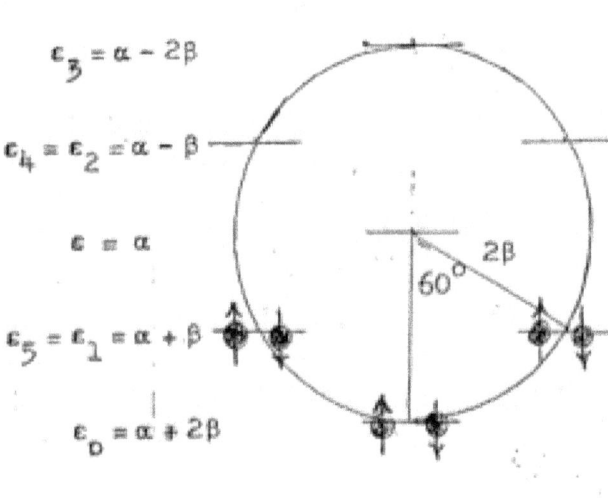

$$\varepsilon = 0$$
$$\varepsilon_3 = \alpha - 2\beta$$
$$\varepsilon_4 = \varepsilon_2 = \alpha - \beta$$
$$\varepsilon = \alpha$$
$$\varepsilon_5 = \varepsilon_1 = \alpha + \beta$$
$$\varepsilon_0 = \alpha + 2\beta$$

A hexigon with a ring of pi bonds would seem to be a better representation.

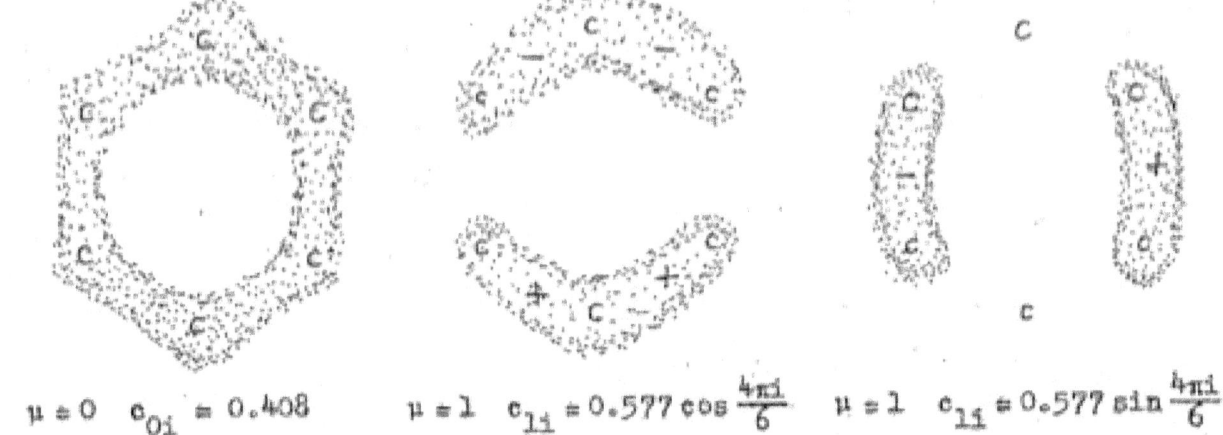

$$\mu = 0 \quad c_{0i} = 0.408 \qquad \mu = 1 \quad c_{1i} = 0.577 \cos\frac{4\pi i}{6} \qquad \mu = 1 \quad c_{1i} = 0.577 \sin\frac{4\pi i}{6}$$

PROBLEM 37-3: In the same manner, show the shapes of the anti-bonding pi orbitals for benzene.

PROBLEM 37-4: Using the method from page 37-22, find the energy levels for linear hexatriene, C_6H_8. Put two electrons in each of the lowest energy states. Remembering that β is negative, add these energies to get the energy of the ground state pi bonds for C_6H_8. Do the same thing for the bonding pi electrons for benzene. Which molecule has the lower ground state energy? Which is the more stable?

FOR FURTHER READING:

Szabo and Ostland MODERN QUANTUM CHEMISTRY
older but interesting: Pullman and Pullman QUANTUM BIOCHEMISTRY

The discussion of benzene opens the
door into a vast realm of modern
science and technology. We are
familiar with graphite being a form
of carbon with sheets of carbon
atoms that can slide over each other
making it useful for writing instru-
ments called pencils. The picture
below shows the sheets of carbon
atoms holding lithium ions in a
lithium ion battery. In 2004, Andre
Geim and Konstantin Novoselov at the
University of Manchester managed to
isolate a plane of graphite carbon
which turned out to be a vast sheet
of benzene rings shown to the right.
Being halfway between benzene and
graphite, they called it GRAPHENE.

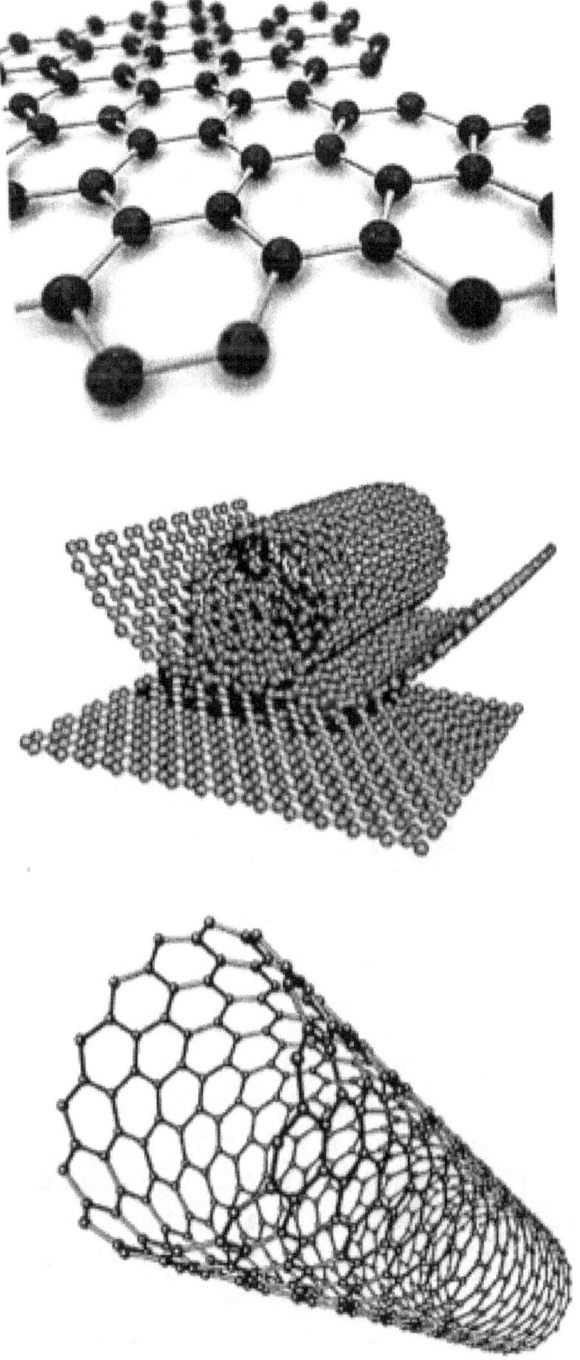

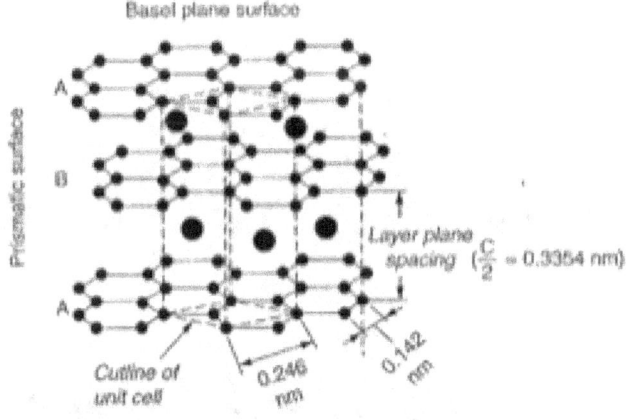

Basel plane surface

Layer plane
spacing ($\frac{C}{2}$ = 0.3354 nm)

Cutline of
unit cell

0.246
nm

0.142
nm

If you could take a sheet of graphene
and roll it into a cylinder you would
get another structure called a carbon
NANOTUBE. This was discovered by Sumio
Iigima at the NEC Corporation Laboratory in 1991. The two structures are
related this way, but produced by
different processes.

Obviously, benzene related structures are going to be very interesting.
Diameter of carbon nanotubes (CNT) is around 10 nanometers (nm), one
nanometer being 10^{-9} meter. Most of them being made are around a micron
in length (1.0 micron is 10^{-6} meter). As they are made of carbon in a
benzene-like structure, let us return to the discussion of benzene.
Each atom of carbon has three nearest neighbors. Each atom is bonded
to its neighboring atoms by hybrids of it $2s$, $2p_x$ and $2p_y$ orbitals
in a sigma bond structure. The remaining $2p_z$ orbital goes into a
pi molecular orbital, which is spread out through all carbons in the
CNT. This results in a sea of conducting electrons.

CARBON NANOTUBE TECHNOLOGY

When we get more and more transistors (sometimes several million) on
a computer chip so that the computer can do more and more, things get
rather small. Transistors must be electrically connected, and that
requires conducting "wires". Normally the connections are made of
metal, but when the diameters get to be only a few atoms wide, there
are problems which do not occur with normal wires. Metallic materials
are not single crystals but are composed of many crystals and there
is some scattering of flowing electrons from intercrystaline boundaries
which contributes to resistance. When the cross sectional area of the
conducting wire gets very small, the electron current per unit area
gets very big creating a high wind of electrons. Just as a tornado does
large wind damage, this electron wind can even displace atoms and carry
them along. This electromigration is not a great problem in normal
wires, but when the wire is only a few atoms wide, the results can
thin the wire in some places and make it bulge in others, now greatly to
vary the cross sectional area.

At this point chip designers start to look at carbon nanotubes (CNT).
As long as they can be made, they have a uniform structure of continuously connected benzene rings. Thus no crystal boundaries to create
scattering of electrons. As the tube is essentially graphene, it is a
two dimential structure and not three dimensional. Without a third
dimension, electrons can only move forward or only sidewise in one
other dimension. As the chemical bond in graphene is stronger than
metallic bond, atoms cannot be moved around. Hence, chip designers
look to carbon nanotubes to Interconnect transistors.

The next step has been to use carbon
nanotubes to make transistors. Also, IEEE
Spectrum of October 2019 reported that Max
Shulaker and his group at MIT developed a
microprocessor using 15,000 carbon nanotube
(CNT) transistors. Carbon nanotubes unlike
silicon can be easily stacked in layers for
3 dimensions. Engineers at Analog Devices
and Skywater Technology Foundry helped
implement the fabrication, which can be done
with existing chip manufacturing methods. P
and N type CNT are gotten with layers of
dielectrics as hafnium dioxide manipulated
to add or subtract electrons from the
nanotube.

THE STRUCTURE OF SOLID MATTER

When there are very many atoms in the solid phase, they most often arrange themselves in an orderly structure called a crystal lattice. Some basic pattern of atoms, sometimes simple but often quite complicated, is repeated over and over again in an orderly manner. An explanation of this involves the shapes of atomic electron orbitals developed in chapter 17.

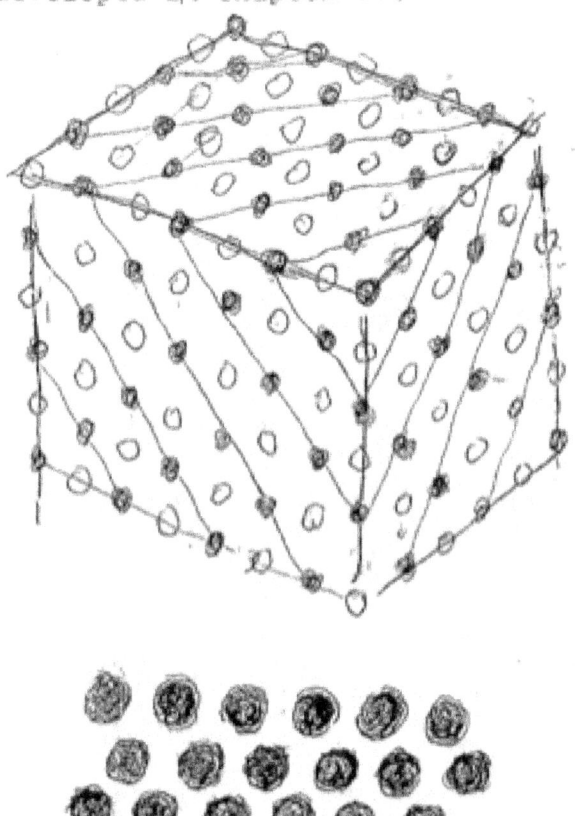

An example of a crystal is MnO. The black atoms are manganese and the white atoms are oxygen in the picture.

Two dimensional structures can be stacked to form a three dimensional solid.

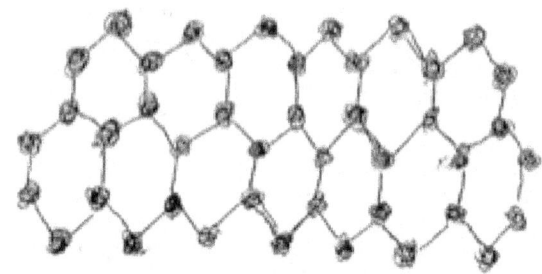

Graphene, being a vast extension of benzene rings. When rolled into a cylinder it makes carbon nanotubes. When stacked in layers it makes graphite.

Hexagonal close packed

In some materials, NaCl for example, the atoms are charged as positive and negative ions. Here, the bonds are electrostatic attraction of opposite charges.

A typical ionic crystal is shown in the picture. MnO also has this structure.

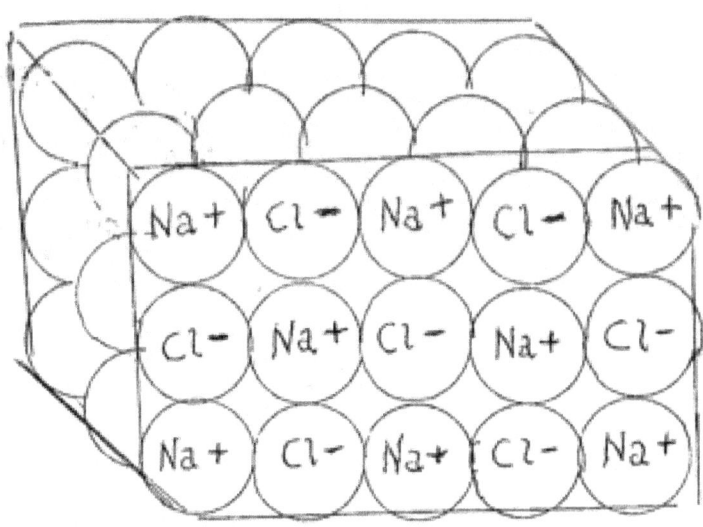

BONDING IN METALS AND THEIR COMPOUNDS

Consider transition metals   Sc Ti V Cr Mn Fe Co Ni Cu Zn
Here we are concerned with the 3d atomic orbitals. In those atoms
where more than one 3d orbital is occupied, when one electron is
given to the band of conduction electrons, the remaining 3d electrons
are available for bonding.

Consider a lattice of
metal atoms and let the
$3d_{x^2-y^2}$ orbitals be
contributing to the bond.
Notice the overlap
regions where the
exchange integrals
make a major contra-
bution to the strength
of the bond.

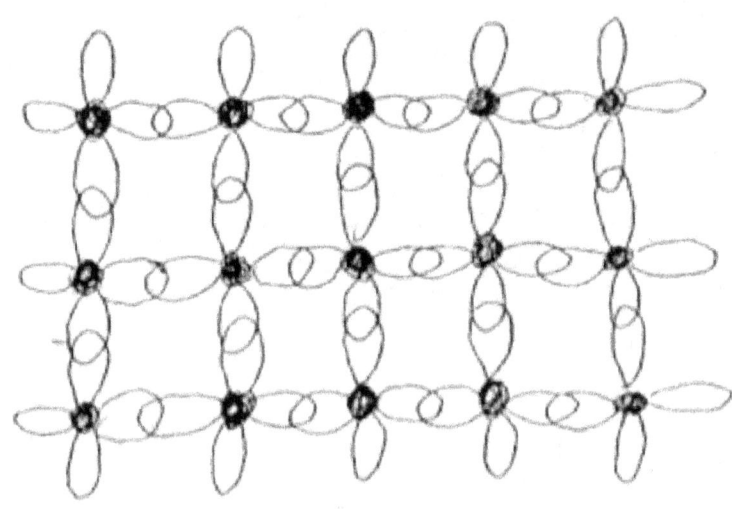

The $3d_{xz}$ and $3d_{yz}$ and
$3d_{z^2}$ orbitals are
involved in bonds to the
layers of atoms above
and below this one.

Metal oxides are much
involved in some
exotic modern science.
Consider a material
with alternating planes
of metal and oxygen
atoms, the planes seen
here end on, Oxygen atoms
are bonded to metal by
their $2p_x$ orbitals and
to other oxygen by $2p_y$.

Oxygen $2p_z$ orbitals are
used to bond to oxygen in
the next layer of atoms.

Because oxygen uses all
of its electrons in the
bonds, it has none left
for conducting. The
oxygen layer is non-
conducting.

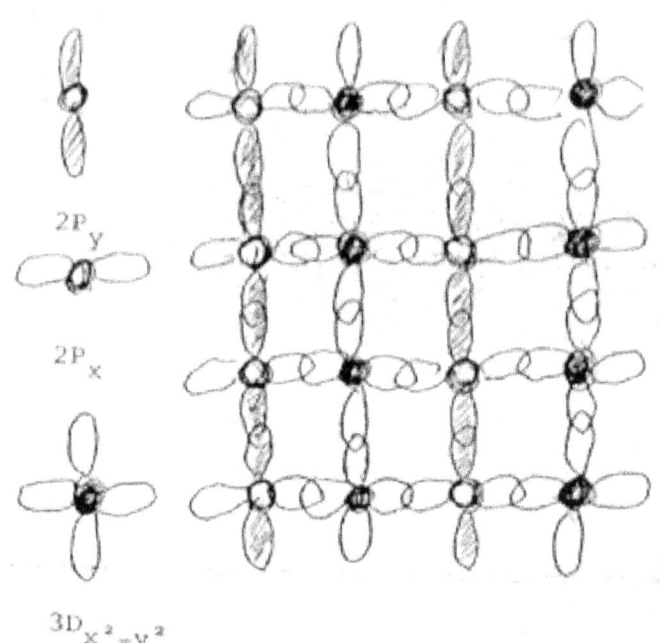

2P_y

2P_x

$3D_{x^2-y^2}$

FOR FURTHER READING:

M. T. Dove STRUCTURE AND DYNAMICS: AN ATOMIC VIEW OF MATERIALS Oxford U.P
W. A. Harrison ELECTRONIC STRUCTURE AND THE PROPERTIES OF SOLIDS Dover
P. A. Cox TRANSITION METAL OXIDES Oxford U. P.

Krzysztof Iniewski: NANOELECTRONICS, NANOWIRES, MOLECULAR ELECTRONICS,
                AND NANODEVICES McGraw Hill (2011)

CHAPTER TWENTY: STATISTICAL THERMODYNAMICS OF THE STATES OF MATTER

Behold whenever
The sun´s light and rays, let in, pour down
Across dark halls of houses; thou wilt see
Dust particles in many a manner mixed
Amid the void in the very light of the rays,
And battling on, as in eternal strife,
And in battalions contending without halt,
In meetings, partings, harried up and down,
From this thou mayest conjecture of what sort
The ceaseless tossing of primordial atoms
Amid the mightier void - at least so far
As small affair can for a vaster serve -
And by example put thee on the trail
Of knowledge. For this reason too ´tis fit
Thou turn thy mind the more unto these bodies
Which here are witnessed tumbling in the light;
Namely because such tumblings are a sign
That motions also of the primal stuff
Secret and viewless lurk beneath, behind.
For thou wilt mark here many a speck, impelled
By various blows, to change its little course,
And beaten backwards to return again
Hither and thither in all directions round.
Lo, all their shifting movement is of old.
From the primordial atoms, for the same
Atoms of things first move of self,
And then molecules minute of atoms conjoined
And nearest, as it were, unto the powers
Of primeval atoms, are stirred up
By impulse of those atoms´ unseen blows,
And these thereafter goad the next in size;
Thus motion ascends from the primevals on,
And stage by stage emerges to our sense
Until those objects also move which we
Can mark in sunbeams, though it not appears
What blows do urge them on.

                    Titus Lucretius Carus (96-55 BC)

Twenty centuries ago, the idea of what we now call the kinetic theory of gases (gas made of molecules in random motion) was brought forth as an explanation of what we now call the Brownian motion of dust particles dancing around in a sun beam. If thinking was so advanced at that time, why then did modern science not develop? Mathematics had to get to where it could put speculation in precise language and it took a while for the method of experimental proof to catch on.

This chapter will develop a theory of the states of matter: gas, liquid and solid, and the transitions (melting, boiling, etc.) between them. Starting with entropy (missing information) the primary ideas will be energy and temperature. The primary method will be statistical mechanics where we start with the quantized nature of the particles of matter and develop a description of the bulk matter of our everyday experience.

ENTROPY, THERMODYNAMICS AND THE STATES OF MATTER

Starting with the idea of missing information, or entropy, this chapter develops the formal theory of heat phenomena and applies that to the study of the relation between the gas, liquid and solid states of matter.

A couple mathematical relations are needed for the theory of information and entropy. These are developed in the first pages of the chapter. If you are familiar with them, you can skip too the discussion of entropy.

The first relation comes under the mathematical category called combinatorics. You are familiar with the question, in flipping a coin seven times, how many orders are there where four heads and three tails occur?

Suppose we consider six different and distinct items which are to be placed in six different boxes that are arranged in a row.

Let the items be labeled A, B, C, D, E, and F. They can be arranged in six boxes in the order: $A\,B\,C\,D\,E\,F$; or in the order: $D\,B\,C\,A\,E\,F$; which is a different arrangement, obtained from the first by exchanging the A and the D. We now ask how many different arrangements there can be of six different things in six boxes. This is easy to answer. First take six empty boxes and then take item A. There are 6 choices of places to put item A, for each one of which there are 5 remaining choices of places to put item B, giving 30 choices of combinations of places to put items A and B. For each of these 30 combinations there are 4 remaining choices of places to put item C, and then 3 choices for items D, and 2 choices for item E, and finally 1 choice of a place to put item F. All six boxes are now full.

The total number of different combinations of 6 different things in 6 boxes is, then, $6 \times 5 \times 4 \times 3 \times 2 \times 1 = 6!$. From this it is evident that there will be $N!$ different arrangements of N different items in N boxes.

Now, suppose we have six coins. On one side they are labeled with different numbers. On the other side they are labeled only with A or B.

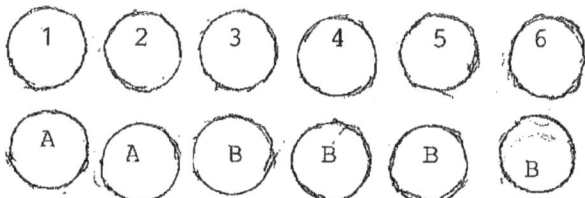

If we reverse the two A's the letter pattern doesn't change but the numbers do change. If we reverse two B's, the letter pattern doesn't change but the number pattern does change. If we reverse an A and a B, both patterns change. Now, how many obviously different arrangements are there for 2 A's and 4 B's in six places?

Although we have six items, not all 6 arrangements will be observably different from each other. So, let the symbol, P, (for permutations) stand for the number of observably different arrangements. For each one of these there are $2!$ re-arrangements of the 2 A's that will not change the patterns and $4!$ re-arrangements of the 4 B's that will not change the patterns. Thus, for each distinct pattern of the items, there will be $2! \times 4!$ arrangements of the A's and the B's that give the same pattern. Thus, the total number of arrangements of the items, distinguishable or not, is $P \times 2! \times 4!$, which gives the number of observably different arrangements of 6 items, 2 of one kind and 4 of another kind, in 6 places to $P = \dfrac{6!}{2! \, 4!}$.

Following the same argument, we obtain the generalization that if in N places we put N items, K of them being of one kind and $(N-K)$ being of another kind, then the total number of distinguishable combinations will be $P(K, N-K) = \dfrac{N!}{K!(N-K)!}$.

This can be looked at in another way. If you flip a coin N times, $P(K, N-K)$ gives the total number of patterns in which K heads and $N-K$ tails could appear in the order of flipping of the coin. That, you see, is the same question as that in which we want to know the number of different arrangements of K H's and $N-K$ T's in N boxes.

DERIVATION OF STIRLING´S FORMULA

In many situations in Statistics and Statistical Physics, we need a simple way to evaluate $y = ln\,(x!)$, especially with x being a big number.

To get such a formula, consider $y = ln\,(x!)$ to be a continuous function that hits the values of $ln\,(x!)$ where x is an integer.

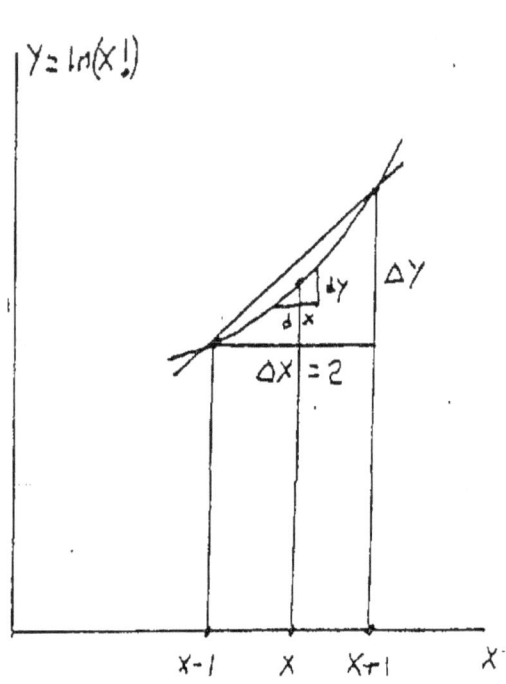

Assume that near the point $y(x)$ the slope of the curve equals the steepness of the line between $y(x - 1)$ and $y(x + 1)$. Thus,
$$\frac{dy}{dx} = \frac{ln\,(x+1)! - ln\,(x-1)!}{2}.$$

But, $ln\,(x + 1)! = ln\,(x - 1)! + ln\,x + ln\,(x + 1)$.

Thus, $\dfrac{dy}{dx} = \dfrac{ln\,x + ln\,(x+1)}{2}$, but for very large x, $ln\,(x + 1) = ln\,x$ to the limit of our ability to calculate it. Hence, $y = ln\,(x!)$ has the slope,
$$\frac{dy}{dx} = ln\,x.$$

Recall that $\dfrac{d}{dx}(x\,ln\,x) = ln\,x + 1$, but 1 is $\dfrac{dx}{dx}$, so $\dfrac{d}{dx}(x\,ln\,x - x) = ln\,x$. Therefore $\dfrac{dy}{dx} = ln\,x$ is satisfied by $y = x\,ln\,x - x + constant$.

When the constant is evaluated at small values of x it is found to be too small to be bothered with at large values of x. Thus, we have the approximation, $ln\,(x!) = x\,ln\,x - x$, which is STIRLING´S FORMULA.

INFORMATION, MISSING INFORMATION AND ENTROPY

We start this topic with a question whose answer is a basis for the discussion of information (an important topic today) and of entropy. You are on your way to visit friend in a part of the city built up with apartment buildings. There are B buildings in the area, each having A apartments. As with most modern cities, everything is identical with everything else, and very boring.

When you get to the area, you discover that you have left the memorandum with the building address and the apartment number back home on your desk. How much information is lying back on your desk? How much information are you missing? In this discussion it will be most useful to define a Missing Information Function.

Suppose N is the number of things through which you are searching, only one of which is the thing which you actually want. We wish to obtain a missing information quantity, S, which is a function of the number, N, of things to be searched through. We write $S = S(N)$ to say that the relation exists. Obviously, $S(N)$ increases with increasing N.

With regard to the number, $B\,A$, of apartments to be searched through in finding you friend, the total missing information is $S(B\,A)$. But this is the sum of missing information in two statements, $S(B)$ as missing information about the buildings to be searched and $S(A)$ as missing information about apartment number. Hence, we write $S(B\,A) = S(B) + S(A)$.

This may allow us to find a formula for the function. In Chapter Eight it is shown that there is a natural number, $e = 2.718...$ and that the natural logarithm for any number is the exponent to which e is a raised to give that number. We write $A = e^{ln\,A}$, $B = e^{ln\,B}$ and $A\,B = e^{ln\,A\,B}$.

Obviously, $e^{ln\,A\,B} = e^{ln\,A}\,e^{ln\,B} = e^{ln\,A + ln\,B}$ or $ln\,A\,B = ln\,A + ln\,B$.

For the function, $S(B\,A) = S(B) + S(A)$, obviously, $ln\,(A\,B) = ln\,A + ln\,B$ is a strong candidate to be used in constructing a formula for missing information. To get the most useful formula for the missing information about the number, N, of things or places to be searched through, we will take $ln\,N$ and multiply it by some quantity, k, which will be arbitrary but constant throughout any discussion using the concept of missing information.

In the field of Physics, the concept of missing information is called ENTROPY. In this application, it is convenient for k to be the Boltzmann constant which is a gas constant per molecule.

Then, $k = 1.38 \times 10^{-23}$ Joule/molecule-°K.

Hence, the formula for missing information is $S(N) = k \ln N$.

An interesting and important application of the concept of entropy is to the mixing of two dissimilar fluids. Consider a tank with a removable partition. On one side let there be N_w white molecules in N_w molecules-sized spaces each having volume, V. On the other side let there be N_b black molecules in N_b molecule-sized spaces each with volume, v.

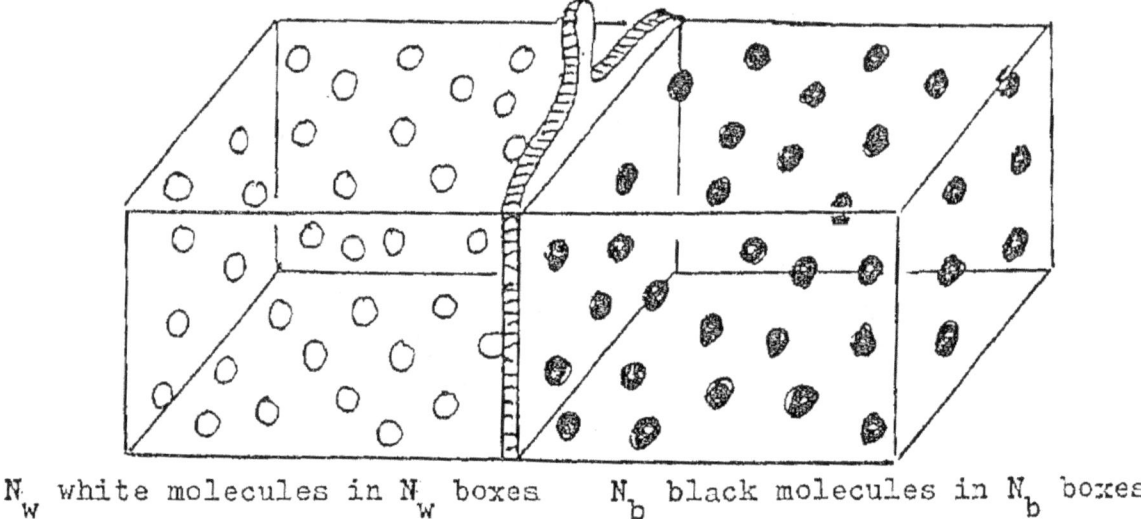

N_w white molecules in N_w boxes N_b black molecules in N_b boxes

To consider the entropy of the system, we must define the STATE of the system. A state of this system will be a unique pattern of molecules arranged in available molecule-sized spaces or boxes. If two white molecules exchange places, the pattern or state does not change. If two black molecules exchange places, the state does not change. If a white molecule exchanges places with a black one, the pattern changes, the state changes.

Initially, with the partition in, there is only one state to be found for the system. Every space on the left is white and every space on the right is black. Only one pattern has to be looked through to find the initial state. Hence the initial entropy, designated by subscript $_o$, is $S_o = k \ln 1 = 0$.

With the partition in, there is no missing information. You know exactly what pattern of molecules to expect.

Now, remove the partition. This allows white and black molecules to exchange places. White molecules move into the black region and black molecules move into the white region. This process is called DIFFUSION. If we wait for a little while, the entire tank becomes gray; although possibly little white regions or little black regions might appear momentarily. We now do not know what pattern or arrangement of molecules that we should expect to see at any instant. The entropy has increased.

To determine the entropy of the mixed system of molecules, it is necessary that we know the number of patterns or states of the molecules, out of which one exists uniquely at any instant of time. The number of patterns of N_w white things and N_b black things in $N_w + N_b$ boxes is exactly the same as the number of arrangements or patterns of N_h heads and N_t tails of coins that result from $N_h + N_t$ flips, hence the number of patterns of molecules is

$$P = \frac{(N_w + N_b)!}{N_w! \, N_b!}.$$

The entropy of the mixed system is the missing information as to what specific pattern of molecules will exist at a designated instant of time.

Hence, $S = k \ln \left[\dfrac{(N_w + N_b)!}{N_w! \, N_b!} \right]$.

Although this looks like a disaster at the end of a bad trip, with a few tricks it can be made to be very simple. First as $\ln(AB) = \ln A + \ln B$, we also notice that $\dfrac{A}{B} = \dfrac{e^{\ln A}}{e^{\ln B}} = e^{\ln A} - e^{\ln B} = e^{\ln A - \ln B}$.

Hence, $\ln\left(\dfrac{A}{B}\right) = \ln A - \ln B$ which gives the entropy of mixing:

$$S = k \ln(N_w + N_b)! - k \ln N_w! - k \ln N_b!$$

Next, if we can see the liquids at all, N_w and N_b will be very large numbers. There are, approximately, 10^{25} molecules in a cup of water.

Actually, this allows the formula for entropy to be further simplified. Previously in this chapter, we derived Stirling's Formula, whereby if x is a very large number, we have $\ln x! = x \ln x - x$.

Then, $S = k[(N_w + N_b) \ln (N_w + N_b) - (N_w + N_b) - N_w \ln N_w + N_w - N_b \ln N_b + N_b]$, from which you can do some simple algebra as an exercise to show that $S = k N_w \ln \left[\frac{N_w + N_b}{N_w}\right] + k N_b \ln \left[\frac{N_w + N_b}{N_b}\right]$ in which $N_w + N_b$ is the total number of molecules.

If each molecule occupies a volume, v, then $N_w v = V_w$ is the part of the total volume occupied by white fluid and $N_b v = V_b$ is the part of the total volume occupied by black fluid.

Then, $\dfrac{N_w + N_b}{N_w} = \dfrac{(N_w + N_b) v}{N_w v} = \dfrac{V_w + V_b}{V_w}$ and similarly for the other ratio.

The entropy of two mixed fluids is $S = k N_w \ln \left[\frac{V_w + V_b}{V_w}\right] + k N_b \ln \left[\frac{V_w + V_b}{V_b}\right]$.

At this point we can get a precise answer for a most intriguing question. What is the probability that one day we shall look at the tank only to find that all of the black molecules are on the left side and all of the white molecules are on the right side? At that moment we can put in the partition and the fluids will be totally separated.

Let us think about this. Although the separated state appears unique, it is only one of all the allowed states of the mixed system. It is no more nor less probable than any other state. It is only more interesting.

Hence, the probability of the separated system to exist is the same as the probability for any other state or pattern to exist. Now, recall that if there are P different patterns or states possible for the mixed system with one of them existing at any instant of time, then the entropy of the mixed system is given by $S = k \ln P$, from which $\dfrac{S}{k} = \ln P$, or $P = e^{S/k}$ is the number of patterns.

Now the probability that one of P patterns may occur is merely $\dfrac{1}{P}$.

$$\begin{bmatrix} \text{the probability} \\ \text{of one pattern} \end{bmatrix} = \frac{1}{\begin{bmatrix} \text{the number} \\ \text{of patterns} \end{bmatrix}} = \frac{1}{e^{S/k}} = e^{-S/k}$$

This is the most important interpretation of the entropy function.

Return to the entropy of the mixture. If we had equal number, N, of each type of molecule, so that $N_w = N$ and $N_b = N$, the entropy becomes $S = k N \, ln \, 2 + k N \, ln \, 2 = 2 \, k N \, ln \, 2 \, .$

This then gives: $\begin{bmatrix} \text{probability for} \\ \text{one pattern} \end{bmatrix} = e^{-2N \, ln \, 2}$

This can be further simplified if we recall that when e is raised to a product of exponents it can be raised to them in their order, one by one. Thus, $e^{-2 \, N \, ln \, 2} = (e^{ln \, 2})^{-2 \, N} = 2^{-2 \, N}$, from which we get the result that if a cup of white fluid and a cup of black fluid are allowed to diffuse into each other, the probability that they will naturally separate into their original states is $P = 2^{-2 \times 10^{25}}$, which is not very likely!

ENTROPY AND IRREVERSIBILITY, ORDER, DISORDER AND CHAOS

What have we learned from this discussion? A system of white and black
fluids separated by a partition had zero entropy as to what it would look
like as we had no missing information about that matter. When the partition
was removed the entropy went up as a result of our having less and less
information about what the system would look like. Also increasing entropy
is associated with the system going from an orderly separated state to a
disordered mixed condition. The more the entropy went up the less probable
it was that the orderly initial state would occur again by natural
processes. For large numbers of molecules, this means that the diffusion
process is not-reversible.

We now generalize on this result. If a physical process happens in such a
way that the entropy of the system and its environment does not change, we
know that some sort of orderliness is preserved. If the process happens with
constant entropy, the process can be reversed naturally. Such processes are
considered ideal.

If a process happens in which the entropy of a system and its environment
increases, the process involves going from order to disorder or chaos. As
the entropy increases, the probability of reversing the process by natural
means decreases. Reversal requires some sort of work to be done. Real
processes in nature happen this way.

A process which requires the entropy of a system and its environment to
decreases by natural means is improbable to the point of being impossible.

These statements constitute what is known as the SECOND LAW OF
THERMODINAMICS.

WE NEXT CONSIDER A BOX FULL OF MOLECULES

Initially the box will be loaded by a
molecular beam. Every molecule in the beam
has the same velocity in direction and
speed. There is only one state of the
molecules in the beam, and the entropy is
$S = k \, ln \, 1$ which is zero entropy. But as
they bounce off the wall of the box they
come off in many directions and when they
begin to bounce off each other they get a
bunch of different speeds. The entropy
rises.

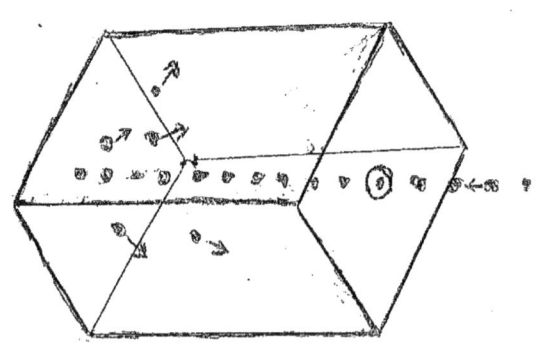

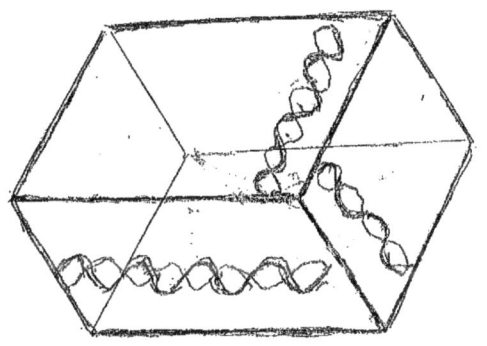

When the molecules settle down in the box, they settle into quantum states which are standing waves. Each dimension of the box is an integer number of wavelengths.

$$L_x = n_x \lambda_x \quad L_y = n_y \lambda_y \quad L_z = n_z \lambda_z$$

Now, momentum is $p = h/\lambda$ and a molecule's energy is $E = \frac{1}{2M}\left[p_x{}^2 + p_y{}^2 + p_z{}^2\right]$.

With a box full of wavefunctions, forming three dimensional standing waves, each momentum is determined by the wavelength of the standing wave in its direction, and each wavelength is determined by the dimension of the box as $\lambda_x = L_x/n_x$ and $p_x = h/\lambda_x$ and the same in the other directions. The energy of each quantum state is $E_i = E_{n_x,n_y,n_z} = \frac{h^2}{2M}\left[\frac{n_x{}^2}{L_x{}^2} + \frac{n_y{}^2}{L_y{}^2} + \frac{n_z{}^2}{L_z{}^2}\right]$, where n_x, n_y and n_z are three integers.

IN THE NEXT DISCUSSION, WE WILL DEFINE THE SUBSCRIPT i AS A GENERALIZED QUANTUM NUMBER AND REDEFINE n AS A NUMBER OF PARTICLES.

The state of a particle is defined by its quantum numbers. The state of a system of molecules is given by the distribution of molecules among the allowed quantum states. For example, a state of a system would be to have some particular number, n_i molecules, to be in any particular energy state, E_i.

The total number of molecules is $N = \sum_i n_i$ and $U = \sum_i n_i E_i$ is the total energy of the system which is called the internal energy. If the molecules were laid out in some line and randomly assigned their energy states, the number of patterns of energies among molecules is $P = \dfrac{N!}{n_1!\,n_2!\,n_3!...}$ and $S = k\,ln\,P$ would be the entropy.

This gives $S = k\,[ln\,N! - \sum_i ln\,n_i!]$. But using Stirling's formula and $N = \sum_i n_i$,
$$S = k\,[N\,ln\,N \cancel{-N} - \sum_i (n_i\,ln\,n_i \cancel{-n_i})] = -N\,k\,\sum_i \left(\frac{n_i}{N}\right) ln\left(\frac{n_i}{N}\right).$$

If the distribution is described by $f_i = \dfrac{n_i}{N}$, the fraction of the molecules in each energy state, E_i, then we get:

$$N = N \sum_i f_i \qquad U = N \sum_i f_i E_i \qquad \text{and} \quad S = -N\,k\,\sum_i f_i\,ln\,f_i$$

We will consider the system to be closed with definite boundaries separating it from its environment. In our approach to a definition of equilibrium we will consider that the system can be isolated with no heat energy transport across the boundaries, and that boundaries do not change. In the latter condition, if L_x, L_y, and L_z are constant, E_i is constant for each set of its quantum numbers.

Molecular collisions change the distribution among quantum states. Changes are more likely towards distributions with more, rather than less, quantum states. Hence, with collisions happening, the entropy tends to increase. We will define equilibrium as that configuration with the greatest entropy. With any fluctuation away from equilibrium, the tendency will always be to return to the state of maximum entropy.

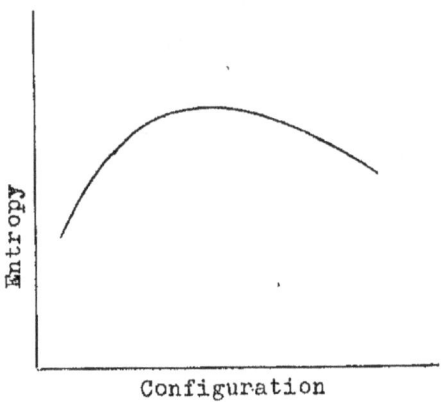

Let δf_i be a fluctuation in f_i. If the number of molecules is constant, $\delta N = N \sum_i \delta f_i = 0$

If the internal energy is constant and each E_i is constant, $\delta U = N \sum_i E_i \delta f_i = 0$.

If entropy is continuously varying near its maximum, $\delta S = -N \, k \sum_i (\ln f_i + 1) \, \delta f_i = 0$.

We can simultaneously state the condition that these three quantities are zero by adding them; but if we add them their units will be made the same as the units of entropy (whatever those are). We multiply δN by some a and δU by some b. Now adding them we get:

$$\delta S + a \, \delta N + b \, \delta U = N \sum_i \left[-k \ln f_i - k + a + b \, E_i \right] \delta f_i = 0$$

If the set δf involves arbitrary fluctuations, the condition for equilibrium for a closed system of molecules is $\ln f_i = -1 + \dfrac{a}{k} + \dfrac{b}{k} E_i$ or $f_i = e^{-1 + a/k} \, e^{\left(\frac{b}{k}\right) E_i}$.

As the f_i must decrease as energy, E_i, increases, we set $\dfrac{b}{k} = -\beta$; a negative number. As f_i, being a fraction, must have a value between zero and one, we set $-1 + \dfrac{a}{k} = -\alpha$, another negative number. Thus, when an isolated system of molecules is in equilibrium, the fraction of molecules in the ith quantum state is:

$$f_i = e^{-\alpha} e^{-\beta E_i}$$

Now, if $\sum_i f_i = 1$, we get the result that $e^{+\alpha} = \sum_i e^{-\beta E_i}$, which will be such an important tool in the following analysis that we give it a symbol, Z, and call it the PARTITION FUNCTION.

Interactions can change the distributions of molecules among the E_i. If we decrease the f_i for lower E_i and increase the f_i for higher E_i the internal energy rises. If we decrease the f_i for higher E_i and increase the f_i for lower E_i the internal energy decreases. Now we can consider changes in internal energy. $dU = N \sum_i E_i \, df_i + N \sum_i f_i \, dE_i = 0$.

In the first term we keep the size of the box, and hence the E_i, constant and change the occupation of the quantum states. If the first term is positive, the internal energy goes up; and vice versa. Let dQ represent heat (thermal energy) transferred into the gas. If dQ is positive, U goes up; and vice versa.

Thus, we make the identification, $N \sum_i E_i \, df_i = dQ$.

Next, $N \sum_i f_i \, dE_i$ keeps the occupation of quantum states constant but changes the E_i which means changing the size of the box. This term is positive if the volume decreases. And negative if the volume increases. What happens here is obvious. The pressure that the gas exerts on its walls produces a force, $F = PA$, on a surface of area, A. If the surface moves back by dx, the work done by gas is positive.

$dW = P A \, dx = P \, dv$ and the internal energy goes down. Now, $N \sum_i f_i \, dE_i$ is associated with work and is positive when dW is negative and negative when dW is positive. We write $N \sum_i f_i \, dE_i = -dW$.

Thus, we have $dU = dQ - dW$ which represents the conservation of energy which is the FIRST LAW OF THERMODYNAMICS.

Take another look at $f_i = e^{-\alpha - \beta E_i}$.

The fraction of molecules in energy
states decreases with energy. The
larger β the more rapidly the fraction
decreases with energy. The smaller β the
less rapidly f_i decreases with energy, E_i.
Now consider temperature, T. The larger
the temperature, the greater the fractions
at higher energy. The lower the
temperature the less molecules will have
higher energies.

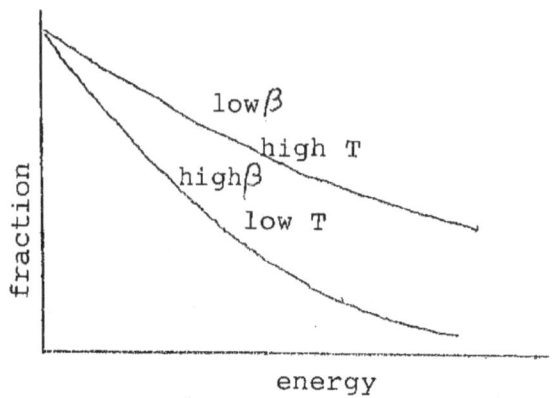

This suggests a thermodynamic temperature with $\beta = \dfrac{1}{kT}$, k being some
constant.

Now consider changes in entropy, $S = -Nk\sum_i f_i \ln f_i$. If $\sum_i f_i = 1$, $\sum_i df_i = 0$.
$dS = -Nk\sum_i \left[\ln f_i\, df_i + f_i \dfrac{1}{f_i} df_i\right]$ in which the second summation is zero.

If $f_i = e^{-\alpha - \beta E_i}$, $\ln f_i = -\alpha - \beta E_i$ giving $dS = Nk\alpha \sum_i df_i + Nk\beta \sum_i E_i\, df_i$.

Here the first sum is zero and if $\beta = \dfrac{1}{kT}$, we get $dS = \dfrac{dQ}{T}$.

Using $f_i = e^{-(\alpha + \beta E_i)}$ gives $S = -Nk\sum_i f_i \ln f_i = -Nk\sum_i f_i(-\alpha - \beta E_i)$. But, $\sum_i f_i = 1$
and $\sum_i N f_i E_i = U$, giving the entropy to be $S = Nk\alpha + k\beta U$, where $Z = e^{\alpha}$ or
$\alpha = \ln Z$. It is now evident that calculating the partition function,
$Z = \sum_i e^{-\beta E_i}$, is the link between the quantized world of individual molecules
and the collective world of the thermal behavior of matter consisting of
many molecules.

If $\beta = \dfrac{1}{kT}$ then $S = Nk\alpha + \dfrac{kU}{kT}$, since we haven't said what either constant T was
equal to, let them be equal to each other. Now, multiply S by T. We get
$ST = NkT\alpha + U$. In traditional thermodynamics there is a quantity called
Helmholtz free energy, $F = U - TS$.

What we have just derived gives us Helmholtz free energy, $F = -NkT\alpha$.

But $\alpha = \ln Z$, giving $F = -NkT \ln Z$. If $F = U - TS$, $dF = dU - TdS - S\,dT$; but
$dU = dQ - P\,dV$ and $dQ = TdS$.

$dF = T\,dS - P\,dV - T\,dS - S\,dT$ or $dF = -P\,dV - S\,dT$.

Obviously, Helmholtz free energy is a function of volume and temperature. From simple calculus we can write $dF = \frac{\partial F}{\partial V}dV + \frac{\partial F}{\partial T}dT$.

This gives $P = -\frac{\partial F}{\partial V}$ and $S = -\frac{\partial F}{\partial T}$, where $F = -N\,k\,T\,\ln Z$.

For a monatomic gas, there is only translational energy, ϵ_{tr}, and

$$Z = Z_{tr} = \textstyle\sum_{n_x\,n_y\,n_z} e^{-\beta\,\epsilon_{tr}}, \quad \text{where } \epsilon_{tr} = \frac{\hbar^2}{2\,m}\left[\frac{n_x{}^2}{L_x{}^2} + \frac{n_y{}^2}{L_y{}^2} + \frac{n_z{}^2}{L_z{}^2}\right].$$

The sum over quantum states is:

$$Z = \left[\textstyle\sum_{n_x=0}^{\infty} exp\left(\frac{-\beta\,\hbar^2}{2\,m\,L_x{}^2}\,n_x{}^2\right)\right]\left[\textstyle\sum_{n_y=0}^{\infty} exp\left(\frac{-\beta\,\hbar^2}{2\,m\,L_y{}^2}\right)n_y{}^2\right]\left[\textstyle\sum_{n_z=0}^{\infty} exp\left(\frac{-\beta\,\hbar^2}{2\,m\,L_z{}^2}\right)n_z{}^2\right]$$

Ignoring the labels, x, y, z, you will notice that the summations in the boxes are all equal, so we shall evaluate one of them and Z will be the cube of the result.

If the mass of a molecule is of the order of 10^{-27} kilogram and $\hbar = 10^{-34}$ Joule-second, it is obvious that the spectrum of energies is almost a continuum and the summations can be done by integration. For the summation over n_x notice that $dn_x = 1$. Now look at $\sum_{n_x=0}^{\infty} exp\left(\frac{-\beta\,\hbar^2}{2\,m\,L_x{}^2}\,n_x{}^2\right)dn_x$ and let

$u = \frac{\hbar}{L_x}\sqrt{\frac{\beta}{2\,m}}\,n_x$. Then, $dn_x = \frac{L_x}{\hbar}\sqrt{\frac{2\,m}{\beta}}\,du$.

The summation can now be written $\frac{L_x}{\hbar}\sqrt{\frac{2\,m}{\beta}}\int_{u=0}^{\infty} e^{-u^2}\,du = \frac{L_x}{\hbar}\sqrt{\frac{2\,m}{\beta}}\sqrt{\frac{\pi}{2}}$ in which we used an integral from chapter 4.

Doing the same with y and z gives $Z = \frac{L_x\,L_y\,L_z}{\hbar^3}\beta^{-3/2}\sqrt{\frac{m\,\pi}{2}} = V\,\beta^{-3/2}$ (constant) as the partition function for a monatomic gas where $V = L_x\,L_y\,L_z$ is the volume of the containing box.

This gives $\alpha = \ln Z = \ln V - \frac{3}{2}\ln\beta + \ln(constant)$, $\ln\beta = \ln\left(\frac{1}{k\,T}\right) = -\ln T - \ln k$.

We now get: $F = -N\,k\,T\left[\ln V + \frac{3}{2}\ln T + constant\right]$, $P = -\frac{\partial F}{\partial V}$, $S = -\frac{\partial F}{\partial T}$.

$\frac{\partial F}{\partial V} = -N\,k\,T\left[\frac{1}{V}\right]$ giving $P = \frac{N\,k\,T}{V}$ or $P\,V = N\,k\,T$ but $P\,V = R\,T$.

For one mole of gas N_A is Avogrado's number and $N_A k = R$, the molar gas constant, so $k = R/N_A$ is the Boltzmann constant, $k = 1.38 \times 10^{-23} \frac{Joule}{molecule \cdot K}$ and T is absolute temperature in (degrees) Kelvin.

Cancelling the temperatures in the last term and lumping all constants into a S_o we get for entropy, $S = N k \ln V + \frac{3}{2} N k \ln T + S_o$.

For internal energy, $dU = T \, dS - P \, dV$, but $dS = N k \frac{dV}{V} + \frac{3}{2} N k \frac{dT}{T}$. $\frac{N k T}{V} = P$. $dU = \frac{3}{2} N k \, dT$ and $U = \frac{3}{2} R T$.

In partial derivatives, we take derivatives of functions with respect to single variables, keeping other variables constant. Partial derivatives in thermodynamics will have different forms depending on which other variables are kept constant. Subscripts on partial derivatives will specify which other variables are kept constant.

The (MOLAR) HEAT CAPACITY of a substance is the amount of heat energy necessary to change the temperature of one kg-mole of the substance by one Kelvin. The heating can be done at constant volume or at constant pressure. At constant pressure, the volume changes and work is done. The symbol for heat capacity is C. Recall that heat transfer is $dQ = T \, dS = dU + P \, dV$.

The heat capacity at constant volume is $C_v = \left(\frac{\partial Q}{\partial T}\right)_v = T \left(\frac{\partial S}{\partial T}\right)_v = \left(\frac{\partial U}{\partial T}\right)_v$.

At constant pressure $C_p = T \left(\frac{\partial S}{\partial T}\right)_p = \left(\frac{\partial U}{\partial T}\right)_p + P \left(\frac{\partial V}{\partial T}\right)_p$.

For monatomic gas, $U = \frac{3}{2} R T$ and $V = \frac{R T}{P}$. $C_v = \frac{3}{2} R$ and $C_p = \frac{3}{2} R + P \left(\frac{R}{P}\right) = \frac{5}{2} R$.

Before we get too far along with changes in thermodynamics, recall that the distribution, $f_i = e^{-\alpha - \beta E_i}$, held for the state of equilibrium where entropy, $S = -N k \sum_i f_i \ln f_i$, had a maximum value for a closed system. The partition function and the formulas derived from it describe a system in equilibrium. If all this is to be valid during processes involving changes in the system, the changes would have to be done in the following manner. A change, dT, dP, dV, dQ etc. would be made. The system would be allowed to re-establish equilibrium. Then new changes would be made.

Then equilibrium re-established; and so on. Changes would occur in such a slow manner that the system would always be infinitesimally close to equilibrium. Such changes would always be reversible. We call this a QUASI-STATIC PROCESS.

Obviously, such processes seldom happen in the practical world; especially in the heat engines discussed in chapter 10 where there are explosions, shock waves and turbulence which make even temperature a concept of doubtful validity. Thermodynamics is a necessary and useful theory in that it comes sufficiently close to describing processes involving energy in bulk matter. But we should remember its limitations.

The manner in which changes in a system occur are assigned various classifications.

An ISOTHERMAL process occurs with no temperature change. $dT = 0$

An ISOBARIC process occurs with no change in pressure. $dP = 0$

An ADIABATIC process occurs with no heat transfer. $dQ = 0$

An ISENTROPIC process is a reversible adiabatic process. $dQ = T\,dS = 0$

Consider and adiabatic process with an idea gas: $PV = RT$ and $U = U(T)$. Now, $dQ = T\,dS = dU + P\,dV$. But $PV = RT$ and $P\,dV = R\,dT - V\,dP$.

So $dQ = T\,dS = dU + R\,dT - V\,dP$.

$C_v = T\left(\frac{\partial S}{\partial T}\right)_v = \frac{\partial U}{\partial T}$ and for an ideal gas, $dU = C_v\,dT$ as U has no other variables.

$C_p = T\left(\frac{\partial S}{\partial T}\right)_p = \frac{\partial U}{\partial T} + R = C_v + R$.

Restating the equations for heat transfer, $T\,dS = C_v\,dT + P\,dV$, and $T\,dS = C_p\,dT - V\,dP$.

For an adiabatic process, $dQ = T\,dS = 0$ giving $C_p\,dT = V\,dP$ and $C_v\,dT = P\,dV$. Dividing gives $\frac{C_p}{C_v} = -\frac{V\,dP}{P\,dV}$. Let $\gamma = C_p/C_v$ be the heat capacity ratio. This gives $\gamma P\,dV + V\,dP = 0$, or $\gamma\frac{dV}{V} + \frac{dP}{P} = 0$, or $d(\gamma \ln V + \ln P) = 0$.

For an adiabatic process $PV^\gamma = constant$. For an adiabatic process, show that $TV^{\gamma - 1} = constant$ and $T^\gamma V^{1 - \gamma} = constant$.

For n moles of a gas, the gas law is $PV = nRT$ and is fairly well followed by most gases under normal conditions. For many substances the gas law is not too legal. For water (H_2) the gas or vapor phase is called steam. In the curves below for water, pressure is plotted against volume for several temperature values.

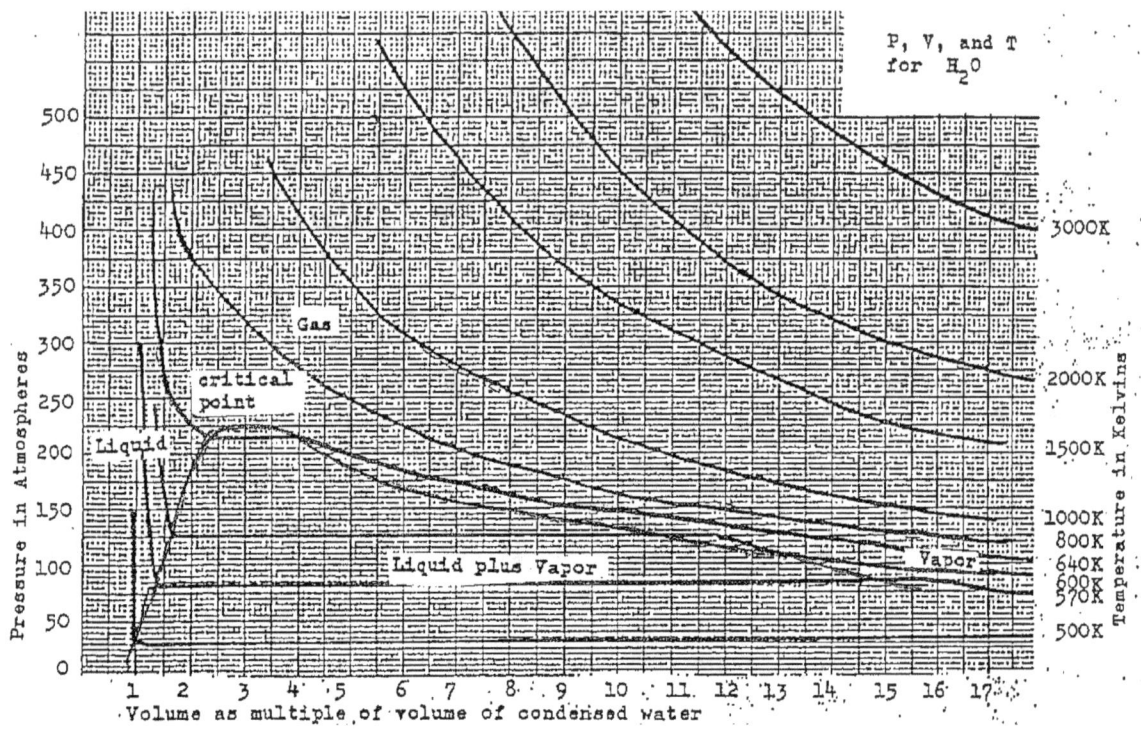

You would expect that $P = \dfrac{RT}{V}$ would, keeping temperature constant, give a plot of pressure vs. volume to be a hyperbola, as is the case at $3000°\,K$. But at lower temperatures a totally different behavior occurs. But recall that all pots of boiling water at atmospheric pressure had the same temperature. So long as liquid and vapor co-existed (two phases of matter) the temperature at atmospheric pressure was uniquely defined. For the contours below $640°\,K$, the horizontal lines are for boiling two-phase systems of liquid and vapor. For each contour the temperature is the boiling-point temperature at the pressure shown. Obviously, the higher the pressure the higher the boiling temperature which is the principle of the pressure-cooker. To the left of the boiling region, a great change of pressure produces only a small change of volume. There we have a pure liquid (or condensed) phase. To the right of the boiling region, a small change of pressure makes a large change in volume. There we have a vapor phase. Notice that at temperature $640°\,K$ there is a point above the liquid-vapor transition which does not occur. This is called the critical point. For all temperatures below this, we refer to the vapor and liquid phases.

We noticed that for a boiling mixture with liquid and vapor co-existing, that at each temperature there was a unique pressure at which boiling occurred, up to a critical temperature and pressure, above which the phase transition no longer happened. Boiling point temperatures and pressures for water are plotted in the figure to the right.

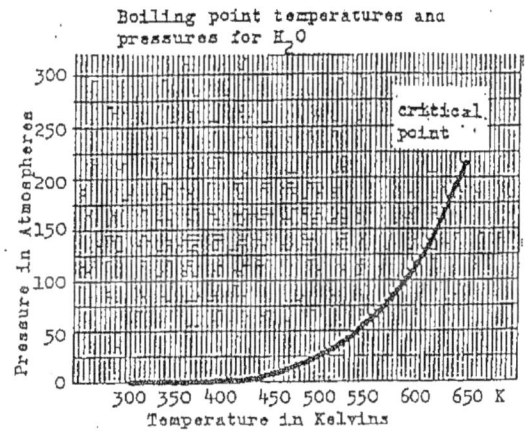

The gas law, $PV = RT$, assumes non-interacting molecules banging into each other and into the walls of their container. The first modification notices that there is a part of the volume that is occupied by the molecules themselves which is on the order of the volume of the condensed liquid. What is compressed with the gas or vapor is the vacuum between the molecules. The actual volume of one mole of molecules is given the symbol, b, and the volume of the inter-molecular vacuum is $(V - b)$ giving the gas law, $P(V - b) = RT$. There is experimental and theoretical evidence that the two hydrogen atoms are located to one side of the oxygen atom with the resulting effect on bonding electrons that the hydrogen end is positively charged and the oxygen end is negative. Having two charges (poles) such a molecules is called a dipole and is represented by charges $+q$ and $-q$ separated by a distance, d. The figures below show two dipoles, a distance, x, apart.

We label each charge in the top dipole by its energy related to the charges in the bottom dipole.

$$\frac{q^2}{x} - \frac{q^2}{x + d}$$

$$\frac{q^2}{x} - \frac{q^2}{x - d}$$

The total energy is $2q^2\left[\frac{1}{x} - \frac{x}{x^2 + d^2}\right]$. This is $\frac{2q^2}{x}\left[1 - \frac{1}{1 - d^2/x^2}\right]$.

For small a, $(1 + a)(1 - a) = 1 - a^2$.

Ignoring a^2, $\frac{1}{(1 - a)} = (1 + a)$.

Dipole-dipole energy is $-\frac{2q^2 d^2}{x^3}$.

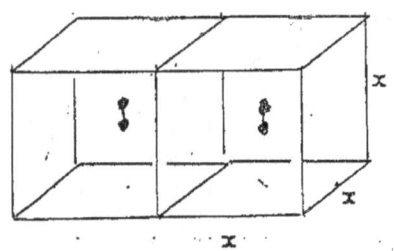

If N molecules are in a box of volume, V, and each molecule occupies a little box of dimension, x, then $x^3 = \dfrac{V}{N}$ and the energy is $-\dfrac{2 N q^2 d^2}{V}$.

With only one energy number, its partition function is $Z = e^{-2 N q^2 d^2 / V}$.

The total partition function now is $Z = (const)\, V\, T^{3/2}\, e^{-2 N q^2 d^2 / V k T}$.

Free energy is $F = -N k T \ln Z$ or $F = -N k T \left[\ln V + \dfrac{3}{2} \ln T - 2 N q^2 d^2 / V k T \right]$.

If $F = -N k T \ln V + \dfrac{3}{2} N k T \ln T + 2 N q^2 d^2 / V$, pressure is $P = -\left(\dfrac{\partial F}{\partial V} \right)_T$, giving

$$P = \frac{N k T}{V} - \frac{2 N^2 q^2 d^2}{V^2}.$$

We abbreviate $a = 2 N^2 q^2 d^2$ and under $N k T$, modify V for b which we take to be the volume of the molecules.

Then $P + \dfrac{a}{V^2} = \dfrac{N k T}{V - b}$, which is the van der Waals equation.

What, now, is the behavior of a system described by the van der Waal formula? Notice that the units for PV are:

(Newtons/meter^2) $\times$ (meter^3/kg-mole)

or

Newton-meter/kg-mole=Joule/kg-mole

We take:
$R = 8.31 \times 10^3$ Joule/(Kelvin, kg-mole)

Strictly, P is in Newtons/meter^2 but it is often reported in atmospheres where the standard reference atmosphere is
1 atm $= 1.016 \times 10^5$ Newton/meter^2.

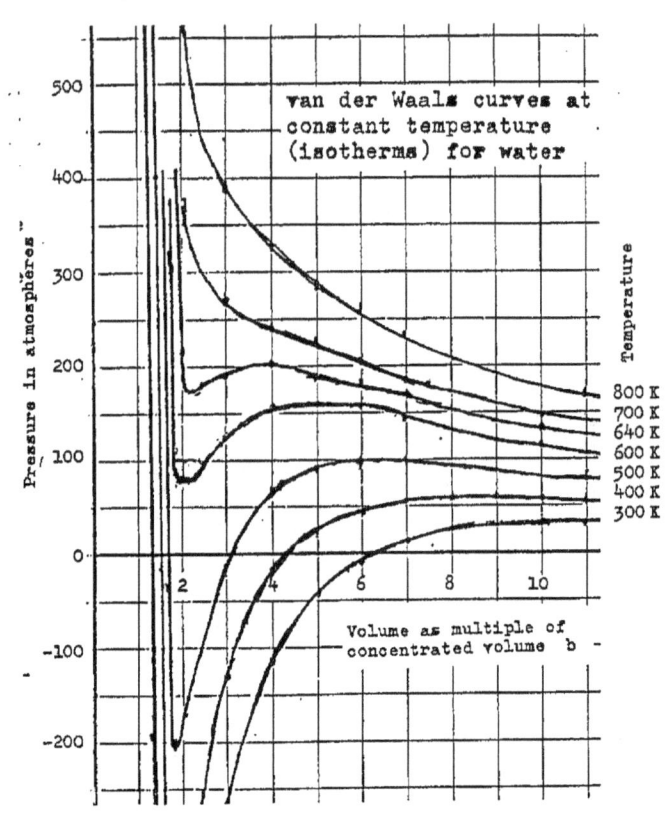

van der Waals curves at constant temperature (isotherms) for water

Pressure in atmospheres

Temperature

800 K
700 K
640 K
600 K
500 K
400 K
300 K

Volume as multiple of concentrated volume b

Then, b, which is the
volume of the condensed
phase of molecules, is
in meter^3/kg-mole. Then,
a, has the units of
Joules-meter^3/kg-mole^2.
The values, $b = 0.03$ and
$a = 5.5 \times 10^5$ should give
a description of water,
but the resulting
curves look absurdly un-
physical, even to the
point of requiring
negative pressures.

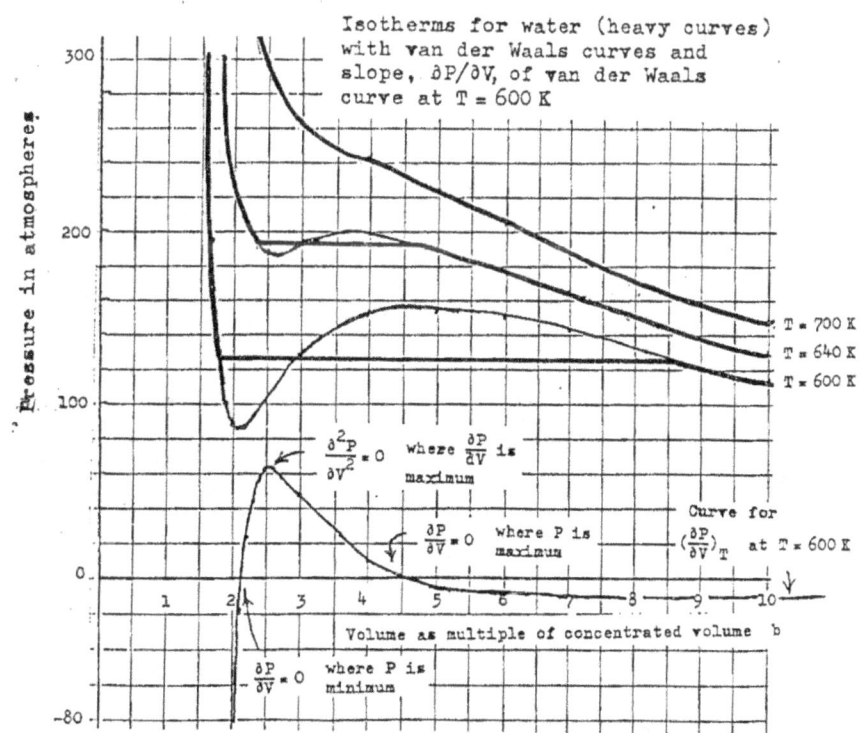

But notice that the
curves at $700°$K and
$800°$K do give gas-phase
behavior and at lower
temperatures they do
give vapor behavior at
volumes slightly larger
than b.

Putting the van der Waals curves on top of the constant temperature curve
for water shows that the misbehavior occurs in the two-phase region where
the pressure is constant at given temperature.

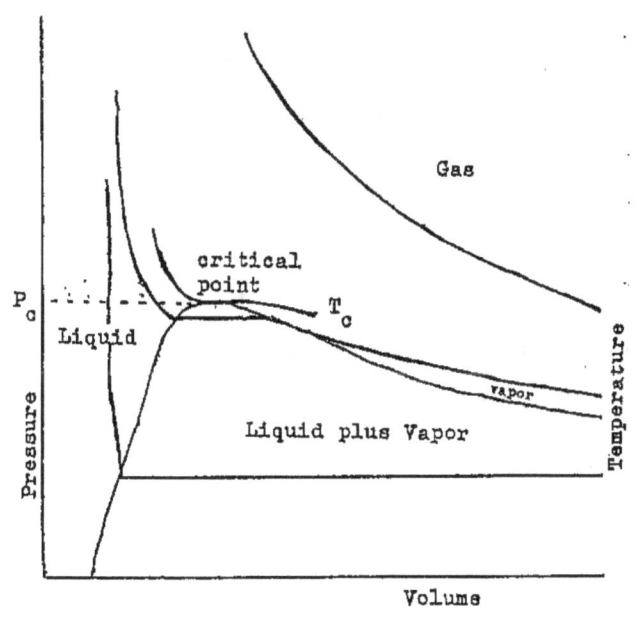

Compare, now, the graph above to the
curves at the top of page 20-16. The
van der Waals equation is giving a
realistic description of the pure
vapor phase and of the liquid phase.
Where it does goofy things, it is
not needed as the system remains at
a constant pressure for a given
temperature when vapor and liquid
are both present. Although the van
der Waals equation appears to do
non-physical things in this region,
it is here that it can be most
interesting and useful.

For an isotherm (curve at constant temperature) the interesting part is the slope of the curve, dP compared with dV. Here we introduce a slightly new notation. P now varies with both V and T and we want the rate of change of P with V at constant T. To remind ourselves that the variable, T, is kept constant we round off the d's denoting change.

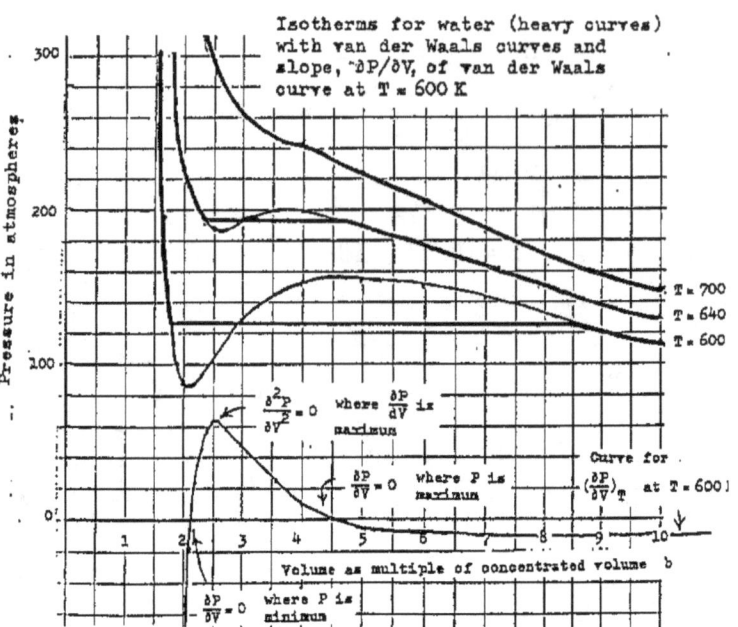

Isotherms for water (heavy curves) with van der Waals curves and slope, $\partial P/\partial V$, of van der Waals curve at T = 600 K

Instead of dP we write ∂P and instead of dV we write ∂V and often outside of a parenthesis we indicate which other variable is constant. Thus, $\left(\dfrac{\partial P}{\partial V}\right)_T$ is called the partial derivative of P with V at constant T. In what follows, we shall omit the extra reminder and write $\dfrac{\partial P}{\partial V}$.

In the figure above, we have plotted $\dfrac{\partial P}{\partial V}$ for the curve at $600°$ K temperature. The pressure-volume curve has a low point and a high point and you will see that at both of these, $\dfrac{\partial P}{\partial V} = 0$.

Then, $\dfrac{\partial P}{\partial V}$ plotted against V has a high point where its slope, $\dfrac{\partial\left(\frac{\partial P}{\partial V}\right)}{\partial V} = \dfrac{\partial^2 P}{\partial V^2}$, is zero (using the standard notation for second derivative). With the isotherm at $640°$ K the points where $\dfrac{\partial P}{\partial V} = 0$ approach where $\dfrac{\partial^2 P}{\partial V^2} = 0$. At the critical point, $\dfrac{\partial P}{\partial V}$ and $\dfrac{\partial^2 P}{\partial V^2}$ are zero at the same volume.

For any substance, the critical temperature, T_c, and critical pressure, P_c, can be determined experimentally. The van der Waals equation allows us to extract additional information from these measurements.
Recall, $P = \dfrac{RT}{(V-b)} - \dfrac{a}{V^2}$, where $a = 6\,k\,(q\,d)^2 N_a{}^2$ for one kilogram-mole.
$R = 8.31 \times 10^3$ Joule/(Kelvin-kg-mole); $\quad k = 9 \times 10^9$ N-m^2/Coulomb^2;
$N_a = 6.023 \times 10^{26}$ Molecules/kg-mole

$q\,d$ for the charges $+q$ separated by distance, d, is called the electric dipole moment of the molecules. Its standard symbol is a Greek mu, $\mu = q\,d$, (Unfortunately, the Greek letter mu will much overworked as a standard symbol for other things in the following discussions. You must determine what it means from the equations in which it is used.)

Next, we see how we can determine the dipole moment of a substance's molecules from the critical temperature, T_c, and critical pressure, P_c, of that substance.

Notice the curves for pressure as a function of volume at constant temperature (isotherms). Where the curves are steep we have liquid, where the curves have a very small slope we have a vapor, and where the curves are flat we have a liquid-vapor phase transition at constant pressure and temperature. At a given pressure, the temperature of the phase transition is called the boiling point.

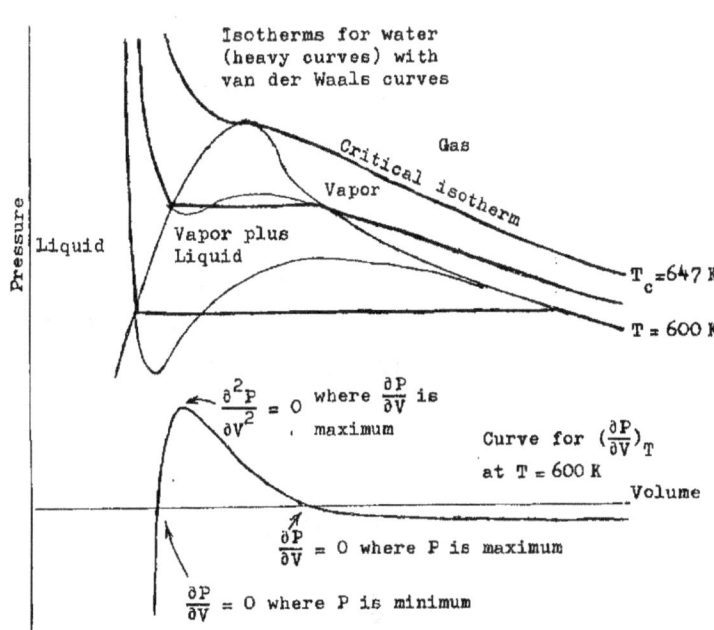

The point where an isotherm only barely gives a hint of phase transition is called the CRITICAL POINT. T_c and P_c for that point are the critical temperature and critical pressure.

Where there is only a single phase, liquid or vapor, the isotherms follow the van der Waals curves. Where the two phases coexist, the isotherms are straight lines at constant temperature and pressure. In this region, the van der Waals curves have maximum and minimum values where $\left(\frac{\partial P}{\partial V}\right)_T = 0$, and a point of inflection where $\left(\frac{\partial P}{\partial V}\right)_T$ is at a maximum and $\left(\frac{\partial^2 P}{\partial V^2}\right)_T = 0$.

At the critical point, all of these conditions exist, where $T = T_c$, $P = P_c$ and $V = V_c$.

If $P = \dfrac{RT}{(V-b)} - \dfrac{a}{V^2}$, $\left(\dfrac{\partial P}{\partial V}\right)_T = -\dfrac{RT}{(V-b)^2} + \dfrac{2a}{V^3}$, and $\left(\dfrac{\partial^2 P}{\partial V^2}\right)_T = \dfrac{2RT}{(V-b)^3} - \dfrac{6a}{V^4}$.

At the critical point, both of these derivatives are zero, giving
$\frac{R T_c}{(V_c - b)^2} = \frac{2 a}{V_c^3}$ and $\frac{2 R T_c}{(V_c - b)^3} = \frac{6 a}{V_c^4}$. If $\frac{6 a}{V_c^4} = \frac{3}{V_c}\left(\frac{2 a}{V_c^3}\right)$, then $\frac{2 R T_c}{(V_c - b)^3} = \frac{3 R T_c}{V_c (V_c - b)^2}$, which gives $V_c = 3 b$.

Putting that result into $\frac{2 a}{V_c^3} = \frac{R T_c}{(V_c - b)^2}$ gives $\frac{a}{b} = \frac{27}{8} R T_c$.

Putting all this into $P_c = \frac{R T}{(V - b)} - \frac{a}{V^2}$ gives $P_c = \frac{R T_c}{8 b} = \frac{3 R T_c}{8 V_c}$.

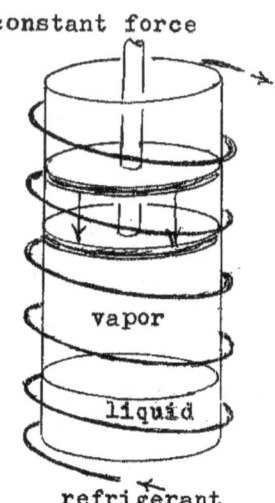

constant force

vapor

liquid

refrigerant

In the process of liquefying a vapor, it would be assumed that one would use a piston in a cylinder to compress the vapor into a liquid at constant pressure as a refrigerant is passed through cooling coils wrapped around the cylinder to remove heat to keep the temperature below that of any convenient coolant. Suppose we want to liquefy nitrogen at 77.4° K which is −195.8° C. We will discuss how to do that with a throttling process.

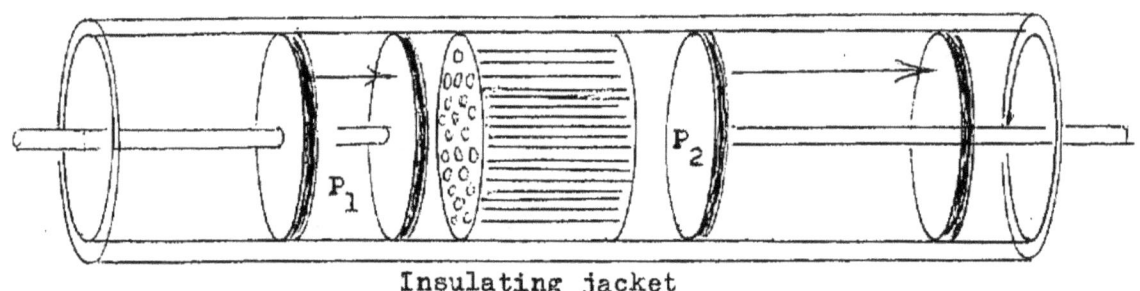

Insulating jacket

Two pistons move in two cylinders that are separated by a porous plug. The whole assembly is surrounded by an insulating jacket so that there is no heat transfer. Maintaining a constant pressure, P_1, the left-hand piston moves in pushing the gas through the porous plug, while the right hand piston moves out also maintaining a constant pressure, P_2. The volume of the gas in the left-hand cylinder goes from V_1 to zero and the volume in the right hand cylinder goes from zero to V_2.

The work done by the right-hand piston is $+P_2 V_2$ while $-P_1 V_1$ is the work done on the left-hand piston. The net work done on the external environment is $P_2 V_2 - P_1 V_1$. With no heat transfer, $dQ = dU + P\, dV = 0$.

Hence, $U_2 - U_1 + P_2 V_2 - P_1 V_1 = 0$. Notice that $U_2 + P_2 V_2 = U_1 + P_1 V_1$ and the quantity, $U + PV$, is conserved here. This quantity is sufficiently important in thermodynamics that we give it a symbol and a name.

Define ENTHALPY as $H = U + PV$.

We are interested in the change of temperature of the gas undergoing this process, being the change of temperature with pressure at constant enthalpy. This given by the JOULE-THOMSON coefficient, $\mu_{JT} = \left(\frac{\partial T}{\partial P}\right)_H$.

For using this process to cool and liquefy a gas, we are interested in the conditions where temperature decreases with decreasing pressure, where μ_{JT} is positive. But before we can derive a formula for μ_{JT}, say from the van der Waals equation, it is necessary that we have some results of formal thermodynamic theory.

For a reversible process, $dQ = T\,dS$. This gives $C_v = T\left(\frac{\partial S}{\partial T}\right)_V$ and $C_p = T\left(\frac{\partial S}{\partial T}\right)_P$.

A change in internal energy is $dU = T\,dS - P\,dV$.
If enthalpy is $H = U + PV$, then $dH = dU + P\,dV + V\,dP$ or $dH = T\,dS + V\,dP$.

Obviously, $C_v = \left(\frac{\partial U}{\partial T}\right)_V$ and $C_p = \left(\frac{\partial H}{\partial T}\right)_P$.

Enthalpy is heat transfer at constant pressure, which makes it of interest to the chemist as many of his processes take place at constant pressure. Now recall that entropy is derived from the partition function,
$S = N\,k\,\ln Z + \frac{U}{T}$.

This makes $N\,k\,T\,\ln Z = T\,S - U$.

The HELMHOLTZ FREE ENERGY is defined as $F = U - TS$ giving $F = -N\,k\,T\,\ln Z$.

Then $dF = dU - T\,dS - S\,dT$, but $dU = T\,dS - P\,dV$, so $dF = -S\,dT - P\,dV$.

It is evident that, at constant temperature, dF is the negative of work done by the system, or equals the work done on the system. F remains constant for processes happening at constant temperature and volume.

Another classical thermodynamic function is the GIBBS FREE ENERGY,
$G = H - TS$.

$dG = dH - T\,dS - S\,dT$. But $dH = T\,dS + V\,dP$. Thus, $dG = -S\,dT + V\,dP$.

In discussing the van der Waals equation, it was observed that a liquid-vapor phase transition for a single substance or component took place at constant temperature and pressure. While the two phases are present, the Gibbs free energy is constant during the phase transition. For multi-component systems things are not so simple. Never the less, the Gibbs free energy has a major role in studying phase transitions.

We just now saw that the major link between the partition function, $Z = \sum_i e^{-\epsilon_i/kT}$, and thermodynamics is going to be the Helmholtz free energy, $F = -N\,kT\,\ln Z$.

If $dF = -S\,dT - P\,dV$, then as $dF = \left(\frac{\partial F}{\partial T}\right)_V dT + \left(\frac{\partial F}{\partial V}\right)_T dV$, we get

$S = -\left(\frac{\partial F}{\partial T}\right)_V = \left(\frac{\partial}{\partial T} N\,kT\,\ln Z\right)_V$ and $P = \left(\frac{\partial}{\partial V} N\,kT\,\ln Z\right)_T$.

If F and its first derivatives are continuous functions of T and V, then the second derivatives are independent of the order, $\frac{\partial}{\partial T}\left(\frac{\partial F}{\partial V}\right) = \frac{\partial}{\partial V}\left(\frac{\partial F}{\partial T}\right)$, keeping each variable constant while taking the derivative with the other.

Now, $\left(\frac{\partial S}{\partial V}\right)_T = \frac{\partial}{\partial V}\left(-\frac{\partial F}{\partial T}\right) = \frac{\partial}{\partial T}\left(-\frac{\partial F}{\partial V}\right) = \left(\frac{\partial P}{\partial T}\right)_V$.

Thus, $dF = -S\,dT - P\,dV$ gives $\left(\frac{\partial S}{\partial V}\right)_T = \left(\frac{\partial P}{\partial T}\right)_V$.

Then $dG = -S\,dT + V\,dP$ gives $\left(\frac{\partial S}{\partial P}\right)_T = -\left(\frac{\partial V}{\partial T}\right)_P$

and $dU = T\,dS - P\,dV$ gives $\left(\frac{\partial T}{\partial V}\right)_S = -\left(\frac{\partial P}{\partial S}\right)_P$

and $dH = T\,dS + V\,dP$ gives $\left(\frac{\partial T}{\partial P}\right)_S = \left(\frac{\partial V}{\partial S}\right)_P$.

These equations were given by James Clerk Maxwell in his <u>Theory of Heat</u> (1871). They are very important in the development of thermodynamic theory. First, they allow us to develop the two important $T\,dS$ equations.
Recall that $C_v = T\left(\frac{\partial S}{\partial T}\right)_V$ and $C_p = T\left(\frac{\partial S}{\partial T}\right)_P$.

If $S = S(T,V)$, then $dS = \left(\frac{\partial S}{\partial T}\right)_V dT + \left(\frac{\partial S}{\partial V}\right)_T dV$. But $\left(\frac{\partial S}{\partial V}\right)_T = \left(\frac{\partial P}{\partial T}\right)_V$.

Thus, $T\,dS = C_v\,dT + T\left(\frac{\partial P}{\partial T}\right)_V dV$.

If $S = S(T,P)$, then $dS = \left(\frac{\partial S}{\partial T}\right)_P dT + \left(\frac{\partial S}{\partial P}\right)_T dV$. But $\left(\frac{\partial S}{\partial P}\right)_T = -\left(\frac{\partial V}{\partial T}\right)_P$.

Thus, $T\,dS = C_p\,dT + T\left(\frac{\partial V}{\partial T}\right)_P dV$.

These two $T\,dS$ equations allow us to treat heat transfer in terms of the heat capacities and the equation of state.

Our first use of the second of the $T\,dS$ equations will be to discuss the Joule-Thomson (or Joule-Kelvin, as William Thomson was raised to the peerage as Lord Kelvin) coefficient.

$\mu_{JT} = \left(\frac{\partial T}{\partial P}\right)_H$, where $dH = T\,dS + V\,dP$. Now, $T\,dS = C_p\,dT - T\left(\frac{\partial V}{\partial T}\right)_P dP$.

Hence, $dH = C_p\,dT - \left[T\left(\frac{\partial V}{\partial T}\right)_P - V\right]dP$ but with $dH = 0$, $dT = \left[T\left(\frac{\partial V}{\partial T}\right)_P - V\right]dP$.

At constant H, $\mu_{JT} = \left(\frac{\partial T}{\partial P}\right)_H = \frac{1}{C_p}\left[T\left(\frac{\partial V}{\partial T}\right)_P - V\right]$.

For an ideal gas, $V = \frac{RT}{P}$ and $\mu_{JT} = \frac{1}{C_p}\left[T\left(\frac{\partial V}{\partial T}\right)_P - V\right] = 0$, which is no surprise.

For a van der Waals gas, $P = \frac{RT}{V-b} - \frac{a}{v^2}$ and $\left(\frac{\partial V}{\partial T}\right)_P$ is not easy to do directly.

But if $P = P(T,V)$, then $dP = \left(\frac{\partial P}{\partial T}\right)_V dT + \left(\frac{\partial P}{\partial V}\right)_T dV$ and if P is constant, $dP = 0$,

giving $\left(\frac{\partial V}{\partial T}\right)_P - \frac{\left(\frac{\partial P}{\partial T}\right)_V}{\left(\frac{\partial P}{\partial V}\right)_T}$.

EXERCISE: Show that for a van der Waals gas, $\mu_{JT} = \frac{1}{C_p}\left[\frac{RT(V-b)}{RT - \frac{2a(V-b)^2}{v^3}} - V\right]$.

$\mu_{JT} = \left(\dfrac{\partial T}{\partial P}\right)_H$ must be positive if temperature is to decrease with decreasing pressure in a throttling process. Show that $\mu_{JT} \geq 0$ requires the condition:

$$-RTb \geq -2a\left(\frac{b}{v}\right)^2$$

If $\dfrac{1}{V} = \dfrac{1}{RT}$, approximately

$$-RTb \geq -2a + \frac{4abP}{RT} - \frac{2ab P^2}{R^2 T^2}$$

If the last term is dropped this gives $\mu_{JT} = 0$ when

$P \leq \dfrac{RT}{2b} - \dfrac{R^2 T^2}{4a}$. This means

that $\left(\dfrac{\partial T}{\partial P}\right)_H \geq 0$, the condition for cooling will exist for those temperatures and pressures to the left of the

curve, $P = \dfrac{RT}{2b} - \dfrac{R^2 T^2}{4a}$.

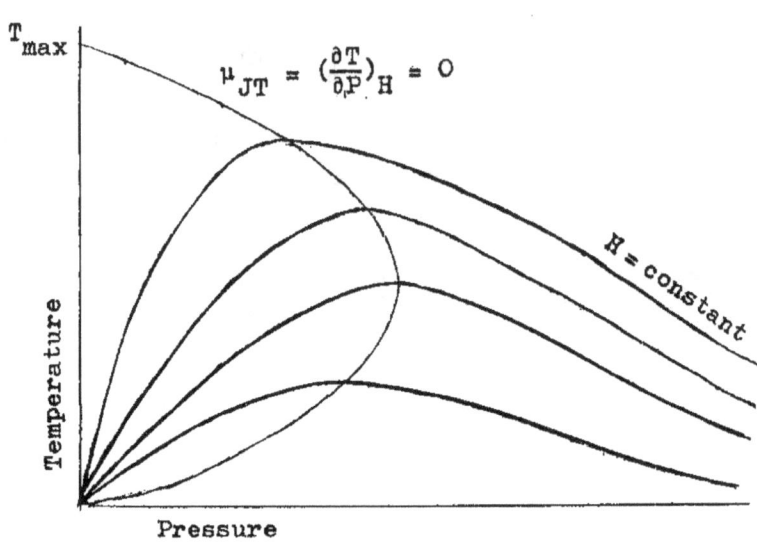

This is called the INVERSION CURVE. Where $P = 0$ at the top of this curve is the highest temperature at which cooling can be started by throttling a van der Waals gas. We get $T_{max} = \dfrac{2a}{Rb}$. On page 20-24, $\dfrac{a}{b} = \dfrac{27}{8} R T_c$, T_c being the critical temperature.

EXERCISE: The van der Waals equation has given us a model explaining the cooling of a gas by passing it through a porous plug in hopes that recycling it through the process will cool it to the point of liquification. But how good a guide is the van der Waals equation when predicting numbers? The maximum inversion temperature is the highest temperature at which cooling can be started by throttling the gas. The critical temperature is the highest temperature that the liquid state can exits. For several gases, T_c and T_{max} are given from measurements for several gases. Check their agreement according to the predictions from the van der Waals equation.

Gas	T_c	T_{max}
Helium	5.3 K	40 K
Hydrogen	33.3	202
Nitrogen	126.1	621
Argon	151.2	723

For an ideal gas, internal energy depends only upon temperature, but for a non-ideal gas as one described by the van der Waals equations, the internal energy may depend upon other variables of the state of the system.

Now, $dU = T\,dS - P\,dV$, but on page 20-27, $T\,dS = C_v\,dT + T\left(\frac{\partial P}{\partial T}\right)_V dV$.

Thus, $dU = C_v\,dT + \left[T\left(\frac{\partial P}{\partial T}\right)_V - P\right]dV$.

For an ideal gas, $P = \frac{RT}{V}$, $\left(\frac{\partial P}{\partial T}\right)_V = \frac{R}{V}$, and $dU = C_v\,dT$.

From the van der Waal equation, $P = \frac{RT}{V-b} - \frac{a}{v^2}$ and $\left(\frac{\partial P}{\partial T}\right)_V = \frac{R}{V-b}$.

Then, $dU = C_v\,dT + \left[\frac{RT}{V-b} - \frac{RT}{V-b} + \frac{a}{v^2}\right]dV = C_v\,dT + \frac{a}{v^2}\,dV$,

so $U = \int C_v\,dT - \frac{a}{v} + constant$ is the ENERGY EQUATION for a van der Waals gas.

For a single component system, the change of phase from liquid to vapor takes place at constant temperature and pressure.

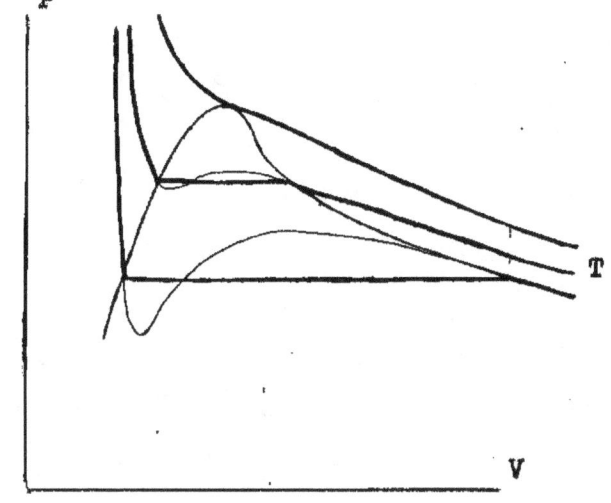

From $dG = -S\,dT + V\,dP$, it is obvious the Gibbs free energy remains constant during the phase transition.

From $dH = T\,dS + V\,dP$, it is obvious that the heat of vaporization is given by the change in enthalpy.

We define the LATENT HEAT OF VAPORIZATION:

$L = H_{vap} - H_{liq}$ and $dL = dH = T\,dS$ at constant P. From page 20-26,

$T\,dS = C_v\,dT + T\left(\frac{\partial P}{\partial T}\right)_V dV$, but $dT = 0$ here.

In this case, T is the boiling point temperature at a pressure, P. $\left(\frac{\partial T}{\partial P}\right)_V$ would be the rate of change of boiling point temperature with pressure going up or down between horizontal isotherms in the phase change region.

With T and P constant, $dL = T\left(\frac{\partial P}{\partial T}\right)_V dV$ and $L = T\left(\frac{\partial P}{\partial T}\right)_V \left(V_{vap} - V_{liq}\right)$.

If at boiling temperature, T, V_{vap} and V_{liq} and L are measurable,

$\left(\frac{\partial P}{\partial T}\right)_V = \frac{L}{T\left(V_{vap}-V_{liq}\right)}$, which is the CLAPEYRON EQUATION.

At the critical point, $V_{vap} = V_{liq}$ and $L = 0$.

For further reading on the topics just covered:
M.W. Zemansky, Heat and Thermodynamics, 5th edition, McGraw Hill (1968)

For the next topics refer to:
E.A. Guggenheim, THERMODYNAMCS: An Advanced Treatment for Chemists and
                  Physicists, North Holland Publishing Co. (1950)
E.A. Guggenheim, Mixtures, Oxford Univ. Press (1952)
Alexander Findlay, The Phase Rule, 9th edition, Dover (1951).

The methods here used to describe phenomena in multi-component systems with multi-phases are greatly influenced by the works of E. A. Guggenheim. The treatment of the chemical potential, μ_i, will involve a significant departure from this and the standard convention where it has the units of energy per mole. Because interactions at the molecular level play such an important part in this development we shall give chemical potential the units of energy per molecule. Multiply our chemical potential by Avogadro´s number to get what appears in the standard treatment.

For a multicomponent system, for the ith component, let N_i be the number of molecules and n_i the number of moles. Then, $N_i = n_i N_A$, where N_A is Avogadro´s number. Obviously, $\sum_i N_i = N$, the total number of molecules. For the ith component, we define the MOLE FRACTION, $X_i = \frac{N_i}{N}$; $\sum_i X_i = 1$.

For an ideal gas, $P V = N k T = \sum_i N_i k T$. If $p_i = \frac{N_i k T}{V}$ is the PARTIAL PRESSURE of the ith component, then $P = \sum_i \frac{N_i k T}{V} = \sum_i p_i$ and $p_i = X_i P$.

At this point we are talking about a multicomponent system where only the gaseous or vapor phases are present.

With a fixed number of molecules of one component, $dU = T dS - P dV$.

If there are several components, changes will involve not only heat transfer and work but also variations of the chemical composition. If, for example, we removed nitrogen molecules and added helium molecules there would be a loss of internal energy as nitrogen molecules rotate and vibrate but helium atoms don´t. To account for this, we introduce the CHEMICAL POTENTIAL, μ_i, for each type of molecule in the mixture.

Now, $dU = T\,dS - P\,dV + \sum_i \mu_i\,dN_i$, where $\mu_i = \left(\dfrac{\partial U}{\partial N_i}\right)_{S,V,N_{j\neq i}}$.

If $H = U + PV$, then $dH = T\,dS + P\,dV + \sum_i \mu_i\,dN_i$.

If $F = U - TS$, then $dF = -S\,dT - P\,dV + \sum_i \mu_i\,dN_i$.

If $G = H - TS$, then $dG = -S\,dT + V\,dP + \sum_i \mu_i\,dN_i$.

At this point notice that there are two types of quantities in these equations; U, S, V and $N_i = X_i N$ depend on the size or extent of the system and are called EXTENSIVE VARIABLES. On the other hand, T, P and μ_i describe the quality or nature of the system and are called INTENSIVE VARIABLES.

Now notice that in $dU = T\,dS - P\,dV + \sum_i \mu_i\,dN_i$, the changing variables all depend on the extent of the system. Let λ be a variable giving the extent or size of the system. Let U_1, S_1, V_1 and N_{i1} be the values of these quantities when $\lambda = 1$. Now vary λ giving a set of comparable systems having identical intensive variables; for which notice that $T = \left(\dfrac{\partial U}{\partial S}\right)_{V,\,N_i}$ $-P = \left(\dfrac{\partial U}{\partial V}\right)_{S,\,N_i}$ and $\mu_i = \left(\dfrac{\partial U}{\partial N_i}\right)_{S,V,N_{j\neq i}}$.

According to the way λ, U_1, S_1, V_1 and N_{i1} were defined, $U = \lambda U_1$, $S = \lambda S_1$, $V = \lambda V_1$ and $N_i = \lambda N_{i1}$ with $\dfrac{\partial U}{\partial \lambda} = U_1$, $\dfrac{\partial S}{\partial \lambda} = S_1$, etc.

If $U = U(S,V,N_i)$, then $\dfrac{\partial U}{\partial \lambda} = \dfrac{\partial U}{\partial S}\dfrac{\partial S}{\partial \lambda} + \dfrac{\partial U}{\partial V}\dfrac{\partial V}{\partial \lambda} + \sum_i \dfrac{\partial U}{\partial N_i}\dfrac{\partial N_i}{\partial \lambda}$,

from which $U_1 = T\,dS_1 - P\,V_1 + \sum_i \mu_i\,N_{i1}$, which when being multiplied by λ, gives $U = T\,S - P\,V + \sum_i \mu_i\,N_{i1}$.

But, $H = U + PV$, so $H = T\,S + \sum_i \mu_i\,N_i$.

But, $F = U - TS$, so $F = -P\,V + \sum_i \mu_i\,N_i$.

But, $G = H - TS$, so $G = \sum_i \mu_i\,N_i$.

On the top of this page, $dG = -S\,dT + V\,dP + \sum_i \mu_i\,dN_i$, but now, $dG = \sum_i \mu_i\,dN_i + \sum_i N_i\,d\mu_i$, which when combined, gives $S\,dT - V\,dP + \sum_i N_i\,d\mu_i = 0$, which is called the GIBBS-DUHEM equation.

Starting with what we said about U, H, F and G was general thermodynamic theory and did not depend on the system being in the gaseous phase. The theory then could apply to matter in any phase (gas, liquid, or solid) or to matter co-existing in several phases. We now take a closed system of several types of molecules existing in several phases. If the system is in equilibrium, temperature and pressure are uniform throughout the system; hence, in going from one phase to another, $dT = 0$ and $dP = 0$. Let i denote a chemical component and φ denote a phase. Each extensive variable will have a value in each phase. S^φ is the entropy of the φth phase. Likewise, V^φ and N_i^φ. With units of energy per molecule we might expect μ_i^φ to have a different value in each phase and would change as molecules go from one phase to another.

The total value of any extensive variable would be the sum of its values in each of the coexisting phases. Thus, $S = \sum_\varphi S^\varphi$, $V = \sum_\varphi V^\varphi$ etc.

For multicomponents in multiphases, the Gibbs-Duhem equation becomes
$$\sum_\varphi S^\varphi \, dT - \sum_\varphi V^\varphi \, dP + \sum_i \left[\sum_\varphi N_i^\varphi \, d\mu_i^\varphi \right] = 0 \,.$$

But if the system is in equilibrium, $dT = 0$ and $dP = 0$ in going from one phase to another. If the result is to hold for any mixture of components then for each component, $\sum_\varphi N_i^\varphi \, d\mu_i^\varphi = 0$.

But N_i^φ can have any value in any phase with the result that $d\mu_i^\varphi = 0$ as a molecule goes from one phase to another.

As a result, for a closed system of several components in equilibrium in several phases, the chemical potential for each component has the same value in each and all of the phases.

We will now use this result to study the liquid-vapor phase transition for a system with several components. With the liquid and vapor at equilibrium in a closed system, let us study the vapor phase first. The Gibbs-Duhem equation is $S \, dT - V \, dP + \sum_i N_i \, d\mu_i = 0$.

At equilibrium $dT = 0$, but although $dP = 0$ we recall $P = \sum_i p_i$, where we shall assume that the ideal gas law, $p_i V = N_i k T$, holds approximately. Here, V is the volume of the vapor phase.

If we write the Gibbs-Duhem equation, $\sum_i N_i \, d\mu_i - \sum_i V \, dp_i = 0$, and let $V = \dfrac{N_i kT}{p_i}$,

we get $\sum_i N_i \left[d\mu_i - \dfrac{kT}{p_i} dp_i \right] = 0$.

Thus, for each component in the vapor phase, $d\mu_i = \dfrac{kT}{p_i} dp_i$.

Now let $\mu_i{}^o$ and $p_i{}^o$ be the chemical potential and the vapor pressure of pure i^{th} component alone. When the components are mixed, their chemical potentials will change, and their pressures become partial pressures. Integrating the changes $d\mu_i$ gives $\mu_i - \mu_i{}^o = kT \, ln \, \dfrac{p_i}{p_i{}^o}$.

If we can determine the chemical potentials in the liquid phase, then if the chemical potentials are the same in both phases, we can determine the partial pressures in the vapor phase. This will lead to comparing mole fractions in the two phases and give a complete picture of the process of the phase transition. We will analyze a system with two components.

In the liquid phase, let each molecule be surrounded, on the average by Z nearest neighbors. For a two-component system let there be molecules of type A and of type B. Let their interactions extend only to nearest neighbors. If a molecule of type A interacts with another molecule of type A, let their energy (attractive) be $-2\,\epsilon_A$, and assign the energy, $-\epsilon_A$, to each molecule.

Let two molecules of type B interact with their attractive energy being $-2\epsilon_B$ for the pair and assign $-\epsilon_B$ to each molecule.

When a type A molecule is paired with a type B, let the pair have an energy, $-\epsilon_A - \epsilon_B + w$, and give type A the energy, $-\epsilon_A + w/2$, and give to type B the energy, $-\epsilon_B + w/2$, giving half the perturbation energy to each.

Of the Z molecules surrounding any molecule, $\dfrac{N_A Z}{N_A + N_B}$ are of type A and $\dfrac{N_B Z}{N_A + N_B}$ are of type B, on the average.

The average energy of an A molecule is $\dfrac{N_A Z}{N_A + N_B}(-\epsilon_A) + \dfrac{N_B Z}{N_A + N_B}(-\epsilon_A + w/2)$.

The average energy of a B molecule is $\dfrac{N_B Z}{N_A + N_B}(-\epsilon_B) + \dfrac{N_A Z}{N_A + N_B}(-\epsilon_B + w/2)$.

The total internal energy of N_A and N_B molecules in the liquid phase is

$$-\frac{N_A{}^2}{N_A+N_B} Z \epsilon_A - \frac{N_B{}^2}{N_A+N_B} Z \epsilon_B + \frac{N_A N_B}{N_A+N_B} Z(-\epsilon_A - \epsilon_B + w) \quad \text{or} \quad U = -N_A Z \epsilon_A - N_B Z \epsilon_B + \frac{N_A N_B}{N_A+N_B} Z w.$$

From page 20-8, the entropy is $S = -N_A k \ln \dfrac{N_A}{N_A+N_B} - N_B k \ln \dfrac{N_B}{N_A+N_B}$.

The Helmholtz free energy, $F = U - TS$, for the liquid phase is

$$F = -N_A Z \epsilon_A - N_B Z \epsilon_B + \frac{N_A N_B}{N_A+N_B} Z w + N_A k T \ln \frac{N_A}{N_A+N_B} + N_B k T \ln \frac{N_B}{N_A+N_B}.$$

If $dF = -S\, dT - P\, dV + \mu_A\, dN_A + \mu_B\, dN_B$, then $\mu_A = \left(\dfrac{\partial F}{\partial N_A}\right)_{T,V,N_B}$ and $\mu_B = \left(\dfrac{\partial F}{\partial N_B}\right)_{T,V,N_A}$.

EXERCISE: Show that for component A, $\mu_A = -Z \epsilon_A + \dfrac{N_B{}^2 Z w}{(N_A+N_B)^2} + k T \ln \dfrac{N_A}{N_A+N_B}$;

and for component B, $\mu_B = -Z \epsilon_B + \dfrac{N_A{}^2 Z w}{(N_A+N_B)^2} + k T \ln \dfrac{N_B}{N_A+N_B}$.

If $\ln 1 = 0$, show that for pure A in the liquid phase, $\mu_A{}^o = -Z \epsilon_A$, and for pure B in the liquid phase, $\mu_B{}^o = -Z \epsilon_B$.

Going back to the vapor phase, $k T \ln \dfrac{p_A}{p_A{}^o} = \mu_A - \mu_A{}^o$.

If the chemical potentials are the same in both phases,

$$k T \ln \frac{p_A}{p_A{}^o} = k T \ln \frac{N_A}{N_A+N_B} + \left[\frac{N_A}{N_A+N_B}\right]^2 Z w.$$

Let $X_A{}^L = \dfrac{N_A}{N_A+N_B}$ be the mole fraction of A in the liquid phase.

Then, $X_B{}^L = \dfrac{N_B}{N_A+N_B} = 1 - X_A{}^L$.

Then, $\ln \dfrac{p_A}{p_A{}^o} = \ln X_A{}^L + \left(1 - X_A{}^L\right)^2 \dfrac{Z w}{k T}$

or $p_A = X_A{}^L p_A{}^o \exp\left[\left(1 - X_A{}^L\right)^2 \dfrac{Z w}{k T}\right]$ and $p_B = X_B{}^L p_B{}^o \exp\left[\left(1 - X_B{}^L\right)^2 \dfrac{Z w}{k T}\right]$.

If the two components are sufficiently similar that $w = 0$, this gives RAOULT'S LAW, $p_A = X_A{}^L p_A{}^o$ and $p_B = X_B{}^L p_B{}^o$, in which you will notice that the vapor pressures in the vapor phase are given by the mole fractions in the liquid phase.

An example of a system obeying Raoult´s law is a mixture of benzene and ethylene chloride. In the graph, pressure is given in the height of a column of mercury with a vacuum above it that would be supported by the pressure. The pressure of a standard atmosphere is:

One atmosphere $= 760$ mm.Hg.
$= 1.013 \times 10^5 \, \text{N/m}^2$

The left-hand axis are data for pure ethylene chloride. The diagonal lines are the partial pressures for ethylene chloride (lower left to upper right) and benzene (upper left to lower right), and the line at the top is the total pressure of the mixtures specified by the mole fraction of ethylene chlorine. All data are at a temperature of 50° C.

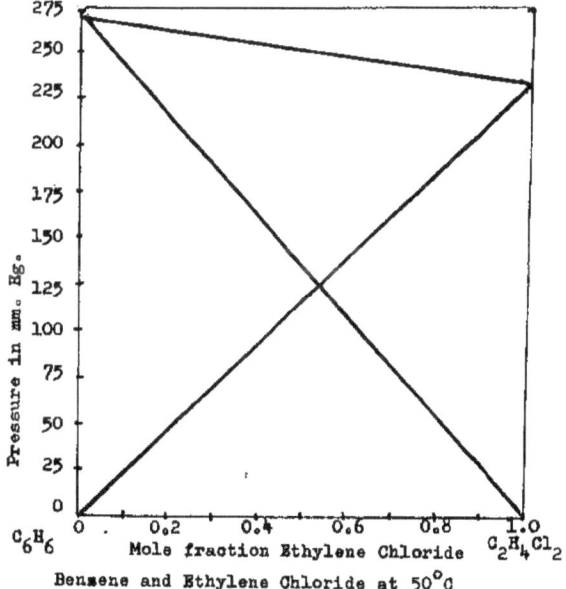

Benzene and Ethylene Chloride at 50°C

Because of their similar molecular structure, we should expect CCl_4 and $SnCl_4$ to obey Raoult´s law. The two curves in the top figure give saturated vapor pressure for the pure vapor and liquid in equilibrium for each substance. To get the Raoult´s law behavior for any mixture take a temperature and find the vapor pressure for the pure materials.

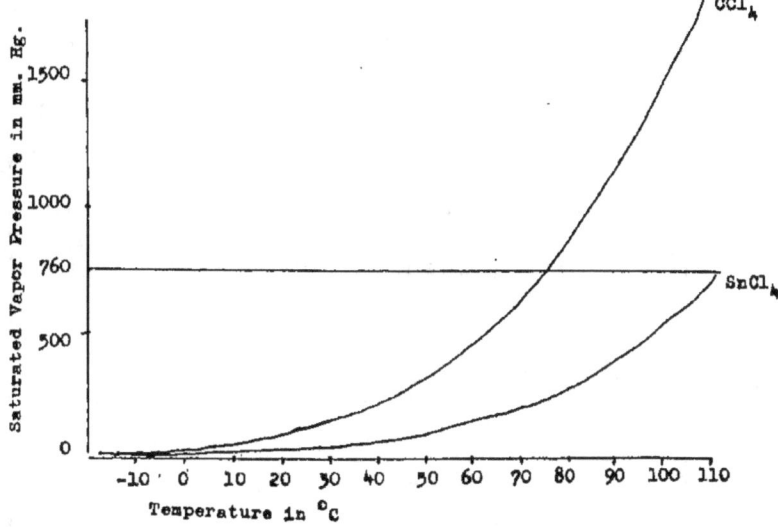

Now plot vapor pressure against liquid mole fraction of $SnCl_4$.

Locate the pressure of pure CCl_4 on the left axis and of pure $SnCl_4$ on the right axis. If Raoult´s law holds, the vapor pressure of any mixture should lie on a straight line between these two points.

With the data in the previous graph we can do this at a number of temperatures. For any temperature, the liquid mole fraction with an equilibrium vapor pressure of **760** mm.Hg. should boil at a standard atmospheric pressure of **760** mm.Hg. This gives the boiling point temperature at standard atmospheric pressure. (This can be done for any pressure.)

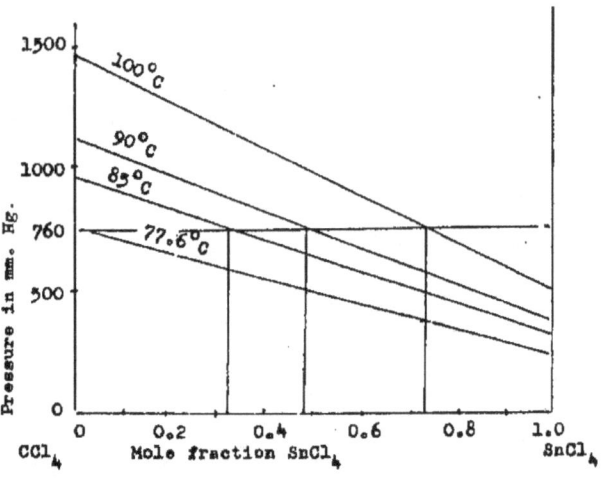

Next, we can plot the boiling point temperatures at one atmosphere against liquid mole fraction SnCl$_4$. That will give the curve labeled "liquid" in the next figure.

Now we raise the question about the composition of the vapor coming off the liquid at any temperature. Let "A" denote CCl$_4$ and "B" denote SnCl$_4$. By Raoult's law $p_A = X_A{}^L p_A{}^o$ and $p_B = X_B{}^L p_B{}^o$

Then the mole fraction of "B" in the vapor phase would be $X_B{}^V = \dfrac{p_B}{p_A + p_B}$.

Let $p_A{}^o = p_B{}^o + \delta p$. Then,

$p_B + p_A = X_B{}^L p_B{}^o + X_A{}^L p_B{}^o + X_A{}^L \delta p$

and $X_B{}^V = \dfrac{X_B{}^L}{1 + \left(X_A{}^L \delta p / p_A{}^o\right)}$.

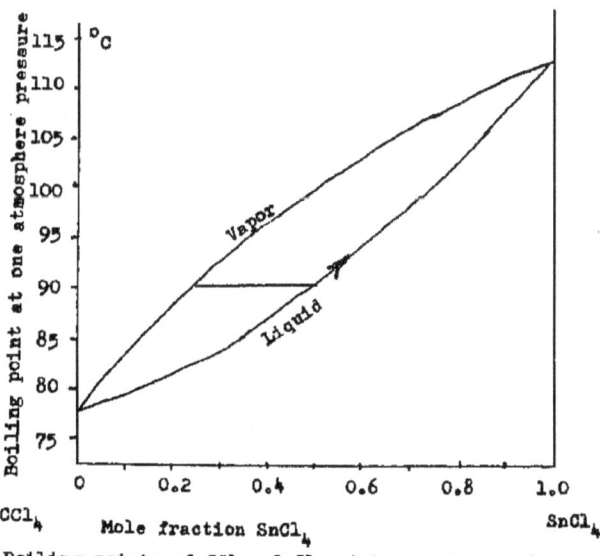

Boiling points of CCl$_4$ - SnCl$_4$ mixtures at one atmosphere pressure assuming Raoult's law behavior

In this case, where "A" is CCl$_4$ and "B" is SnCl$_4$, $\delta p = p_A{}^o - p_B{}^o$ is positive as at any temperature CCl$_4$ is more volatile than SnCl$_4$. Thus, for any liquid mole fraction, $X_B{}^L$, of SnCl$_4$, the vapor mole fraction, $X_B{}^V = \dfrac{X_B{}^L}{1 + \left(X_A{}^L \delta p / p_A{}^o\right)}$, giving the vapor to be poorer in SnCl$_4$ the less volatile component and richer in CCl$_4$ the more volatile component. Thus, the upper curve gives the composition of the vapor in equilibrium with the liquid given by the lower curve. At **90°** the liquid is **0.5** SnCl$_4$ and the vapor is **0.23** SnCl$_4$.

At one atmosphere, starting here, the boiling point temperature will rise as the liquid becomes more concentrated in $SnCl_4$.

For non-ideal, or regular, solutions, if the energy of two A molecules is $-2\,\epsilon_A$ and of two B molecules, $-2\,\epsilon_B$; but of a pair of A with B the energy was $-\epsilon_A - \epsilon_B + w$. Then,

$$p_A = X_A{}^L\, p_A{}^o\, exp\left[\left(1 - X_A{}^L\right)^2 \frac{Z\,w}{k\,T}\right]$$ and
$$p_B = X_B{}^L\, p_B{}^o\, exp\left[\left(1 - X_B{}^L\right)^2 \frac{Z\,w}{k\,T}\right].$$

If w is negative, each molecule is more strongly attracted to its opposite than to its own kind. The quantity in the bracket is negative and the vapor pressures are less than Raoult´s law behavior. As opposite types are more strongly attracted than same types, mixtures are less volatile than the pure substances.

When w is positive, each molecule is less strongly attracted to it opposite than to its own kind, acting as a slight repulsion. We would now expect the mixture to be more volatile than the pure substances and that turns out to be the case. When w is positive, the quantities in the brackets are positive and the vapor pressures are greater than the Raoult´s law behavior.

For regular solutions where $w \neq 0$ notice that when either component is dilute, X is very small and $(1 - X)$ is approximately unity and its behavior is close to Raoult´s law. The richer of the two components in the mixture producing most of the deviation.

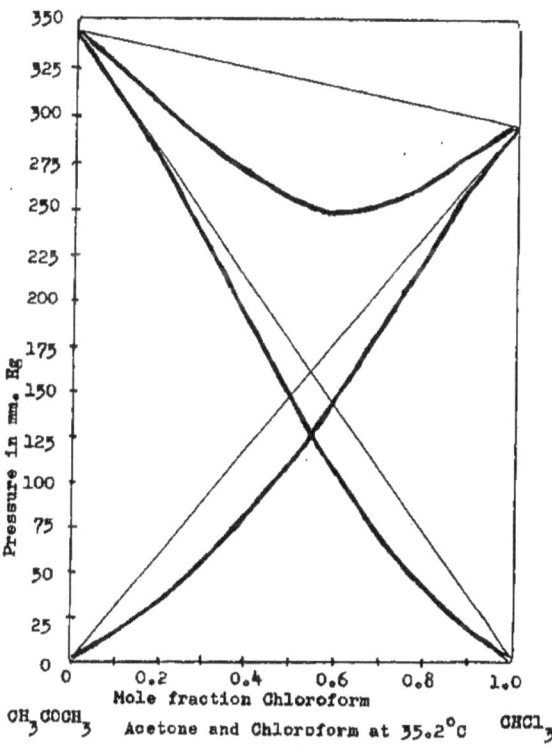

Acetone and Chloroform at 35.2°C

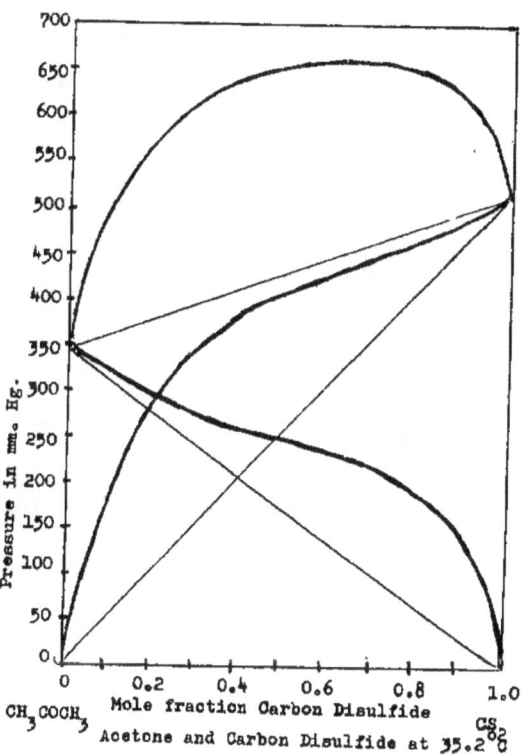

Acetone and Carbon Disulfide at 35.2°C

Next, we shall investigate the liquid and vapor compositions at the boiling points of various mixtures of two components when w is positive. For components "A" and "B" vapor pressure is plotted against liquid mole fractions of B for several temperatures.

At one atmosphere pressure, pure A has a boiling temperature of T_A^o and pure B boils at T_B^o. Vapor pressure curves are given at four other temperatures. Notice that at each temperature T_4, T_3, and T_2, only one mixture of A and B will boil at one atmosphere. At temperatures below T_B^o, as for example T_1, there are two mixtures of A and B that boils at one atmosphere.

If we plot the boiling point temperatures at one atmosphere against mole fraction of B we get the curve labeled "liquid" on graph on the right. At each temperature, the composition of the vapor in equilibrium with the liquid is given by the curve labeled "vapor".

At a minimum in the curves, the vapor and liquid have the same composition. This constant boiling point mixture is called an azeotrope and acts here as a pure substance. To the right of the azeotropic mixture, the vapor is richer than the liquid in A and the liquid approaches pure B as boiling continues. To the left of the azeotropic mixture, the mixture becomes richer in B and the liquid approaches A as boiling continues. An example of this behavior is the water-propyl alcohol mixture on the right.

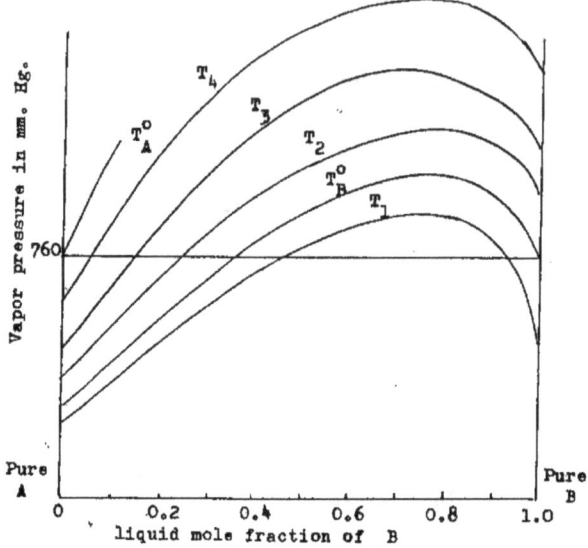

Vapor pressure of mixtures of "A" and "B" at temperatures T_1, T_B^o, T_2, T_3, T_4, T_A^o

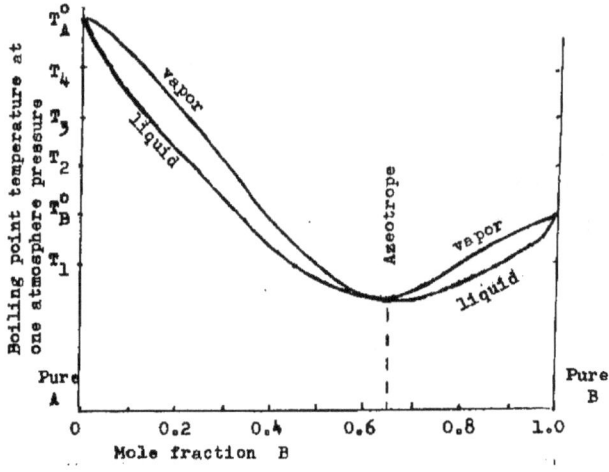

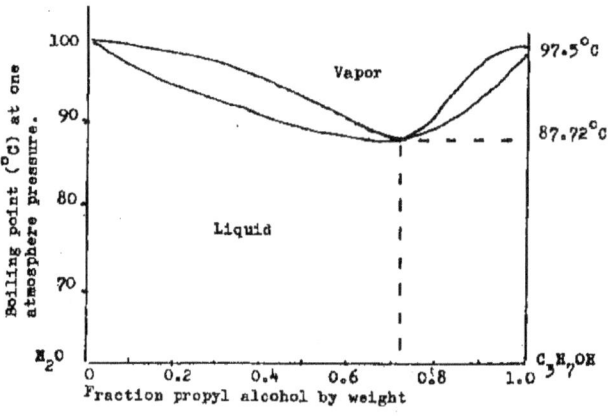

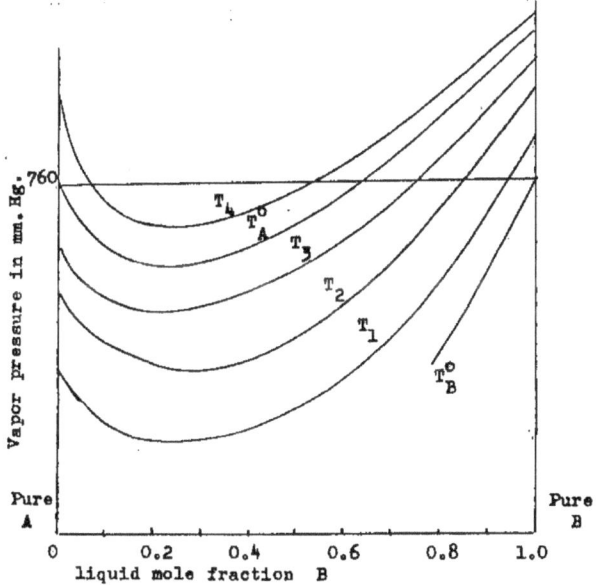

EXERCISE: Go through the same analysis in the case where *w* is negative, shown in the graph on the left. Where is the azeotropic mixture? Again, the vapor composition curves lie above the liquid. To either side of the azeotrope, what happens to the liquid composition as boiling continues?

The behavior of non-reacting components in several phases is generalized by the GIBBS PHASE RULE. If several components of a closed system exist in several phases we may assume that because of the structural differences between phases, we can identify the extent of the properties and variables of each phase. Let components be denoted by the subscript, *i*, and let the phases be denoted by the superscript, α.

For an isolated system in several phases coexisting in equilibrium:

$$\sum_{\alpha} U^{\alpha} = constant \qquad \sum_{\alpha} V^{\alpha} = constant \qquad \sum_{\alpha} N_i^{\alpha} = constant$$

the last condition holding for each component for a non-reacting system. Hence,

$$\sum_{\alpha} dU^{\alpha} = 0 \qquad \sum_{\alpha} dV^{\alpha} = 0 \qquad \sum_{\alpha} dN_i^{\alpha} = 0$$

For an isolated system, for shifts to occur reversibly, it is necessary that $\sum_{\alpha} S^{\alpha} = $ maximum for all states, hence $\sum_{\alpha} dS^{\alpha} = 0$.

But for each phase, $T^{\alpha} dS^{\alpha} = dU^{\alpha} + P^{\alpha} dV^{\alpha} + \sum_i \mu_i^{\alpha} dN_i^{\alpha}$, giving

$$\sum_{\alpha} \left[dS^{\alpha} - \frac{dU^{\alpha}}{T^{\alpha}} - \frac{P^{\alpha} dV^{\alpha}}{T^{\alpha}} - \sum_i \frac{1}{T^{\alpha}} \mu_i^{\alpha} dN_i^{\alpha} \right] = 0.$$

But introducing Lagrangian multipliers *a*, *b*, and c_i, we add the previous summations: $\sum_{\alpha} [dS^{\alpha} - a \, dU^{\alpha} - b \, dV^{\alpha} - \sum_i c_i \, dN_i^{\alpha}] = 0$

Subtracting the second sum from the first gives:

$$\sum_{\alpha} \left(a - \frac{1}{T^{\alpha}} \right) dU^{\alpha} + \sum_{\alpha} \left(b - \frac{P^{\alpha}}{T^{\alpha}} \right) dV^{\alpha} + \sum_{\alpha} \sum_i \left(c_i - \frac{\mu_i^{\alpha}}{T^{\alpha}} \right) dN_i^{\alpha} = 0$$

This requires that $\frac{1}{T^\alpha} = a$ for all phases, so at equilibrium, $T^\alpha = T$ is the same temperature in all phases. Then if $\frac{P^\alpha}{T} = b$, $P^\alpha = P$ is the same pressure in all phases. Finally, $\frac{\mu_i^\alpha}{T} = c_i$ for each component requiring that for each component, μ_i has the same value in all phases.

Now, how many variables are needed to describe the system and how many variables can be manipulated independently?

The variables are T, P, and all of the mole fractions, X_i^α. If there are C components, there are C values for it. If there are Φ phases, then for each i, there are Φ values for α. Thus, there are $C\Phi$ mole fractions which with T and P give a total of $C\Phi + 2$ variables.

The number of these which are independent is limited by the number of equations in which they appear. $\sum_i X_i^\alpha = 1$ for each α gives Φ equations. $\mu_i^1 = \mu_i^2 = \cdots = \mu_i^\Phi$ gives $\Phi - 1$ equations for each of C components.

The total number of equations is $\Phi + C(\Phi - 1)$. This gives the total number of independent variables, or DEGREES OF FREEDOM, to be
$$F = C\Phi + 2 - \Phi - C(\Phi - 1) \text{ or } F = C - \Phi + 2,$$
which is the GIBBS PHASE RULE.

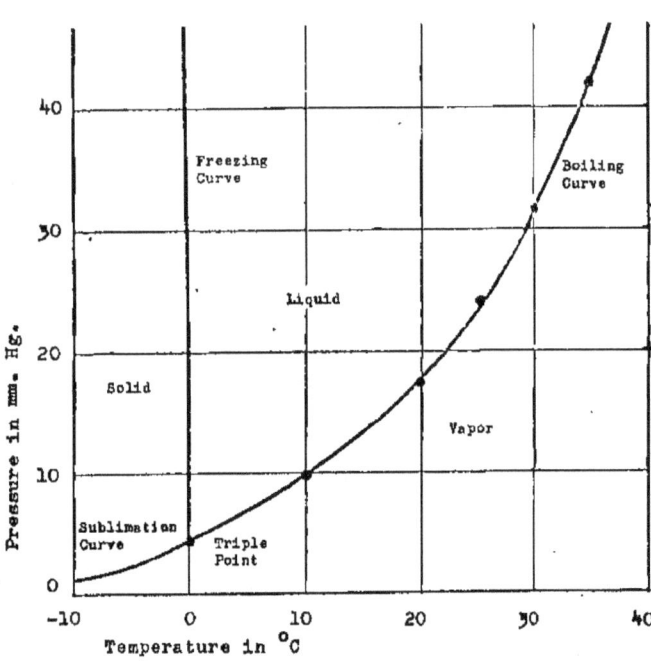

For a system of one component, $F = 3 - \Phi$. The phases of water are shown on a diagram where pressure is plotted against temperature. On the left is the solid phase. At the bottom and right of the diagram is the vapor phase; and the upper middle is the liquid phase.

Where there is only one phase, $F = 2$, allowing pressure and temperature to vary independently. Where there are two phases (e.g. liquid and vapor), $F = 1$. Once we have picked a pressure or temperature, the other variable is uniquely determined as on the boiling curve.

If all three phases are to exist in equilibrium, $F = 0$ and that state happens only at the triple point.

Next consider two components; now $F = 4 - \Phi$. When one phase exists, $F = 3$, allowing free variation of pressure, temperature and mole fraction of one of the components.

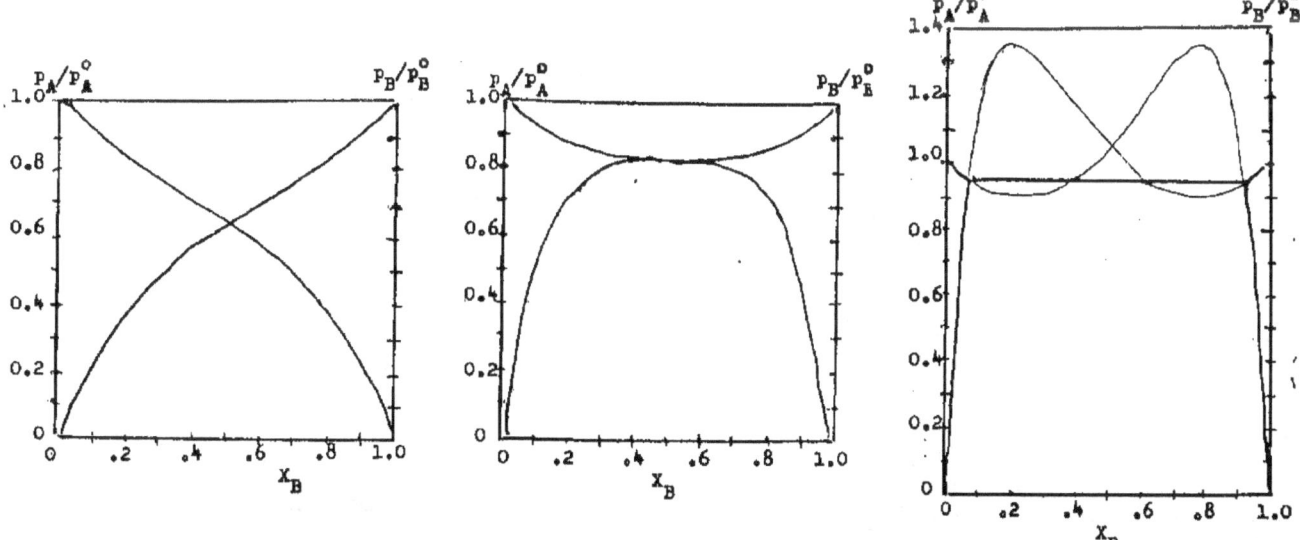

When two phases exist, $F = 2$. Then if temperature and mole fraction are picked freely, pressure is uniquely specified. The equations for partial pressure on page 20-34 will allow us to study the effect of temperature on partial pressure for the liquid vapor interface. We will re-write them as the ratio of partial pressure to pressure of the pure component. Thus,

$$\frac{p_A}{p_A{}^o} = X_A{}^L \, exp\left[(1 - X_A{}^L)^2 \, \frac{z\,w}{k\,T}\right] \quad \text{and} \quad \frac{p_B}{p_B{}^o} = X_B{}^L \, exp\left[(1 - X_B{}^L)^2 \, \frac{z\,w}{k\,T}\right].$$

In the first graph, $\dfrac{p_A}{p_A{}^o}$, goes from upper left to lower right corners and $\dfrac{p_B}{p_B{}^o}$ goes from lower left to upper right corners. The total pressure is the sum of the partial pressures, $P = p_A + p_B$.

EXERCISE: The effect of w and temperature, T, can be studied by varying the quantity $\dfrac{z\,w}{k\,T}$. When $\dfrac{z\,w}{k\,T} = 0$, we get Raoult's law.

Take the values for $\dfrac{z\,w}{k\,T}$ to be -2, $+1$, $+2$, and $+3$ and in each case plot (or have your computer plot) the curves for $\dfrac{p_A}{p_A{}^o}$ and $\dfrac{p_B}{p_B{}^o}$.

Referring to the 3 graphs on the previous page, if the left-hand curves result you have one liquid phase, and one vapor phase in equilibrium. As the temperature goes down, $\dfrac{Z\,w}{k\,T}$ goes up. If the right-hand curves result, you have one vapor phase but at the lower temperature you now have two liquid phases with unique concentrations at the end of the straight line. The curves between these two points are not followed.

With two components and three phases there remains one degree of freedom, temperature. Starting with the right-hand figure, if the temperature is increased the length of the solid line decreases, the compositions of the two liquid phases approach each other until the curves in the middle figure happen. Here the compositions of the two liquid phases become equal giving what is called the point of critical mixing at the temperature of critical mixing. Above this temperature there is one liquid phase, below there are two liquid phases.

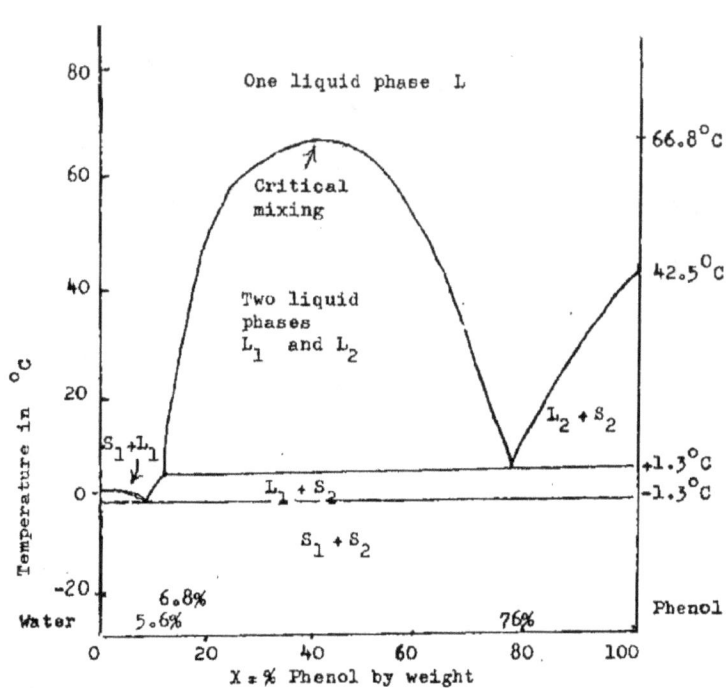

Phenol and water at one atmosphere pressure

S_1 = solid water phase S_2 = solid phenol phase

This is demonstrated in the water-phenol mixtures. Here we select one pressure, opting one degree of freedom, and when one phase exists we are left with two remaining degrees of freedom, temperature and mole fraction. The closed system of water and phenol has no vapor phase but will have a solid as well as liquid phase.

Above 66.8° C there is one liquid phase. Critical mixing occurs at 66.8° C and the curves coming down from that point give the concentrations of the two liquid phases into which mixtures with compositions lying between the curves will separate. For example, at 50° C, mixtures between the curves separate into one phase with about 20% phenol and 80% water with the other phase being about 60% phenol and 40% water. When phenol is less than 20% it is miscible in water giving one phase and when water is less than 40% it is miscible in phenol. As the temperature continues to drop from 50° C, the substances become less miscible in each other.

Referring to the previous graph, on the right-hand axis, pure phenol goes from a liquid to solid phase at 42.5°C. Below that temperature, pure solid phenol comes out of water-phenol mixtures leaving a liquid phase whose composition follows the far right-hand curve of L_2 until +1.3°C. Below that solid phenol continues to come out, now leaving a L_1 phase until −1.3°C at which the remaining water comes out as pure solid water. There are two solid phases, crystals of pure water or crystals of pure phenol.

Copper and nickel are miscible in liquid and solid phases. If a mixture 30% Ni / 70% Cu is cooled from the molten phase, when the temperature reaches the liquidus curve, the composition of the crystals is given by the solidus curve. As the temperature drops the liquid becomes richer in Cu until the last crystals are pure copper. The outside of a casting is richer in nickel. The inside richer in copper.

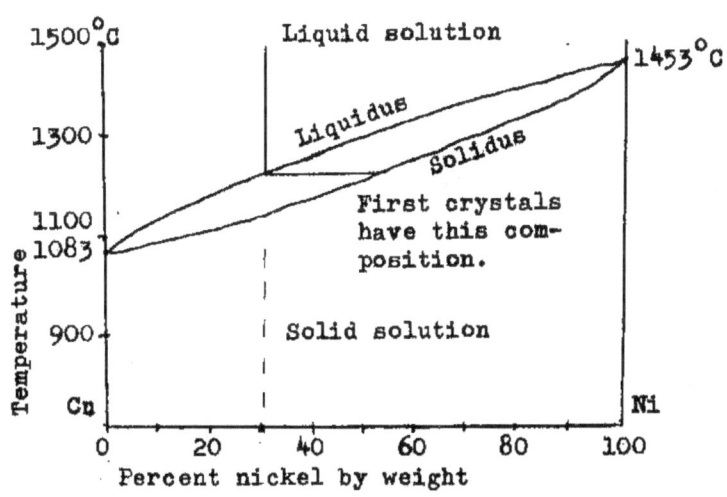

Silver and copper are totally mixable in the liquid phase. The melting point of pure silver is 961°C, and 1083°C is the melting point of copper. Below the melting point of silver, there are two curves. As a molten mixture rich in silver is cooled, when the temperature reaches the top of the curves, an alloy, α, becomes a solid phase. Its composition is given by the left-hand curve. As cooling continues, the molten metal gets richer in copper as more of the silver-rich solid comes out.

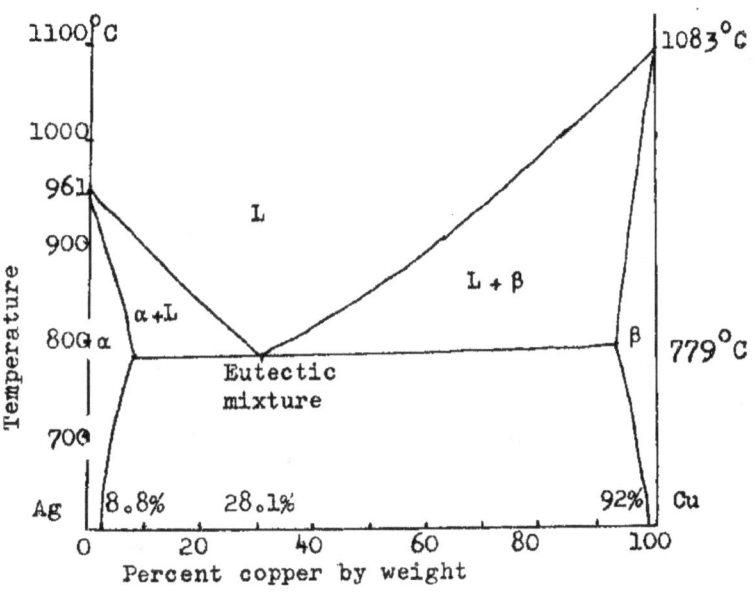

When the temperature reaches 779°C, there is now a β solid phase with 92% copper as well as an α phase with 8.8% copper. With three phases, liquid, α, and β, and two components, the phase rule gives us one degree of freedom which is opted when we select the pressure. Hence the temperature is constant while both α and β crystals solidify. This point is called the EUTECTIC POINT.

With copper-rich molten mixtures, the same behavior happens except that there is now liquid plus β crystals until the eutectic point is reached.

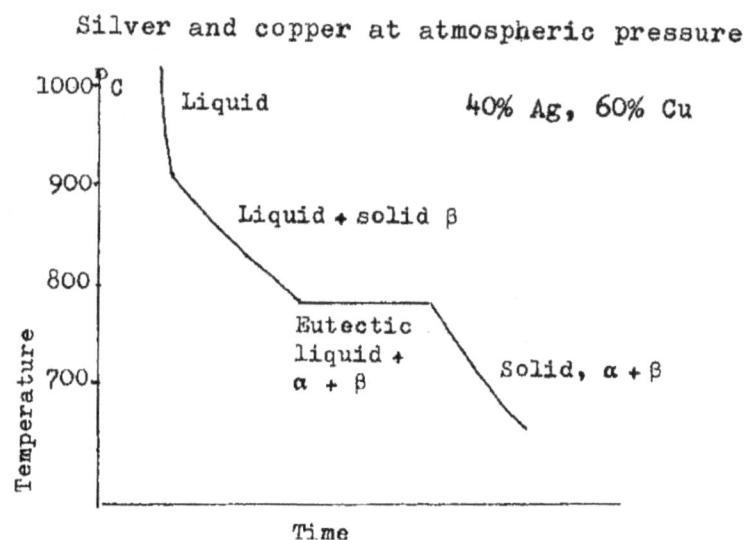

Silver and copper at atmospheric pressure

When liquid 40% Ag and 60% Cu is cooled, plotting temperature against time, the liquid cools rapidly until β crystals start to come out. Cooling slows until the eutectic point is reached where temperature is constant until the liquid is gone.

With zinc-magnesium mixtures there is the possibility of $MgZn_2$ which behaves as a pure component. Zinc-rich mixtures cool as L+Zn or $L+MgZn_2$. Mg-rich mixtures cool as L+Mg.

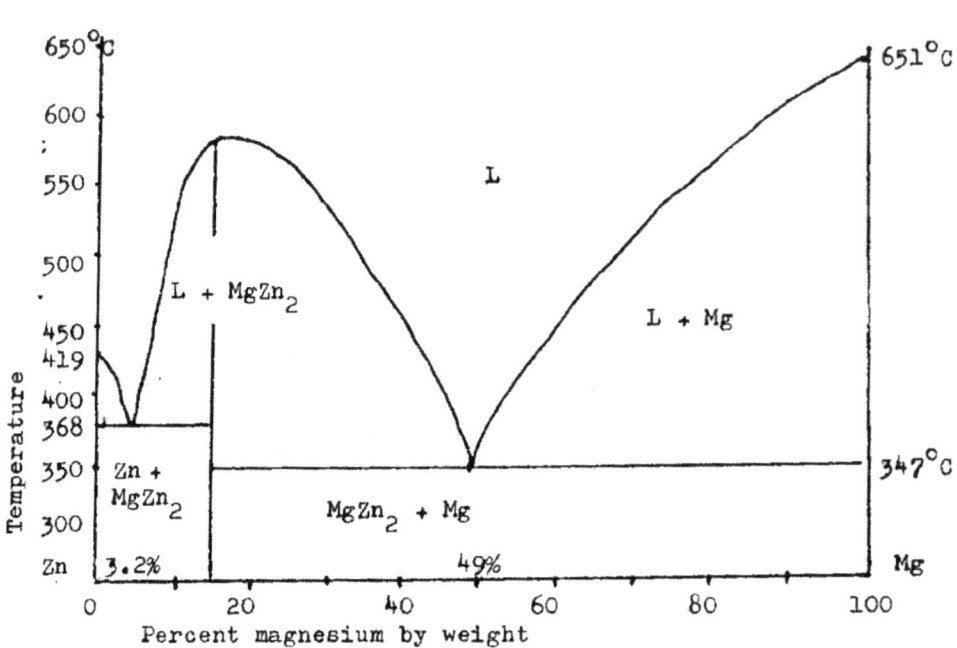

APPENDIX: ENTROPY, INFORMATION, MARKOV PROCESSES AND TIME

When I was wandering around Brazil, every morning I would be served a tray
with a pot of very black coffee, a pot of hot milk and a pot of hot water
and I could do whatever I wanted with them. (I still miss that piece of
papaya with a wedge of lime). If I mixed the coffee and milk, I got cafe
au lait. But I couldn't go back to separate pots of coffee and milk.
When a process goes from order to dis-order, there is a quantity called
ENTROPY involved which increases in the processes.

Let us demonstrate. Suppose we have some blocks
of wood with letters on them, and suppose we
find them as in picture no.1. Suppose, with
some effort, we re-arranged them in picture
no.2, we would have word of the English
language, but they wouldn't make any sense.
Next, go from get the famous slogan of the
philosopher Rene Descartes. But, if we give the
blocks in picture no.3 a swift kick, they
quickly go back to something like picture no.1.
Notice that going from order to dis-order is
easy. Going from dis-order to order takes
thought and effort. But, notice that in going
from dis-order to order gave us information.

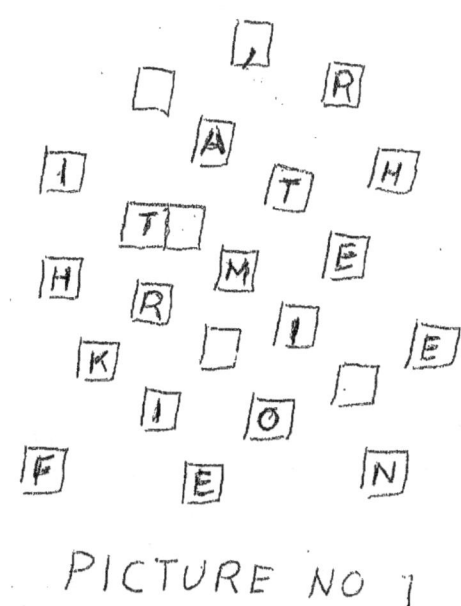

PICTURE NO 1

Let us think about the arithmetic. Your friend lives on a certain street.
You start driving up the street but by mistake pass his house which at **5**
miles and go **3** miles farther. You have gone **+8** miles up the street but must
go back **3** miles. In going back, you add **−3** miles to **+8** miles. If increasing
or going forward is a positive quantity, decreasing or going backwards adds
a negative quantity. If increasing dis-order (making a mess) involves
increasing entropy, going from dis-order to order (working to clean up your
mess) involves decreasing the entropy, or adding negative entropy.

Going from random
letters to an
arrangement whereby
they contain
information involves
negative entropy.
Therefore, we call
INFORMATION itself
NEG-ENTROPY.

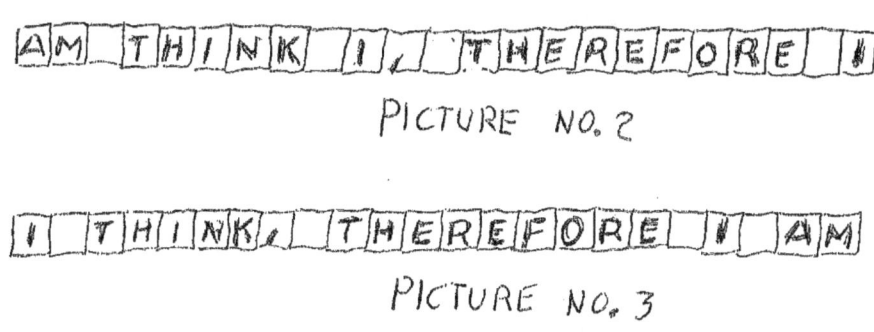

PICTURE NO. 2

PICTURE NO. 3

Let us see where else we can find increasing entropy. Let us consider optical refraction where we trace a light ray going through a slab of plate glass. After we are through tracing, it is evident that if you turn the picture upside down, it looks the same. If light worked the way we traced it, it goes forward and backward equally well. Time can be reversed. But, now look through a window, You see vague images of things behind you. The window is a bad mirror, but a mirror just as well. It is now definite which way the direction of light cannot be reversed. Entropy is increasing.

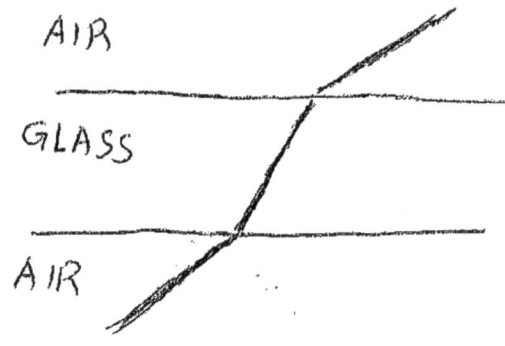

This raises a question about time itself In locating things in space, we can go east or west, north or south, up or down. But the clock only goes in one direction.

Let us first look at space. We don't see space. We see things. And we have the perception of relative location. If we are at one thing, we have to bestir ourselves and move to be at another thing.

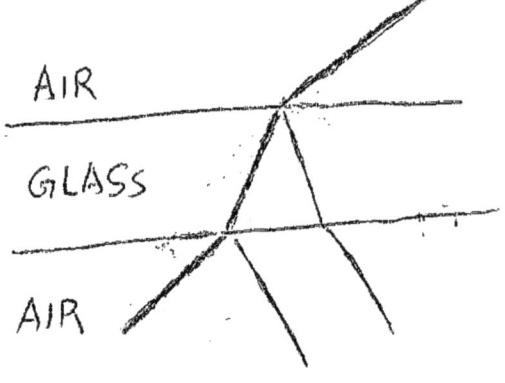

Space is, then, a mental idea which we invent to describe location. With reference to some things, we define directions and then take some object (a meter stick?) to enable to reduce our perception if a distance to number. Notice that our reality involves things. Next, we notice that things change. Change their nature and change their location but change always seems to go in one direction. If a fire burns a log, the ashes will not go back to being a log. We invent time to reduce our perception of change to number. But notice that in doing this, we are merely comparing changes of one thing to changes in another thing. The reference is always to the motion of things seen in the sky, or to hands going around a circular dial. But it is things that change. Time is a word applied to the numbers that we use to describe change.

This now explains why tie appears to go only one direction. By ENTROPY changes can go only in one direction. So, the basic PHYSICAL quantity in change is ENTROPY and time is our mentally defined idea which allows us to reduce our perception of change to number.

This leads us to an intellectual mystery. How is it that for things that we ourselves can experience happening in the universe, we can make our mathematical description describe what happened in the past as well as in the future? Newton´s can postdict as well as predict.

The statistical basis for thermodynamics, or the theory of heat phenomena, will give us a clue. If we start with a gas, its molecules are total disorder. But the ideal gas law $PV = RT$ is absolute order.

At start of this chapter, we go through the details of this reasoning. The logic is found in an amazing branch of mathematics called probability and statistics. The particles of matter are governed by quantum mechanics which tells us the states that they are allowed to be in. But the vast number of particles in a box of gas are scattered with total randomness among these states, and when two particles interact, the distribution changes. But mathematically, we can derive a formula for what any distribution might be, and also the entropy for that distribution. Given any distribution of particles among quantum states, any change will most likely be in the direction of increasing entropy (a more random distribution). But there is a finite number of molecules to play with and a constant amount of energy to be shared among them. It is then assumed that the existing conditions of the gas will be in the neighborhood of maximum entropy.

This gives something called the partition function out of which we immediately get a Helmholtz free energy which then gives all of the thermodynamics describing the thermal behavior of the gas without any reference to what the particles are doing. (The nature of the molecules themselves doe have an important effect as with the heat capacity ratio). Mathematically, there is a bridge between the total randomness at the small-scale level of the universe and our predictable large-scale behavior. The philosophical question is always: how real is the mapping of the details of the mathematical model onto the details of the actual universe?

Going to chapter 21, we find that a process governed by entropy is a MARKOV PROCESS. That is a random process where if the future can be predicted, the past is unknown. Given the present state of a system, we have no idea how it got there. Diffusion (or random walk) is such a process. It is shown in chapter 21 that the Schrödinger equation is a complex valued diffusion equation, and not a wave equation as it is usually called. The evolution of a quantized system would appear to be a Markov process. How much this perception requires the re-writing of standard quantum mechanics book has to be thought about.

At this point a physicist discovers biology. (When a physicist does this sort of thing, he totally messes the order field up for those people who have been working in the other field.) A living cell is a very complicated thing Among other things, it needs very big molecules to run its functions. A talk with bio-chemists comes up with an estimate of around 100 amino acids in a peptide chain that would be capable of enzyme activity. A simple calculation gives 10^{130} different peptide chains of that length to be possible. How, out of that number of choices, did the first cell find just the right enzyme molecules to run its functions? But, next the cell must reproduce itself, and that requires DNA. Now, the DNA encodes the information (neg-entropy) used to form the cell from basic chemicals in its beginning. Although we may not here be able to calculate this, we know that it is going to be an outrageous number.

We turn to the talk of monkeys on type writers writing Hamlet. It doesn't take too much arithmetic to show that no amount of monkeys or no amount of typewriters can do this. It takes the outside intelligence of a Shakespeare to do this. Although human intelligence may in the distant future be able to make a living cell, no human intelligence was around when the first cell happened.

Once life got going, it differentiated into all sorts of animals, plants, bugs and other things. And even in one species such as dog, a Scottish terrier doesn't produce a German shepherd. The DNA of a species must be in sufficient length to encode just that species and no other. This gives some idea of the amount of information needed for biology, and the amount of intelligence needed to reduce the entropy to produce it.

Now, before there was life, there had to be a universe with the right conditions for life to happen at all. Recently, in the past 50 years, physicists have run into some strange facts to give an example, if the masses of the proton, neutron and electron were not exactly what they are, there would be different types of universes. One thing about them all, life would be impossible in all of them. Most other constants of physics are weird like that. But aside from that, all the information needed to reduce the randomness of matter at the big bang to give us our universe had to be somehow encoded in the big bang.

Now, Hamlet did not write his own play. Some intelligence, Shakespeare, on outside had to do it. With what we know about physics, the universe could not have reduced its entropy by natural processes. By natural processes, entropy goes up. There is no getting around that reduce the entropy of the universe, there had to be an intelligence outside of the physical universe. At this point, you can draw your own conclusions.

CHAPTER TWENTY-ONE: FERMI-DIRAC AND BOSE-EINSTEIN QUANTUM STATISTICS THE GROSS-PITAEVESKII EQUATION AND THE GINZBURG-LANDAU THEORY OF SUPERCONDUCTION

> The Dao that can be stated cannot be the eternal Dao. The Name that can be named cannot be the eternal Name. The Nameless is the origin of the Universe. The Named is the mother of all matter.
>
> Lao Zi

> From our experience of the history of physics, we can safely assume that none of the current physical theories are completely correct. (According to Lao Zi, the theory that can be written down cannot be the eternal theory, because it is limited by mathematical symbols that we use to write down the theory.) The problem is to determine in which ways the current theories are wrong and how to fix them. Here we need a lot of imagination and stimulation.
>
> Xiao-Gang Wen

The phenomenon of superconductivity was first discussed in chapter 13. The first superconducting material was lead, discovered by Kammerling-Onnes in 1911, which was superconducting below a critical temperature of $4\,K$. By 1970, Nb_3Ge was found to be superconducting at $23\,K$. Just what was going on was not very evident. In 1956, John Bardeen, Leon Cooper and Robert Schrieffer (BCS) published a theory that accounted for most observed aspects of the phenomenon. Then in 1986, J. G. Bednorz and K. A. Müller in Zurich got $LaBaCuO$ to be superconducting at $35\,K$. The phenomenon has been observed at critical temperatures up to $135\,K$. But CuO is an antiferromagnetic insulator called a Mott insulator. This didn't make any sense at all and the BCS theory did not help here. For a new starting point, physicists went back to a model put forth by V. L. Ginzburg and L. D. Landau in 1950. At the time this model had little theoretical background but was put together because it worked.

In this chapter, we will derive a variation of the Schrödinger equation called the Gross-Pitaevskii equation and from this derive the Ginzburg Landau model. This will give a physical interpretation of the terms in that model and, then, raise further questions about the physics involved.

In the October-December 2017 issue of the Reviews of Modern Physics, Xiao-Geng Wen of MIT states that to fully understand high temperature superconducting (and the fractional quantum Hall effect of chapter 16) we must start all over again. The new starting point, he says, is the new field of topological quantum matter. In the same issue of The Reviews of Modern Physics, J. M. Kosterlitz and F. D. M. Haldame discuss their contributions, along with those of D. Thouless, to this new field of theoretical physics. As the level of that subject is beyond the level of this textbook, we must stop with Ginzburg-Landau. But if you wish to further pursue the topic, hold on for an exciting ride.

In Chapter 13, we discussed systems of particles which did not obey the Pauli exclusion principle, where any number of particles could occupy the same quantum state. These particles obeyed Bose-Einstein statistics and were called bosons. An important example of these are the quanta of electromagnetic radiation called photons; discussed at the end of Chapter 13. It is important that photons are emitted and absorbed by charges and hence, their number is not conserved.

Particles which obey the Pauli exclusion principle, at most one particle of a system could occupy a quantum state, are described by Fermi-Dirac statistics and are called fermions. Electrons are fermions and their number in a closed system is conserved.

In Chapter 20, entropy was defined as the missing information about all patterns of configuration of things in a system distributed among their allowed states, which pattern happens to be the one existing at the present moment. For a system not restricted by Pauli's exclusion principle, if f_i were the fraction of particles in energy state, ϵ_i, the entropy was $S = -k \sum_i f_i \ln f_i$ with $\sum_i f_i = 1$ and $U = \sum_i f_i \epsilon_i$ being internal energy.

Equilibrium for a closed system involved maximum entropy subject to constant internal energy and sum of fractions. Introducing Lagrangian multipliers, $\delta\left(\frac{S}{k}\right) + b\,\delta U + a\,\delta(1) = 0$ gives $\sum_i(-1 - \ln f_i + b\,\epsilon_i + a)\delta f_i = 0$ giving $f_i = e^{a-1}\,e^{b\,\epsilon_i}$.

If ϵ_i increases with increasing index, i, and f_i must decrease with increasing, b must be negative, call it $b = -\beta$. To assure that f_i is less than one, $a - 1$ must be negative. Call it $a - 1 = -\alpha$.

If $\sum_i f_i = 1$, then $e^\alpha = \sum_i e^{-\beta\,\epsilon_i}$. Call $Z = \sum_i e^{-\beta\,\epsilon_i}$ the partition function, giving $\alpha = \ln Z$.

If $f_i = e^{-\alpha - \beta\,\epsilon_i}$, then $S = -k \sum_i(-\alpha - \beta\,\epsilon_i)f_i = k\,\alpha + k\,\beta\,U$.

Notice the limit, $\lim_{\beta \to \infty} f_i = 0$, for all i except the ground state. This suggests β involves the inverse of temperature.

From $S = k\alpha + k\beta U$, if we allow $\beta = \dfrac{1}{kT}$, we get $TS = kT\alpha + U$ or $kT\alpha = TS - U$. But in Chapter 19, the Helmholtz free energy was $F = U - TS$. Hence, $F = -kT \ln Z$.

Now $dF = dU - T\,dS - S\,dT$. But $T\,dS = dU + P\,dV$ is conservation of energy. Hence, $dF = dU - T\,dS - S\,dT$, giving $P = -\left(\dfrac{\delta F}{\delta V}\right)_T$ and $S = -\left(\dfrac{\delta F}{\delta T}\right)_V$. Thermodynamics starts with finding a partition function, $Z = \sum_i e^{-\beta \epsilon_i}$.

ELECTRONS AND FERMI-DIRAC STATISTICS

For particles obeying Fermi-Dirac statistics, a quantum state is either filled with one particle, or it is empty. This makes the entropy function easy to come by. For any energy, ϵ_i, let there be Γ_i quantum states. If ν_i of these are filled then $(\Gamma_i - \nu_i)$ are empty.

For example, let the boxes be quantum states having the same energy, and let the black circles be particles.

The number of patterns of filling quantum states for ϵ_i is $\dfrac{\Gamma_i!}{\nu_i!\,(\Gamma_i - \nu_i)!}$ and when the particles are distributed with ν_i particles in energy states, ϵ_i, the product of all of these is the total number of patterns, or distributions. The entropy is $S = k\,ln\,\Pi_i\,\dfrac{\Gamma_i!}{\nu_i!\,(\Gamma_i - \nu_i)!}$, or using Stirling's formula and cancelling a few terms, $S = k\sum_i[\Gamma_i\,ln\,\Gamma_i - \nu_i\,ln\,\nu_i - (\Gamma_i - \nu_i)\,ln\,(\Gamma_i - \nu_i)]$.

If the distribution of occupation numbers is varied, the variation of the entropy is $\delta\left(\dfrac{S}{k}\right) = \sum_i[-ln\,\nu_i + ln\,(\Gamma_i - \nu_i)]\,\delta\nu_i$.

Again, we maximize entropy for the equilibrium distribution, subject to the energy, $U = \sum_i f_i\,\epsilon_i$, and the total number of particles, $N = \sum_i \nu_i$, being constant.

$\delta\left[\dfrac{S}{k} - b\,U - a\,N\right] = 0$ gives $\sum_i[-ln\,\nu_i + ln\,(\Gamma_i - \nu_i) - b\,\epsilon_i - a]\,\delta\nu_i$ giving $\dfrac{\Gamma_i}{\nu_i} - 1 = e^{b\,\epsilon_i + a}$ or $\nu_i = \dfrac{\Gamma_i}{1 + e^{b\,\epsilon_i + a}}$.

As ν_i must decrease as ϵ_i increases, so b is positive and from past experience we set $b = \beta = \dfrac{1}{kT}$. What about a? As T approaches zero, all quantum states with energy below the Fermi level, ϵ_f, are full, and all states with energy above ϵ_f are empty. Hence, $\lim\limits_{T \to 0}(\nu_i) = \Gamma_i$ when $\epsilon_i < \epsilon_f$ and $\lim\limits_{T \to 0}(\nu_i) = 0$ when $\epsilon_i > \epsilon_f$. This is accomplished if $a = -\dfrac{\epsilon_f}{kT}$ and $\nu_i = \dfrac{\Gamma_i}{1 + e^{(\epsilon_i - \epsilon_f)/kT}}$.

We shall apply this to conduction electrons in metals. The simplest approach is the free electron model which ignores the lattice of positively charged nuclei and filled electron shells except to assume an overall distribution of positive charge so that the electrons move in an environment that is electrically neutral. For most aspects of conductivity in metals and semiconductors it is necessary to go to the band theory of electrons moving in a periodic potential from the lattice of nuclei. For that you are referred to any standard book on solid-state physics. For certain phenomena, the free electron model is an adequate description.

For electrons in a box, what will be the number of quantum states with energies between E and $E + dE$?

We now take all standing waves in a resonant box of dimensions, L_x, L_y, and L_z. If there are n_x wavelengths in length, L_x, then $\lambda_x = L_x/n_x$ is the wavelength in that direction. The x-component of photon momentum is $p_x = h/\lambda_x = n_x\, h/L_x$.

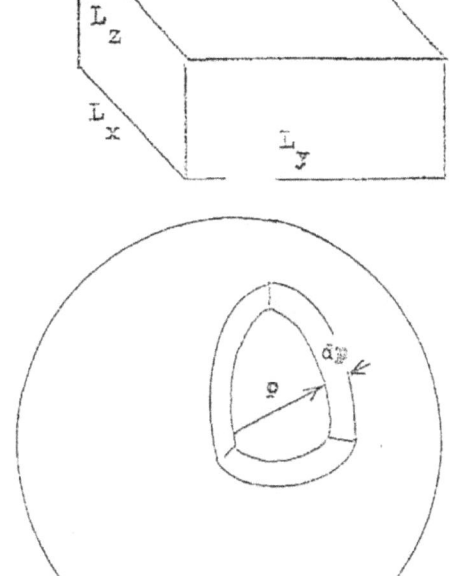

Likewise, the momentum states in the y and z directions are $p_y = n_y\, h/L_y$ and $p_z = n_z\, h/L_z$. The jumps between quantum numbers are $\Delta n_x = \Delta n_y = \Delta n_z = 1$. The jumps between momentum values are $\Delta p_x = h/L_x$, $\Delta p_y = h/L_y$, and $\Delta p_z = h/L_z$.

We now construct a momentum space. Each point in this space is a value of photon momentum. All points on a sphere of momentum radius, p, will have the same energy. Take a momentum shell of thickness, dp, about the radius, p.

The momentum-volume of this shell is $4\pi p^2\, dp$. Now, the volume of the momentum space per quantum state is $\Delta p_x\, \Delta p_y\, \Delta p_z = \dfrac{h^3}{L_x\, L_y\, L_z} = \dfrac{h^3}{V}$, where V is the volume of the resonant cavity.

The number of quantum states in a shell of radius, p, and thickness, dp, is $\dfrac{4\pi p^2\, dp}{h^3/V}$. But we are interested in energy rather than momentum.

If $p^2 = 2\,M\,E$, then $p\, dp = M\, dE$ and $p = \sqrt{2\,M\,E}$. The number of quantum states between E and $E + dE$ is $\dfrac{4\pi V}{h^3}\sqrt{2\,M^2 E}\; dE$.

Electrons are spin one half particles, they have two spin orientation states. Hence the number of momentum states is the same and the number of quantum states with momentum between p and $p + dp$ is:

$$\Gamma(p)\,dp = \frac{8\,\pi\,V\,p^2\,dp}{h^3}.$$

But we can treat electrons at normal temperatures as non-relativistic particles. $E = p^2/2\,M$, $p^2 = 2\,M\,E$, $2\,p\,dp = 2\,dE$, and $p = (2\,M\,E)^{1/2}$ giving $\Gamma(E)\,dE = \frac{4\,\pi\,V\,(2\,M)^{3/2}\,\sqrt{E}\,dE}{h^3}$, from which, let $A = 4\,\pi\,(2\,M)^{\frac{3}{2}}/h^3$. When the temperature is much above $T = 0$, we replace ϵ_f in the formula for ν_i by a function, μ, which may vary with T. $\lim_{T \to 0} \mu = \epsilon_f$. The number of electrons with energies between E and $E + dE$ is now:

$$\nu(E)\,dE = \frac{A\,V\,\sqrt{E}\,dE}{1 + e^{(E-\mu)/kT}}.$$

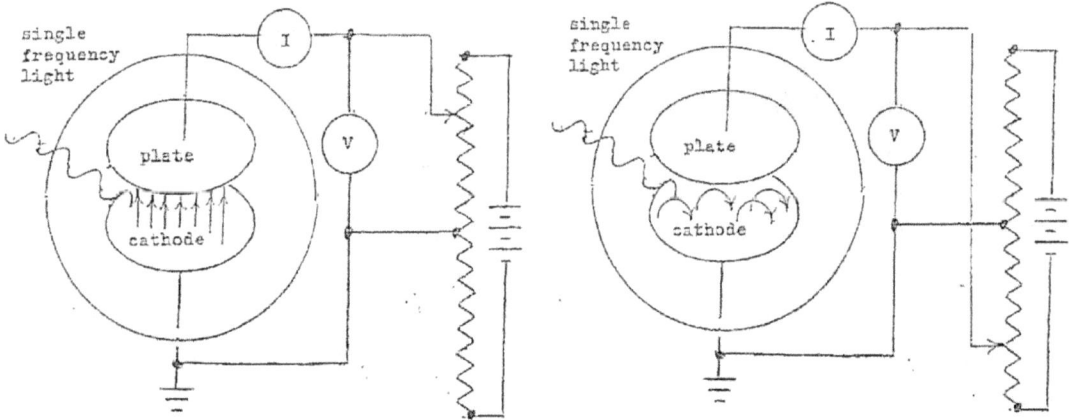

This apparatus demonstrates the photoelectric effect. When light with a single wavelength and frequency falls upon a metal surface called a cathode, electrons are emitted. When another surface, called a plate, is at a positive voltage with respect to the cathode, electrons will be drawn to the plate and a current flows in the circuit shown.

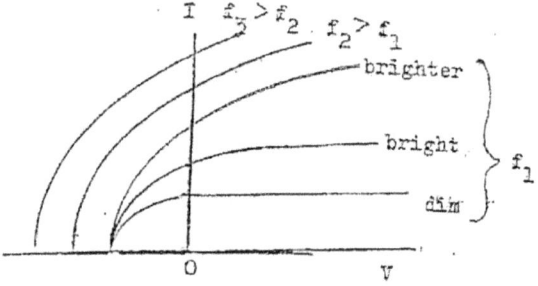

If the voltage at the plate is very negative, electrons will be repelled back to the cathode. If the plate is only slightly negative, current will still flow, but the current will approach zero at a particular negative voltage.

This is interpreted as, the electrons emitted from the surface when they absorb light have an energy when escaping. When the plate voltage is $-V$, the electrons will lose an energy, $-Ve$, on travelling to the plate. At the cut-off voltage, call it V_{co}, the electrons will have lost all of their escape energy and just barely get to the plate. It is now noticed that, keeping the frequencies constant, increasing the intensity of the light will increase the electron current to the plate but will not change the cut-off voltage.

However, increasing frequency of the light will result in the cut-off voltage becoming more negative. Decreasing the frequency reduces the magnitude of the cut-off voltage until the cut-off voltage is zero. Below this frequency, no electrons will be photo-emitted. A positive voltage at the plate would itself pull some electrons from the cathode. Plotting the magnitude of cut-off voltage against frequency might look like the plot shown here.

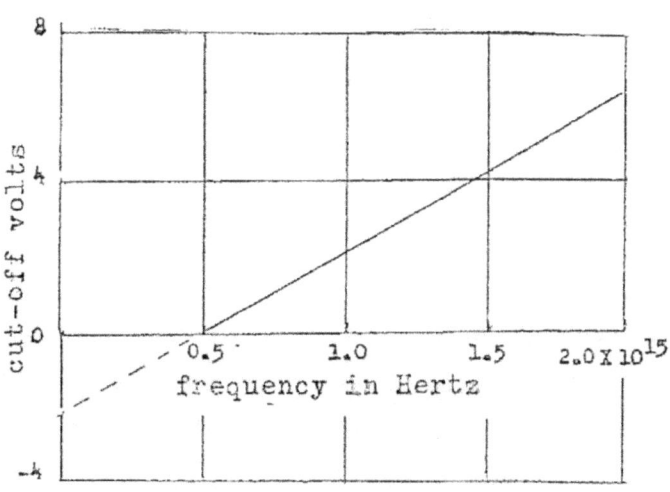

For a frequency below 0.5×10^{15} Hertz ($\lambda = 6000\,\text{Å}$) no electrons are photo-emitted. The cut-off voltage gives the energy of emitted electrons. At room temperature, the most energetic electrons in the metal are at the Fermi level.

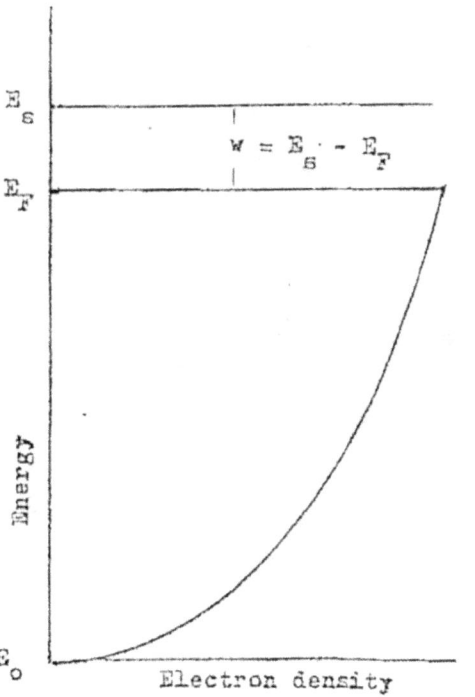

Room temperature is close enough to $T = 0$, that $v(E)\,dE = A V \sqrt{E}\,dE$ gives the electron density among energy states below the Fermi level, and $v(E) = 0$ for states above the Fermi level. The figure gives the occupation of states near $T = 0$. Let E_o be the lowest quantum state for free electrons. E_F is the Fermi energy and E_s is the energy an electron needs to escape from the surface. An electron at the Fermi level needs an energy, $w = E_s - E_F$, in order to just emerge from the surface. This is called the work function.

In the photo-electric effect, if an electron at the Fermi level absorbs a photon of energy, hf, it will emerge from the surface with an energy,

$$\frac{1}{2}mv^2 = hf - w.$$

When the cut-off voltage exists between the cathode and the plate, this energy is all absorbed by the electric field by the time the electron gets to the plate. $eV_{co} = hf - w$.

PROBLEM: Use the data at the bottom of the previous page.
   1. What is the work function for the surface in volts?
   2. What value of Plank's constant in Joule-sec do you get from these data?

In the next set-up, the cathode is a metal wire heated to a temperature of, maybe, 2000° K. The energy distribution of electrons extends to higher energies than E_F. Energies higher than E_s can be emitted from the surface. This is called thermionic emission. In working out the current of emitted electrons, we shall consider those electrons moving in the direction perpendicular to the surface with energy, $p_z^2/2m$, greater than E_s.

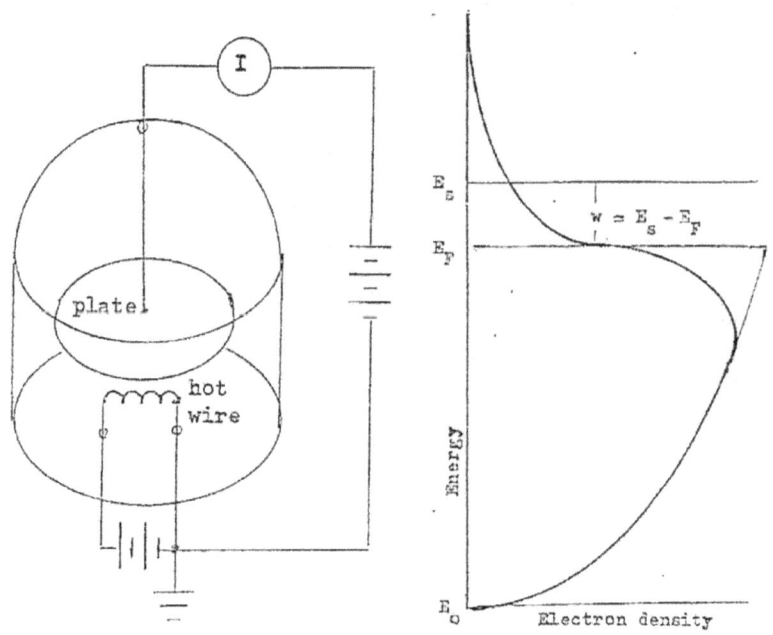

There are $\dfrac{2V}{h^3} dp_x\, dp_y\, dp_z$ momentum-spin quantum states in the volume $dp_x\, dp_y\, dp_z$ of momentum space. Then, $v\, dp_x\, dp_y\, dp_z = \dfrac{2V\, dp_x\, dp_y\, dp_z}{h^3\left[1 + e^{E - \mu/kT}\right]}$, gives the electrons occupying those states.

BOSE-EINSTEIN STATISTICS AND BOSE-EINSTEIN CONDENSATION

Here we treat particles which do not obey the Pauli exclusion principle. In addition to ^{4}He, these include photons and phonons. For photons and phonons, the quantized energy states are shown in chapters 13 and 16 to be $E_n = (n + 1/2)\, h f$ with $n = 0,\ 1,\ 2, \ldots$

The harmonic oscillator energy, $\dfrac{p^2}{2M} + \dfrac{1}{2} M \omega^2 Q^2 = E$, has quantum states, $E_n = (n + 1/2)\, h f$. The partition function is $Z = \sum_n e^{-h f / 2 k T_e}\, e^{-n h f / k T}$. For the moment, let $e^{-h f / k T} = x$, which is less than one. Next, notice that $\dfrac{1 - x^N}{1 - x} = 1 + x^2 + x^3 + x^4 + \cdots + x^{N-1}$. Then, $\lim\limits_{N \to \infty} x^N = 0$ if $x < 1$. With the result that $1 + x^2 + x^3 + x^5 \ldots = \dfrac{1}{1-x}$, $Z = e^{-h f / 2 k T} \sum_{n=0}^{\infty} \left(e^{-h f / k T} \right)^n = \dfrac{e^{-h f / 2 k T}}{1 - e^{-h f / k T}}$. If N is the number of particles, $F = -N k T \ln Z$ with $S = -\dfrac{\partial F}{\partial T}$ and $U = F + T S$. The internal energy is $U = -N k T \ln Z + T \dfrac{\partial}{\partial T} (-N k T \ln Z) = N k T^2 \dfrac{\partial}{\partial T} (\ln Z)$.

$\ln Z = -\dfrac{h f}{2 k T} - \ln \left(1 - e^{-h f / k T} \right)$ gives $U = N \dfrac{1}{2} h f + \dfrac{N h f e^{-h f / k T}}{1 - e^{-h f / k T}}$. $N \dfrac{1}{2} h f$ is the quantum ground state energy.

At absolute zero temperature, $T = 0$. Notice that $\lim\limits_{T \to 0} e^{-h f / k T} = e^{-\infty} = 0$. Here the internal energy is $U = N \dfrac{1}{2} h f$. At $T = 0$ all particles have collapsed into the lowest quantum state. Compare this with the distribution of particles over quantum states below the Fermi level in the case of Fermi-Dirac statistics.

SUPERCONDUCTIVITY, THE GROSS-PITAEVSKII EQUATION AND THE GINZBURG-LANDAU THEORY

There are two prominent theories of superconductivity. A more elegant model was published by J. Bardeen, L. N. Cooper, and J. R. Schrieffer (BCS) in 1957. A more naïve model by V. I. Ginzburg and L. D. Landau was published in 1950. As it depicted better the actual phenomenon, we shall build our discussion on the latter model. The Ginzburg-Landau theory is put on a more rigorous basis by an equation developed independently be E. P. Gross and L. P. Pitaevskii in 1961.

The phenomenon of superconductivity was discussed in chapter 13. This discussion will continue from the topics discussed in chapters 13 and 14, so you might review that material.

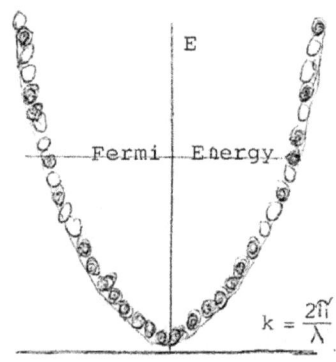

Consider electrons in a box of length, L, to have wavelength, $\lambda = L/n$, with n being an integer. Plotting energy of their quantum states against wavenumber as a parabola, we get the pictures shown here.

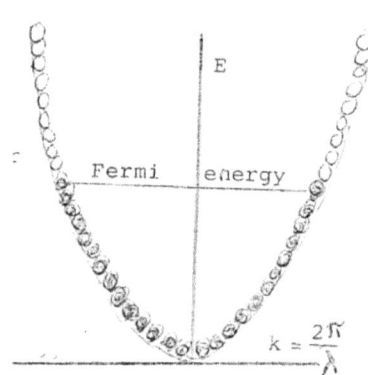

Being fermions, only one electron is in any state. In the picture, white states are empty and black states are singly occupied.

Superconductivity belongs to the class of phenomena called Bose-Einstein condensation (BEC) and a discussion of that starts with a consideration of heat capacity. At room temperature, electrons are randomly distributed among quantum states and the discussion of heat capacity in Chapter 19 can be taken to hold here. However, at absolute zero all states below the Fermi level are full and states above the Fermi level are empty. At $T = 0$, the internal energy is constant, and the heat capacity is zero. Somewhere between room temperature and $T = 0$, the heat capacity starts to decrease with temperature, continuing to go down to $c = 0$ and $T = 0$.

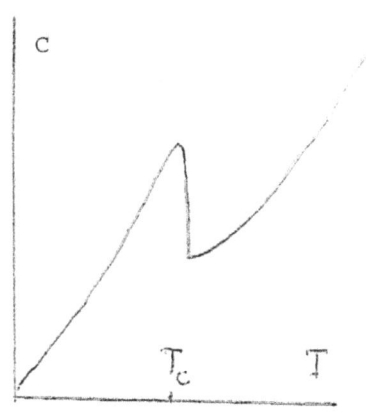

The expected heat capacity would behave as the picture to the right. But, with liquid helium, ^{4}He, and super-conducting electrons near $T = 0$, measurements give the picture to the left. T_c is some sort of critical temperature.

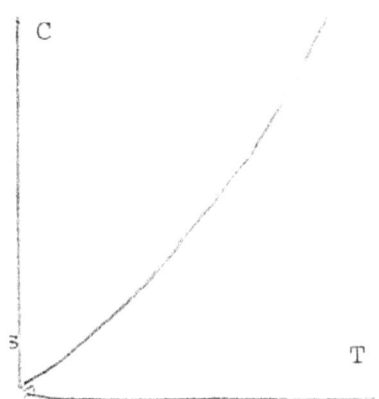

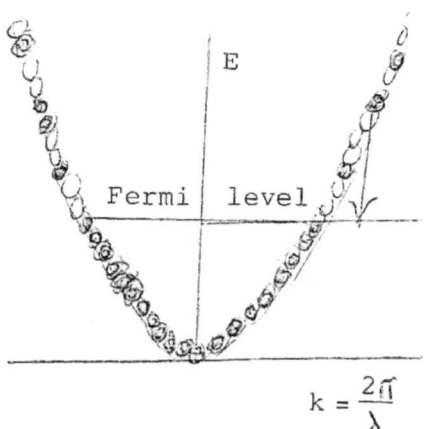

Look at this question for electrons. They obey Fermi statistics with only one electron in a quantum state. Hence, for an electron in an excited state, the most energy it can lose is to fall to the Fermi level. This is shown in the left-hand picture.

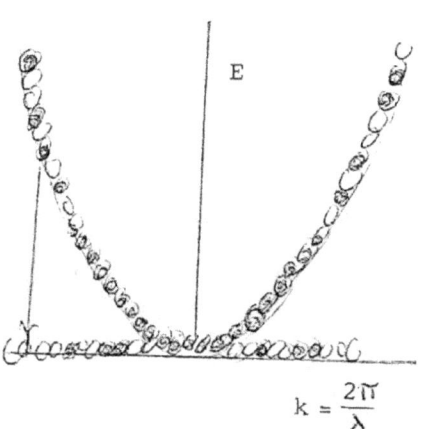

Suppose somehow two electrons with opposite spins combine to form a composite particle of spin-zero. That would be a boson, not obeying the Pauli exclusion principle. Now, an excited state electron could drop all the way to the ground state, losing much more energy. If this happened at the critical temperature, it would produce a big increase in the heat capacity. But this would have to overcome the Coulomb repulsion of the charges on the electrons. This idea was explored by L. N. Cooper in 1956 and such electron pairs, if they happen, are called Cooper pairs.

Because of its shape at the critical temperature, that point in the heat capacity against wave number curve is called the lambda point. We will use the Ginzburg-Landau theory to explore what goes on at that point. But as a foundation for that discussion, we will develop the Gross-Pitaevskii equation, and that needs a description of electron-electron interactions in quantum mechanics.

To start, look at an electron wave function. Take an electron in a 2px state. Assume that it exists most strongly in the x-direction. Its algebraic sign is both positive and negative, making it in itself not suitable for giving probability, but if we square the wavefunction, we get a function of the same shape, but having a positive value everywhere. (The algebraic sign is on the function. It does not imply charge.)

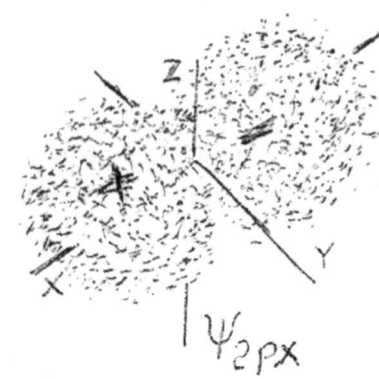

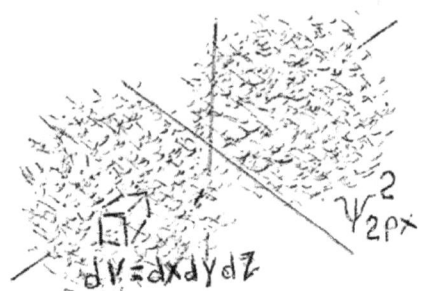

This is interpreted in the following manner. The square of the wave function is taken to be a density of probability of the electron being in any element of volume, $dV = dx\,dy\,dz$. The probability density of the electron is $\psi^2 dV$, requiring $\int \psi^2\, dV = 1$.

The electron's charge, $-e$, is interpreted as being distributed as the probability, thus $-e\,\psi^2$ is the electron's charge density. The next task is to describe the interaction between two electrons.

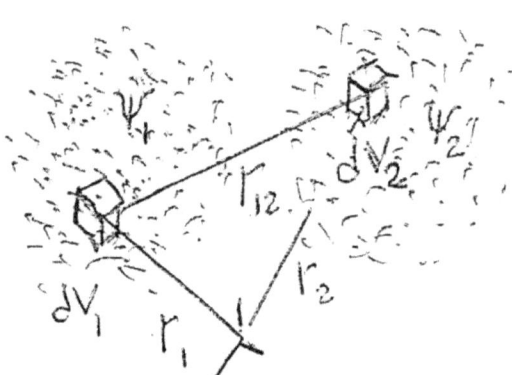

Let electron 1 be distributed according to wave function, $\psi_1(r_1)$, and electron 2 have wavefunction, $\psi_2(r_2)$. Let dV_1 be a volume element for electron 1 and dV_2 be a volume element for electron 2. The potential energy is $\dfrac{k\left(-e\,\psi_1{}^2 dV_1\right)\left(-e\,\psi_2{}^2 dV_2\right)}{r_{12}}$, where $k = \dfrac{1}{4\pi\epsilon_o}$ and r_{12} is the distance from dV_1 to dV_2.

$$\iint\limits_{r_1\,r_2} \frac{k\,e^2\,\psi_1{}^2\,\psi_2{}^2 dV_1\, dV_2}{r_{12}}$$ is the potential energy.

For electron 1 in dV_1, $\left[\displaystyle\int\limits_{r_2} \frac{k\,e^2\,\psi_2{}^2\, dV_2}{r_{12}}\right]\psi_1{}^2\, dV_1$

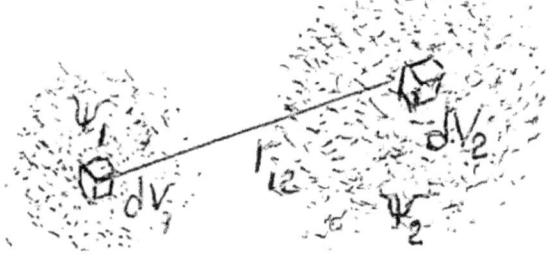

is its potential energy with respect to all of the second electron.

Now, if the electrons interacted with some other interaction, $U(r_{12})$,

$\left[\displaystyle\int\limits_{r_2} U(r_{12})\,\psi_2{}^2\, dV_2\right]\psi_1{}^2\, dV_1$

is the potential energy of electron 1 in dV_1, repelled by electron 2. This is a positive quantity.

The electrons have been denoted by the lattice atoms which are now positive ions. The electrons are now attracted to the positive atoms with a negative energy. For the particular lattice, let this have a net value, $-w$.

The amount of this energy in dV_1 is $-w\,\psi^2 dV_1$.

The fraction of kinetic energy in dV_1 is $\dfrac{p^2}{2M}\psi_1{}^2\,dV_1$ and $E\,\psi_1{}^2\,dV_1$ is the fraction of the total energy in dV_1. This gives

$$\frac{p^2}{2M}\psi_1{}^2\,dV_1 - w\,\psi_1{}^2\,dV_1 + \int_{r_2} U(r_{12})\,\psi_2{}^2\,dV_2\,\psi_1{}^2\,dV_1 = E\,\psi_1{}^2\,dV_1.$$

When this is divided by dV_1 we get electron 1 to have energy density

$$\frac{p^2}{2M}\psi_1{}^2 - w\,\psi_1{}^2 + \int_{r_2} U(r_{12})\,\psi_2{}^2\,dV_2\,\psi_1{}^2 = E\,\psi_1{}^2.$$

This is our form of the Gross-Pitaevskii equation.

Next, let both electrons be in the same quantum state with opposite spins, $\psi_1(r) = \psi_2(r) = \psi(r)$.

Then consider that there are two electrons to interact only when they are in contact with each other, that is at the same point. We describe the contact potential with something called the Dirac delta "function".

$\delta(r) = 0$ when $r \neq 0$ and $\int f(r)\,\delta(r)\,dr = f(0)$ for some function, $f(r)$.

For a contact potential, we write $U(r_{12}) = U\,\delta(r_{12})$.

Then: $\displaystyle\int_{r_2} U\,\delta(r_{12})\,\psi^2(r_2)\,dV_2 = U\,\psi^2(r_1)$

The energy density is now $\dfrac{p^2}{2M}\psi^2 - w\,\psi^2 + U\,\psi^4 = E\,\psi^2.$

To simplify the energy density, let $\frac{p^2}{2M} - w = a$ and $U = \frac{1}{2}b$ giving
$a\,\psi^2 + \frac{1}{2}b\,\psi^4 = E\,\psi^2$.

Ginzburg and Landau invented a density of Helmholtz free energy, $f = \frac{F}{V}$, and let f_n be normal phase and f_s be superconducting.

They let the energy density be the difference between the free energies of the two phases, $E\,\psi^2 = f_s - f_n$, giving $a\,\psi^2 + \frac{1}{2}b\,\psi^4 = f_s - f_n$.

This is in the form of a fourth order parabola, $a\,x^2 + \frac{1}{2}b\,x^4 = y$.

Such a curve has a local maximum or minimum value when $\frac{dy}{dx} = 0$.

Then, $\frac{dy}{dx} = 2\,a\,x + 2\,b\,x^3 = 0$. $(a + b\,x^2)\,x = 0$ for $x = 0$, $x^2 = -\frac{a}{b}$.

Ginzburg and Landau took a and b to be arbitrary terms, to be interpreted in the following manner.

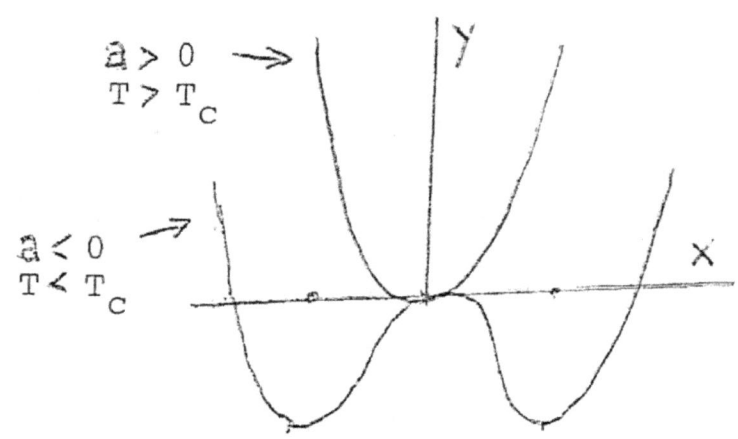

They took b to be positive and a to be either positive or negative.

With being positive, $\frac{dy}{dx} = 0$ happens for $x = 0$ or $x = \sqrt{-\frac{a}{b}}$. With a minimum at only one real valued point, the curve of y against x is the parabola shown above the horizontal axis in the picture. With $a = -|a|$, $\frac{dy}{dx} = 0$, and $x = \pm\sqrt{\frac{|a|}{b}}$.

This is interpreted by the curve below the horizontal axis. The local maximum at $x = 0$ is an unstable point. At $x^2 = -\frac{a}{b}$, the local minimum points are stable.

Following the standard treatment of Ginzburg-Landau theory, this gives $f_s - f_n = -\frac{a^2}{2b}$. Free energy density is $f = \frac{F}{V}$. The physical meanings of a and b were $a = \frac{p^2}{2M} - w$ and $b = 2U$. $-w$ was the interaction with lattice ions and U was the electron-electron repulsion energy.

At this point, let T be the temperature for an electron. (In this scheme, each electron has its own temperature.) Call this a kinetic energy with $kT = \frac{p^2}{2M}$. Now, $a = kT - w$.

When kT is greater than w, a is positive and when kT is less than w, the a is negative. Obviously, $kT_c = w$ gives the critical temperature. Free energy is now $F_s - F_n = \frac{k^2 V (T-T_c)^2}{4U}$ and entropy is $S = -\left(\frac{\partial F}{\partial T}\right)_V$. For $T > T_c$, $S_s - S_n = 0$. For $T < T_c$, $S_s - S_n = \frac{k^2 V (T-T_c)}{2U}$.

Once we have entropy, we have heat capacity, $C_V = T \left(\frac{\partial S}{\partial T}\right)_V$.

For $T > T_c$, $C_{V_s} - C_{V_n} = 0$. For $T < T_c$, $C_{V_s} - C_{V_n} = \frac{k^2}{2U} V T$.

In the right-hand picture at the top of page 20-11, we show the normal heat capacity. What does the result that we just now have derived look like? They look like the left-hand picture. For starters, the Ginzburg-Landau model gives the lambda point.

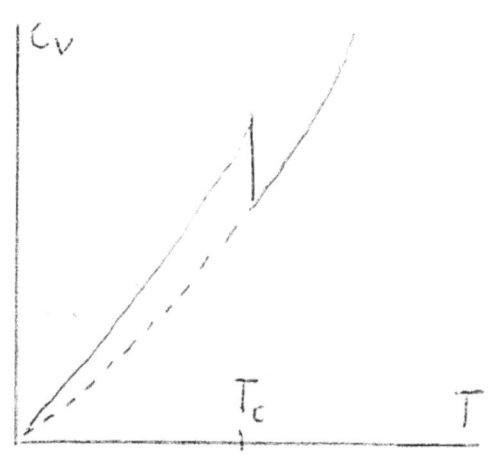

What happens at the critical temperature?

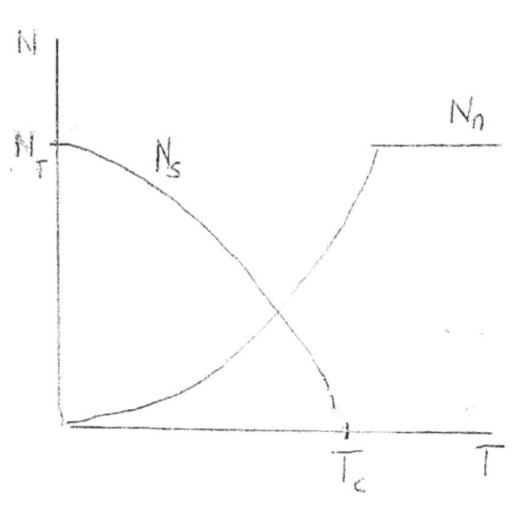

Let n be the number of conducting particles. N_n are the number of normal particles and N_s are the number of superconducting particles and N_T are the total number of conducting particles.

Superconductivity requires that all particles condense into their lowest quantum state. We have seen that electrons by themselves obey the Pauli exclusion principle and cannot do this. If two of them with opposite spin combined to form a spin-zero particle, they would obey Bose-Einstein statistics and be able to drop to their lowest quantum state. Such a material can become superconducting. In 1956, Leon Cooper published a paper describing how this could happen, and thus such composite objects are called Cooper pairs.

Let us return to the Ginzburg-Landau model as derived from the Gross-Pitaevskii equation. $+U$ was the generalized energy of electron-electron repulsion and $-w$ was the generalized energy of electrons to the positive ions of the crystal lattice. T was the temperature of an electron's kinetic energy. For a volume, V, of material, F_s and F_n were the free energies of the superconducting and normal states of matter. When $kT > w$, $F_s - F_n = 0$. When $kT < w$, $F_s - F_n = \dfrac{V(kT-w)^2}{4U}$.

Obviously, $kT_c = w$ gave a critical temperature. Superconductivity appears to result from an interaction of electrons with positive ions in the crystal lattice. When $T > w/k$, the kinetic energy is too large for this to happen. What might happen when $T < w/k$? In this model, in the neighborhood of lattice ions, their positive charges might shield electrons from the electron-electron repulsion. This would allow the weak magnetic dipole-dipole attraction of opposite spins to pull together electrons in the same quantum state to form Cooper pairs. The strength of the electron-electron repulsion, U, is in the denominator, and would weaken the resulting effect.

The mutual energy of a pair of dipoles is $U = -\dfrac{(q\,h)^2}{4\pi\,\epsilon_o}\left[\dfrac{3\cos^2\theta - 1}{r^3}\right]$.

For the magnetic case, replace $\dfrac{(q\,h)^2}{4\pi\,\epsilon_o}$ by $\dfrac{\mu_o\,(I\,A)^2}{4\pi}$.

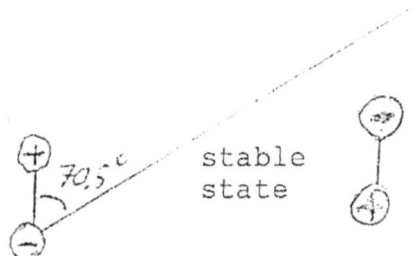

This assumes that the dipoles are in the same direction and parallel. This is stable (or negative) when $\cos\theta$ is greater than $\sqrt{1/3}$, or θ is less than 70.5 degrees. When θ is greater than 70.5 degrees, this gives a positive number (unstable). For these directions, the stable state is when the dipoles are in opposite directions.

Look at a superconducting material. In a lattice of atoms, let each atom donate an electron to a negatively charged "gas", leaving a lattice of positive ions.

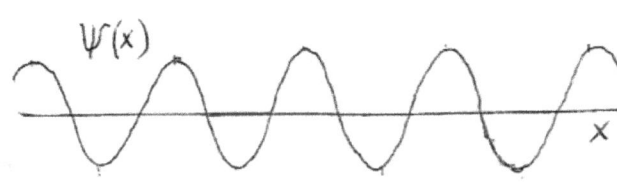

Left alone, the free electrons would have standing wave functions giving a uniform "gas" shown in the left-hand picture. But the potential wells of the positive ions would concentrate the wave functions as shown in the next picture.

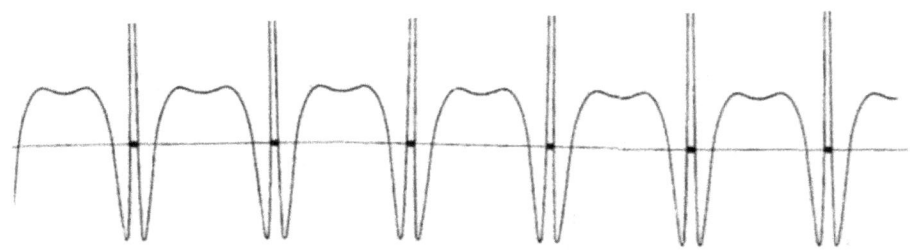

The positive charge on each lattice atom creates a potential well, and that interrupts the free space electron wave function into the wave function of a particle trapped in a potential well. If the square of a wave function is the probability density of the particle, then most of the electron's charge is trapped in the potential well of the positive ion which buffers the coulomb repulsion of its negative charge. The effect of the lattice atom ions is to let electrons weak dipole-dipole interaction to bring electrons of opposite spin to form compound particles with zero spin. These particles obeying Bose-Einstein statistics are now free to all drop to their lowest quantum state forming a superconductor.

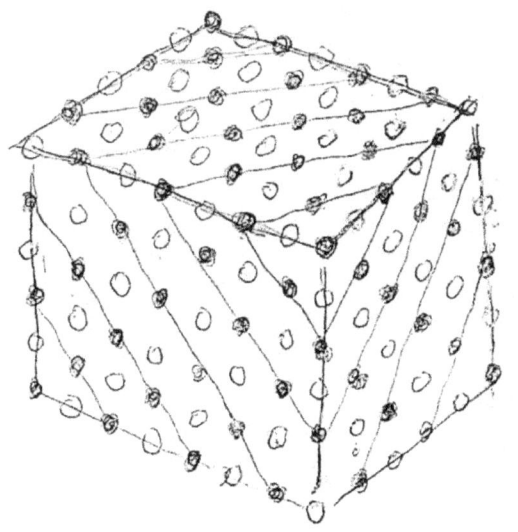

This discussion will deal with those materials whose critical temperatures are above $25\,\text{K}$. These involve primarily transition metal oxides, primarily copper oxides. The picture shows the crystal structure of such a material. The black atoms are copper, and the white atoms are oxygen. In its pure form, such a material will not conduct electricity and is called a Mott insulator. But when certain impurity atoms are inserted, it can go all the way to become a superconductor.

Let us take a plane of atoms perpendicular to a face of the crystal shown above. Then each line of atoms would be perpendicular to this page. Each plane of copper atoms is separated from the next plane of copper atoms by a plane of oxygen atoms. Copper atoms in parallel planes have opposite spins making this a candidate for forming Cooper pairs if electrons from parallel copper planes could somehow get together.

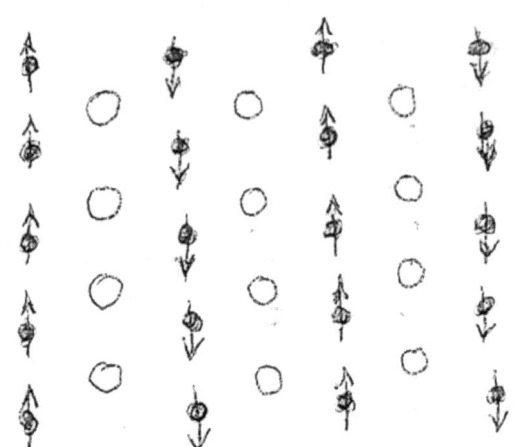

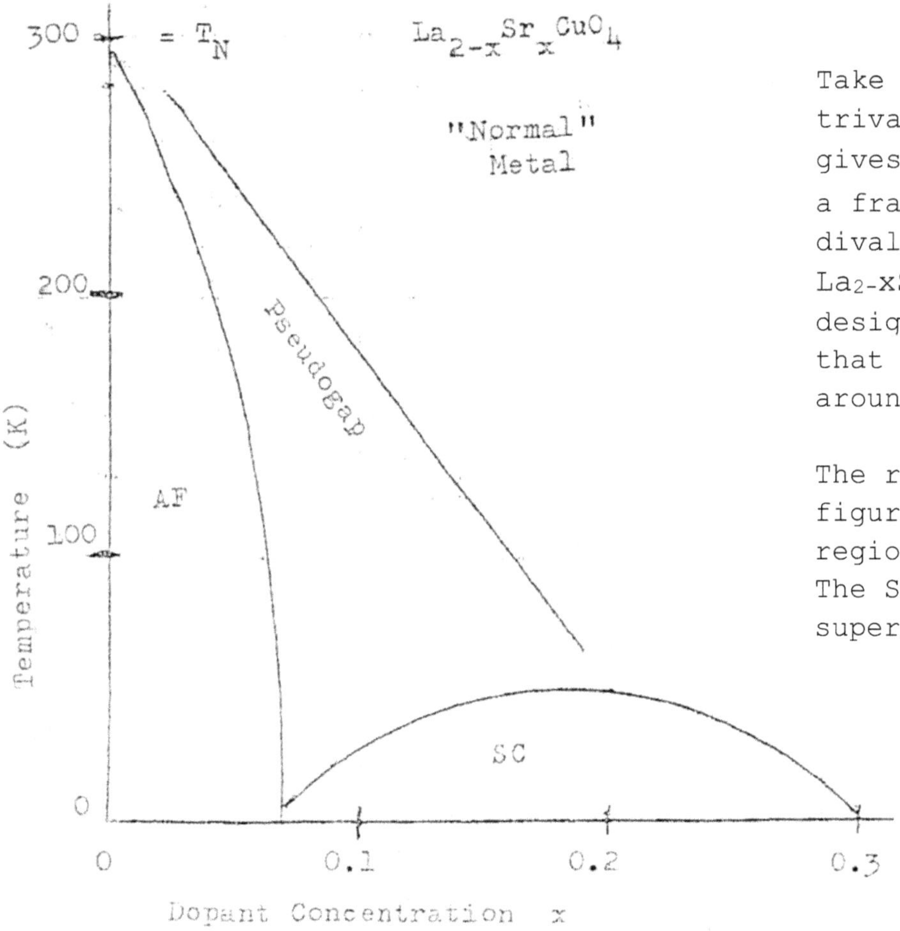

$La_{2-x}Sr_xCuO_4$

"Normal" Metal

Take CuO_4 and introduce trivalent lanthanum. This gives La_2CuO_4. Then replace a fraction, x, of La with divalent strontium to form $La_{2-x}Sr_xCuO_4$. This is designed to provide holes so that the bosons can move around.

The results are shown in the figure to the left. The AF region is antiferromagnetic. The SC region is superconducting.

If these are taken to be phases, there is not a normal phase transition between them, but rather a strange pseudogap. This does achieve superconductivity at temperatures higher than attained with metals.

We next compare the phases with two doped cuprates, the one on the right
introducing holes into the lattice, the one on the left introducing
electrons.

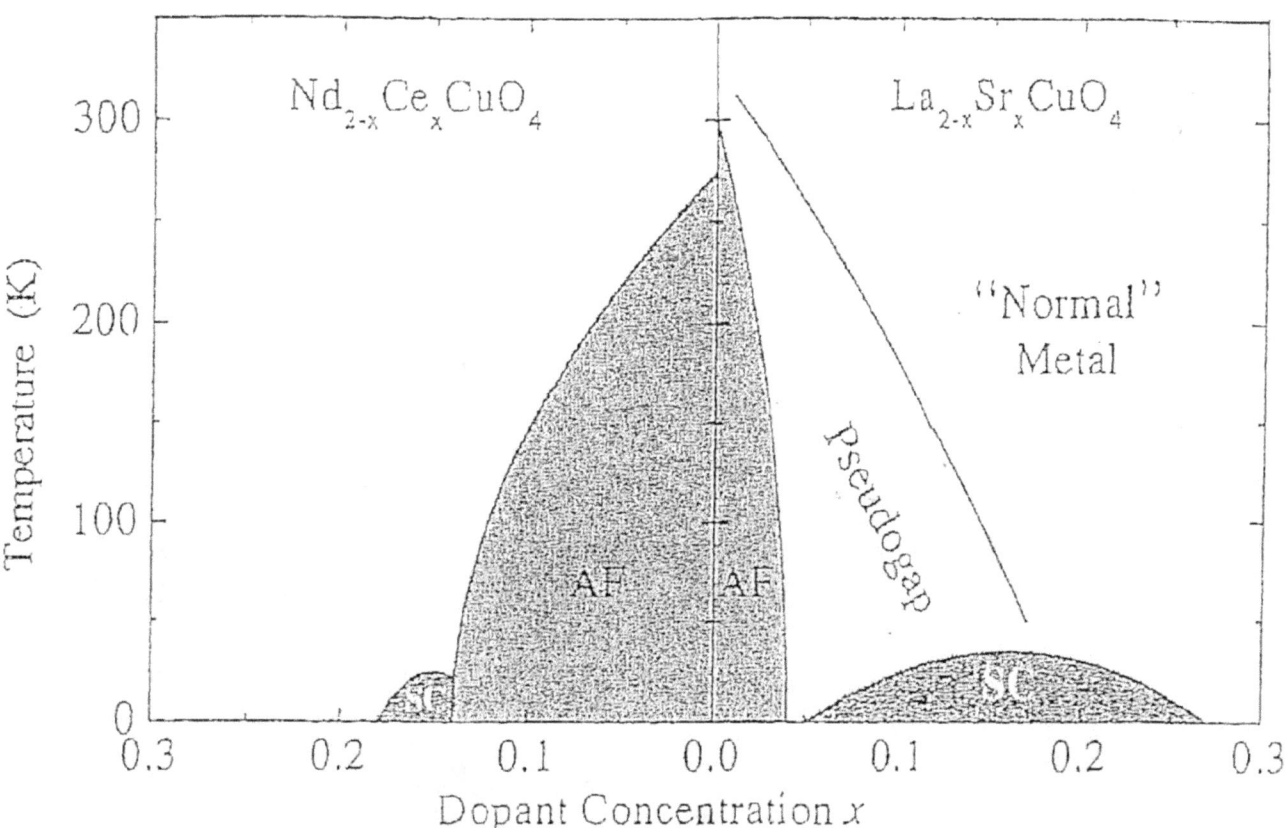

These data tell us several things. Somehow or other there is electron
pairing to produce bosons. The CuO_4 is a nonconducting Mott insulator, just
like pure semiconductors. The introduction of impurities with holes lets the
bosons move about just like with p-type semiconductors. Impurities
introducing electrons do not lead to conduction.

PHYSICS AT HIGH PRESSURE

"The unforeseen may sometimes end conditions which are perfect"
Shih Nai-an, <u>All men are Brothers</u> or <u>Water Margin</u>

The metric unit of pressure is the Pascal with units,
$1.0\,Pa = 1.0$ Newton/meter^2. $10^5\,Pa$ is the pressure of the atmosphere at the
earth's surface. 362×10^9 is estimated to be the pressure at the earth's
center. 10^9 Pascals is 1.0 GPa or gigapascal. The diamond anvil cell was
invented to measure such pressures. See H.K. Mao and P.M. Bell, Science 200,
1145 (1978).

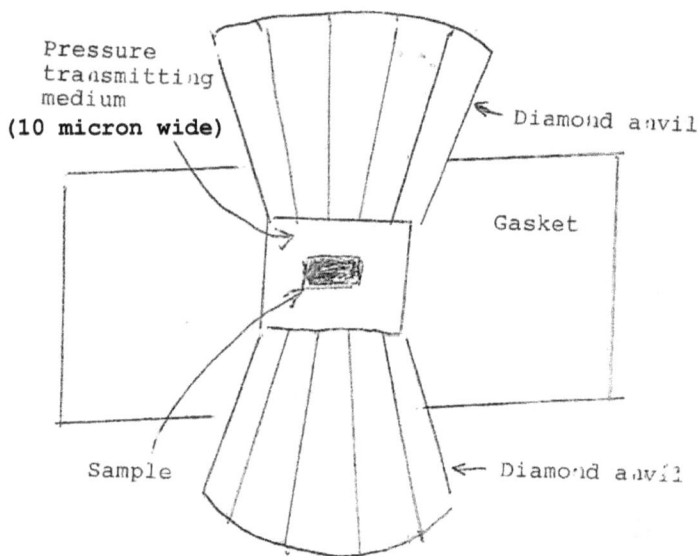

In the diamond anvil cell (DAC), two cut-off diamond cones are pressed on to a pressure transmitting medium in a cylindrical cavity in what is called a gasket. The gasket material is selected for its strength and inertness. The pressure transmitting material is selected for its ability to exert a uniform pressure on to all surfaces of the sample.

In the January-to-March (2018) issue of the Reviews of Modern Physics (RMP)
an article by H.K. Mao and others, SOLIDS, LIQUIDS, and GASES UNDER HIGH
PRESSURE, summarizes the current research in that area. Also in the same
issue of RMP is an article by L.P. Girkov and V.Z. Kresen on HIGH PRESSURE
AND ROAD TO ROOM TEMPERATURE SUPERCONDUCTIVITY.

At high pressure, the study of superconductivity of hydrogen-rich compounds
has been of great interest. Under high pressure, the reaction $3H_2S \rightarrow 2H_3S+S$
appears to take place. Data on superconductivity of sulfur hydride are
reported by:

A.P. Drozdov, et al in Nature (London) 525, 72 (2015)

M.I. Ermets & A.P. Drozdov in Uspechi 57, 1154 (2016)

M. Einaga, et al in Nat. Phys 12, 835 (2016)

The results are shown in the graph on the right.

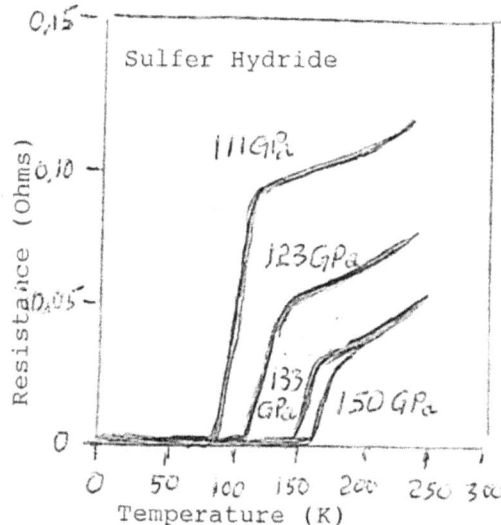

The first thing that you see is the effect of pressure on the critical temperature of transition from the normal phase to the superconducting phase. This phenomenon has been observed in superconductivity in normal laboratory conditions.

The importance of these studies of hydrogen-rich compounds is that they point the way to verifying the predicted metallic hydrogen.

The phenomenon of superconductivity will be an active area of research for a long time to come. There will continue to be surprises and bewilderment. If you desire to study it further, the following are sources for further study.

J.F. Annett: SUPERCONDUCTIVITY, SUPERFLUIDS AND CONDENSATES, Oxford Univ. Press (2004)

P.A. Cox: TRANSITION METAL OXIDES, Oxford (1992)

R.P. Huebener: CONDUCTORS, SEMICONDUCTORS, SUPERCONDUCTORS, Springer International (2015)

A.J. Leggett: QUANTUM LIQUIDS: BOSE CONDENSATION AND COOPER PARIING IN CONDENSED MATTER SYSTEMS, Oxford (2006)

M. Tinkham: INTRODUCTION TO SUPERCONDUCTIVITY, McGraw-Hill (1996)

For a non-mathematical description of the phenomenon, you might start with:

Steven Blundell: SUPERCONDUCTIVITY: A VERY SHORT INTRODUCTION, Oxford (2009)

Lev Landau, laureate of a Lenin and a Nobel Prize, would like to write secondary-school textbooks on physics and mathematics.

"The first thing I will do when I get well is to try to set up a special commission at the Academy of Sciences to draw up new school curricula for eight-year instruction," he told a *Komsomolskaya Pravda* correspondent. "We are teaching children as we were taught 20-30 years ago. New textbooks are needed in many subjects."

Yesterday (22 Jan.) he was 60 and was awarded the Order of Lenin in recognition of his great contribution to the advancement of physical science.

"No, I am not a many-sided scientist," Landau said. "I am just a theoretical physicist. I am really interested only in natural phenomena as long as they are unknown. To investigate them is no work for me; it is great pleasure, satisfaction, tremendous joy, which cannot be compared to anything."

The scientist regards Isaak Pomeranchuk, the physicist who died last year at the age of 53, as his favorite disciple ("How terrible that he died."). He also mentioned Arkadi Migdal, a nuclear physicist: "He is a very, very gifted man but lazy at times."

PRESENTATION OF THE NOBEL PRIZE to Landau by the Swedish ambassador to the USSR (December 1962). Those present at the ceremony are (from left) Mrs. Landau, Igor Tamm, Mstislav Keldish, Nikolai Semenov and Peter Kapitsa.

Landau's favorite writer is Nikolai Gogol. He also eagerly reads Byron in the original and singles out Konstantin Simonov among Soviet prose writers. "However, this has no bearing on my work," he said. "The world of science and the world of art are not in any way connected for me."

Recalling his training abroad in the early 1930's, Landau spoke with great warmth about Niels Bohr, in whose institute many young, and now world famous, scientists had worked.

"Nearly every day we gathered at his institute in Copenhagen and engaged in endless discussions. Incidentally, these were not discussions as such, but a form of creativity, perhaps one of its highest forms."

Asked whether he had spoken with Einstein, Landau replied, "Yes, but little. It was difficult to speak with him since I did not interest him. Nobody interested him. He was too preoccupied with himself."

NOVOSTI PRESS AGENCY (APN)

Unfortunately, Landau died before he could get going with his textbook project. It would be interesting to know what he might have produced. A curriculum that was 20-30 years out of date in 1968 is 50-60 years out of date today. The same thing might be said of the undergraduate physics curriculum in this country today. A glance at freshman physics books now being used will show the same syllabus that I studied in freshman physics in 1947; only we now have colored pictures and higher prices.

CHAPTER TWENTY-TWO: PATH INTEGRALS, DIFFUSION, FUNCTIONAL INTEGRALS, AND
                    QUANTUM RANDOM WALK

This chapter honors Norbert Wiener, Mark Kac and Edward Nelson; three
great American mathematicians. 21st century physics owes a great debt to
their work. Their influence is scattered throughout this chapter.

INTRODUCTION TO PATH INTEGRALS IN PHYSICS

We now move into a very old but very
new realm of thought where the work
of pure mathematicians produces an
insight into physical reality that
cannot be gotten from physics alone.

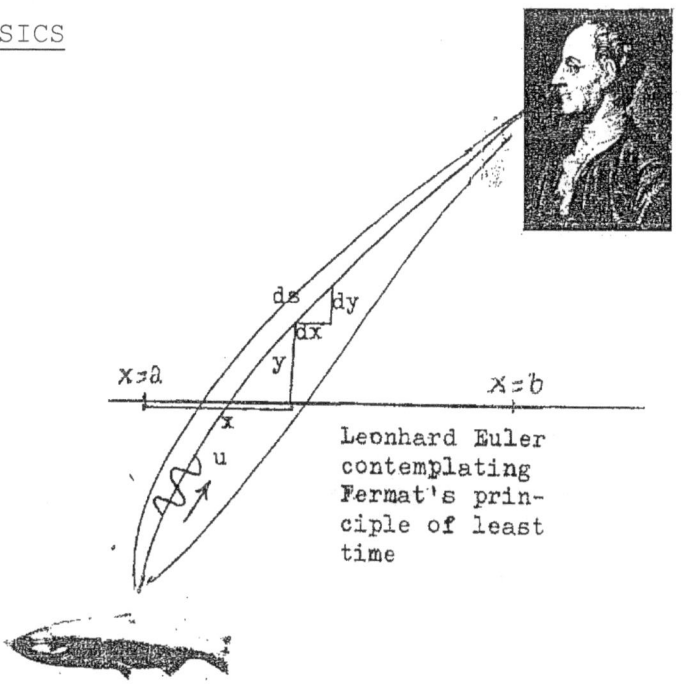

Leonhard Euler
contemplating
Fermat's prin-
ciple of least
time

The French mathematician Pierre
de Fermat (1601-1665) postulated
that light, a wave, travelled
from one point to another over
such a path that the time of
propagation was a minimum.

Let the wave velocity be $u(x,y)$.
The time to traverse an arc,

$$ds = \sqrt{(dx)^2 + (dy)^2} = \sqrt{1 + \left(\frac{dy}{dx}\right)^2}\ dx$$

is $dt = \frac{ds}{u}$.

For convenience, we adopt the notation, $y' = \frac{dy}{dx}$, $y'' = \frac{d^2y}{dx^2}$.

Then, $dt = \frac{1}{u(x,y)}\sqrt{1 + (y')^2}\ dx = f(x,y,y')\ dx$.

Then the time for the wave to move over any path is:

$$t = \int_{x=a}^{b} f(x,y,y')\ dx$$

We are looking for the path, $y = y(x)$, which makes this integral have a
minimum value. Let $y_o(x)$ be whatever path that minimizes t. Let $w(x)$ be any
function such that $w(a) = w(b) = 0$. Let ϵ be a parameter of variation. With
each function, $w(x)$, form a family of paths such that $y(x) = y_o(x) + \epsilon\, w(x)$.
$y(a) = y_o(a)$ and $y(b) = y_o(b)$.

For all paths starting at $x = a$, $y = y(a)$ and ending at $x = b$, $y = y(b)$, we are looking for the path, $y_o(x)$, which causes an integral, $I = \int_{x=a}^{b} f(x, y, y')\, dx$, to have a minimum (or maximum sometimes) value.

In this case, I is time of wave propagation between the fixed points. We pick some function, $w(x)$, and a parameter, ϵ, and construct a family of paths, $y(x) = y_o(x) + \epsilon\, w(x)$, which will give values for $f(x, y, y')$ which now, also, becomes a function of ϵ. For each value of ϵ the integral I will have a value. If I is a continuous function of ϵ and has a minimum value for $y = y_o(x)$ and hence when $\epsilon = 0$, then when $\epsilon = 0$, $\dfrac{dI}{d\epsilon} = 0$.

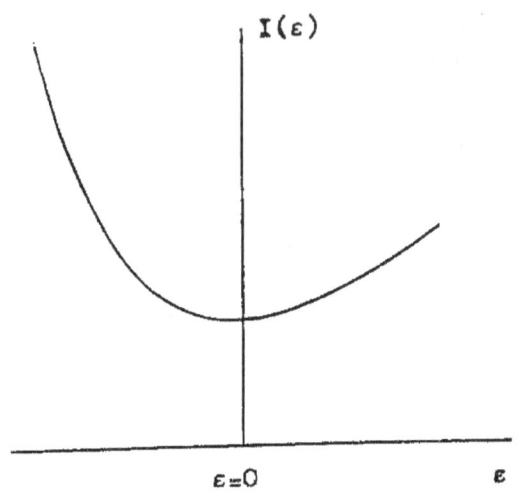

This procedure is called the CALCULUS OF VARIATIONS and will lead to a method for finding the function, $y_o(x)$.

Although the integral, $I(\epsilon)$, is a function only of ϵ, the integrand, $f(x, y, y')$, is a function $y(x, \epsilon)$ and $y'(x, \epsilon)$ so the total derivative, $\partial/\partial\epsilon$, becomes a partial derivative, $\partial/\partial x$, inside the integral. Recall that the integral is a sum, and the derivative of a sum is the sum of the derivatives.

If $y(x, \epsilon) = y_o(x) + \epsilon\, w(x)$, then $\dfrac{\partial y}{\partial \epsilon} = w(x)$, and if $y'(x) = y_o'(x) + \epsilon\, w'(x)$, then $\dfrac{\partial y'}{\partial \epsilon} = w'(x)$.

If $I = \int_{x=a}^{b} f\, dx$, then $\dfrac{dI}{d\epsilon} = \int_{x=a}^{b} \left[\dfrac{\partial f}{\partial y}\dfrac{\partial y}{\partial \epsilon} + \dfrac{\partial f}{\partial y'}\dfrac{\partial y'}{\partial \epsilon}\right] dx = \int_{x=a}^{b} \left[\dfrac{\partial f}{\partial y} w + \dfrac{\partial f}{\partial y'}\dfrac{\partial w}{\partial x}\right] dx$.

We integrate the second integral by parts and note that because of the fixed endpoints on $y(a)$ and $y(b)$, $w(a) = w(b) = 0$.

Thus,

$$\int_{x=a}^{b} \dfrac{\partial f}{\partial y'}\, dw = \dfrac{\partial f}{\partial y'}\Big|_{y=b} w(b) - \dfrac{\partial f}{\partial y'}\Big|_{y=a} w(a) - \int_{x=a}^{b} w(x)\dfrac{d}{dx}\left(\dfrac{\partial f}{\partial y'}\right) dx = -\int_{x=a}^{b} w(x)\dfrac{d}{dx}\left(\dfrac{\partial f}{\partial y'}\right) dx$$

Thus, $\dfrac{dI}{d\epsilon} = \int_{x=a}^{b} \left[\dfrac{\partial f}{\partial y} - \dfrac{d}{dx}\left(\dfrac{\partial f}{\partial y'}\right)\right] w(x)\, dx = 0$ when $\epsilon = 0$ and $y(x) = y_o(x)$.

If this must happen for any function $w(x)$, then when $y = y_0$, $\dfrac{\partial f}{\partial v} - \dfrac{d}{dx}\left(\dfrac{\partial f}{\partial v'}\right) = 0$ which is not zero for any other $y(x)$.

For a while in this discussion we shall be concerned only with those functions which minimize the integral I and $y(x)$ will stand only for those functions allowing us to drop the subscript o.

If we write the result as $\dfrac{d}{dx}\left(\dfrac{\partial f}{\partial y'}\right) - \dfrac{\partial f}{\partial y} = 0$, you will notice that it looks like Lagrange's equation of mechanics, except that the independent variable is x rather than t. In the Calculus of Variations, this result is called the Euler-Lagrange equation.

Recalling our discussion of the Lagrangian function, let us investigate the function $h = \left(\dfrac{\partial f}{\partial y'}\right) y' - f$ recalling that $y' = \dfrac{dy}{dx}$ and $y'' = \dfrac{d^2 y}{dx^2}$.

$$\frac{dh}{dx} = \left(\frac{\partial f}{\partial y'}\right) y'' + y' \frac{d}{dx}\left(\frac{\partial f}{\partial y'}\right) - \left(\frac{\partial f}{\partial y}\right) y' - \left(\frac{\partial f}{\partial y'}\right) y'' - \frac{\partial f}{\partial x} \quad \text{or} \quad \frac{dh}{dx} = -\frac{\partial f}{\partial x}.$$

Suppose our wave moves through a horizontally stratified medium, where the wave velocity is a function of y only. That is $u = u(y)$. When we have discussed waves previously in this treatise, we saw that the frequency remained constant and the velocity and wavelength changed together in $u = f\lambda$. If we let t_w be the time for a wave to move from one point to another, then as the wave goes through an arc, ds, the

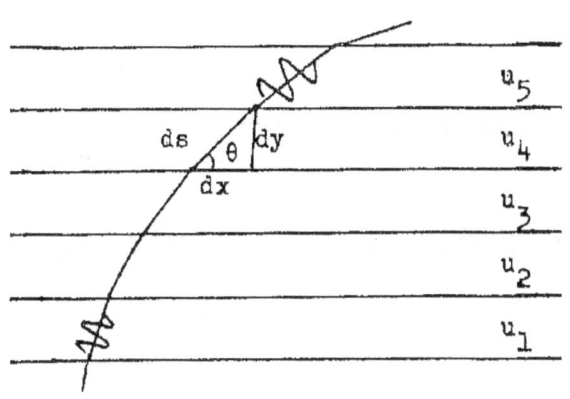

interval of wave time is $dt_w = \dfrac{ds}{u} = \dfrac{1}{u(y)}\sqrt{1 + \left(\dfrac{dy}{dx}\right)^2}\, dx$.

If $t_w = \int dt_w = \int f(y, y')\, dx$, then $f(y, y') = \dfrac{1}{u(y)}\sqrt{1 + (y')^2}$. If t_w is a minimum for the path traversed, then $\dfrac{d}{dx}\left(\dfrac{\partial f}{\partial y'}\right) - \left(\dfrac{\partial f}{\partial y}\right) = 0$.

This gives a second order differential equation to be solved for $y(x)$. But that bother can be avoided by forming $h = \left(\dfrac{\partial f}{\partial y'}\right) - f$. Then, $\dfrac{dh}{dx} = -\dfrac{\partial f}{\partial x}$.

But in the case of a horizontally stratified medium, $u = u(y)$ and neither f nor h is explicitly a function of the letter, x. Hence, $\frac{\partial h}{\partial x} = 0$ and $\frac{dh}{dx} = 0$.

EXERCISE: Starting with the $f(y, y')$ for this case, show that

$$h = \frac{-1}{u(y)\sqrt{1+(y')^2}}$$ is a constant. If h is constant, and

$$y' = \frac{dy}{dx} = \tan\theta, \text{ show that } \frac{\cos\theta}{u(y)} = constant.$$

This is the condition for the path making the wave-time to be a minimum and you will notice that it is identical to Snell's law discussed in chapter 13, pages 13-20 et seq.

In chapter 13 we considered the de Broglie rule that an object with momentum, $m\,v$, has wavelength, $\lambda = \frac{h}{m\,v}$. This allowed us to analyze successfully the motion of baseballs and planets under gravity using Snell's law rather than Newton's second law; giving first order rather than second order differential equations.

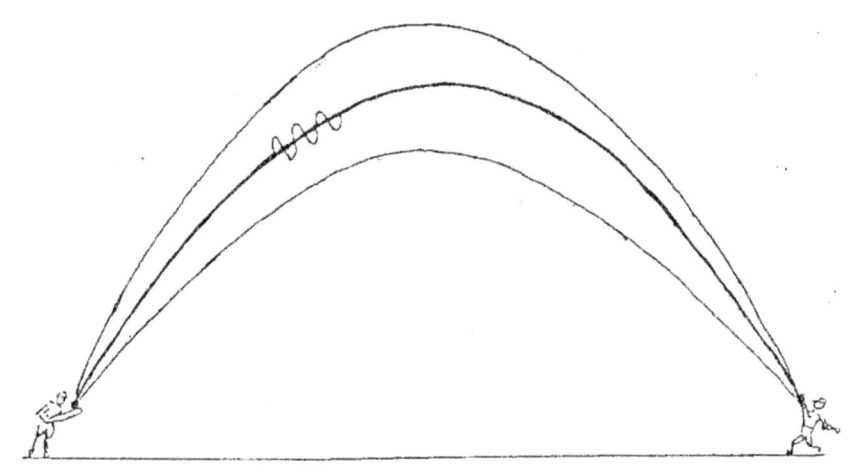

We took an object starting at a definite observed point, (x_1, y_1, z_1), and object time, $t_o = t_1$; and move to a definite point, (x_2, y_2, z_2), and object time, t_2. The definite observed object time of the motion was $T_o = t_2 - t_1$. If the object's velocity is v, its wave velocity is $u = f\lambda = \frac{hf}{m\,v}$, which is obviously different from the object velocity. If the object were a baseball, all we know is that the ball was batted at a definite object time, t_1, and caught at a definite object time, t_2. All the possible paths of motion have the same object time but differing wave times.

As the wave properties of a wave are independent of what is waving, we will assume that Fermat's principle of least time applies to all types of waves. We therefore assume the path for the wave - and hence also the object - will be the one that gives the minimum wave time of transit in the fixed particle time.

Going through an element, ds, of path length, the wave time is $dt_w = \dfrac{ds}{u}$ and the object time is $dt_o = \dfrac{ds}{v}$. If the wave velocity is $u = \dfrac{hf}{mv}$ and $ds = v\,dt_o$, then the wave and object intervals of time are related by $dt_w = \dfrac{mv^2}{hf}\,dt_o$.

Summing over intervals of fixed object time, $T_w = \int_{t_o=t_1}^{t_2} dt_w = \int_{t_o=t_1}^{t_2} \dfrac{mv^2}{hf}\,dt_o$ is the time for the wave to go over a possible path. Let ϵ be a parameter with a value associated with each possible path. Then $d\epsilon$ would shift us from one path to a neighboring path. Then in the neighborhood of the path minimizing T_w, $\dfrac{dT_w}{d\epsilon} = 0$.

Thus, in the neighborhood of the path of wave or object motion,
$\dfrac{d}{d\epsilon}\int_{t_o=t_1}^{t_2} \dfrac{mv^2}{hf}\,dt_o = 0$ or $\dfrac{d}{d\epsilon}\int_{t_o=t_1}^{t_2} mv^2\,dt_o = 0$, after factoring out the constant $\dfrac{1}{hf}$.

The quantity $mv^2\,dt_o$ has the units of what is called Action in physics, and this result is a statement of the PRINCIPLE OF LEAST ACTION as formulated by Sir William Rowan Hamilton. An alternate statement of the principle of least action can be derived from this in the following manner. Recall that the total energy is $E = \frac{1}{2}mv^2 + V$. Now $mv^2 = \frac{1}{2}mv^2 + \frac{1}{2}mv^2 = \frac{1}{2}mv^2 + E - V$. From this we get $\dfrac{d}{d\epsilon}\int_{t_o=t_1}^{t_2}\left[\frac{1}{2}mv^2 - V\right]dt_o + \dfrac{d}{d\epsilon}\int_{t_o=t_1}^{t_2} E\,dt_o = 0$.

But E is constant and $t_2 - t_1$ is constant so $\dfrac{d}{d\epsilon}\int_{t_o=t_1}^{t_2} E\,dt_o = \dfrac{d}{d\epsilon}E(t_2 - t_1) = 0$.

So $\dfrac{d}{d\epsilon}\int_{t_o=t_1}^{t_2}\left[\frac{1}{2}mv^2 - V\right]dt_o = 0$ gives us the path of motion with the principle of least action of Euler and Lagrange. You will recognize $L = \frac{1}{2}mv^2 - V$ as the Lagrangian function.

Let us pretend that we have never seen Lagrangian mechanics and see what happens when we apply the calculus of variations to this problem. Let $L(q_n, \dot{q}_n)$ be a function of a set of coordinates and their velocities, $\dot{q}_n = \dfrac{dq_n}{dt}$. Each point on a path is given by the values of the $q_n(t)$'s at a moment of time. Let the $q_{no}(t)$'s give the coordinates of the path minimizing $\int_{t=t_1}^{t_2} L(q_n, \dot{q}_n)\,dt$, in which t denotes particle time and we omit the subscript o.

Let $w_n(t)$ be any functions for which $w(t_2) = w(t_1) = 0$. We construct a set of paths, $q_n(t) = q_{no}(t) + \epsilon\, w_n(t)$, all starting from the same point and time, $q_n(t_1) = q_{no}(t_1)$, and ending at the same point and time, $q_n(t_2) = q_{no}(t_2)$.

For each path we have a set of velocities, $\dot{q}_n(t) = \dot{q}_{no}(t) + \epsilon\, \dot{w}_n(t)$.

Notice that $\dfrac{\partial q_n}{\partial \epsilon} = w_n$ and $\dfrac{\partial \dot{q}_n}{\partial \epsilon} = \dot{w}_n$. The condition for the action integral to have a minimum value is $\dfrac{d}{d\epsilon} \int_{t=t_1}^{t_2} L\, dt = \int_{t=t_1}^{t_2} \sum_n \left[\dfrac{\partial L}{\partial q_n} \dfrac{\partial q_n}{\partial \epsilon} + \dfrac{\partial L}{\partial \dot{q}_n} \dfrac{\partial \dot{q}_n}{\partial \epsilon} \right] dt = 0$ when $\epsilon = 0$.

EXERCISE: In a double summation, the summations can be taken in any order. Recall that $w(t_2) = w(t_1) = 0$. Then follow the procedure of the calculus of variations from page 22-2 and integrate the second integral by parts to show that:
$$\frac{d}{d\epsilon} \int_{t=t_1}^{t_2} L\, dt = \sum_n \int_{t=t_1}^{t_2} \left[\frac{\partial L}{\partial q_n} - \frac{d}{dt}\left(\frac{\partial L}{\partial \dot{q}_n} \right) \right] w_n(t)\, dt = 0 \text{ when } \epsilon = 0.$$

As the $w_n(t)$ are an arbitrary set of functions, this requires that $\dfrac{\partial L}{\partial q_{no}} - \dfrac{d}{dt}\left(\dfrac{\partial L}{\partial \dot{q}_{no}} \right) = 0$ for the minimizing path, $q_{no}(t)$.

If we assume that we are dealing with the minimizing path, dropping the subscript, o; then taking the negative of this result we get the condition, $\dfrac{d}{dt}\left(\dfrac{\partial L}{\partial \dot{q}_n} \right) - \dfrac{\partial L}{\partial q_n} = 0$, which we recognize as Lagrange's equation for Newton's second law of motion. But we got them from Fermat's principle of least time, using the calculus of variations without any direct mention of Newton' laws except $E = \dfrac{1}{2} m v^2 + V$.

For motion described by a set of variables q_n and a Lagrangian function, $L(q_n, \dot{q}_n) = T(q_n, \dot{q}_n) - V(q_n)$, the momentum gotten from $p_n = \dfrac{\partial L}{\partial \dot{q}_n}$ is called the momentum, Canonically Conjugate to q_n. This is the official definition of momentum in theoretical physics.

Lagrange's equation is primarily a convenient way to find Newton's second law using a difficult set of coordinates which may, however, be convenient for setting up the problem. We are still left with second order differential equations which may be rather difficult to solve.

Some help may come from a contribution from Hamilton - which is also fundamental to the basic structure of theoretical physics.

We define a Hamiltonian function, $H = \sum_n p_n \dot{q}_n - L(q_n, \dot{q}_n)$.

In taking derivatives of H, the partial derivative with respect to a variable takes the derivatives only where the symbol for that variable is found as a letter. The total derivative with respect to a variable takes the derivative where that variable appears as a letter and also where that variable affects any other variable appearing as a letter in the function.

First notice that $\dfrac{\partial H}{\partial \dot{q}_n} = p_n - \dfrac{\partial L}{\partial \dot{q}_n} = p_n - p_n = 0$.

This means that in spite of the manner in which it is defined, the Hamiltonian H is not a function explicitly of the velocities, $\dot{q}_n$.

Now that we have established that $H = H(q_n, p_n)$:

$$\frac{\partial H}{\partial q_n} = -\frac{\partial L}{\partial q_n} = -\dot{p}_n \quad \text{from Lagrange's equation,} \quad \frac{d}{dt}\left(\frac{\partial L}{\partial \dot{q}_n}\right) - \frac{\partial L}{\partial q_n} = 0$$

$$\frac{\partial H}{\partial p_n} = \dot{q}_n$$

$$\frac{dH}{dt} = \sum_n \left[\frac{\partial H}{\partial q_n}\frac{dq_n}{dt} + \frac{\partial H}{\partial p_n}\frac{dp_n}{dt}\right] = 0$$

$H(q_n, p_n)$ is a quantity which has the units of energy and is constant in time. Obviously, the Hamiltonian is the total energy as a function of position and momentum.

Let us now review what has been said about the PRINCIPLE OF LEAST ACTION and the Euler-Lagrange equation that minimizes the action integral.

If y is a position variable and $v = \dfrac{dy}{dt}$ is its velocity, an object subject to a potential energy, $V(y)$, has a Lagrangian function, $L(y, v) = \frac{1}{2}m v^2 - V(y)$, and the condition, $\dfrac{d}{dt}\left(\dfrac{\partial L}{\partial v}\right) - \dfrac{\partial L}{\partial y} = 0$, makes the integral, $S = \int_{t_1}^{t_2} L(y, v)\, dt$, have a minimum value.

$S = \int_{t_1}^{t_2} L(y, v)\, dt$ is called the ACTION INTEGRAL.

To show what can be done with the Lagrangian function, let us consider one dimensional motion under gravity at the Earth's surface. There, $V = m g y$, and $L(y, v) = \frac{1}{2} m v^2 + m g y$ where $v = \frac{dy}{dt}$. Separately we can take the rate-of-change of $L(y, v)$ with either v or y. When there are several variables in a function and we take the rate-of-change of the function with only one of them, the others being treated as constants, we round off the d, for change, to ∂.

Thus, $\frac{\partial L}{\partial v} = m v$ and $\frac{\partial L}{\partial y} = -m g$. Recalling Newton's second law, $\frac{d(m v)}{dt} = -m g$, we can write that as $\frac{d}{dt} \left(\frac{\partial L}{\partial v} \right) = \frac{\partial L}{\partial y}$ or $\frac{d}{dt} \left(\frac{\partial L}{\partial v} \right) - \frac{\partial L}{\partial y} = 0$.

This is known as the Euler-Lagrange equation. Our demonstration was a very cheap trick to derive a very powerful tool for theoretical mechanics.

WHAT DOES THIS HAVE TO DO WITH QUANTUM MECHANICS?

Combining sine and cosine functions, consider a simple wave of the form, $\psi = A e^{i(k x - \omega t)}$. A point of constant phase, $\psi = A e^{i \alpha}$, moves so that $\alpha = (k x - \omega t)$ is constant giving $x = \frac{\alpha + \omega t}{k}$ with a wave velocity, $u = \frac{dx}{dt} = \frac{\omega}{k}$.

As is usual with vibrations, $\omega = 2 \pi f$, and the wave velocity is $u = f \lambda$ giving $f \lambda = \frac{2 \pi f}{k}$ or $k = \frac{2 \pi}{\lambda}$.

This gives a wave function, $\psi = A e^{i 2 \pi \left(\frac{x}{\lambda} - f t \right)}$.

If we use de Broglie's rule, $(m v) \lambda = h$, and define frequency by Planck's rule, $f = \frac{E}{h}$, we get $\psi = A e^{i (2 \pi / h) (p x - E t)}$.

For a constant energy, E, and velocity, v, we get $p x = M v (v t) = M v^2 t$. Then, $E t = \frac{1}{2} M v^2 t + V t$. Thus, $p x - E t = M v^2 t - \frac{1}{2} m v^2 t - V t$, $p x - E t = \frac{1}{2} M v^2 t - V t$. But, $\frac{1}{2} M v^2 - V = L$, the Lagrangian and $S = \int L \, dt$ is the action over the time of the event. $\psi = A e^{i S / \hbar}$.

WHAT DOES THAT MEAN?

PATH INTEGRALS FOR RANDOM WALK, DIFFUSION AND QUANTUM MECHANICS

For a mist of water droplets or a cloud of dust particles dancing in a sun beam, the density of particles is a measure of the probability of finding a particle at some location. When the density of dust particles varies with location, particles tend to drift from higher density locations to where the density is lower. This is called diffusion. Although the solution $\psi(x,t)$ of the Schrödinger equation is called a wave function, it seldom looks like a wave. We have interpreted $\psi(x)^2\,dx$ as the probability of finding the particle so described as being in the dx neighborhood of location, x.

We shall see in this section that the Schrödinger equation is not a wave equation at all, but a diffusion equation. The development of the ψ function in time is given by the functional path integrals first introduced by Norbet Wiener in 1921 and further developed by Mark Kac (pronounced Katz) and Edward Nelson. As we will be talking about probabilities, you might review probability related to statistics at the end of chapter 4 and related to entropy at the start of chapter 19.

MATHEMATICAL PRELIMINARY I: GAUSSIAN INTEGRALS

The integral $\int_{-\infty}^{\infty} e^{-a\,x^2 + b\,x}\,dx$ will occur often in the following discussions. Multiply by $1 = e^{-b^2/4a}\,e^{+b^2/4a}$ to complete the square in the exponent. Then,

$$\int_{-\infty}^{\infty} e^{-a\,x^2 + b\,x}\,dx = e^{b^2/4a} \int_{-\infty}^{\infty} e^{-a\,(x - b/2a)^2}\,dx$$

Let $u = x - b/2a$ and $du = dx$. Then, $\int_{-\infty}^{\infty} e^{-a\,u^2}\,du = \sqrt{\dfrac{\pi}{a}}$.

Hence, $\int_{-\infty}^{\infty} e^{-a\,x^2 + b\,x}\,dx = e^{b^2/4a}\sqrt{\dfrac{\pi}{a}}$.

MATHEMATICAL PRELIMINARY II: RECALL THE FOURIER TRANSFORM

It is easy to show that $\int_{-L}^{L} e^{i\,\alpha\,\pi\,x/L}\,e^{-i\,\beta\,\pi\,x/L}\,dx = 2L\,\delta_{\alpha\beta}$. $\psi(x) = \sum_{\alpha=-\infty}^{+\infty} C_\alpha\,e^{i\,\alpha\,\pi\,x/L}$ has a period of $2L$.

Then, $\int_{-L}^{L} \psi(x)\,e^{-i\,\beta\,\pi\,x/L}\,dx = \sum_{\alpha=-\infty}^{+\infty} C_\alpha \int_{-L}^{L} e^{i\,\alpha\,\pi\,x/L}\,e^{-i\,\beta\,\pi\,x/L}\,dx = 2L\,C_\beta$.

Let $x = \xi$ in this integral and substitute for C_α in the series for $\psi(x)$.

$$\psi(x) = \sum_{\alpha=-\infty}^{+\infty} \frac{1}{2L} \int_{\xi=-L}^{+L} \psi(\xi)\,e^{i\,\alpha\,\pi\,x/L}\,e^{-i\,\alpha\,\pi\,\xi/L}\,d\xi$$

If $\psi(x)$ is a non-periodic function, but where $\int_{x=-\infty}^{+\infty} \psi(x)\,dx$ converges, we get a Fourier representation if we let $k = \alpha\pi/L$, $dk = \pi/L$. Substituting this into the integral and letting $L \to \infty$ gives:

$$\psi(x) = \frac{1}{2\pi} \int_{k=-\infty}^{+\infty} dk \int_{\xi=-\infty}^{+\infty} d\xi\, \psi(\xi)\, e^{ikx}\, e^{-ik\xi}$$

From which it is evident that $\dfrac{1}{2\pi} \int_{k=-\infty}^{+\infty} e^{ikx}\, e^{-ik\xi}\, dk = \lim\limits_{k\to\infty} \dfrac{\sin k\,(x-\xi)}{\pi\,(x-\xi)} = \delta(x-\xi)$.

e^{ikx} has a wavelength, λ, where $e^{ik\lambda} = e^{i2\pi}$. Hence, $k = 2\pi/\lambda$. If this is a wave function for a particle whose momentum is p, then $\lambda p = h$ and $k = 2\pi p/h$. Let $|p\rangle$ be a momentum eigenvector, $P|p\rangle = p|p\rangle$. Then, $\langle x|p\rangle = e^{ipx/h}$ and $\langle p|\xi\rangle = e^{-ip\xi/h}$.

From the bottom of the previous page,

$$\psi(x) = \int_{p=-\infty}^{+\infty} \int_{\xi=-\infty}^{+\infty} \psi(\xi)\, e^{ip\frac{x}{h}}\, e^{-ik\frac{\xi}{h}}\, d\xi\, \left(\frac{dp}{2\pi h}\right)$$

In Dirac notation, $\psi(x) = \langle x|\psi\rangle = \int_{p=-\infty}^{+\infty} \int_{\xi=-\infty}^{+\infty} \langle x|p\rangle\, \langle p|\xi\rangle\, \langle \xi|\psi\rangle\, d\xi\, \left(\frac{dp}{2\pi h}\right)$.

Then, $\psi(p) = \langle p|\psi\rangle = \int_{\xi=-\infty}^{+\infty} \langle p|\xi\rangle\, \langle \xi|\psi\rangle\, d\xi$.

The Fourier transform is the same function vector, $|\psi\rangle$, but with a rotation of basis in Hilbert space.

INTRODUCTION TO DIFFUSION

Consider a fluid flowing without drag in a tube with cross section area, A. Let ρ (rho) be density in kg/m³. Let x be distance along the tube. The volume of a piece dx of the tube is $A\,dx$ and it contains a mass, $dM = \rho\,A\,dx$. If all of that mass crosses the right-hand boundary in time, dt, the mass per second is $\dfrac{dM}{dt} = \dfrac{\rho\,A\,dx}{dt}$, but $\dfrac{dx}{dt} = v$ is the velocity of the flow.

$\dfrac{dM}{dt} = \rho\,A\,V$. Now define $J = \rho\,v$ in (kg/sec)/meter². $\dfrac{dM}{dt} = J\,A$.

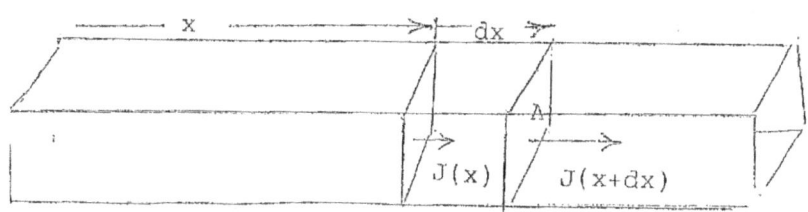

With the volume, $A\,dx$, the flow out is $J(x+dx)\,A$. The flow in is $J(x)\,A$. If the net flow is out, the mass is going down. dM/dt is negative.

For $dM = \rho\,A\,dx$, it is the density now changing. If the flow out is positive and $\dfrac{dM}{dt} = \dfrac{d\rho}{dt}A\,dx$ is negative, $-\dfrac{d\rho}{dt}A\,dx = [\,J(x+dx)\,A - J(x)\,A\,]$,

$\dfrac{d\rho}{dt} = -\left[\dfrac{J(x+dx) - J(x)}{dx}\right]$, or $\dfrac{d\rho}{dt} = -\dfrac{dJ}{dx}$.

This is the equation for a conserved fluid where mass is neither gained nor lost.

Next let the fluid be a mist of particles in motion with no transport flowing motion. Let the density of particles be increasing with x.

If the particles are in random motion, we would expect a drift from greater to lesser density.

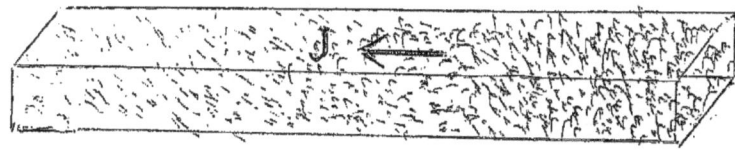

This motion is called DIFFUSION. The diffusion current would be opposite to the direction of increasing density, and proportional to the rate of increase of density with position. The number giving the amount of flow for a rate of density increase is the DIFFUSION CONSTANT, D.

We get $J = -D\dfrac{d\rho}{dx}$, but if the fluid is conserved we get $\dfrac{d\rho}{dt} = D\dfrac{d^2\rho}{dx^2}$.
This is the diffusion equation.

But there are cases where particles are not conserved. In a mist of water droplets, we could have evaporation. If one droplet is as likely to evaporate as another, the rate of loss of density would be proportional to the density by a fraction, f. But the fraction could vary from place to place if, say the temperature varies. We then have $\frac{d\rho}{dt} = D\frac{d^2\rho}{dx^2} - f(x)\rho$. This is the Bloch equation.

Now stop and look at time-dependent Schrödinger equation.

$$i\hbar\frac{d\psi}{dt} = -\frac{\hbar^2}{2m}\frac{d^2\psi}{dx^2} + V(x)\psi$$

Re-writing this gives. $\frac{d\psi}{dt} = \frac{i\hbar}{2m}\frac{d^2\psi}{dx^2} - \frac{i}{\hbar}V(x)\psi$.

Hence, $D = \frac{i\hbar}{2m}$ is an imaginary diffusion constant for the Schrödinger equation. We shall see this again later.

Solving the one-dimensional diffusion equation for a conserved "mist" gives some tricks that will be used later.

$\frac{d\rho}{dt} = D\frac{d^2\rho}{dx^2}$ suggests $\rho = e^{\alpha t}e^{\beta x}$ with $\alpha = D\beta^2$. As ρ must not diverge at infinite x we require $\beta^2 = -k^2$ and $\alpha = -k^2 D$. $\rho = C e^{-Dk^2 t}e^{ikx}$.

We don't know what k represents, but let it be a continuous parameter. Then $C = C(k)$, and $\rho(x,t) = \int_{k=-\infty}^{+\infty} C(k)e^{-Dk^2 t}e^{ikx}\,dk$.

At $t = 0$, $\rho(x,0) = \int_{k=-\infty}^{+\infty} C(k)e^{ikx}\,dk$ is a Fourier transform.
So, $C(k) = \frac{1}{2\pi}\int_{\xi=-\infty}^{+\infty} \rho(\xi,0)e^{-ik\xi}\,d\xi$.

$C(k)$ is the inverse Fourier transform of the initial density distribution.

Thus, $\rho(x,t) = \frac{1}{2\pi}\int_{\xi=-\infty}^{+\infty} \rho(\xi,0)\,d\xi \int_{k=-\infty}^{+\infty} e^{-Dk^2 t}e^{ik(x-\xi)}\,dk$.

In the integral over k; t, x, and ξ remain constant. Hence, we have the Gaussian integral from page 22-9, where $a = Dt$ and $b = i(x-\xi)$.

Hence, $\rho(x,t) = \frac{1}{2\pi}\sqrt{\frac{\pi}{Dt}}\int_{\xi=-\infty}^{+\infty} \rho(\xi,0)e^{-(x-\xi)^2/4Dt}\,d\xi$.

For a point source at $t = 0$, $\rho(\xi, 0) = \delta(\xi)$, $\rho(x, t) = \sqrt{\frac{1}{4\pi D t}} \, e^{-x^2/4Dt}$.

If we put the quantum diffusion constant into the solution of the diffusion equation, we get $\psi(x, t) = \sqrt{\frac{m}{ibt}} \, exp\left[-\frac{\pi m}{ibt} x^2\right]$, where $b = 2\pi h$, which is exactly what Feynman got with his path integral. But what does the exponent mean? With zero potential energy the momentum is constant. Hence, $p = m x/t$ and $-\frac{\pi m x^2}{iht} = i - \frac{p x}{2h}$ which is $i\frac{p x}{h} - i\frac{p x}{2h}$ but $\frac{p x}{2h} = \frac{m v^2 t}{2h} = E t/h$.

So $exp\left[-\frac{\pi m}{ibt} x^2\right] = e^{i(p x - E t)/h}$ which is the expected wave function for a particle with zero potential energy. Next, what about a solution to the BLOCH EQUATION, $\frac{d\rho}{dt} = D\frac{d^2\rho}{dx^2} - f\rho$.

Again, let $\rho(x, t) = C e^{\alpha t + ikx}$. $\alpha = -k^2 D - f$. Integrating over continuous $C(k)$, $\rho(x, t) = \int_k G(k) e^{(-Dk^2 - f)t} e^{ikx} dk$.

Again, $C(k) = \frac{1}{2\pi}\int_x \rho(x, 0) e^{-ikx} dx$ and if $\rho(x, 0) = \delta(x)$, $C(k) = \frac{1}{2\pi}$. Reduced to the solution of the diffusion equation, $\rho(x, t) = \sqrt{\frac{1}{4\pi D t}} \, e^{-x^2/4Dt} e^{-ft}$, which is no surprise.

Another version of diffusion is the FOKKER-PLANK EQUATION. This involves diffusion, as in a cloud, moving with an overall velocity, v_0. Here in one dimension, $\frac{\partial\rho}{\partial t} = -\frac{\partial J}{\partial x}$, with $J = -D\frac{\partial\rho}{\partial t} + \rho v_0$ or $\frac{\partial\rho}{\partial t} = D\frac{\partial^2\rho}{\partial x^2} - v_0\frac{\partial\rho}{\partial x}$.

Now, $\rho(x, t) = C e^{\alpha t + ikx}$ gives $\alpha = -k^2 D - i v_0 k$.

Again, if $\rho(x, 0) = \delta(x)$, $C(k) = \frac{1}{2\pi}$, and $\rho(x, t) = \frac{1}{2\pi}\int_k e^{-Dtk^2 + i(x - v_0 t)k} dk$.

Compared with $\int e^{-ak^2 + bk} dk = \sqrt{\frac{\pi}{a}} \, e^{b^2/4a}$, $\rho(x, t) = \sqrt{\frac{1}{4\pi D t}} \, e^{-(x - v_0 t)^2/4Dt}$, which is no surprise.

The literature on diffusion, random walk and stochastic processes is vast.
The following are several references from the author´s library:

A. Einstein INVESTIGATION ON THE THEORY OF BROWNIAN MOTION Dover(1956)

M. Chaichian, A. Demichev PATH INTEGRALS, VOLUME I: STOCHASTIC PROCESSES
                          AND QUANTUM MECHANICS Institute of Physics (2001)
R.P. Feynman and A.R. Hibbs QUANTUM MECHANICS AND PATH INTEGRALS
                          McGraq Hill (1965)
R.P. Feyman STATISTICAL MECHANICS W.A. Benjamin (1972)

DIFFUSION AS RANDOM WALK

A particle (or drunken college student)
moves randomly in one dimension starting
at point zero and taking a step to the
right or left from which point taking a
step to the right or left with equal
probability. Let m be his location from
point zero after n steps. To make things
easy, let us consider n to be even
integers. If l is the distance between
occupied locations, the displacement will
be $x = m\,l$. After n steps, the number of
paths to the mth location will be

$$P_n(m) = \frac{n!}{\left(\frac{n}{2}+m\right)!\left(\frac{n}{2}-m\right)!}.$$

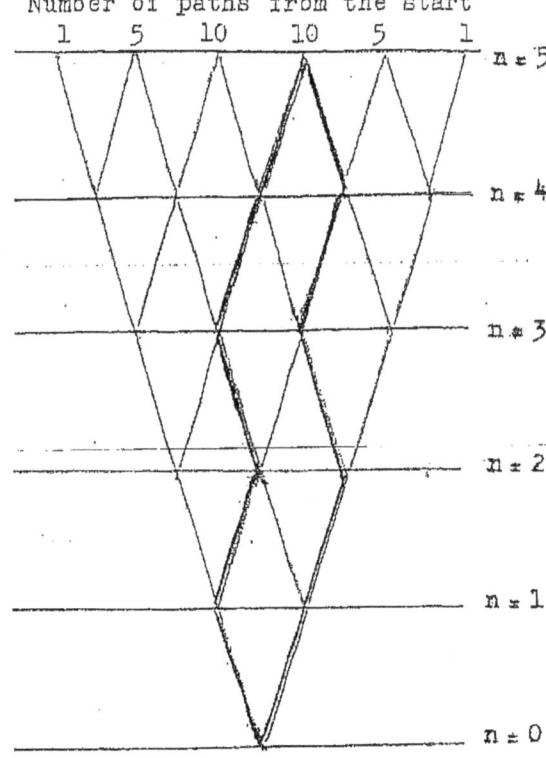

Number of paths from the start
1   5   10   10   5   1

$n = 5$

$n = 4$

$n = 3$

$n = 2$

$n = 1$

$n = 0$

To get a simpler formula for P as a
function of x let us look at

$$\frac{dP}{dx} = \frac{1}{2\,l}\big[P(m+1) - P(m-1)\big].$$

A little algebra shows that

$$\frac{dP}{dx} = \frac{P_n(m)}{2\,l}\left[\frac{\frac{n}{2}-m}{\frac{n}{2}+m+1} - \frac{\frac{n}{2}+m}{\frac{n}{2}-m+1}\right] = \frac{P_n(m)}{2\,l}\left[\frac{-2\,n\,m}{\left(\frac{n}{2}\right)^2 - m^2}\right].$$

For large $n/2$ we ignored the 1's in the denominators. After a large number of steps let us take only those positions m that have a significant probability of being occupied, that is $\left(\dfrac{n}{2}\right)^2$ is much bigger than m^2. Then letting $m = x/l$, we get $\dfrac{dP}{dx} = -P(x)\dfrac{1}{2\,l^2} = \left(\dfrac{8\,x}{n}\right)$ or $\dfrac{dP}{P} = -\dfrac{4\,x\,dx}{n\,l^2} + 0$. From this we get $P_n(x) = C\exp(-2\,x^2/n\,l^2)$. Compared with $\rho(x,t) = \sqrt{\dfrac{1}{4\pi D t}}\,\exp(-x^2/4\,D\,t)$, suggests that $D = n\,l^2/8\,t$.

For a system of particle diffusing by collisions, $\varepsilon = t/n$ is the time between collisions. Notice carefully that $L = l/2$ is the mean free path between collisions. The diffusion constant is then $D = L^2/2\,\varepsilon$.

FUNCTIONAL INTEGRALS WITH WIENER MEASURE

We now follow Norbert Wiener in generalizing on a random walk. We refer to:

Norbert Wiener (1921) Proc. Natl. Acad. Sci. 7, 256.
Norbert Wiener (1921) Proc. Natl. Acad. Sci. 7, 294.
Norbert Wiener (1923) J. Math. Phys. Sci. Natl. 2, 132.
Norbert Wiener (1924) Proc. Lond. Math. Soc. 22, 454.
Norbert Wiener (1930) Acta. Math. 55, 304.
Edward Nelson (1967) Dynamical Theories of Brownian Motion Princeton U.
              Press
Edward Nelson (1985) Quantum Fluctuations Princeton U. Press

Return to the diffusion function on page 22-12. Let the elapsed time, t, be divided into n intervals with each interval being $dt = t/n$. Let epsilon (ϵ) be the duration of an interval, $\epsilon = t/n$. $t_n = n\epsilon$ is the time after n intervals. Let x_n be the location of things happening at the nth time. All x's are the same position domain but each gives position at a particular moment of time. Let x_o instead of ξ be locations at time, $t = 0$. The equation for density at locations at time, t, was:

$$\rho(x_1, \epsilon) = \sqrt{\frac{1}{4\pi D t}} \int_{x_o=-\infty}^{+\infty} \rho(x, 0)\, e^{-(x-x_o)^2/4Dt}\, e^{-ft}\, dx_o$$

In the picture on the right, time advances toward the top. Density of the mist or dust cloud varies with location, x, left-right.

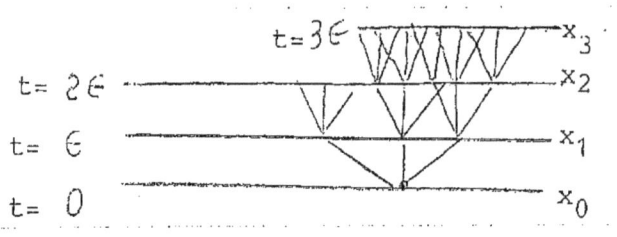

We are not showing the values of a density function; we are merely showing the points where the density has a value.

From each position at $t = 0$, the paths show the steps in the calculation of density at time $t = \epsilon$, and from there the steps in the calculation of density at time, $t = 2\epsilon$. We use the above formula, but only step by step as time advances, so at $t = \epsilon$, density is:

$$\rho(x_1, \epsilon) = \sqrt{\frac{1}{4\pi D \epsilon}} \int_{x_o=-\infty}^{+\infty} \rho(x_o, 0)\, e^{-(x_1-x_o)^2/4D\epsilon}\, e^{-f\epsilon}\, dx_o$$

Looking at the first exponent of e, $\frac{(x_1-x_0)^2}{4 D \epsilon} = \left[\frac{x_1-x_0}{\epsilon}\right]^2 \frac{\epsilon}{4 D}$. Let $v_1 = \frac{x_1-x_0}{\epsilon}$ be a velocity going from x_0 to x_1 in time, ϵ.

$$\rho(x_1,\epsilon) = \sqrt{\frac{1}{4 \pi D \epsilon}} \int_{x_0} dx_0 \, \rho(x_0,0) \, e^{-[(v_1^2/4 D + f)]\epsilon}$$

It will be understood that the summation over position goes from $x_n = -\infty$ to $+\infty$, and we will put the change, dx_n, immediately after the summation over x_n at $t = n\epsilon$.

At time $t = 2\epsilon$, $\rho(x_2, 2\epsilon) = \sqrt{\frac{1}{4 \pi D \epsilon}} \int_{x_1} dx_1 \, \rho(x_1,\epsilon) \, e^{-[(v_2^2/4 D + f)]\epsilon}$.

Putting in $\rho(x_1,\epsilon)$ gives:

$$\rho(x_2, 2\epsilon) = \frac{1}{4 \pi D \epsilon} \int_{x_1} dx_1 \int_{x_0} dx_0 \, \rho(x_0,0) \, e^{-[(v_2^2/4 D + f)]\epsilon} \, e^{-[(v_1^2/4 D + f)]\epsilon}$$

It is evident that after n jumps in time, the density at points, x_n, being position at time, $t = n$, will be:

$$\rho(x_n, n\epsilon) = \left[\frac{1}{4 \pi D \epsilon}\right]^{n/2} \int_{x_{n-1}} dx_{n-1} \int_{x_{n-2}} dx_{n-2} \cdots \int_{x_0} dx_0 \, e^{-\Sigma_{k=1}^{n}[(v_k^2/4 D + f)]\epsilon} \, \rho(x_0,0)$$

Remember that x_n is the x-axis on which the density is calculated at the time, $t = n\epsilon$. In the picture we show some of the infinitely many paths by which the density of a diffusing dust cloud or mist of water drops. At each instant of time, density is calculated from only what it was at the just previous instant of time. The process is merely a vast random walk.

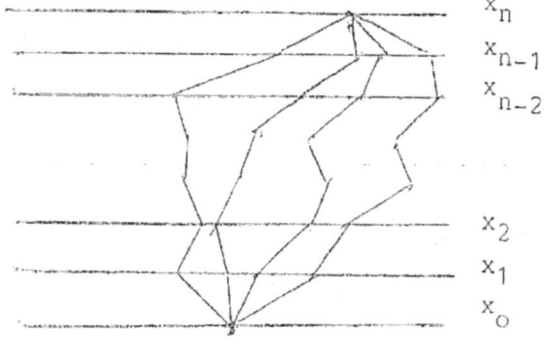

At each stage, the density function depends only on what it was at the previous stage, without knowing how it got to the previous stage. $t = 0$ could be arbitrarily at any time. A process that starts with its nature at a particular instant of time and evolves into the future without any knowledge of how it got to its starting point is called a MARKOV PROCESS.

The integral for $\rho(x_n, n\,\epsilon)$ was developed by Norbert Wiener in a set of papers published between 1921 and 1930. It is called the Wiener functional integral or the Wiener path integral. For a physicist its importance is its insight into diffusion as a random walk. To the mathematician its importance was that it fell off the cliff of known mathematics and forced the development of new mathematical ideas.

The product of changes, $dx_{n-1}\, dx_{n-2}\, dx_{n-3}\, ... \, dx_3\, dx_2\, dx_1\, dx_o$, is the product of position elements on the same x-axis but taken at different times. That said, however, to a mathematician it looks like a volume element in an n-dimensional space. The problem is that the dimension is arbitrary and undefined.

If the integral at the top of this page is to be used often in a text, we don't want to write the whole thing out every time it is used. Once we know what it is, there is short way of referring to it. We write:

$$\int \mathcal{D}\,x = \left[\frac{1}{4\,\pi\,D\,\epsilon}\right]^{n/2} \int\limits_{x_{n-1}} dx_{n-1} \int\limits_{x_{n-2}} dx_{n-2} \, ... \int\limits_{x_2} dx_2 \int\limits_{x_1} dx_1 \int\limits_{x_o} dx_o$$

$$\rho(x,t) = \int \mathcal{D}\,x \; e^{\int_t (v^2/4\,D + f)\,dt} \; \rho(x_o, 0)$$

The concept of MEASURE refers, roughly speaking, to volume in a given space. This is called a FUNCTIONAL INTEGRAL with WIENER MEASURE.

We have seen that the Schrödinger equation is identical mathematically to the Bloch equation except for different letters. We solve for $\psi(x,t)$ rather that $\rho(x,t)$. The quantum diffusion constant is $D = \frac{i\,\hbar}{2\,M}$ and $f = \frac{i\,V}{\hbar}$. $\frac{1}{D} = -\frac{i\,2\,M}{\hbar}$ as $\frac{1}{i} = -i$. The quantity in the exponent is now $-[(v^2/4\,D + f)] = i\left[\frac{M\,v^2}{2\,\hbar} - \frac{V(x)}{\hbar}\right]$.

But $\frac{1}{2}M\,v^2 - V(x) = L(x,v)$, the Lagrangian and $S = \int L(x,v)\,dt$ is the action integral. $\sqrt{\frac{1}{4\,\pi\,D\,\epsilon}}\,dx$ is replaced by $\sqrt{\frac{M}{i\,2\,\pi\,\hbar\,\epsilon}}\,dx$. The quantum "wave function" evolves in time as $\psi(x,t) = \int \mathcal{D}\,x \; e^{i\,S/\hbar}\,\psi(x_o, 0)$.

Although this looks like the formula at the bottom of page 22-8, it has a totally new interpretation. Richard Feynman (1918-1987) developed it in his doctoral dissertation at Princeton, independent of the work of Norbert Wiener. The approach taken here is influenced by the work of Mark Kac. It is called the Feynman-Kac path integral or the functional integral with Feynman-Kac measure.

Our interpretation is that it is a quantum random walk. Feynman had another idea, Feynman went back to a principle of wave propagation stated by Christian Huygens (1629-1695). A disturbance in a medium moves away from its origin as a wave. Each point on the resulting wave is itself a disturbance and hence produces a new wavelet. Starting from a configuration of a wave front, pieces of the wavelets moving parallel to the wave front all cancel. Pieces of the wavelets moving in the forward direction are in phase with each other and unite to form the envelope of a new configuration of the waveform. Each piece of the new configuration involving the sum of wavelets from all pieces of the former configuration.

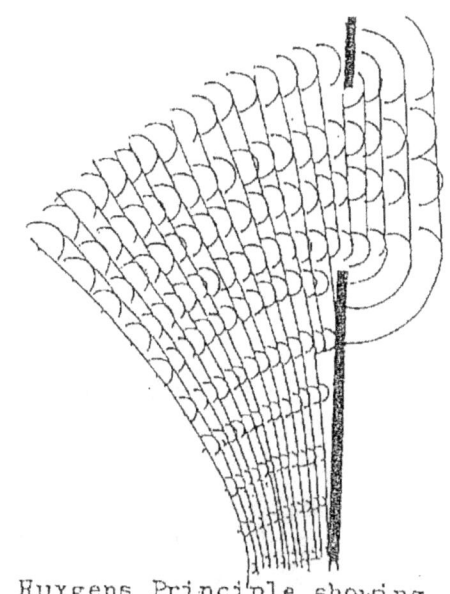

Huygens Principle showing refraction and diffraction

In the Huygens-Feynman picture, wavefront constitutes a disturbance that moves forward to create the next wavefront, it retains no memory of the past as to how it got to its position at the moment. This also has the condition of a Markov process.

The evaluation (getting numbers) for these monstrosities is more than we want to get into now. Even for the "easy" cases it is more like cracking peanuts with a ten-ton press. The importance lies in the profound questions that random (stochastic) Markov processes raise for our understanding of what physics is all about.

The following is list of some books explaining or using path integrals.

Feynman and Hibbs: QUANTUM MECHANICS AND PATH INTEGRALS
        McGraw-Hill (1965) 0-07-020650-3

R. P. Feynman: STATISTICAL MECHANICS
        W. A. Benjamin (1972) 0-805-325650-3

Hagen Kleinert: PATH INTEGRALS IN QUANTUM MECHANICS, STATISTICS, POLYMER
        PHYSICS, AND FINANCIAL MARKETS
        World Scientific Publishing (2004) 981-238-107-4 paperback
        1468 page report on the evolution of path integrals and their
        application since Kleinert's collaboration with Feynman

L. S. Schulman: TECHNIQUES AND APPLICATIONS OF PATH INTEGRATION
        Dover Publications (2005) 0-486-44528-3
        Contains hundreds of references. Excellent introduction to the
        literature

G. W. Johnson and M. L. Lapidus: THE FEYNMAN INTEGRAL AND FEYNMAN'S
        OPERATIONAL CALCULUS, Oxford Univ. Press (2000) 0-19-853574-0
        Two mathematicians try to make sense out of just what did Feynman
        do? Lapidus worked with Feynman

M. Chaichian, A. Demichev: PATH INTEGRALS IN PHYSICS
        V. I. STOCHASTIC PROCESSES AND QUANTUM MECHANICS
        V. II. QUANTUM FIELD THEORY, STATISTICAL PHYSICS AND OTHER MODERN
        APPLICATIONS, Institute of Physics (2001) 2 vol. 0-7503-0713-7

Harold J. W. Mueller-Kirsten: INTRODUCTION TO QUANTUM MECHANICS, SCHRODINGER
        EQUATION AND THE PATH INTEGRAL
        World Scientific (2006) 981-256-692-9

J. Zinn-Justin: PATH INTEGRALS IN QUANTUM MECHANICS
        Oxford (2005) 0-19-856674-3

Ashok Das: FIELD THEORY: A PATH INTEGRAL APPROACH
        World Scientific (2006) 981-256-848-4

Appendix A:

We review the Lagrangian mechanics

If q_i and $\dot{q}_i = \dfrac{dq_i}{dt}$ with $L(q_i, \dot{q}_i)$ is the Lagrangian for a phenomenon, the

conjugate momentum is $p_i = \partial L / \partial \dot{q}_i$ and $\dfrac{d}{dt}(\partial L / \partial \dot{q}_i) - \partial L / \partial q_i = 0$ is the

Euler-Lagrangian equation.

Let us apply all this to a particle moving in an electro-magnetic field. We shall need to recall some things from chapter 32.

In electric and magnetic fields $\overline{E} = -\overline{\nabla}\phi - \dfrac{\partial \overline{A}}{\partial t}$ and $\overline{B} = \overline{\nabla} \times \overline{A}$

as derived from the scalar and vector potentials $\phi(x,y,z,t)$ and $\overline{B}(x,y,z,t)$ a particle with charge q and mass M will obey Newton's second law

$$M\dfrac{d\overline{v}}{dt} = q(\overline{E} + \overline{v}\times\overline{B}) \quad\text{or}\quad M\dfrac{d\overline{v}}{dt} = q\left[-\overline{\nabla}\phi - \dfrac{\partial\overline{A}}{\partial t} + \overline{v}\times(\overline{\nabla}\times\overline{A})\right]$$

Taking the x-component of the last term

$$\left[\overline{v}\times(\overline{\nabla}\times\overline{A})\right]_x = v_x\left(\dfrac{\partial A_z}{\partial z} - \dfrac{\partial A_x}{\partial x}\right) - v_y\left(\dfrac{\partial A_z}{\partial y} - \dfrac{\partial A_y}{\partial x}\right)\ \text{to which add}\ \ v_x\dfrac{\partial A_x}{\partial t} - v_x\dfrac{\partial A_x}{\partial x} = 0$$

and recall that the velocity may vary as the particle moves from position to position but does not have an explicit formula with position for taking partial derivatives - so $\partial v_x/\partial x = 0$, etc.

$$\left[\overline{v}\times(\overline{\nabla}\times\overline{A})\right]_x = v_x\dfrac{\partial A_x}{\partial x} + v_y\dfrac{\partial A_y}{\partial x} + v_z\dfrac{\partial A_z}{\partial x} - v_x\dfrac{\partial A_x}{\partial x} - v_y\dfrac{\partial A_x}{\partial y} - v_z\dfrac{\partial A_x}{\partial z}$$

Generalizing to three dimensions $\overline{v}\times(\overline{\nabla}\times\overline{A}) = \overline{\nabla}(\overline{v}\cdot\overline{A}) - (\overline{v}\cdot\overline{\nabla})\overline{A}$

giving $\quad M\dfrac{d\overline{v}}{dt} = q\left[-\overline{\nabla}\phi + \overline{\nabla}(\overline{v}\cdot\overline{A}) - (\overline{v}\cdot\overline{\nabla})\overline{A} - \dfrac{\partial\overline{A}}{\partial t}\right]$

Now $\dfrac{\partial\overline{A}}{\partial t}$ is the partial derivative of $\overline{A}$ with respect to t, operating only

on t only when that letter appears in the formula for $\overline{A}$. But

$$(\overline{v}\cdot\overline{\nabla})\overline{A} + \dfrac{\partial\overline{A}}{\partial t} = \dfrac{\partial\overline{A}}{\partial x}\dfrac{dx}{dt} + \dfrac{\partial\overline{A}}{\partial y}\dfrac{dy}{dt} + \dfrac{\partial\overline{A}}{\partial z}\dfrac{dz}{dt} + \dfrac{\partial\overline{A}}{\partial t} = \dfrac{d\overline{A}}{dt}$$

is the total derivative of $\overline{A}$ with t in any way that t can make a variation of $\overline{A}$.

Thus $M\dfrac{d\overline{v}}{dt} = \overline{\nabla}(q\overline{v}\cdot\overline{A} - q\phi) - q\dfrac{d\overline{A}}{dt}$ or $\dfrac{d}{dt}(M\overline{v} + q\overline{A}) = \overline{\nabla}(\overline{v}\cdot(q\overline{A}) - q\phi)$

With $\overline{v}\cdot q\overline{A} = qv_x A_x + qv_y A_y + qv_z A_z$, $\dfrac{\partial}{\partial v_x}(\overline{v}\cdot q\overline{A}) = qA_x$, $\dfrac{\partial}{\partial v_y}(\overline{v}\cdot q\overline{A}) = qA_y$, etc

The x-component of Newton's second law is $\dfrac{d}{dt}(Mv_x + qA_x) = \dfrac{\partial}{\partial x}(\overline{v}\cdot q\overline{A} - q\phi)$

If there were a Lagrangian for a particle in an electromagnetic field, then $\dfrac{d}{dt}(\dfrac{\partial L}{\partial v_x}) = \dfrac{\partial L}{\partial x}$ etc. suggesting that $L = \frac{1}{2}Mv^2 + q\overline{v}\cdot\overline{A} - q\phi$

Associated with the variable x is the conjugate momentum $p_x = \dfrac{\partial L}{\partial v_x} = Mv_x + qA_x$

giving a momentum $\overline{p} = M\overline{v} + q\overline{A}$.

Appendix B:

The AHARONOV-BOHM EFFECT demonstrates the results derived on the previous page. Consider a toroidal winding (much longer than shown) enclosing a magnetic field. Assume that a vector potential may be produced in the space outside the torus.

$$\oint_C \bar{A} \cdot d\bar{r} = \iint \bar{\nabla} \times \bar{A} \cdot \bar{n} dS$$

But $\bar{\nabla} \times \bar{A} = \bar{B}$ C S bounded by C

Then $\oint_C \bar{A} \cdot d\bar{r} = \iint_{S \text{ in } C} \bar{B} \cdot \bar{n} dS$ the magnetic flux enclosed by C

An electron with charge $q = +e$ moving in the region will have a momentum $\bar{p} = m\bar{v} - e\bar{A}$ but the electron's wavelength is $\lambda = h/p$.

In the second figure, two electrons, initially with momentum $\bar{p}_0 = m\bar{v}$ enter a region around a confined column of magnetic flux. The near-electron will have a wavelength $\lambda = \dfrac{h}{mv - eA}$

and $\lambda = \dfrac{h}{mv + eA}$ is the wavelength of the electron on the opposite side, with the results as shown. This is the Aharonov-Bohm effect from Phys. Rev. 115, 485 (1959).

In a beautifully written paper, RMP 59, 639 (1987) Akira Tonomura discusses the fascinating phenomenon of electron holography. He lets an electron beam pass through a magnetized square torus and interfere with a reference beam showing wavelength displacements, as shown.

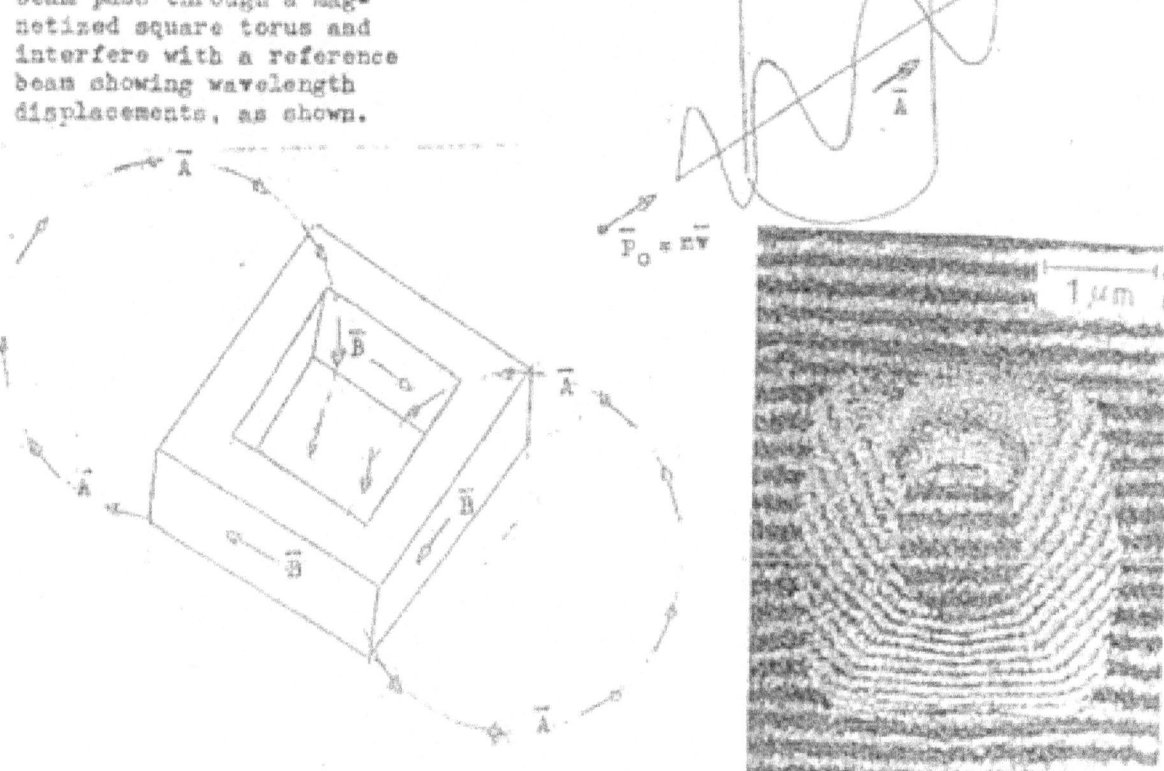

Now are they all undone, the ancient laws,   Aeschylus
                                                   "The Eumenides"

The components of the classical angular momentum vector are related to
the components of the position and linear momentum according to

$$L_x = yp_z - zp_y \quad , \quad L_y = zp_x - xp_z \quad , \quad L_z = xp_y - yp_x \; .$$

You may recall that in the harmonic oscillator problem we related the energy
operator to the position and momentum operators by replacing the terms in
their classical relation by the equivalent operators. To construct the
angular momentum operators, we shall procede to do the same thing. Thus, if
X, Y, and Z are taken to be the three position operators and P_x, P_y, and P_z
are the three linear momentum operators, we write the three angular momentum
operators $L_x = YP_z - ZP_y$, $L_y = ZP_x - XP_z$, $L_z = XP_y - YP_x$.

The commutation relations of the angular momentum operators are derived
from the commutation relations of the position and linear momentum operators.
In three dimensions, the fourth axiom of the first lesson states that the
position and momentum operators will commute if they are in <u>different</u> directions and will <u>not</u> commute if they are in the <u>same</u> direction. Hence.

$$XP_x - P_x X = YP_y - P_y Y = ZP_z - P_z Z = i\hbar \quad \text{and} \quad XP_y - P_y X = XP_z - P_z X = \text{etc} = 0$$

In addition to this, we state that the position operators all commute with
each other, $XY - YX = \text{etc} = 0$, and the momentum operators will commute with
each other, $P_x P_y - P_y P_x = \text{etc} = 0$. Thus, in products of position and momentum
operators, we must be especially careful of the order of multiplication only
when the position and momentum operators are in the <u>same</u> direction.

The quantum rules for angular momentum will be derived from the commutator rules for the operators. Thus

$$L_x L_y - L_y L_x = (YP_z - ZP_y)(ZP_x - XP_z) - (ZP_x - XP_z)(YP_z - ZP_y)$$
$$= YP_z ZP_x - YP_z XP_z - ZP_y ZP_x + ZP_y XP_z - ZP_x YP_z + ZP_x ZP_y + XP_z YP_z - XP_z ZP_y$$
$$= (YP_x - XP_y)(P_z Z) - (YP_x - XP_y)(ZP_z)$$
$$= (YP_x - XP_y)(P_z Z - ZP_z) = (-L_z)(-i\hbar)$$

Hence $L_x L_y - L_y L_x = i\hbar L_z$ and in the same manner $L_z L_x - L_x L_z = i\hbar L_y$ and
$L_y L_z - L_z L_y = i\hbar L_x$.

Instead of the length of the angular momentum vector, we shall consider
an operator for the square of the length of the vector $L^2 = L_x^2 + L_y^2 + L_z^2$.
We shall now show that L^2 commutes with any one of the components. Let us
work out the commutator or L^2 and L_z. We will consider each term of
L^2 seperately. First notice that $L_z L_z^2 - L_z^2 L_z = 0$. Next, multiply
$L_z L_x - L_x L_z = i\hbar L_y$ by L_x on the right and by L_x again on the left and add.

$$L_z L_x^2 - L_x L_z L_x + L_x L_z L_x - L_x^2 L_z = i\hbar(L_y L_x + L_x L_y) \quad \text{or} \quad L_z L_x^2 - L_x^2 L_z = i\hbar(L_y L_x + L_x L_y)$$

Similarly with $L_y L_z - L_z L_y = i\hbar L_x$ we get $\qquad L_z L_y^2 - L_y^2 L_z = -i\hbar(L_x L_y + L_y L_x)$

If we put these three commutators together we get $L_z L^2 - L^2 L_z = 0$.
Likewise $L_x L^2 - L^2 L_x = L_y L^2 - L^2 L_y = 0$. Hence L_x , L_y , and L_z do not commute
with each other, but they do all commute with L^2. According to the relation
between the Heisenberg uncertainty principle and non-commuting operators
only one component of the set L_x , L_y , and L_z can be measured in any
observation, but no matter which component of the vector we choose to measure, the square of the length of the vector — and, hence, the length — can
be determined in the same observation.

This result isn't really so strange, if you think about it. An object
in atomic physics that has angular momentum will generally have the property
of charge, giving it a magnetic moment vector either parallel or anti-parallel to the angular momentum. Observations of either vector will, therefore,
involve in many cases a constant magnetic field — usually taken to be along
the z-axis. As a result, the z-component of the angular momentum and the
length of the vector will be constant, but the x and y components will be
precessing around the magnetic field. In any determination of the angular
momentum of an object, we shall consider that we can know the magnitude
of the vector, or its square, and the projection of the vector onto the
permanent magnetic field, usually taken to be L_z .

In the process of determining the eigenvalues and eigenvectors of L_z and
L^2 we shall find it convenient to use a related set of operator symbols,
J_x , J_y , and J_z gotten by dividing L_x , L_y , and L_z by $\hbar$. Hence $L_x = \hbar J_x$
$L_y = \hbar J_y$, and $L_z = \hbar J_z$. Then $L^2 = \hbar^2 J^2$. The J's have the commutators
$$J_x J_y - J_y J_x = i J_z$$
$$J_z J_x - J_x J_z = i J_y \qquad J_z J^2 - J^2 J_z = etc = 0$$
$$J_y J_z - J_z J_y = i J_x$$

Because J_z and J^2 can be determined in any single observation, the
eigenkets for this problem will be eigenkets of both J_z and J^2. We will
write the eigenkets $|j,m\rangle$ with two sets of quantum numbers, j and m.
With each eigenket there will be associated two eigenvalues, ... for J_z and
J^2. We shall write $J_z |j,m\rangle = m|j,m\rangle$ and $J^2 |j,m\rangle = \lambda_j |j,m\rangle$ where m and
λ_j are to be determined. Now that we have given J^2 and J_z something to do,
what about J_x and J_y ? Well, just to keep them from feeling left out and
useless, we will give them something to do. Let us define two new operators
$J_+ = J_x + i J_y$ and $J_- = J_x - i J_y$. These will turn out to be raising and
lowering operators on the eigenvalues m . Because all components of angular
momentum commute with J^2, it evident that $J_+ J^2 - J^2 J_+ = 0$ and $J_- J^2 - J^2 J_- = 0$.
We now need the commutation relations for J_z, J_+, and J_- .

The commutator relations for J_+ and J_- follow from the commutators for J_x, J_y, and J_z. Hence

$$J_+J_- - J_-J_+ = (J_x + iJ_y)(J_x - iJ_y) = J_x^2 + J_y^2 + i(J_yJ_x - J_xJ_y) - J_x^2 - J_y^2 + i(J_yJ_x - J_xJ_y)$$

$$J_+J_- - J_-J_+ = 2i(-iJ_z) = 2J_z$$

In the same manner we get

$$J_zJ_+ - J_+J_z = J_zJ_x + iJ_zJ_y - J_xJ_z - iJ_yJ_z = iJ_y + i(-iJ_x) = J_x + iJ_y = J_+$$

and by the same reasoning, $J_zJ_- - J_-J_z = -J_-$

The commutation relations for J_z, J_+, and J_- and also J^2 are summarized by

$$J_zJ^2 - J^2J_z = J_\pm J^2 - J^2 J_\pm = 0$$

$$J_+J_- - J_-J_+ = 2J_z \qquad \text{and} \qquad J_zJ_\pm - J_\pm J_z = \pm J_\pm$$

Now recall that $|j,m\rangle$ is an eigenket of J_z with eigenvalue m, which is written $J_z|j,m\rangle = m|j,m\rangle$.

Now, if we take the identity $J_zJ_+ = J_+J_z + J_+$ and multiply it by the ket $|j,m\rangle$ we have $J_zJ_+|j,m\rangle = J_+J_z|j,m\rangle + J_+|j,m\rangle = (m+1)J_+|j,m\rangle$ which says that the new ket $J_+|j,m\rangle$ is an eigenket of J_z with the eigenvalue $(m+1)$. Hence, if m is an eigenvalue of J_z, $(m+1)$ is also an eigenvalue.

Next, taking the identity $J_zJ_- = J_-J_z - J_-$ and multiplying by $|j,m\rangle$ we get $J_zJ_-|j,m\rangle = J_-J_z|j,m\rangle - J_-|j,m\rangle = (m-1)J_-|j,m\rangle$. This says that $J_-|j,m\rangle$ is an eigenket of J_z with the eigenvalue $(m-1)$. Thus if m is an eigenvalue of J_z, then $(m-1)$ also is an eigenvalue.

Our next task is to show that the set of eigenvalues m has an upper and a lower limit, from which we can determine the values allowed for λ_j and the particular sets of numbers that are permitted for m. For this job we need the identities $J_+J_- = J_x^2 + J_y^2 + i(-iJ_z) = J^2 - J_z^2 + J_z$

and $\qquad\qquad J_-J_+ = J_x^2 + J_y^2 + i(iJ_z) = J^2 - J_z^2 - J_z$

in which we have used $J^2 = J_x^2 + J_y^2 + J_z^2$.

Now, for the operator J_+, J_- is the operator complex conjugate and vice versa. Hence, if $|\Psi\rangle$ is any vector, $\langle\Psi|J_-J_+|\Psi\rangle \geq 0$.*

- -

*Strictly speaking, for the ket $(J_x + iJ_y)|\Psi\rangle$, the associated bra is $\langle\Psi|(J_x^* - iJ_y^*)$ where J^* designates the hermitian adjoint of J. But in quantum mechanics, the operators for all observables are hermitian or self adjoint operators, hence, $J^* = J$. Thus for the ket $J_+|\Psi\rangle$, we have the associated bra $\langle\Psi|J_-$. Then $\langle\Psi|J_-J_+|\Psi\rangle$ is a real number, greater than or equal to zero, being the square of the magnitude of $J_+|\Psi\rangle$.

- -

By the same arguments, $\langle\Psi|J_+J_-|\Psi\rangle \geq 0$.

For a particular eigenvalue λ_j of J^2 we would expect to find a set of eigenvalues m of J_z giving the permitted projections of the angular momentum onto a magnetic field, which is taken to define the z-axis. We expect the set of m's to have a least upper bound and a greatest lower bound with m going from one to the other in steps of unity. These bounds will be found next along with the relation between λ_j and the quantum number j for the total angular momentum. We shall expect the angular momentum eigenkets to have unit length, $\langle j,m|j,m\rangle = +1$. Recall that if $|\Psi\rangle$ is any vector, $\langle\Psi|J_+J_-|\Psi\rangle \geq 0$ and $\langle\Psi|J_-J_+|\Psi\rangle \geq 0$.

In terms of the state $|j,m\rangle$ the expectation value of $J_+J_- = J^2 - J_z^2 + J_z$ is
$$\langle j,m|J_+J_-|j,m\rangle = \langle j,m|J^2 - J_z^2 + J_z|j,m\rangle = \left[\lambda_j - m^2 + m\right]\langle j,m|j,m\rangle$$
where we let the operators operate on $|j,m\rangle$ and factored out the eigenvalues which are just numbers. Because $\langle j,m|J_+J_-|j,m\rangle \geq 0$ and $\langle j,m|j,m\rangle \geq 0$ we get $\lambda_j - m^2 + m \geq 0$ for all m belonging to λ_j. By the same reasoning
$$\langle j,m|J_-J_+|j,m\rangle = \langle j,m|J^2 - J_z^2 - J_z|j,m\rangle = \left[\lambda_j - m^2 - m\right]\langle j,m|j,m\rangle \text{ giving}$$
$\lambda_j - m^2 - m \geq 0$. Hence, for all m belonging to a <u>particular</u> λ_j

$$\lambda_j \geq m(m-1) \qquad \text{and} \qquad \lambda_j \geq m(m+1)$$

Hence, for a particular λ_j there will be a maximum value of the m's which can be denoted by $\bar{m}$ and which we shall take to be a positive number. Thus, $\lambda_j \geq \bar{m}(\bar{m}+1)$ which, of course, is greater than $\bar{m}(\bar{m}-1)$. If $\bar{m}$ is the greatest value in a set of m's, then $J_+|j,\bar{m}\rangle$ would give the eigenvector for $\bar{m}+1$ which does not exist. Thus, $J_+|j,\bar{m}\rangle$ does not exist and this is indicated by $J_+|j,\bar{m}\rangle = 0$. Now, using the normalized eigenvectors $\langle j,\bar{m}|j,\bar{m}\rangle = 1$. Now, from the expectation value of J_-J_+ for the state $|j,\bar{m}\rangle$
$$\left[\lambda_j - \bar{m}^2 - \bar{m}\right] = \langle j,\bar{m}|J_-J_+|j,\bar{m}\rangle = 0 \qquad \text{because} \quad J_+|j,\bar{m}\rangle = 0$$
This gives the important result that $\lambda_j = \bar{m}(\bar{m}+1)$.
Up to this point there has been no indication as to what numbers we get for the eigenvalue λ_j of J^2 or its quantum number j. At this point it is convenient to let $j = \bar{m}$ the greatest member of the particular set of m's, then $\lambda_j = j(j+1)$.

Now, if the angular momentum of an object is a finite length vector, its projection on the z-axis will have a minimum value as well as a maximum value. Let $\underline{m}$ be the least member of the set of all m for a particular j. Then $J_-|j,\underline{m}\rangle = 0$ because there is no state with $m = \underline{m}-1$. Then, if $\langle j,\underline{m}|j,\underline{m}\rangle = 1$ and $\lambda_j = j(j+1)$ we have, by arguments identical to those above $\left[j(j+1) - \underline{m}^2 + \underline{m}\right] = \langle j,\underline{m}|J_+J_-|j,\underline{m}\rangle = 0$ from which we get $\underline{m}(\underline{m}-1) = j(j+1)$ which is satisfied by either $\underline{m} = j+1$ or $\underline{m} = -j$.

Now, the idea that $\underline{m = j+1}$ is preposterous, so we must settle for the least member of the set of all m's for a particular j is $\underline{m = -j}$.

So far, the quantum theory of angular momentum has given the following results. If $|j,m\rangle$ is an eigenvector of both J_z and J^2, then $J_z|j,m\rangle = m|j,m\rangle$ and $J^2|j,m\rangle = j(j+1)|j,m\rangle$, where for a particular j the values of m range from $-j$ to $+j$, consecutive values of m differing by unity. It is easy to see that all this works very well if j is a positive integer or zero. Hence, if $j=0$ then $m=0$, and if $j = 1$ then $m = -1, 0, +1$, and if $j = 2$ then $m = -2, -1, 0, +1, +2$, and so with the rest of the positive integers.

It is demonstrable, but maybe not so evident, that j can be half of a positive odd integer. However, if $j = \frac{1}{2}$ then $m = -\frac{1}{2}, +\frac{1}{2}$, and if $j = \frac{3}{2}$ then $m = -\frac{3}{2}, -\frac{1}{2}, +\frac{1}{2}, +\frac{3}{2}$, and so forth for j being any other half positive integer. You can show from this that associated with any value of j there will be $2j + 1$ values of m.

The physical significance of all this is easy to show. For the angular momentum operators, recall that $L_z = \hbar J_z$ and $L^2 = \hbar^2 J^2$ with the result that the quantized angular momentum will be a vector with a magnitude of $\sqrt{j(j+1)}\,\hbar$ and having a projection onto the z-axis of $m\hbar$. Thus, if $j = 1$ for a rotating object,

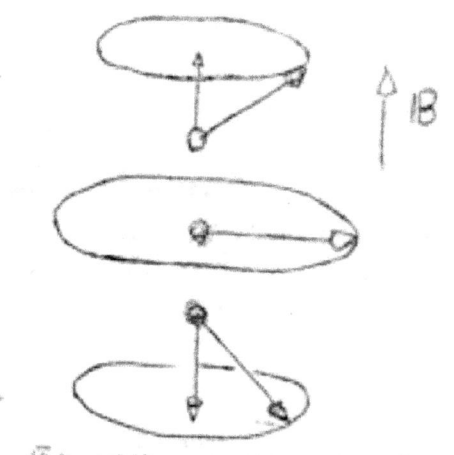

$j=1, m=+1$

$j=1, m=0$

$j=1, m=-1$

its angular momentum will have a magnitude $L = \sqrt{2}\,\hbar$ with z-components of $L_z = -\hbar, 0, +\hbar$. These are shown in the above figure. If there is a magnetic field in the region, directed along the z-axis, and if the rotating object has a charge, the angular momentum will be caused to precess about the magnetic field. If the object has a mass M and a charge q then it will have a magnetic moment vector $\boldsymbol{\mu} = g\frac{q}{2M}\mathbf{L}$ where the g-factor arises from the charge and mass not necessarily having the same distribution over the volume of the object. The potential energy of the magnetic moment $\boldsymbol{\mu}$ in the magnetic field $\mathbf{B}$, taken to be along the z-axis, is $U = -\boldsymbol{\mu}\cdot\mathbf{B}$ which, for the quantized angular momentum, comes out to be $U = -m\frac{gq\hbar B}{2M}$. In all of this, rationalized MKS units have been assumed; for formulas in Gaussian units substitute q/c for q. If the charge and mass of the object have the same distribution, $g = 1$.

Now that the eigenvalues of angular momentum have been found, the next job involves finding the eigenvectors $|j,m\rangle$.

Angular momentum is an observable quantity, hence its operators must be hermitian or self adjoint to assure that the eigenvalues will be real numbers. This also assures that the eigenvectors must be orthogonal — or unitary if they have complex components in any representation. Since the eigenvectors only define directions in Hilbert space, we may specify that their magnitudes be unity. Thus $\langle j,m | j',m' \rangle = \delta_{jj'} \delta_{mm'}$.

It was shown on page (4) that, with the substitution of $\lambda_j = j(j+1)$,
$$\langle j,m | J_- J_+ | j,m \rangle = \left[j(j+1) - m(m+1) \right] \langle j,m | j,m \rangle \qquad \text{and}$$
$$\langle j,m | J_+ J_- | j,m \rangle = \left[j(j+1) - m(m-1) \right] \langle j,m | j,m \rangle$$
On page (3) it was shown that $J_+ |j,m\rangle$ was an eigenket of J_z with an eigenvalue of $(m+1)$. By the rules of unitary spaces, the associated bra vector is $\langle j,m | J_-$. Now, these are not normalized eigenvectors for the eigenvalue $m+1$ of J_z , and since we want a normalized — unit length — set of eigenvectors, we introduce a constant of normalization C, so that $J_+ |j,m\rangle = C |j,m+1\rangle$ and $\langle j,m | J_- = \langle j,m+1 | C$, so that $\langle j,m+1 | j,m+1 \rangle = \langle j,m | j,m \rangle = 1$. Thus we find C from $\langle j,m | J_- J_+ | j,m \rangle = C^2 \langle j,m+1 | j,m+1 \rangle = \left[j(j+1) - m(m+1) \right] \langle j,m | j,m \rangle$ which gives $C^2 = \left[j(j+1) - m(m+1) \right]$ or $J_+ |j,m\rangle = \sqrt{j(j+1) - m(m+1)} \, |j,m+1\rangle$

In a similar manner for the lowering operator J_- , there will be some number C such that $J_- |j,m\rangle = C |j,m-1\rangle$ with the associated bra $\langle j,m | J_+ = \langle j,m-1 | C$. In a manner similar to that above we get
$$\langle j,m | J_+ J_- | j,m \rangle = C^2 \langle j,m-1 | j,m-1 \rangle = \left[j(j+1) - m(m-1) \right] \langle j,m | j,m \rangle$$
giving $C^2 = \left[j(j+1) - m(m-1) \right]$ or $J_- |j,m\rangle = \sqrt{j(j+1) - m(m-1)} \, |j,m-1\rangle$

You will notice that in all these states, j remains the same and we are raising and lowering m. Thus J_+ and J_- relate different states of different z-components of the same total angular momentum vector. In a later set of lectures, raising and lowering operators for j , keeping m constant, will be obtained, but that is not practical to be done here.

The results of our investigation of the quantum theory of angular momentum may be summarized. There is a set of eigenkets of J_z and J^2 such that
$$J_z |j,m\rangle = m |j,m\rangle \qquad\qquad J^2 |j,m\rangle = j(j+1) |j,m\rangle$$
$$J_+ |j,m\rangle = \sqrt{j(j+1) - m(m+1)} \, |j,m+1\rangle \qquad J_- |j,m\rangle = \sqrt{j(j+1) - m(m-1)} \, |j,m-1\rangle$$
These last two relations are more usefully written as
$$J_+ |j,m\rangle = \sqrt{(j-m)(j+m+1)} \, |j,m+1\rangle \qquad\qquad J_- |j,m\rangle = \sqrt{(j+m)(j-m+1)} \, |j,m-1\rangle$$
If we know any $|j,m\rangle$ we can get all other eigenkets for the set of m's corresponding to the particular value of j. In most cases we will relate all kets of the set $|j,m\rangle$ to the ket $|j,0\rangle$ when j is an integer.

To a large extent, the representation or form that the eigenvectors take in this problem will depend on how we label the rows and columns of the matrices of the operators. The Hilbert space spanned by the eigenkets of J_z and J^2 can be broken down into finite dimensional orthogonal subspaces. Two spaces are orthogonal (or mutually perpendicular) subspaces of a Hilbert space if every vector in one space is perpendicular to every vector in the other space. The property of angular momentum, that the eigenvectors span finite dimensioned subspaces of a Hilbert space turns out to be convenient. Many problems are of such nature that when their operators are represented in angular momentum states, the resulting matrices finite dimensioned according to the dimensions of the various orthogonal subspaces. We shall now see how all this works.

Because J_z and J^2 are Hermitian, their eigenkets are orthogonal. Thus $\langle j,m | j',m' \rangle = \delta_{jj'} \delta_{mm'}$. Recall $J_z | j,m \rangle = m | j,m \rangle$ and $J^2 | j,m \rangle = j(j+1) | j,m \rangle$ which give the matrix elements $\langle j',m' | J_z | j,m \rangle = m \, \delta_{j'j} \delta_{m'm}$ and $\langle j',m' | J^2 | j,m \rangle = j(j+1) \, \delta_{j'j} \delta_{m'm}$ which are both diagonal, as you might expect.

The matrices are displayed in the following manner. We label the rows and columns according to <u>increasing</u> values of j , and then for each value of j according to <u>decreasing</u> values of m . Recall that j is permitted to have the values $0, \frac{1}{2}, 1, \frac{3}{2}, 2, \frac{5}{2}, 3, \ldots$ while for each value of j, $m = j, j-1, j-2, \ldots, -j$. Thus we get for J_z , labeling the borders with the j' and m' denoting rows and the j and m denoting columns.

j' \ m'	j:0 m:0	1/2 1/2	1/2 -1/2	1 1	1 0	1 -1	3/2 3/2	3/2 1/2	3/2 -1/2	3/2 -3/2	. . .
0 0	0	0	0	0	0	0	0	0	0	0	. . .
1/2 1/2	0	1/2	0	0	0	0	0	0	0	0	. . .
1/2 -1/2	0	0	-1/2	0	0	0	0	0	0	0	. . .
1 1	0	0	0	1	0	0	0	0	0	0	. . .
1 0	0	0	0	0	0	0	0	0	0	0	. . .
1 -1	0	0	0	0	0	-1	0	0	0	0	. . .
3/2 3/2	0	0	0	0	0	0	3/2	0	0	0	. . .
3/2 1/2	0	0	0	0	0	0	0	1/2	0	0	. . .
3/2 -1/2	0	0	0	0	0	0	0	0	-1/2	0	. . .
3/2 -3/2	0	0	0	0	0	0	0	0	0	-3/2	. . .
. . .	.	.	.	.	.	.	.	.	.	.	. . .

The rows and columns of J_z go on and on and on, and since in this representation J_z is diagonal there are a lot of zeros. But there would be a lot of zeros anyway. For most operators which we shall wish to represent in angular momentum states, the term at the intersection of a row labeled by j' and a column labeled by j will be zero if $j \neq j'$. For this reason we will write matrix elements only for those terms where $j = j'$. This permits us to confine our attention, for each value of j, to the $2j+1$ dimensional square matrices whose rows and columns are labeled in decreasing order by the $2j+1$ eigenvalues m that go with the eigenvalue j.

The representations of the state $|0,0\rangle$ are not very exciting, so we won't bother to write them here — although the state itself is an important state. For $j = 1/2$ we have the matrices $J_z = \begin{pmatrix} 1/2 & 0 \\ 0 & -1/2 \end{pmatrix}$ and $J^2 = \begin{pmatrix} 3/4 & 0 \\ 0 & 3/4 \end{pmatrix}$

which can be written $J_z = \frac{1}{2}\begin{pmatrix} 1 & 0 \\ 0 & -1 \end{pmatrix}$ $J^2 = \frac{3}{4}\begin{pmatrix} 1 & 0 \\ 0 & 1 \end{pmatrix}$

When $j = 1$, we get $J_z = \begin{pmatrix} 1 & 0 & 0 \\ 0 & 0 & 0 \\ 0 & 0 & -1 \end{pmatrix}$ and $J^2 = 2\begin{pmatrix} 1 & 0 & 0 \\ 0 & 1 & 0 \\ 0 & 0 & 1 \end{pmatrix}$

When $j = \frac{3}{2}$, $J_z = \begin{pmatrix} 3/2 & 0 & 0 & 0 \\ 0 & 1/2 & 0 & 0 \\ 0 & 0 & -1/2 & 0 \\ 0 & 0 & 0 & -3/2 \end{pmatrix}$ and $J^2 = \frac{15}{4}\begin{pmatrix} 1 & 0 & 0 & 0 \\ 0 & 1 & 0 & 0 \\ 0 & 0 & 1 & 0 \\ 0 & 0 & 0 & 1 \end{pmatrix}$

All of which follow from the eigenvalues of J^2 being $j(j+1)$ and those of J_z being $m = j, j-1, j-2, \ldots, -j$.

Now it is to be seen how the matrix elements for the other J operators are obtained. Recall that $J_+ = J_x + iJ_y$ and $J_- = J_x - iJ_y$. If the matrices for J_+ and J_- can be found, then $J_x = \frac{1}{2}(J_+ + J_-)$ and $J_y = -\frac{i}{2}(J_+ - J_-)$. The matrices for J_+ and J_- are gotten from

$$\langle j',m'|J_+|j,m\rangle = \sqrt{(j-m)(j+m+1)}\,\langle j',m'|j,m+1\rangle = \sqrt{(j-m)(j+m+1)}\,\delta_{j'j}\delta_{m'm+1}$$
$$\langle j',m'|J_-|j,m\rangle = \sqrt{(j+m)(j-m+1)}\,\langle j',m'|j,m-1\rangle = \sqrt{(j+m)(j-m+1)}\,\delta_{j'j}\delta_{m'm-1}$$

Recall that the rows, m', and the columns, m, are labeled according to m' and m <u>decreasing</u> as you go down and to the right. $\delta_{m'm+1} \neq 0$ only if $m = m'-1$, and $\delta_{m'm-1} \neq 0$ only if $m = m'+1$. Hence, for the states with $j = 1/2$, $\langle \frac{1}{2},\frac{1}{2}|J_+|\frac{1}{2},\frac{1}{2}\rangle = 0$ $\langle \frac{1}{2},\frac{1}{2}|J_+|\frac{1}{2},-\frac{1}{2}\rangle = 1$ $\langle \frac{1}{2},-\frac{1}{2}|J_+|\frac{1}{2},\frac{1}{2}\rangle = 0$

and $\langle \frac{1}{2},-\frac{1}{2}|J_+|\frac{1}{2},-\frac{1}{2}\rangle = 0$ the last term because $\delta_{-1/2,+1/2} = 0$.

For $j = 1/2$, this gives the matrix $J_+ = \begin{pmatrix} 0 & 1 \\ 0 & 0 \end{pmatrix}$ and by similar means $J_- = \begin{pmatrix} 0 & 0 \\ 1 & 0 \end{pmatrix}$

From these we get $J_x = 1/2\begin{pmatrix} 0 & 1 \\ 1 & 0 \end{pmatrix}$ and $J_y = 1/2\begin{pmatrix} 0 & -i \\ i & 0 \end{pmatrix}$

It is easy to show that J_x, J_y, and J_z satisfy the algebra on page (2) and J_z, J_+, and J_- satisfy the algebra on page (3).

For an object with j = 1/2 (which is called spin one-half) we got the
matrix representations of the operators to be

$$J_x = 1/2 \begin{pmatrix} 0 & 1 \\ 1 & 0 \end{pmatrix} \qquad J_y = 1/2 \begin{pmatrix} 0 & -1 \\ i & 0 \end{pmatrix} \qquad J_z = 1/2 \begin{pmatrix} 1 & 0 \\ 0 & -1 \end{pmatrix} \qquad J^2 = 3/4 \begin{pmatrix} 1 & 0 \\ 0 & 1 \end{pmatrix}$$

$$J_+ = \begin{pmatrix} 0 & 1 \\ 0 & 0 \end{pmatrix} \qquad J_- = \begin{pmatrix} 0 & 0 \\ 1 & 0 \end{pmatrix}$$

The angular momentum of a spin one-half object has two projections onto
the z-axis, m = +1/2 which we call 'spin up', and m = -1/2 which we call
spin down. These are specified by the eigenkets $|up\rangle = |\frac{1}{2},\frac{1}{2}\rangle = \begin{pmatrix} 1 \\ 0 \end{pmatrix}$ and
$|down\rangle = |\frac{1}{2},-\frac{1}{2}\rangle = \begin{pmatrix} 0 \\ 1 \end{pmatrix}$. These give us $J_z|up\rangle = +1/2|up\rangle$ and
$J_z|down\rangle = -1/2|down\rangle$. Also $J^2|up\rangle = 3/4|up\rangle$ and $J^2|down\rangle = 3/4|down\rangle$.
The operation $\begin{pmatrix} 0 & 1 \\ 0 & 0 \end{pmatrix}\begin{pmatrix} 0 \\ 1 \end{pmatrix} = \begin{pmatrix} 1 \\ 0 \end{pmatrix}$ is written $J_+|down\rangle = |up\rangle$ and

$\begin{pmatrix} 0 & 0 \\ 1 & 0 \end{pmatrix}\begin{pmatrix} 1 \\ 0 \end{pmatrix} = \begin{pmatrix} 0 \\ 1 \end{pmatrix}$ is written $J_-|up\rangle = |down\rangle$.

That $\begin{pmatrix} 0 & 1 \\ 0 & 0 \end{pmatrix}\begin{pmatrix} 1 \\ 0 \end{pmatrix} = \begin{pmatrix} 0 \\ 0 \end{pmatrix}$ means that there is no spin-1/2 state with $m > +1/2$,

and $\begin{pmatrix} 0 & 0 \\ 1 & 0 \end{pmatrix}\begin{pmatrix} 0 \\ 1 \end{pmatrix} = \begin{pmatrix} 0 \\ 0 \end{pmatrix}$ means that there is no spin-1/2 state with $m < -1/2$.

You may have noticed that the spin-1/2 operators have utilized a set of
matrices that we shall designate by the symbols, sigma, as

$$\sigma_x = \begin{pmatrix} 0 & 1 \\ 1 & 0 \end{pmatrix} \qquad \sigma_y = \begin{pmatrix} 0 & -i \\ i & 0 \end{pmatrix} \qquad \sigma_z = \begin{pmatrix} 1 & 0 \\ 0 & -1 \end{pmatrix}$$

and are called the Pauli spin matrices. They have the following algebra:

$$\sigma_x\sigma_y - \sigma_y\sigma_x = 2i\sigma_z \qquad \sigma_x\sigma_y + \sigma_y\sigma_x = 0 \qquad \sigma_x^2 = \sigma_y^2 = \sigma_z^2 = 1$$
$$\sigma_z\sigma_x - \sigma_x\sigma_z = 2i\sigma_y \qquad \sigma_z\sigma_x + \sigma_x\sigma_z = 0$$
$$\sigma_y\sigma_z - \sigma_z\sigma_y = 2i\sigma_x \qquad \sigma_y\sigma_z + \sigma_z\sigma_y = 0$$

which you can verify without much trouble.

Matrix operators for the angular momentum operators can be derived for
other values of j. Those for j = 1 are given in the problem at the end of
the zeroth lecture.

You will recall that, in terms of the position and momentum operators,
the angular momentum operators were $L_x = YP_z - ZP_y$, $L_y = ZP_x - XP_z$, and
$L_z = XP_y - YP_x$. Recall, also, that the L and J operators were related by $\hbar$
so that $L_z = \hbar J_z$ and $L^2 = \hbar^2 J^2$. Hence, $L_z|j,m\rangle = m\hbar|j,m\rangle$ and
$L^2|j,m\rangle = j(j+1)\hbar^2|j,m\rangle$.

ADDITION OF ANGULAR MOMENTUM VECTORS

Just as orbital angular momentum is quantized as $L^2|\ell,m_\ell\rangle = \ell(\ell+1)\hbar^2|\ell,m_\ell\rangle$
and $L_z|\ell,m_\ell\rangle = m_\ell\hbar|\ell,m_\ell\rangle$ so also the spin $\bar{S} = \frac{\hbar}{2}\bar{\Sigma}$ is quantized as

$S^2|\frac{1}{2},m_s\rangle = \frac{1}{2}(\frac{1}{2}+1)\hbar^2|\frac{1}{2},m_s\rangle$ and $S_z|\frac{1}{2},\frac{1}{2}\rangle = \frac{\hbar}{2}|\frac{1}{2},\frac{1}{2}\rangle$ and $S_z|\frac{1}{2},-\frac{1}{2}\rangle = -\frac{\hbar}{2}|\frac{1}{2},-\frac{1}{2}\rangle$

Consider now the hydrogen atom's single electron. The total angular momentum is the sum of its orbital and spin angular momentum. $\bar{J} = \bar{L} + \bar{S}$ and $J_z = L_z + S_z$

The maximum value of m_ℓ is ℓ and the maximum value of m_s is $\frac{1}{2}$ therefore the maximum value of m_j is $\ell+\frac{1}{2}$. On the other hand, if m_s is $-\frac{1}{2}$ then the maximum value for m_j is $\ell-\frac{1}{2}$. Thus the quantum numbers for j are $j = \ell+\frac{1}{2}$ or $j = \ell-\frac{1}{2}$.

Recall $J_+|j,m_j\rangle = \sqrt{j(j+1) - m_j(m_j+1)}|j,m_j+1\rangle = \sqrt{(j-m_j)(j+m_j+1)}|j,m_j+1\rangle$
and $J_-|j,m_j\rangle = \sqrt{j(j+1) - m_j(m_j-1)}|j,m_j-1\rangle = \sqrt{(j+m_j)(j-m_j+1)}|j,m_j-1\rangle$

Similarly L_+ raises m_ℓ and L_- lowers m_ℓ. Note that $S_+|\frac{1}{2},\frac{1}{2}\rangle = 0$
and $S_-|\frac{1}{2},\frac{1}{2}\rangle = |\frac{1}{2},-\frac{1}{2}\rangle$.

For a given j there are $2j+1$ values for m_j and for a given value of ℓ there are $2\ell+1$ values of m_ℓ.
There are two values for m_s.

Let $|\ell,m_\ell\rangle$ be orbital states and $|\frac{1}{2},m_s\rangle$ and let their diredt product $|\ell,m_\ell\rangle|\frac{1}{2},m_s\rangle$ be spin-orbital state. Then the states $|j,m_j\rangle$ will involve linear combinations of spin-orbital states.

If the hydrogen electron is in an s-state, then $|\ell,m_\ell\rangle = |0,0\rangle$ and $|j,m_j\rangle$ has two states $|\frac{1}{2},\frac{1}{2}\rangle = |0,0\rangle|\frac{1}{2},\frac{1}{2}\rangle$ and $|\frac{1}{2},-\frac{1}{2}\rangle = |0,0\rangle|\frac{1}{2},-\frac{1}{2}\rangle$.

If the electron is in a p-state, $\ell=1$ and the maximum $m_\ell = 1$ so the maximum $m_j = 1+\frac{1}{2}$ going with $j = 1+\frac{1}{2}$.
The state vector for the maximum m_j is $|\frac{3}{2},\frac{3}{2}\rangle = |1,1\rangle|\frac{1}{2},\frac{1}{2}\rangle$
If $J_z = L_z + S_z$ then $J_- = L_- + S_-$ so

$J_-|\frac{3}{2},\frac{3}{2}\rangle = L_-|1,1\rangle|\frac{1}{2},\frac{1}{2}\rangle + |1,1\rangle S_-|\frac{1}{2},\frac{1}{2}\rangle$ or $\sqrt{3}|\frac{3}{2},\frac{1}{2}\rangle = \sqrt{2}|1,0\rangle|\frac{1}{2},\frac{1}{2}\rangle + |1,1\rangle|\frac{1}{2},-\frac{1}{2}\rangle$

or $|\frac{3}{2},\frac{1}{2}\rangle = \sqrt{\frac{2}{3}}|1,0\rangle|\frac{1}{2},\frac{1}{2}\rangle + \sqrt{\frac{1}{3}}|1,1\rangle|\frac{1}{2},-\frac{1}{2}\rangle$

Lowering again $J_-|\frac{3}{2},\frac{1}{2}\rangle = \sqrt{\frac{2}{3}}L_-|1,0\rangle|\frac{1}{2},\frac{1}{2}\rangle + \sqrt{\frac{2}{3}}|1,0\rangle S_-|\frac{1}{2},\frac{1}{2}\rangle + \sqrt{\frac{1}{3}}L_-|1,1\rangle|\frac{1}{2},-\frac{1}{2}\rangle$
Noting $S_-|\frac{1}{2},-\frac{1}{2}\rangle = 0$

Thus $|\frac{3}{2},-\frac{1}{2}\rangle = \sqrt{\frac{1}{3}}|1,-1\rangle|\frac{1}{2},\frac{1}{2}\rangle + \sqrt{\frac{2}{3}}|1,0\rangle|\frac{1}{2},-\frac{1}{2}\rangle$

Lowering once more gives $|\frac{3}{2},-\frac{3}{2}\rangle = |1,-1\rangle|\frac{1}{2},-\frac{1}{2}\rangle$

Thus $j = \frac{3}{2}$ gives $2 \times \frac{3}{2} + 1 = 4$ states for m_j as expected.

Notice that there is a state vector perpendicular to each of $|\frac{3}{2},\frac{1}{2}\rangle$ and $|\frac{3}{2},-\frac{1}{2}\rangle$

Two state vectors are for $j = \frac{1}{2}$ so

$$|\tfrac{1}{2},\tfrac{1}{2}\rangle = \sqrt{\tfrac{1}{3}}|1,0\rangle|\tfrac{1}{2},\tfrac{1}{2}\rangle - \sqrt{\tfrac{2}{3}}|1,1\rangle|\tfrac{1}{2},-\tfrac{1}{2}\rangle$$

$$|\tfrac{1}{2},-\tfrac{1}{2}\rangle = \sqrt{\tfrac{2}{3}}|1,-1\rangle|\tfrac{1}{2},\tfrac{1}{2}\rangle - \sqrt{\tfrac{1}{3}}|1,0\rangle|\tfrac{1}{2},-\tfrac{1}{2}\rangle$$

Six states for $|j,m_j\rangle$ are all of the orthogonal states we can form from the six spin-orbital states that constitute our basis.

PROBLEM 24-4: Starting with $|\frac{2}{2},\frac{5}{2}\rangle = |2,2\rangle|\frac{1}{2},\frac{1}{2}\rangle$, derive the angular momentum states for a d-state, for both $j = \frac{5}{2}$ and for $j = \frac{3}{2}$.

In the case of helium, there are two electrons. If the spins are anti-parallel, both electrons can be in the 1-s orbital state as the Pauli exclusion principle holds for spin-orbital states. This normal helium is called parhelium. If the spins happen to be parallel and one electron is in the 1-s orbital state, the Pauli exclusion principle requires the second electron to be in a 2-s or 2-p state. This is called orthohelium.

Although the Pauli exclusion principle applies to spin-orbital states, when the angular momentum states are combined to give a total angular momentum, spin states couple with spin states and separately orbital states couple with orbital states. This is called Russell-Saunders or L-S coupling. It may be partially violated in some large atoms. The total angular momentum is then gotten by combining the total orbital angular momentum with the total spin as described above.

For further reading you might refer to several classic treatments
E. U. Condon and G. H. Shortley The Theory of Atomic Spectra
M. E. Rose Elementary Theory of Angular Momentum

When combining angular momentum states to get a total angular momentum as

$$|\tfrac{3}{2},\tfrac{1}{2}\rangle = |1,0\rangle|\tfrac{1}{2},\tfrac{1}{2}\rangle\sqrt{\tfrac{2}{3}} + |1,1\rangle|\tfrac{1}{2},-\tfrac{1}{2}\rangle\sqrt{\tfrac{1}{3}}$$

the projection of the resultant vector onto a component vector is called a VECTOR COUPLING COEFFICIENT or a CLEBSCH-GORDON coefficient.

Applied to the hydrogen electron just above discussed, we will use the notation

$$\langle \ell,s,m_\ell,m_s|j,m_j\rangle = \langle \ell,m_\ell|\langle s,m_s|j,m_j\rangle$$

Although this is used by numerous authors, others follow Condon and Shortley using

$$(\ell,s,m_\ell,m_s|\ell,s,j,m_j) = \langle \ell,m_\ell|\langle s,m_s|j,m_j\rangle$$

M. E. Rose and at least one other author use

$$C(\ell,s,j|m_\ell,m_s,m_\ell) = \langle \ell,m_\ell|\langle s,m_s|j,m_j\rangle$$

With the state $|\frac{3}{2},\frac{1}{2}\rangle$, $\langle 1,\frac{1}{2},0,\frac{1}{2}|\frac{3}{2},\frac{1}{2}\rangle = \sqrt{\tfrac{2}{3}}$

When component states $|j_1,m_1\rangle$ and $|j_2,m_2\rangle$ are combined to form $|j_3,m_3\rangle$ then $|j_3,m_3\rangle = \sum_{m_1}\sum_{m_2} |j_1,m_1\rangle|j_2,m_2\rangle\langle j_1,j_2,m_1,m_2|j_3,m_3\rangle$

It has been shown that for a particle with momentum, mv, and wavelength, λ, they were related by $mv\lambda = h$. If the orbit of the hydrogen electron contained n (integer) wavelengths, then $\lambda = \frac{2\pi r}{n}$ and $mv\frac{2\pi r}{n} = h$. Therefore, $mvr = n\frac{h}{2\pi}$.

But, mvr is angular momentum. Giving angular momentum the letter, L, we see that the electron's orbit has a quantized angular momentum, $L_n = n\frac{h}{2\pi}$.

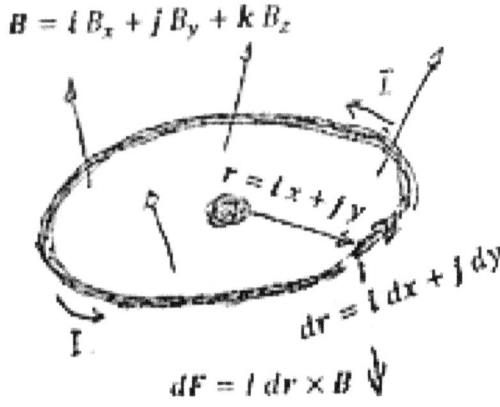

$$B = iB_x + jB_y + kB_z$$

$$r = ix + jy$$

$$dr = i\,dx + j\,dy$$

$$dF = I\,dr \times B$$

If the electron with a charge, $-e$, goes around its orbit in a time, T, it constitutes a current of $I = -\frac{e}{T}$ amperes. The following discussion will involve the energy of such a current loop in a magnetic field.

Look at a current carrying loop in a region with a magnetic field. The field may have differing values at differing locations. The force on a current, I, (taken in the direction positive charge would move) in an arc, dr, of the loop is $dF = I\,dr \times B$ and the sum, $F = \oint I\,dr \times B$, is the total magnetic force.
$$dr \times B = i\left(dy\,B_z - dz\,B_y\right) + j\left(dz\,B_x - dx\,B_z\right) + k\left(dx\,B_y - dy\,B_x\right).$$

In the case we have taken, $dz = 0$. The summation of things around closed loops is done on page 13-27, except this time we have components of magnetic field rather than of velocity.

$$\oint B_z\,dy = \iint \left(\frac{\partial B_z}{\partial x}\right)dx\,dy \qquad \oint B_z\,dx = -\iint \left(\frac{\partial B_z}{\partial y}\right)dx\,dy$$

$$\oint B_y\,dx = -\iint \left(\frac{\partial B_x}{\partial y}\right)dx\,dy \qquad \oint B_x\,dy = \iint \left(\frac{\partial B_x}{\partial x}\right)dx\,dy$$

The double summations are over the pieces $dx\,dy$ of the area bounded by the loop. If the rates of change (derivatives) of the magnetic field are constant over the area, then in each double summation they can be factored out and $\iint dx\,dy = A$, the area of the loop.

The total force on the current loop is:

$$F = I A \left[i \left(\frac{\partial B_z}{\partial x} \right) + j \left(\frac{\partial B_z}{\partial y} \right) - k \left(\frac{\partial B_x}{\partial x} - \frac{\partial B_y}{\partial y} \right) \right]$$

Recall from page 13-15, $\nabla \cdot B = 0$ giving $\frac{\partial B_x}{\partial x} + \frac{\partial B_y}{\partial y} = -\frac{\partial B_z}{\partial z}$.

The force is then $F = I A \left[i \left(\frac{\partial B_z}{\partial x} \right) + j \left(\frac{\partial B_z}{\partial y} \right) + k \left(\frac{\partial B_z}{\partial z} \right) \right]$. But that is the gradient operating on B_z. Therefore, we can rewrite F, $F = I A \nabla B_z$.

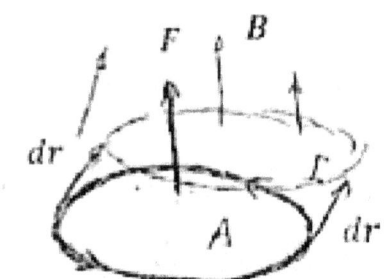

With a force of $F = I A \nabla B_z$ acting, if the ring is displaced, dr, work, $dW = F \cdot dr$, is done on page 13-11, which is $dW = I A \nabla B_z \cdot dr$. But, $\nabla B_z \cdot dr = dB_z$, the change is of B_z with the change of location, dr. This gives $dW = d(I A B_z)$.

If the force has an associated potential energy, $V(x,y,z)$, the work done by the force equals a loss of potential energy, $dW = -dV$. Potential energy changes by $dV = -d(I A B_z)$, suggesting that $V = -I A B_z$. Now, $B = i B_x + j B_y + k B_z$; and $k \cdot i = 0$, $k \cdot j = 0$, and $k \cdot k = 1$.

We can write $V = -I A k \cdot B$. Hanging $I A$ onto k gives a vector, $\mu = I A k$, which we call the magnetic moment of the current loop. Yes, the Greek letter mu is used for too many things in the physics of magnetism, but I am not making this up as I go along, this is standard physics language.

Let the ring be made of something with mass, M, and charge, q, and let it all go around in a time, T. As all the charge passes any point in a time, T, the current is $I = \frac{q}{T}$. If the ring is a circle with radius, R, the area is $A = \pi R^2$. This gives $\mu = \frac{q}{T} \pi R^2$ as the size of the magnetic moment. The rotating mass has an angular momentum, $L = M R v$, but the velocity is $v = \frac{2\pi R}{T}$, giving $L = \frac{M 2 \pi R^2}{T}$, from which $\frac{\pi R^2}{T} = \frac{L}{2M}$.

As result, we hang the magnetic moment onto the angular momentum. $\mu = \frac{q}{2M} L$. And for the hydrogen electron, $L_n = n \frac{\hbar}{2\pi}$.

The energy of a magnetic dipole (call energy E here) in a magnetic field is $E = -\mu \cdot B$. Then, $E = -\frac{q}{2M} L \cdot B$.

We are now ready to study MAGNETISM IN CONDENSED MATTER. Using pictures, we will try to give you physical intuition about the subject. Because the interactions between particles are so much stronger than in the gas phase, the mathematics involved is much more difficult and on top of that, we are not sure that we understand all the physics of what we are about to discuss. On the other hand, the potential applications are so important that industry from computers to medical equipment to electric power distribution is following every bit of research in these areas.

Magnetic moment, μ, was related to angular momentum, L, by $\mu = \frac{q}{2M} L$ and $L = n \frac{h}{2\pi}$ with h being Plank's constant and the quantum number changed by integer jumps. In a magnetic field the particle had energy, $E = \mu \cdot B$.

For the basic particles, the charge is always plus or minus $e = 1.6 \times 10^{-19}$ coulomb. The charge and mass of the particles do not have the same distribution giving rise to a g-factor. $\mu = g\left(\frac{e}{2M}\right) L$. For all particles, the angular momentum has the same quantization and as the proton has about 2000 time the mass of the electron, magnetism is a function of electron behavior.

Statistical mechanics is the branch of physics dealing with randomness. Entropy, or missing information, is defined as $S = k \ln N$ where N is the number of arrangements of a system, one of which exists for your particular sample. In this case, N might be the number of possible paths to a chosen point, one of which was actually taken to get to get there, $S = k \ln P$.

In the light of the next chapter, this suggests a path integral approach to entropy that would give non-equilibrium states as well as equilibrium.

REVIEW OF STATISTICAL MECHANICS

Here, re-define N as the number of particles distributed among energy states E_i. Let n_i be the number of particles having energy E_i. For any particular distribution there are $\frac{N!}{n_1! n_2! n_3! \dots}$ different arrangements.

For Stirling's formula, $\ln(N!) = N \ln N - N$

The entropy is $S = k(\ln N! - \sum_i \ln n_i!) = k(N \ln N - \sum_i n_i \ln n_i)$ with $N = \sum_i n_i$

or $S = -k \sum_i n_i \ln \frac{n_i}{N}$ in which $\frac{n_i}{N} = f_i$ the fraction of particles having E_i.

$S = -Nk \sum_i f_i \ln f_i$

The internal energy is $U = N \sum_i f_i E_i$ and $N = N \sum_i f_i$

We treat an isolated system where the internal energy and the number of particles are constant. From any distribution there a tendency for fluctuations to be toward new distributions with more rather than less arrangements of particles among energy states. Hence the equilibrium state is taken to be that of maximum entropy. For fluctuations in the neighborhood of maximum entropy, $\delta S = 0$ as it is a maximum, and $\delta U = 0$, and $\delta N = 0$ as they are constants. Combining these zeros we introduce Lagrangian multipliers to give consistent units. Thus

$$\delta\left(\frac{S}{k}\right) + a\delta N + b\delta U = 0 \quad \text{which gives}$$

$$N\sum_i \left[-\left(f_i\frac{\delta f_i}{f_i} + \ln f_i\, \delta f_i\right) + a\,\delta f_i + bE_i\delta f_i \right] \quad \text{or} \quad -1 - \ln f_i + a + bE_i = 0$$

Thus $\quad f_i = e^{(a-1)}\, e^{bE_i} \qquad 0 \leq f_i \leq 1$

For f_i less than one, $(a-1) = -a$ must be negative

If f_i decreases with increasing E_i, $b = -\beta$ must be negative

Thus $\quad f_i = e^{-a} e^{-\beta E_i} \quad$ and as $\sum_i f_i = 1$ we get $e^a = \sum_i e^{-\beta E_i}$

We call $\quad Z = \sum_i e^{-\beta E_i} \quad$ the PARTITION FUNCTION $\quad$ and $\quad a = \ln Z$

At equilibrium, $S = -Nk\sum_i f_i \ln f_i = Nka + Nk\beta\sum_i f_i E_i = Nk\ln Z + k\beta U$

as $\ln f_i = -a - \beta E_i \quad$ and $\quad \sum_i f_i = 1$

For each E_i except the ground state, f_i decreases as β increases. But f_i increases as temperature T increases. Hence we define $\beta = \frac{1}{kT}$

Now $\quad S = Nk\ln Z + \frac{U}{T} \quad$ and k has the units of energy per degree per molecule. Thermodynamics defines the Helmholtz free energy $F = U - TS$

Hence $\quad F = -NkT\ln Z \quad$ with $\quad Z = \sum_i e^{-\beta E_i}$

Now $\quad dF = dU - TdS - SdT \quad$ but $\quad dU = TdS - PdV \quad$ from thermodynamics

Hence $\quad dF = -PdV - SdT \quad$ giving $\quad P = -\left(\frac{\partial F}{\partial V}\right)_T \quad$ and $\quad S = -\left(\frac{\partial F}{\partial T}\right)_V$

Given the quantized energy states, we evaluate the partition function from which we get an equation of state and an entropy function for our system.

Let us start with a system of disordered spins at a high temperature. If they
form a ferromagnet, as the temperature
drops below the Curie temperature T_c,

they will align as in the lefthand
figure. With an antiferromagnet when
the temperature drops below the Néel
temperature T_N they will antialign as

in the righthand figure. An important
example of this is the Mott insulator,
N. F. Mott, (1949) Pos. Phys. Soc.
London, Sect. A. 62, 416.

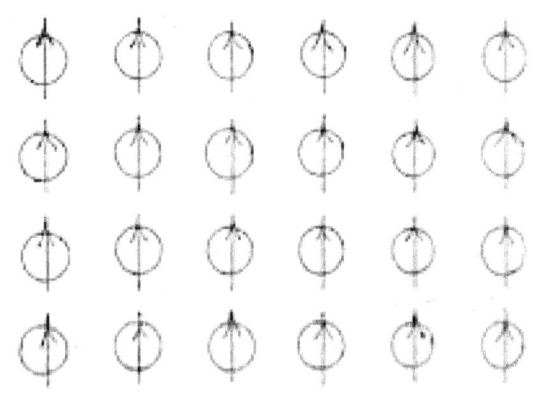

For many properties an
antiferromagnetic sys-
tem can be considered
as two co-existing
ferromagnetic systems.

Many metal oxide solid
crystals are antiferro-
magnetic. In two dimen-
sions, a cuprate Cu-O
lattice is shown in the
lower right-hand figure
with the larger circles
being copper and the
smaller circles being
oxygen.

Some Néel temperatures
are given in the table
to the left.

The importance of Mott
insulators is that with
the inclusion of certain
heavy elements as lanth-
anum, Yttrium, cerium,
and others, they can be
made to be superconduc-
ting at higher tempera-
tures.

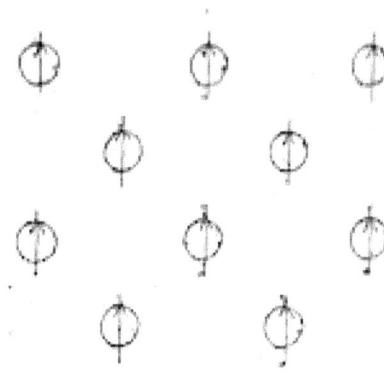

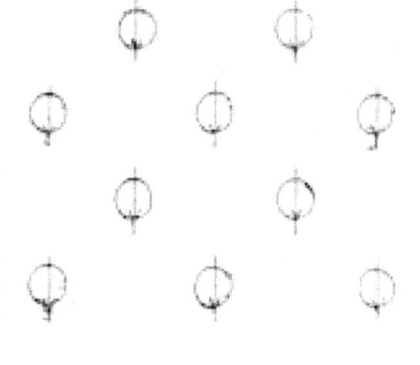

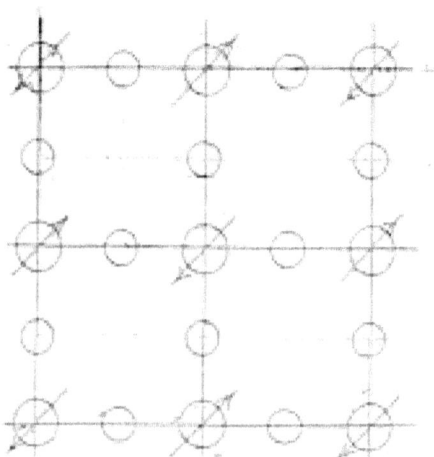

Compound	$T_N(K)$
NiO	515
CuO	300
CoO	291
FeO	185
MnS	165
MnO	122
FeF$_2$	85
NiF$_2$	78
MnO$_2$	56
CoF$_2$	40

PARAMAGNETISM OF SYSTEMS OF PARTICLES WITH MAGNETIC MOMENTS

FIRST A COMMENT ON THERMODYNAMICS

Normally internal energy changes as $dU = TdS - PdV$ but what about magnetic energy. If μ is the average magnetic moment in the direction of B then the average dipole energy is $-\mu B$ and if N is the dipoles per unit volume, $M = N\mu$ is the volume magnetization and thus the magnetic contribution ot the internal energy is $-N\mu B = -MB$. A change in B produces a change $-MdB$.

Now $dU = TdS - PdV - MdB$ and if $F = U - TS$, $dF = -SdT - PdV - MdB$

and $M = -\left(\frac{\partial F}{\partial B}\right)_{T,V}$ $\qquad F = -NkT \ln Z$

CLASSICAL DERIVATION OF LANGEVIN FORMULA

Consider a system of particles, each with a magnetic moment vector μ and place them in a magnetic field $\bar{B}$. The energy of any particle will be $u = -\bar{\mu}\cdot\bar{B} = -\mu B \cos\theta$.

The partition function is $Z = \sum e^{-u/kT} = \int_0^\pi e^{\mu B\cos\theta/kT} 2\pi\sin\theta d\theta$

$\boxed{\begin{aligned} &\text{let } w = \frac{\mu B\cos\theta}{kT} \\ &dw = -\frac{\mu B\sin\theta d\theta}{kT} \\ &\sin\theta d\theta = -\frac{kT}{\mu B}dw \end{aligned}}$

$Z = \frac{2\pi kT}{\mu B}\int_{-1}^{+1} e^w dw = \frac{2\pi kT}{\mu B}\left[e^{\mu B/kT} - e^{-\mu B/kT}\right]$

$\frac{\partial}{\partial B}\ln Z = \frac{1}{Z}\frac{\partial Z}{\partial B} = \frac{\frac{2\pi kT}{\mu B}(\frac{\mu}{kT})(e^{\mu B/kT}+e^{-\mu B/kT})}{\frac{2\pi kT}{\mu B}(e^{\mu B/kT}-e^{-\mu B/kT})} - \frac{\frac{2\pi k^2 T^2}{\mu B^2}(e^{\mu B/kT}-e^{-\mu B/kT})}{\frac{2\pi kT}{\mu B}(e^{\mu B/kT}-e^{-\mu B/kT})}$

$M = NkT\frac{\partial}{\partial B}\ln Z = NkT\left[\frac{\mu}{kT}\coth(\frac{\mu B}{kT}) - \frac{1}{B}\right]$ or $M = N\mu\left[\coth(\frac{\mu B}{kT}) - \frac{kT}{\mu B}\right]$

which is the Langevin formula.

In the following limiting cases, let w now be $w = \mu B/kT$

In the low T or high B case $\displaystyle\lim_{w\to\infty} M = N\mu$

In the high T and low B limit, expand e^w and e^{-w} in power series

$\displaystyle\lim_{w\to 0}\frac{e^w + e^{-w}}{e^w - e^{-w}} = \frac{1}{w}(1 + w^2/2 + \ldots)/(1 + w^2/6 + \ldots) = \frac{1}{w}(1+\frac{w^2}{2})(1-\frac{w^2}{6}) = \frac{1}{w} + \frac{w}{3}$

As a result $\displaystyle\lim_{w\to 0} M = \frac{N\mu^2 B}{3kT}$ which is Curie's law (after Pierre Curie)

Next we let the angular momentum have any total value with quantum number j with a component m parallel to the magnetic field, the $2(j+1)$ values of m going from $-j$ to $+j$, by unit jumps.

The magnetic energy is now for a particle of charge $-e$, $g\mu_B mB$ gives the partition function $\displaystyle\sum_{m=-j}^{+j} e^{g\mu_B mB/kT}$ let $w = g\mu_B B/kT$ where $\mu_B = \frac{eh}{2m_e}$

Then $Z = \displaystyle\sum_{m=-j}^{+j} e^{mw} = e^{-jw}\left[1 + e^w + e^{2w} + \ldots + e^{2jw}\right] = \frac{e^{-jw}\left[e^{(2j+1)w} - 1\right]}{e^w - 1}$

With a bit of manipulation, $Z = \dfrac{e^{(j+1/2)w} - e^{-(j+1/2)w}}{e^{w/2} - e^{-w/2}} = \dfrac{\sinh\left[(j+1/2)w\right]}{\sinh(w/2)}$

Now $F = -NkT\ln Z$ and $M = -\frac{\partial F}{\partial B} = NkT\frac{\partial}{\partial B}\ln Z$

$\ln Z = \ln\sinh\left[(j+1/2)w\right] - \ln\sinh(w/2)$

$$\frac{\partial}{\partial B}(\ln Z) = \left[\frac{(j+1/2)\cosh\left[(j+1/2)w\right]}{\sinh\left[(j+1/2)w\right]} - \frac{\frac{1}{2}\cosh(w/2)}{\sinh(w/2)}\right]\frac{\partial w}{\partial B} \qquad \text{with} \quad \frac{\partial w}{\partial B} = \frac{g\mu_B}{kT}$$

$$M_j = Ng\mu_B\left[(j+1/2)\coth\left[(j+1/2)w\right] - \frac{1}{2}\coth(w/2)\right]$$

At this point we introduce $x = wJ$ and it is customary to use J instead of j

$$M_J = Ng\mu_B J\left[\frac{J+1/2}{J}\coth\left[\left(\frac{J+1/2}{J}\right)x\right] - \frac{1}{2J}\coth\left(\frac{x}{2J}\right)\right]$$

We introduce the Brillouin function $B_J(x) = \frac{2J+1}{2J}\coth\left[\frac{2J+1}{2J}x\right] - \frac{1}{2J}\coth\left(\frac{x}{2J}\right)$

For $J = \frac{1}{2}$ $\quad B_{1/2}(x) = 2\coth(2x) - \coth(x)$

$$B_{1/2}(x) = \frac{2(e^{2x}+e^{-2x})}{e^{2x}-e^{-2x}} - \frac{e^x+e^{-x}}{e^x-e^{-x}} = \tanh(x)$$

This along with $B_{10}(x)$
is plotted in the
accompanying figure

At room temperature
x is very small so
we can approximate

$e^x = 1 + x$

$e^{-x} = 1 - x$

and $B_{1/2}(x) = x$

where $x = \frac{g\mu_B BJ}{kT}$

Hence at room temperature

$$M = \frac{N(g\mu_B J)^2}{kT}B$$

But $M = \chi_m B/\mu_o$
giving Curie's law

$$\chi_m = \frac{N(g\mu_B)^2\mu_o}{4kT} \quad \text{for } J = 1/2$$

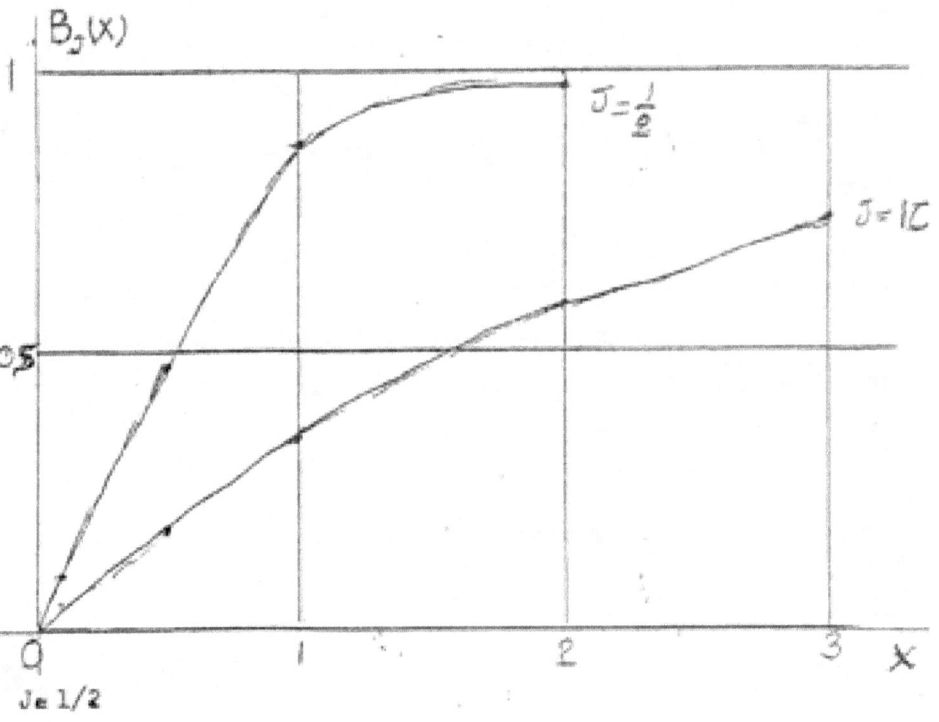

FERROMAGNETISM is an extreme case of paramagnetism where magnetic moments interact with their own magnetization as well as with an external field.

In this case $\quad x = \frac{g\mu_B J}{kT}(B + \alpha M) \qquad$ where α is a measure of the strength of the spin-spin interaction.

At a temperature T, $\quad M(T) = Ng\mu_B J \cdot B_J(x)$

As $T \to 0$, $x \to \infty$ $\quad \lim\limits_{\substack{T \to 0 \\ x \to \infty}} B_J(x) = 1 \qquad$ thus $M(0) = Ng\mu_B J$

The ratio of magnetization at temperature T to the saturated M is $\frac{M(T)}{M(0)} = B_J(x)$

In the formula $x = \frac{g\mu_B J}{kT}(B + \alpha M)$, with no external field (B = 0) we get the self magnetization to be $M(T) = \frac{kTx}{g\mu_B J\alpha}$ but $M(0) = Ng\mu_B J$

Hence $\frac{M(T)}{M(0)} = \frac{kTx}{N(g\mu_B J)^2\alpha}$

represents the effect of magnetic moments interacting with themselves.

This effect plots as a straight line on graph of M(T)/M(0) against x.

For a system of interacting spins, the magnetization at temperature T is given by the point where the two plots intersect.

As the temperature increases, the slope of the self magnetization line increases and the point of intersection with $B_J(x)$ moves to the left with M(T) decreasing until there is a temperature at which M(T) = 0. This is called the Curie temperature T_c and above this temperature there will be no self-organization.

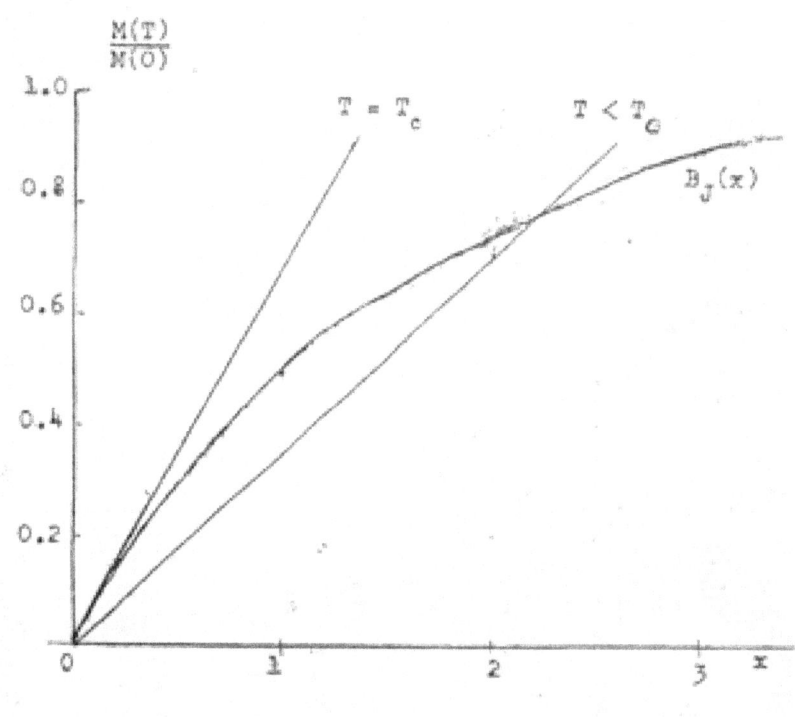

For iron T_c = 1043K, cobalt T_c = 1394K, nickel T_c = 631K.

To evaluate T_c, notice that the straight line for self magnetization at T_c is tangent to the curve for $B_J(x)$ at very small x (x = 0). In taking the limit of $B_J(x)$ for small x notice that for small values of a variable z, we can substitute into $\coth(z) = \frac{e^z + e^{-z}}{e^z - e^{-z}}$

$e^z = 1 + z + \frac{z^2}{2} + \frac{z^3}{6} + \frac{z^4}{24} + \cdots$ and $e^{-z} = 1 - z + \frac{z^2}{2} - \frac{z^3}{6} + \frac{z^4}{24} + \cdots$

and, along the way, using $(1 + \frac{z^2}{2})/(1 + \frac{z^2}{6}) = (1 + \frac{z^2}{2})(1 - \frac{z^2}{6}) = (1 + \frac{z^2}{3})$ approximately, we get for small values of x, $B_J(x) = (\frac{J+1}{J})\frac{x}{3}$

Matching the slope of this with the slope of the self-magnetization line at T_c

$\frac{kT_c}{N(g\mu_B J)^2\alpha} = \frac{1}{3}(\frac{J+1}{J})$ giving $T_c = \frac{J(J+1)N(g\mu_B)^2\alpha}{3k}$

Obviously the stronger the strength, α, of the spin-spin interaction, the more thermal energy that the system can overcome to self-organize.

These pictures show regions of alignment that
that are called domains. In the white areas
the Ising spin variable is S = -1 and in the
black areas S = +1.

If T_c is the Curie temperature, in the top

picture $T = 1.2 T_c$. Here, above the Curie

temperature, we see small domains of aligned
spins, but which are randomly distributed.
Overall, there is no effective long-range
magnetization.

The middle picture shows the system at its
Curie temperature. Now large domains with
magnetization spin up and spin down are
forming.

In the bottom picture, $T = 0.95 T_c$. Now

an over-all magnetization with spin S = +1
dominates the system.

What we have here is spontaneous magnetization. When from the outside we apply a magnetic field B to the domains will be
stimulated to produce an over-all magnetization which will then persist after the
field B is removed. This produces what is
called hysteresis.

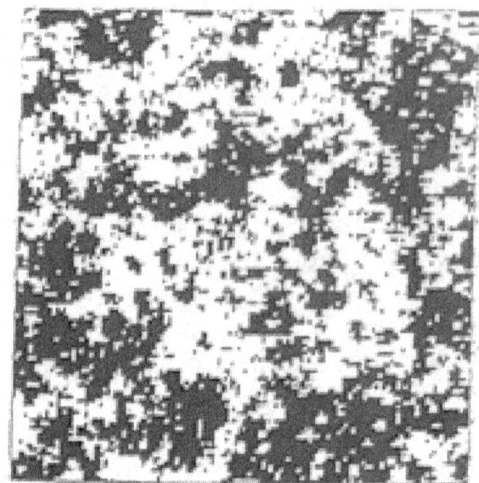

(a)

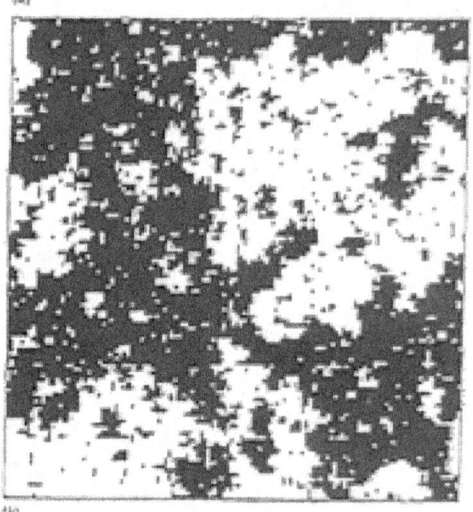

(b)

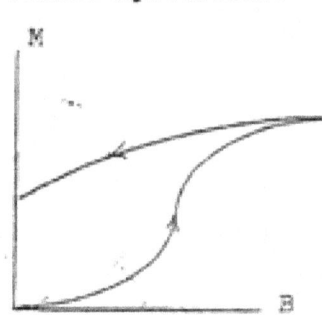

In the accompanying
graph of magnetiza-
tion against magnet-
ic field, we see the
over-all alignement
of domains develop-
ing. But, due to
spin-spin interac-
tions, the aligned
domain structure
persists, after
the applied field
is removed.

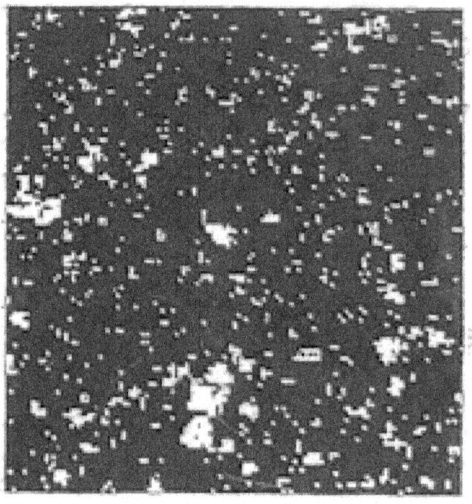

(c)

The energy $E = \sum_i - J_{i,i+1} S_i S_{i+1}$ gave its

lowest value when the spins were all aligned.
The energy $E = \sum_i + J_{i,i+1} S_i S_{i+1}$ gives its

lowest value when adjacent spins are oppositely
directed. This holds for antiferromagnetism.

For further study of critical phenomena as with the van der Waals equation,
order-disorder phase transitions as with the Ising model
some standard references are:

H. E. Stanley INTRODUCTION TO PHASE TRANSITIONS AND CRITICAL PHENOMENA (1971)
Oxford U. P.

S-K Ma MODERN THEORY OF CRITICAL PHENOMENA (1976) W. A. Benjamin

References:

A. H. Morrish **The Physical Principles of Magnetism** John Wiley

R. M. White **Quantum Theory of Magnetism** McGraw Hill

R. Skomski **Simple Models of Magnetism** Oxford University Press

S. Chikazumi **Physics of Ferromagnetism** Oxford University Press

J. Kübler **Theory of Itinerant Electron Magnetism** Oxford University Press

A. Aharoni **Introduction to the Theory of Ferromagnetism** Oxford University
   Press

A Final Word:

At the time this book was published, what might be called topological physics had become established as a major area of physics, especially dealing with condensed matter. Even for many physicists, this is not an easy topic to study. The challenge is: can the concepts of Berry phase and Chern numbers be explained to beginning physics students? An attempt to do this is the article by Ramirez and Skinner, *Dawn of the Topological Age* in <u>Physics Today</u> of September 2020. Serious books on the subject are:

Bernevig and Hughs <u>Topologcial Insulators and Topological Superconductors</u> Princeton University Press

D. Vanderbilt <u>Berry Phase in Electronic Structure Theory</u> Cambridge University Press

M. Nakahara <u>Geometry, Topology and Physics</u> Taylor & Francis

www.ingramcontent.com/pod-product-compliance
Lightning Source LLC
Chambersburg PA
CBHW081553220526
45468CB00010B/2648